REACTIVE
INTERMEDIATES

REACTIVE INTERMEDIATES

A Serial Publication

VOLUME 3

Edited by

MAITLAND JONES JR.
Princeton University

ROBERT A. MOSS
Rutgers University

A WILEY-INTERSCIENCE PUBLICATION

JOHN WILEY & SONS

New York • Chichester • Brisbane • Toronto • Singapore

No Anal

Library of Congress Cataloging in Publication Data:

Reactive intermediates.   v. 3
   New York, John Wiley & Sons, c1978-

   (3) v.   ill.   24 cm.

"A Wiley-interscience publication."
Key title: Reactive intermediates, ISSN 0190-8375.

   1. Chemistry, Physical organic—Collected works.   2. Chemical
reaction, Conditions and laws of—Collected works.
QD476.R414   547'.1'39   79-640411
                                      MARC-S

Library of Congress   78[8311]

ISBN 0-471-01893-7.

Printed in the United States of America

10 9 8 7 6 5 4 3 2 1

CONTRIBUTORS

Edward M. Arnett
Department of Chemistry
Duke University
Durham, North Carolina 27706

Weston Thatcher Borden
Department of Chemistry
University of Washington
Seattle, Washington 95195

Charles P. Casey
Department of Chemistry
University of Wisconsin
Madison, Wisconsin 53706

Peter P. Gaspar
Department of Chemistry
Washington University
Saint Louis, Missouri 63130

Thomas C. Hofelich
Dow Chemical U.S.A.
Midland, Michigan 48640

Maitland Jones, Jr.
Department of Chemistry
Princeton University
Princeton, New Jersey 08544

Leonard Kaplan
Department of Chemistry
Cleveland State University
Cleveland, Ohio 44115

Ronald H. Levin
IBM
Information Products Division
Boulder, Colorado 80302

Walter Lwowski
Department of Chemistry
New Mexico State University
Las Cruces, New Mexico 88003

Robert A. Moss
Department of Chemistry, Rutgers,
The State University of New Jersey
New Brunswick, New Jersey 08903

George W. Schriver
Department of Chemistry
Tulane University
New Orleans, Louisiana 70118

Stuart W. Staley
Department of Chemistry
University of Nebraska-Lincoln
Lincoln, Nebraska 68588

PREFACE TO VOLUME 3

In this third volume of *Reactive Intermediates: A Serial Publication* we hope again to provide authoritative and critical analyses of the recent literature for each major type of reactive organic intermediate. As before, this volume is intended for scientists who are actively involved or contemplating involvement in the chemistry of reactive intermediates.

Authors have been urged to select critically and to evaluate those recent contributions that should be brought to the close attention of researchers and students. Literature coverage in the present volume focuses on 1980–1982, but important contributions from earlier years as well as many references to 1983 have been included at the authors' discretions. As a convenience to the reader, the authors have marked with an asterisk (*) those references that they consider most significant for detailed discussion, analysis, and a re-reading.

The principal authors of the chapters remain the same with one exception. Although we are sorry to see Don Bethell and D. Whittaker retire from the "Carbocations" chapter, we were extraordinarily happy to be able to persuade Ned Arnett and his associates to take over this demanding job.

MAITLAND JONES, JR.
ROBERT A. MOSS

Princeton, New Jersey
New Brunswick, New Jersey

April 1984

CONTENTS

REACTIVE
INTERMEDIATES

1
ARYNES

RONALD H. LEVIN

IBM, Information Products Division Boulder, Colorado 80302

I.   INTRODUCTION

A number of interesting reports on arynes and strained cycloalkynes appeared during the period 1980–1982. The highlights of these studies are considered in the following sections.

II.   THE USE OF *Ortho*-BENZYNE IN ORGANIC SYNTHESIS

The synthetic utility of *o*-benzyne has been markedly expanded by several contributions appearing during this reporting interval. Stevens has amply demonstrated that *o*-benzyne can be used to gain direct and regiospecific access to substituted benzocyclobutenones (Figure 1.1).[1] 1,1-Dimethoxyethylene was employed as a ketene equivalent in the $2 + 2$ cycloaddition and yields in the range of 60–80% were typically achieved. The regioselectivity is in accord with expectations based on inductive and steric effects. The degree of selectivity is impressive and bodes well for further synthetic application. Indeed, Stevens has already applied this methodology to the synthesis of the natural product taxodione.[2]

Wege and Best have employed an alternate tactic to control the regiochemistry in *o*-benzyne Diels–Alder cycloadditions.[3] They have linked the aryne to the cycloaddend; subsequent intramolecular cycloaddition then yields a regiospecific product as illustrated in Figure 1.2. The method worked well for arynes generated from *o*-dihaloaromatics as well as anthranilic acids. The authors used this scheme in their synthesis of mansonone E(4).

Hart and co-workers have employed the concept of diaryne equivalents to gain access to a variety of polycyclic molecules.[4] As depicted in Figure 1.3, upon treating tetrabromo derivatives of 1,4-disubstituted benzenes with *n*-butyllithium in the presence of a cyclic diene, linearly fused polycycles were obtained in good yield. When anthracene or its derivatives were used as the diene component, linearly fused "iptycenes" were produced. Although these

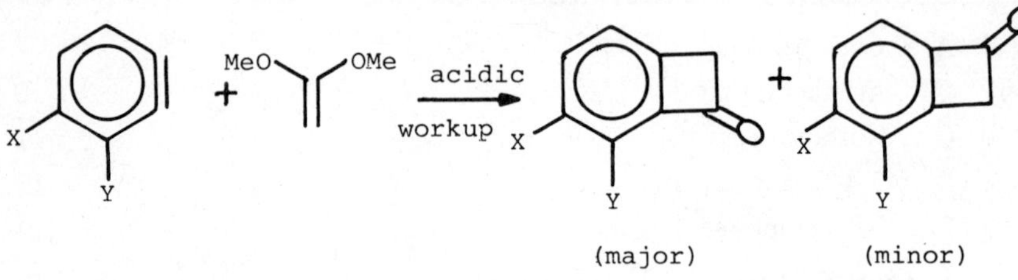

	X	Y	2/3	Yield (%)
a	H	H	–	63
b	H	OMe	∞	70
c	OMe	H	7	64
d	OMe	OMe	∞	73
e	H	Me	3	38
f	H	Cl	∞	76

FIGURE 1.1.   The regioselective cycloaddition of dimethyoxyethylene to substituted benzynes.

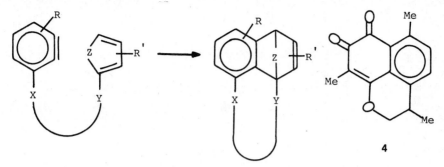

FIGURE 1.2.   Regiospecific intramolecular aryne cycloaddition.

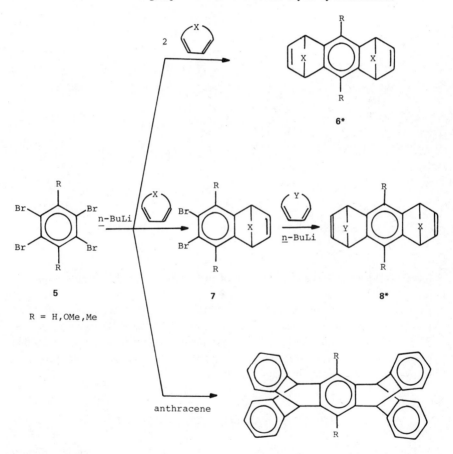

pentiptycene **9**

FIGURE 1.3.   Synthetic application of diaryne equivalents. *Mixture of stereoisomers.

3

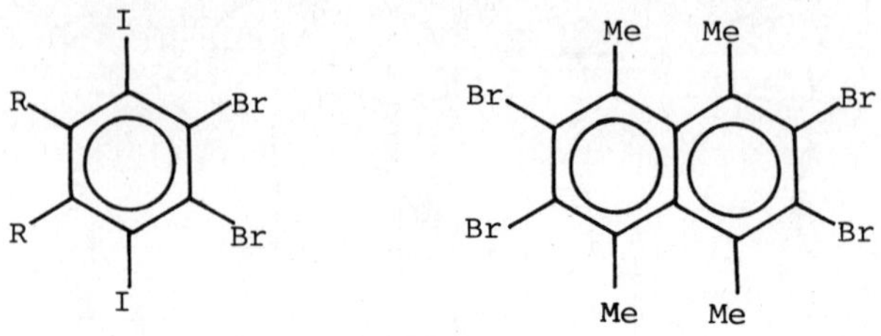

10   **11**

FIGURE 1.4.   Diaryne equivalents.

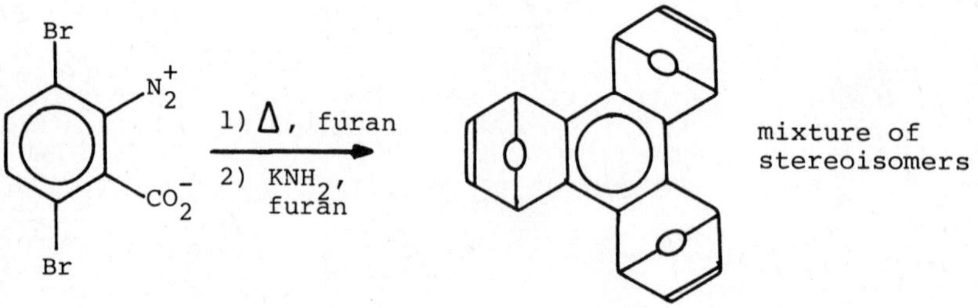

mixture of
stereoisomers

12   **13**

FIGURE 1.5.   Reaction of a triaryne equivalent.

reactions have the formal appearance of proceeding through a bisaryne, experiments revealed that the reaction actually proceeds in a sequential manner through two aryne intermediates. Synthetic advantage was taken of this fact in the preparation of nonsymmetric polycycles such as **8**. Diaryne equivalent **10** provided entry to angularly fused systems, and **11** yielded further extended polycycles (Figure 1.4). Wege has carried matters a step further by using triaryne equivalents as illustrated in Figure 1.5.[5]

III.   THE NATURE OF *Para*-BENZYNE

Bergman has continued his efforts to categorize the various species encountered upon thermolysis of *cis*-enediynes (**14**),[6] shown in Figure 1.6. Early kinetic experiments on the parent system (**14a**) were thwarted by low product yields when the reactions were carried out in solution. These difficulties have been circumvented by derivatizing the parent so as to permit intramolecular

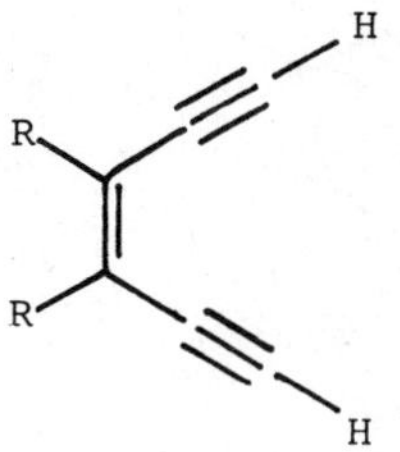

a R = H
b R = Me
c R = Et
d R = n-Pr

14

FIGURE 1.6.   *Cis*-enediynes.

self-trapping. When **14d** was pyrolyzed in the gas phase, three monomeric products (**15**, **16**, and **17**) were obtained. In solution, particularly in the presence of a hydrogen donor, a fourth low molecular weight compound (**18**) was produced. Kinetic and isotopic experiments, in combination, provided strong support for a mechanistic scheme involving a *p*-benzyne intermediate as pressented in Figure 1.7. Through application of activation energies and A factors from appropriate model reactions to this scheme, the authors concluded that the lifetime of *p*-benzyne at 200°C was on the order of 10^{-9} seconds.

Two additional experimental approaches by Bergman corroborated the existence of the *p*-benzyne intermediate and allowed comment on its reactive spin state. In the first of these studies, the authors determined the ratio of cage to escape (*C/E*) reactions from the initial, caged radical pair produced when 1,4-dehydrobenzene biradical **19** abstracts a hydrogen atom from a donor. Within this geminate radical pair there is a proscription against cage reaction (recombination or disproportionation) if the reactants are in the triplet state. In principle, therefore, a *C/E* of zero can be taken as evidence for a triplet reactive state, whereas a ratio greater than zero implies at least some contribution from the singlet. Comparison of the *C/E* value of 0.55 obtained by Bergman with corresponding values from appropriate model systems suggests a high (perhaps 100%) contribution from singlet **19**. This result must be qualified, however, by noting that the reacting species need not be in its ground state (slow intersystem crossing perhaps). Alternatively, if the singlet and triplet were in rapid equilibrium, but reaction from the singlet were kinetically preferred, a misleading *C/E* could again be obtained. Unfortunately, attempts to detect the triplet by varying the concentration of trapping agent, photochemical activation, or heavy atom effects proved inconclusive as they did not produce a shift in the ratio.

The second line of investigation involved CIDNP studies. Upon heating solutions of **14d** in the probe of an nmr spectrometer, emission was observed in some of the side-chain proton signals from the intramolecular product **15**. Because of the complex nature of these signals, a bimolecular reaction product was sought in an effort to simplify the attendant analysis. When **14** was heated

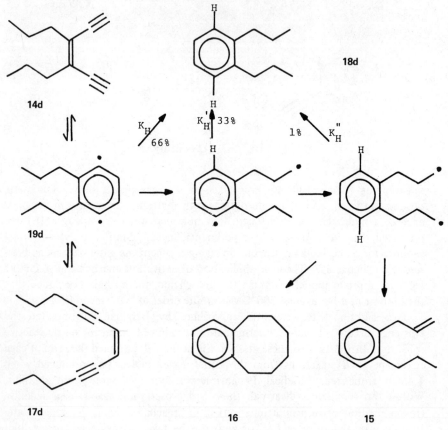

14d

19d

17d

18d

K_H 66%

K_H' 33%

1% K_H''

16

15

FIGURE 1.7.   Mechanism involving a *p*-benzyne intermediate.

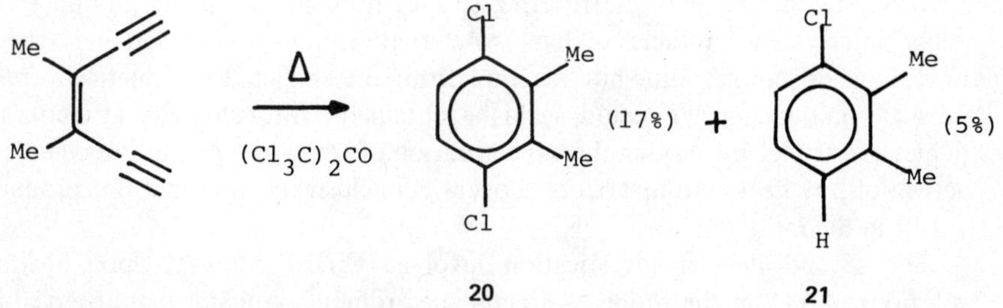

Me

Me

Δ

$(Cl_3C)_2CO$

Cl

Me

Me

Cl

(17%)

20

+

Cl

Me

Me

H

(5%)

21

FIGURE 1.8.   Generation of 1,4-dehydrobenzene in the presence of hexa-chloroacetone.

in the presence of hexachloroacetone, two products, **20** and **21**, resulted, as shown in Figure 1.8. CIDNP emission was observed and tentatively assigned to the aromatic protons in **20**. If this assignment is correct and if the authors' further assumption, that **20** is produced after escape from the radical cage (reasonable arguments were provided to support this position) is correct; then the CIDNP signals provide additional support for reaction by way of the singlet state of biradical **19**. Hence, at least at these conditions of temperature, solvation, and progenitor, a singlet diyl, referred to as 1,4-dehydrobenzene and depicted by **19**, seems to describe adequately the discrete intermediate encountered along this particular reaction coordinate.

IV.   CYCLOPENTYNE AND ANALOGUES

Strained cycloalkynes have received considerable attention of late; even the stability of cyclopropyne has been considered.[7] An interesting body of work relating to cyclopentyne and its derivatives has appeared in the recent literature, and this section serves to review these efforts.

A.   Cyclopentyne

Chapman, Regitz, and co-workers have attempted to apply their cryogenic, *in situ* matrix technique to the generation and isolation of cyclopentyne.[8] The photochemical scheme illustrated in Figure 1.9 was employed. Starting with

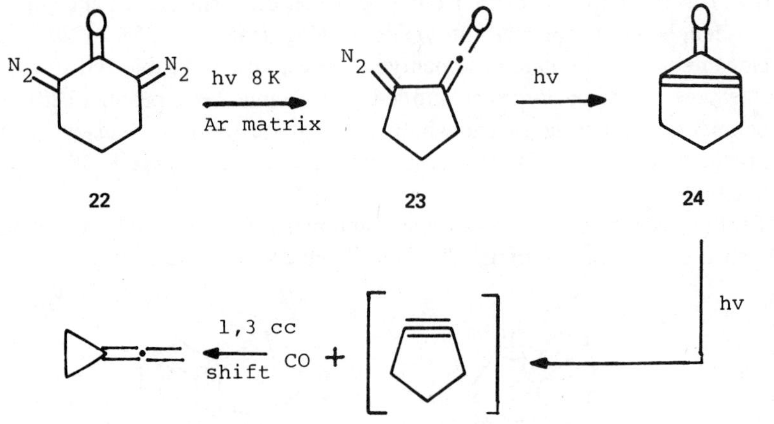

FIGURE 1.9.   Attempt to matrix isolate cyclopentyne.

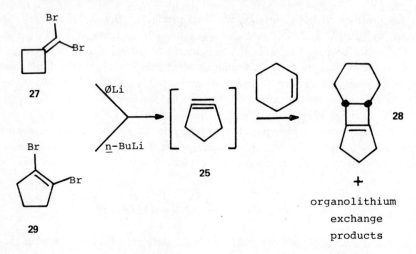

FIGURE 1.10.   Generation and interception of cyclopentyne.

bis-diazo ketone **22**, intermediates **23**, **24**, and **26** were observed, but the anticipated cyclopentyne escaped detection. As it is a logical intermediate in these
interconversions, the authors presumed that its rate of conversion to **26** out-
paced its rate of formation.

In a subsequent report, Fitjer's group claimed to have successfully generated
and trapped cyclopentyne.[9] These workers found that reaction of dibromo-
methylenecyclobutane **27** with phenyllithium in the presence of an olefin led
to the product expected from $2 + 2$ cycloaddition with cyclopentyne (Figure
1.10). The reaction was carried out with seven different olefins and the $2 + 2$
cycloadducts were obtained in yields ranging from 20–35%. 1,2-Dibromo-
cyclopentene, an alternate cyclopentyne source, was also subjected to reaction
in the presence of cyclohexene and led to the same $2 + 2$ product (**28**). These
experiments seem weighty enough to substantiate the intermediacy of cyclopentyne; they also raise some perplexing questions and suggest clear lines for
future investigation.

Striking is the absence of an ene component in these reactions. Owing to
the angular orientation of the in-plane "dehydro" orbitals, there is probably

FIGURE 1.11.   Decreased overlap of the cyclopentyne dehydro orbitals.

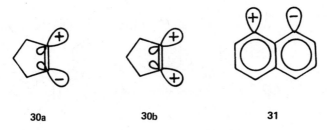

30a   30b   31

FIGURE 1.12.   Antisymmetric/symmetric wave functions.

less overlap between them in cyclopentyne than in o-benzyne, for example
(Figure 1.11). This could serve to either increase the diradical character of 25, or
invert the HOMO-LUMO ordering such that the antisymmetric A combination
30a is located at a lower energy than the symmetric S wave function 30b
(Figure 1.12). Of course the same effect as S/A inversion can be achieved if the
two orbital energies merely approach one another (E_A still greater than E_S)
allowing both orbitals to be populated by way of equilibration, with reaction
then being kinetically preferred from the antisymmetric level. S/A level inver-
sions have been predicted and experimentally verified, 1,8-dehydronaphthalene
(31) being a case in point.[10] Since the generic ene reaction is usually discussed
in terms of a concerted, suprafacial mechanism, its absence here would be
consistent with increased biradical character or S/A orbital inversion (or equilibration) as discussed previously. It is worth noting that ene products are also
absent from the chemistry of 1,8-dehydronaphthalene.[10]

Clearly, then, it would be of interest to understand better the electronic
nature of cyclopentyne. One approach would involve stereochemical analysis of
its 2 + 2 cycloaddition with appropriately labeled olefins; the stereospecificity,
or lack thereof, would allow direct comment on its reactive state. Experiments
aimed at determining the proclivity of 25 toward Diels–Alder cycloaddition
would also be of value. Acyclic dienes would make choice reaction partners as
they would allow for direct observation of the (4 + 2)/(2 + 2) relative rate
differential.

Another point of interest concerns hydrogen abstraction products. Reaction
of 1,8-dehydronaphthalene with dienes often produced such compounds; would
analogous compounds be obtained with cyclopentyne (Figure 1.13)?

A final query concerns the lack of products derived from allene 26 in Fitjer's
study. Perhaps the rearrangement of cyclopentyne to 26 is photochemically
driven, as suggested by Chapman. On the other hand, a thermal mechanism
cannot be dismissed. Chapman's cyclopentyne must almost certainly be produced in a higher vibrational level than Fitjer's cyclopentyne, hence the photochemically derived cyclopentyne should be more prone to thermal unimolecular

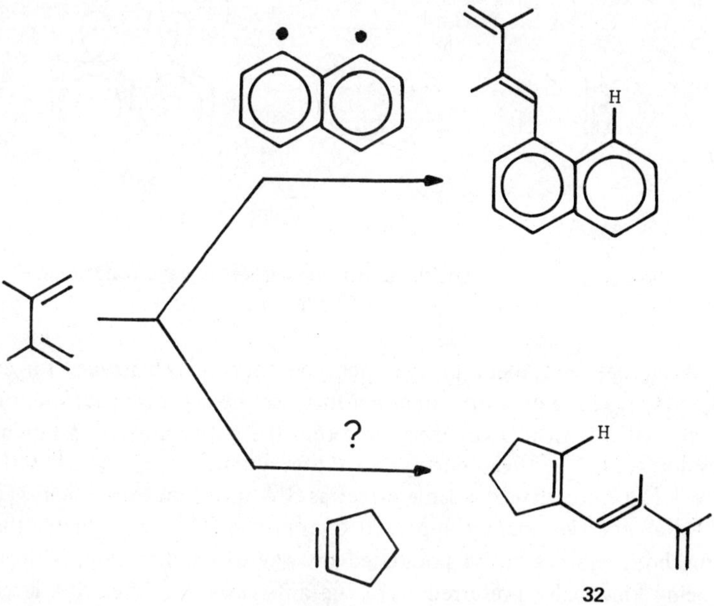

FIGURE 1.13.   Potential reaction by way of hydrogen abstraction.

reorganization. Biradical or concerted mechanisms are both viable candidates for such a rearrangement. It should be noted that if the cyclopentyne anti-symmetric dehydro orbital is the HOMO, then a concerted [1,3]-suprafacial carbon–carbon bond shift with retention of configuration at the migrating terminus is thermally allowed. Again, further experimentation is required to understand better this divergent behavior.

B.   Acenaphthyne

Having been unable to detect cyclopentyne, Chapman and Regitz redirected their efforts toward the generation of acenaphthyne (33).[8] The [1,3]-carbon–carbon bond shift that presumably precluded their observation of cyclopentyne is abjured in this case by the cyclic nature of the system. The same synthetic strategy applied to cyclopentyne was applied here. Irradiation of bis-diazo ketone 34 at 15 K in an argon matrix produced cyclopropenone 35. Subsequent photolysis of 35 generated acenaphthyne. This substance was characterized by both spectral and chemical techniques. The salient chemical trapping reactions are illustrated in Figure 1.14. Nakayama has attempted to generate acenaphthyne starting from the bishydrazone of acenaphthene quinone (36).[12] The reaction was performed in ethylene glycol, but none of the expected adduct from ace-

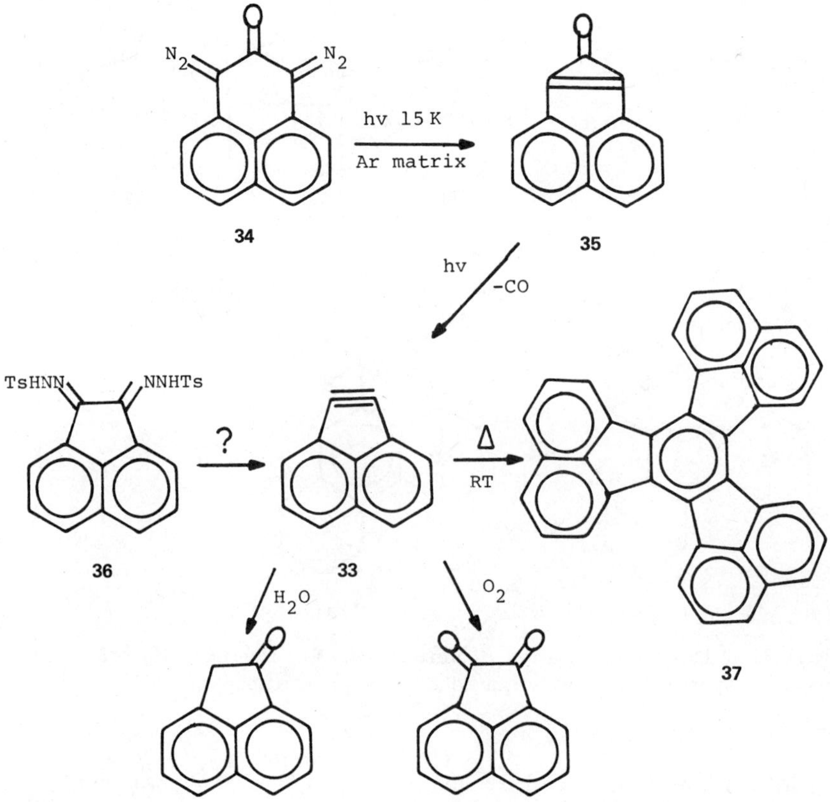

FIGURE 1.14.   Generation and reaction of acenaphthyne.

naphthyne and the glycol was detected. Although decacyclene (**37**) was produced, more work is needed to define the intermediate(s) present in this reaction.

C.   Norbornyne

In a formal sense, at least, norbornyne can be derived from cyclopentyne through incorporation of a two-carbon bridge. This subtle change, with its attendant geometric consequences, could have a significant impact upon the energetic ordering of the dehydro orbitals (S/A; discussed previously) in norbornyne. Since such changes would have a predictable effect upon chemical reactivity, the interception of norbornyne with appropriate trapping agents takes on added interest.

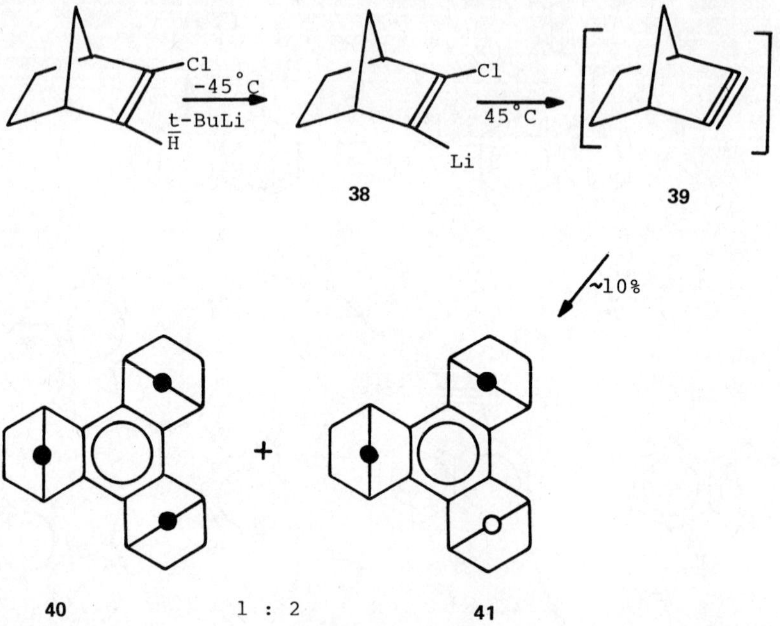

FIGURE 1.15.   Trimerization of norbornyne. ● − Denotes CH_2 bridge above plane of paper; ○ − denotes CH_2 bridge below plane of paper.

Gassman and Gennick[13] have improved upon the earlier procedure[14] for generating norbornyne (Figure 1.15). Lithio halonorbornene **38** is still the precursor, but conditions have been defined that allow it to be prepared without an excess of alkyl lithium reagent. Previously, preparation of **38** and its subsequent decomposition to norbornyne took place in the presence of excess alkyl lithium.[14] Hence, once generated, the norbornyne was rapidly captured by addition of the alkyl lithium reagent. Without this excess of organolithium, other reaction pathways may be observed. In the current study norbornyne reacted with itself to yield the two possible stereoisomeric trimers, **40** and **41**, in the statistical ratio of 1/2.

Experiments using olefins and dienes as trapping agents would be more informative about the electronic properties of the triple bond in norbornyne. Apparently some experiments along these lines have been attempted, but with negative results.[15] One can envision strategies based upon polymer attachment or steric encumbrance that should serve to increase the lifetime of norbornyne by reducing the rate of formation of **40/41**. Such approaches should allow for the interception of norbornyne by reactive olefins and dienes.

Hart has generated the bicyclo[2.2.2]alkyne homologue (**42**) of norbor-

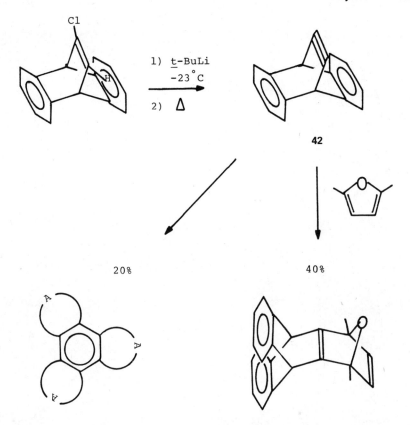

A = 9,10-anthracenyl

FIGURE 1.16.   Generation and reaction of bicycloalkyne **42**.

nyne.[4b] This cycloalkyne also trimerizes but is apparently longer lived, as it can be trapped with dienes (Figure 1.16).

D.   3,4-Thiacyclopentyne

Bolster and Kellogg have generated thiacyclopentyne **43**.[16] The steric congestion caused by the methyl groups and the longer carbon–sulfur bond lengths were expected to stabilize **43** relative to cyclopentyne itself both kinetically and thermodynamically. Compound **43** was produced by oxidation of bishydrazone **44** with lead tetraacetate or manganese dioxide. Figure 1.17 depicts the various reactions that served to characterize **43**. When lead tetraacetate is used to generate **43**, some of the product resulting from the addition of acetic acid,

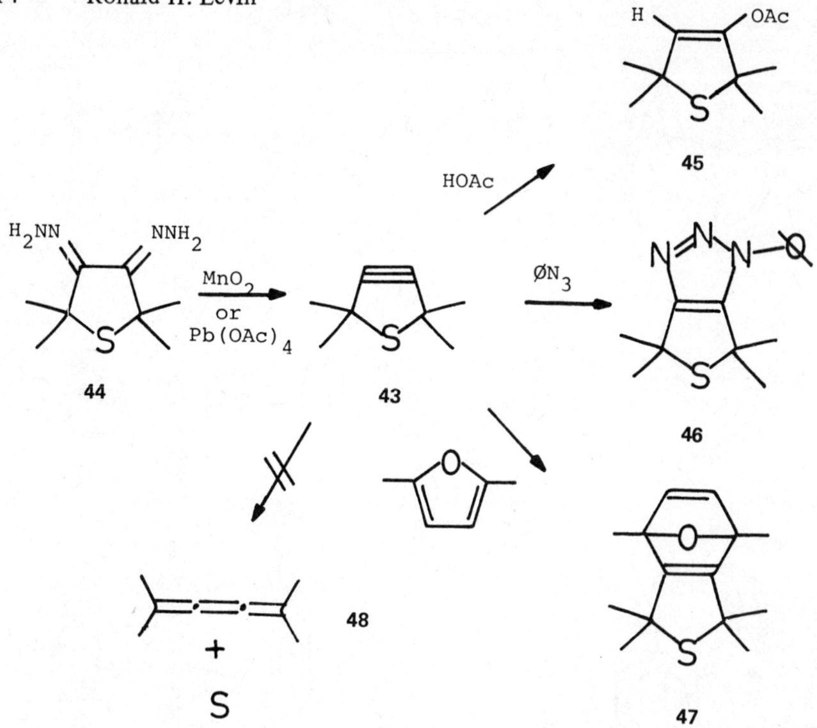

FIGURE 1.17.   Trapping of thiacyclopentyne **43**.

45, was obtained. Even more diagnostic is the obtaining of cycloadducts **46** and **47** from addition of **43** to phenyl azide and 2,5-dimethylfuran, respectively. None of the sulfur extrusion compound, tetramethylbutatriene (**48**), was ever detected. Experiments to characterize the electronic properties of **43**, analogous to those described in the preceding sections, would certainly be of interest.

E.   2,3-Thiophyne

Reinecke's group has continued to survey the chemistry of thiophynes.[17] Flash vacuum pyrolysis (FVP) of anhydride **49** provides a convenient source of 2,3-dehydrothiophene (**50**). Its cycloadditions to 2,3-dimethylbutadiene (**51**) and 2,5-dimethylthiophene (**52**) are of interest because both 4 + 2 and 2 + 2 adducts are obtained.

In the case of **51**, three products were formed as shown in Figure 1.18. The major product, **53**, formed in 13% yield, was the adduct expected from Diels–Alder cycloaddition. The two minor products, **54** and **55** resulted from rearrangement and aromatization of the initial 2 + 2 cycloadduct. Again, the ene reaction

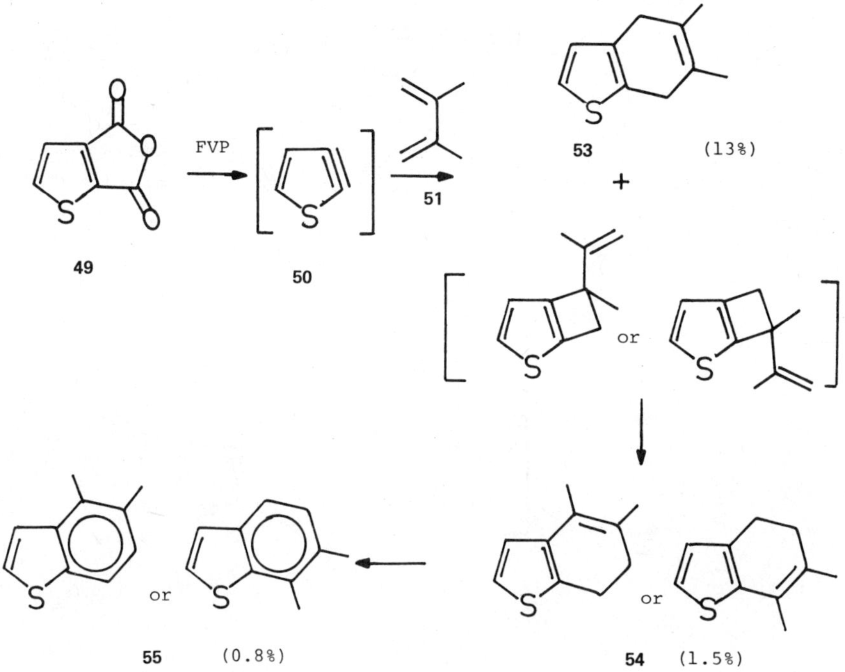

FIGURE 1.18.   Cycloaddition of dehydrothiophene to 2,3-dimethylbutadiene.

product is conspicuous by its absence. The corresponding reaction with o-benzyne also produces 4 + 2 and 2 + 2 cycloadducts, but the major product is formed by way of an ene pathway (Figure 1.19).[18]

When thiophyne **50** was generated in the presence of 2,5-dimethylthiophene (**52**), products resulting from desulfurization of the initial 2 + 2 and 4 + 2 cycloadducts were again produced, as illustrated in Figure 1.20. Two points merit further comment. First, the ratio of the adducts derived from 2 + 2 cycloaddition (**57/58**) is consistent with polarization of thiophyne **50** such that an excess of negative charge resides on the carbon adjacent to the sulfur atom (Figure 1.21). Second, the large preference given to 2 + 2 over 4 + 2

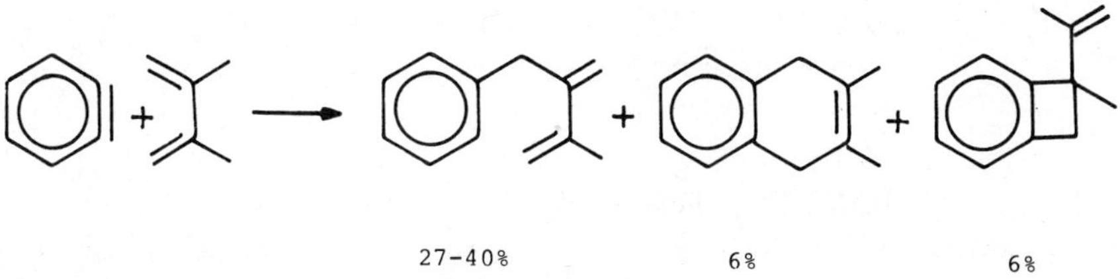

FIGURE 1.19.   Reaction of o-benzyne with 2,3-dimethylbutadiene.

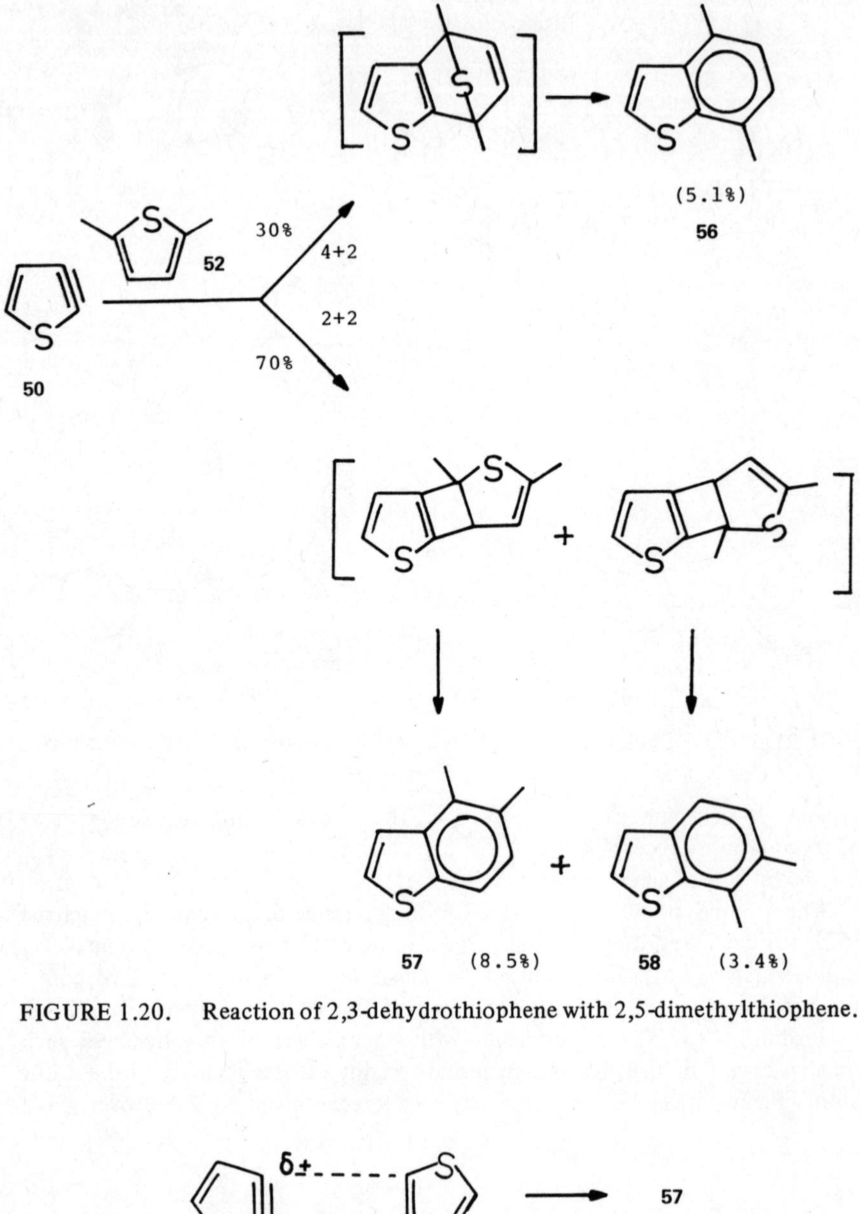

FIGURE 1.20.   Reaction of 2,3-dehydrothiophene with 2,5-dimethylthiophene.

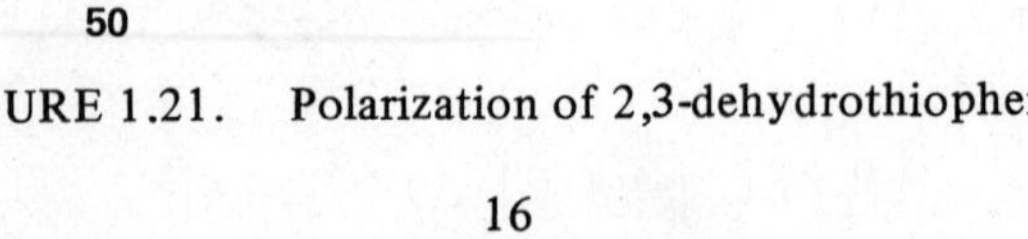

FIGURE 1.21.   Polarization of 2,3-dehydrothiophene.

16

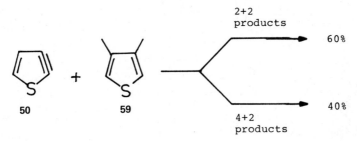

FIGURE 1.22.   Partitioning of reaction pathways with 2,3-dehydrothiophene.

cycloaddition (56) is unusual; the authors note that $4 + 2$ cycloaddition of o-benzyne to generic thiophenes is "substantially preferred" over $2 + 2$ reaction.[17b] In 2,5-dimethylthiophene the methyl groups are geometrically situated so as to retard sterically the $4 + 2$ cycloaddition to a greater extent than $2 + 2$ cycloaddition. Anticipating this complication, the authors also allowed 50 to react with 3,4-dimethylthiophene (59); still the $2 + 2$ reaction was preferred to the Diels—Alder cycloaddition by a 60/40 margin (Figure 1.22).

At the outset of this section, some clear and striking differences between the chemistry of o-benzyne and five-membered cycloalkynes were discussed. In the case of 2,3-dehydrothiophyne the absence of a substantial body of ene chemistry[19] and the increased participation of a pathway leading to $2 + 2$ cycloadducts serve to reinforce these differences. The authors note that the unusual reaction conditions employed, in particular the temperature, may serve to negate the thermodynamic advantage traditionally associated with concerted reactions (e.g., Diels—Alder, ene) and hence lead to an apparent favoring of alternate pathways. Once again, explanations based on increased diradical character or symmetric/antisymmetric molecular orbital inversion (discussed previously) in thiophyne 50 are also plausible, and once again further experimental probing is required.

Reinecke's group has also attempted to generate 3,4-dehydrothiophene (60) by way of pyrolysis of anhydride 61 (Figure. 1.23).[17a] None of the products was consistent with the intervention of the thiophyne intermediate.

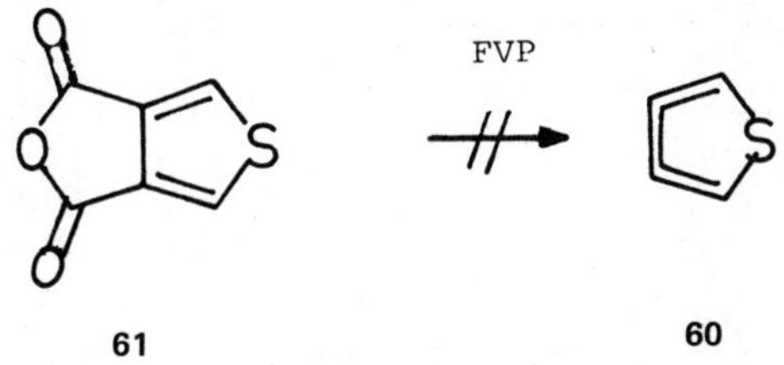

FIGURE 1.23.   Attempted generation of 3,4-dehydrothiophene.

V.   REFERENCES

1.*   R. V. Stevens and G. S. Bisacchi, *J. Org. Chem.*, 47, 2393 (1982).

2.   R. V. Stevens and G. S. Bisacchi, *J. Org. Chem.*, 47 2396 (1982).

3.*   W. M. Best and D. Wege, *Tetradedron Lett.*, 22, 4877 (1981).

4.*   (a) H. Hart, C. Lai, G. Nwokogu, S. Shamouilian, A. Teuerstein; and C. Zlotogorski, *J. Am. Chem. Soc.*, 102 6649, (1980). (b) H. Hart, S. Shamouilian and Y. Takehira, *J. Org. Chem.*, 46, 4427, (1981). (c) H. Hart and S. Shamouilian, *J. Org. Chem.*, 46, 4874 (1981).

5.*   M. B. Stringer and D. Wege, *Tetrahedron Lett.*, 21, 3831 (1980).

6.*   (a) T. P. Lockhart, C. B. Mallon, and R. G. Bergman, *J. Am. Chem. Soc.*, 102, 5976 (1980). (b) T. P. Lockhart, P. B. Comita, and R. G. Bergman, *J. Am. Chem. Soc.*, 103, 4082 (1981). (c) T. P. Lockhart and R. G. Bergman, *J. Am. Chem. Soc.*, 103, 4091 (1981).

7.   P. Saxe and H. F. Schaefer III, *J. Am. Chem. Soc.*, 102, 3239 (1980).

8.*   O. L. Chapman, J. Gano, P. R. West, M. Regitz and G. Maas, *J. Am. Chem. Soc.*, 103, 7033 (1981).

9.*   O. L. Fitjer, U. Kliebisch, D. Wehle, and S. Modaressi, *Tetrahedron Lett.*, 23, 1661 (1982). See also J. C. Gilbert and M. E. Baze, *J. Am. Chem. Soc.*, 105, 664 (1983).

10.   For a discussion of these points as they apply to 1,8-dehydronaphthalene, see R. H. Levin, "Arynes," in M. Jones, Jr., and R. A. Moss, Eds, *Reactive Intermediates*, Vol. 1, Wiley, New York, 1978.

11.   For perspectives on the mechanism of the ene reaction see (a) H. M. R. Hoffmann, *Angew. Chem. Int. Ed. Engl.*, 8, 556; (b) W. Oppolzer and V. Snieckus, *Angew. Chem. Int. Ed. Engl.*, 17, 476 (1978); (c) L. M. Stephenson, M. J. Grdina, and M. Orfanopoulos, *Acc. Chem. Res.*, 13, 419 (1980).

12.   J. Nakayama, T. Segiri, R. Ohya, and M. Hoshino, *J. Chem. Soc., Chem. Commun.*, 791 (1980).

13.*   P. G. Gassman and I. Gennick, *J. Am. Chem. Soc.*, 102, 6863 (1980).

14.   P. G. Gassman and J. J. Valcho, *J. Am. Chem. Soc.*, 97, 4768 (1975).

15.   See footnote 27 in reference 4b.

16.*   J. M. Bolster and R. M. Kellogg, *J. Am. Chem. Soc.*, 103, 2868 (1981).

17.*   (a) M. G. Reinecke, J. G. Newsom, and L-J. Chen, *J. Am. Chem. Soc.*, 103, 2760 (1981). (b) M. G. Reinecke, J. G. Newsom, and K. A. Almqvist, *Tetrahedron*, 37, 4151 (1981).

18.   G. Wittig and H. Dürr, *Justus Liebigs Ann. Chem.*, 672, 55 (1964).

19.   In the reaction of thiophyne 50 with propyne, allene 62 is observed as a product.[17a] This is the only reported potential ene reaction of 50.

50   62

2

CARBANIONS

STUART W. STALEY

Department of Chemistry, University of Nebraska-Lincoln, Lincoln,
Nebraska 68588

I. INTRODUCTION

The field of carbanions has continued a steady growth during recent years. The more widespread availability of high-field NMR spectrometers is making a clear impact on the field. A more detailed description of bonding and charge distribution is available through ^{13}C NMR spectroscopy, and new advances toward understanding ion pair structures are occurring through ^{7}Li and ^{6}Li

NMR spectroscopy. The chemistry and structure of unsolvated or slightly solvated carbanions in the gas phase continues to be investigated by ion cyclotron resonance spectroscopy, Fourier transform mass spectroscopy, and by flowing afterglow. In particular, the use of carbanions in organic synthesis is expanding at an amazing rate. There is clearly an increasing need to gain a better understanding of these species on this basis alone.

This chapter is not a detailed review of the literature — instead, it focuses on various problems of current interest, and particularly on problems yet unsolved. In several cases reinterpretations of literature views are offered. Many of these issues cannot be considered to be settled, and it is hoped that this review will stimulate additional work.

II.   THE STRANGE CASE OF THE PYRENE DIANION

In 1978 Müllen reported that careful reduction of pyrene (1) with sublimed lithium in THF-d_8 at $-40°C$ afforded the corresponding dianion (2) whose ^1H NMR spectrum displayed an upfield shift for the center of gravity ($\Delta \langle \delta \rangle$) of 7.2 ppm, much greater than the value of $ca.$ 1.3 ppm expected from a charge-induced shift alone.[1] The additional shift was reasonably ascribed to a peripheral (16 π-electron) paramagnetic ring current.

Several years later Rabinovitz and co-workers reported that reduction of 1 with sodium in THF-d_8 led not to 2 but to the corresponding *tetraanion* whose ^1H NMR spectrum displayed a $\Delta \langle \delta \rangle$ value of 2.85 ppm relative to 1.[2,3] This interpretation was supported by the results of various quenching experiments. The ^{13}C NMR spectrum of this species was reported to display 16 rather than five different signals, and this was rationalized on the basis of two pyrene tetraanions present in about equal concentration that differ in the location of the counterions (3 and 4).

There are several problems with the foregoing interpretation. First, there appear to be at least 18 or 19 peaks in the ^{13}C NMR spectrum of the sample

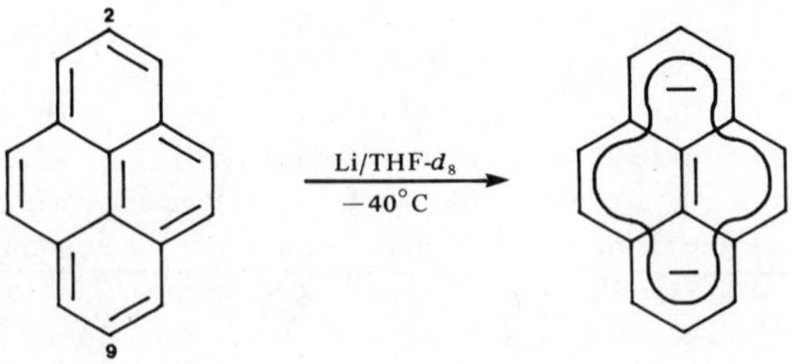

1   2

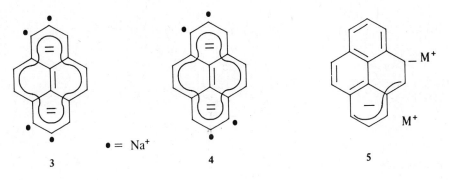

• = Na⁺

3   4   5

attributed to **3** and **4**, of which only 16 are assigned to the species of interest.[3] Second, the center of gravity of the ^{13}C peaks moves upfield only *ca.* 124 ppm on reduction of **1**, whereas a shift of *ca.* 640 ppm would be expected on the basis of the commonly employed proportionality factor of 160 ppm per unit charge.[4] Finally, it does not seem reasonable that the positions of the sodium ions would not be averaged on the NMR time scale.

Recently Tintel et al. reported the generation of the same species (lithium and sodium salts in diethyl ether) and also noted the lack of symmetry in the ^{13}C NMR spectra.[5] Unlike Minsky et al., this group only detected 15 peaks but assumed that an additional peak was buried under a group of peaks at *ca.* 125 ppm. These workers attributed these spectra to a dianion of **1** rather than to a tetraanion. Furthermore, a mixture of dianions differing in the locations of the counterions was rejected on the grounds that a constant 1:1 mixture was unlikely at different temperatures and solvent polarities. Instead, the data were rationalized on the basis of a single dianion (**5**) which is rendered asymmetric by the location of the counterions. As with **3** and **4**, one can again question why the ^{13}C Δ⟨δ⟩ value is more appropriate for a monoanion than for a dianion and also why equivalent positions in **1** do not become averaged on the NMR time scale in **5**.

The uncertainty behind these NMR studies has recently been resolved by Edlund and co-workers, who have attributed these spectra to a monoprotonated product of the pyrene dianion (**6**).[6] The asymmetry in the ^{1}H and ^{13}C NMR spectra is immediately explained by the monoanion, as is the Δ⟨δ⟩ value of the ^{13}C spectrum. Furthermore, the problem of 15 ^{13}C peaks is solved since a sixteenth peak (C_3) is clearly seen at 36 ppm in the C_{sp^3} region. The protons on this carbon are seen as a doublet at δ 4.2 in Figure 1 of Reference 3. Curiously, the adjacent proton (H_2) at *ca.* δ 4.6 has a greatly reduced intensity (see Figure 1 of Reference 3). It appears with its expected intensity in the spectrum of the potassium salt of **6** in THF-d_8.[6] The ^{1}H NMR spectra in Figures 3 and 4 of Reference 3 indicate that 2,9-di-*t*-butylpyrene behaves in exactly the same

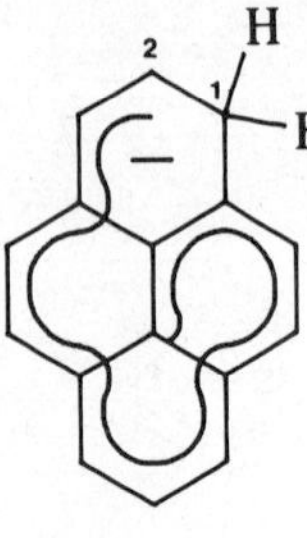

6 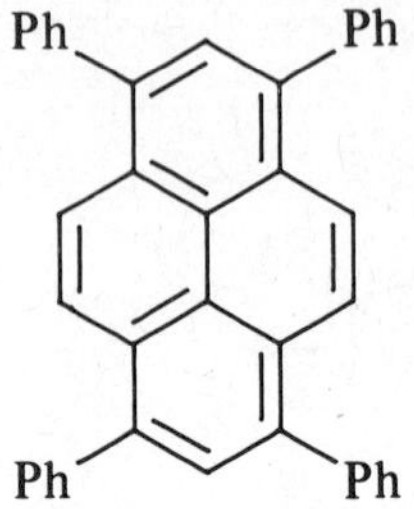   7 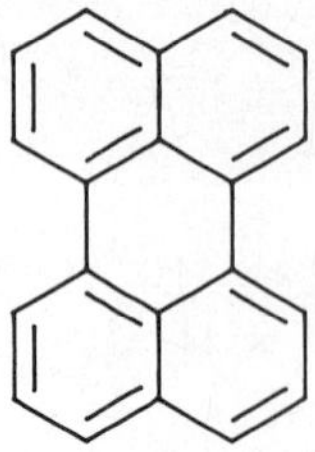   8

way. Furthermore, application of the usual chemical shift/charge correlation suggests that **7** is reduced to a dianion[6] and not to a tetraanion, as claimed.[3]

Edlund et al. believe that **6** may arise by protonation of **2** by residual water in the solvent or by NH_3 when a NH_3/ether mixture is used. Neutral pyrene can possibly provide a proton since it is known that pyrene can be deprotonated by a strong base such as n-BuLi. The pyrene dianion can be observed as a major component in the ^1H and ^{13}C NMR spectra of **6** at $-80°C$, but is unobservable at higher temperatures due to fast electron exchange with pyrene radical anion.[6]

It does not appear that the conclusion of Minsky et al.[3] that Lawler[7] had previously misassigned the ^1H NMR spectrum of the tetraanion of perylene to the dianion is correct. Not only does the ^{13}C chemical shift/charge correlation of this reduction product of **8**[3] favor a dianion assignment, but also the relatively low field positions of the ^1H NMR signals are reasonable for the dianion in view of the relatively large HOMO-LUMO energy gap calculated for this molecule.[8]

The pyrene dianion has proved to be a useful intermediate in the syntheses of various polycyclic aromatic hydrocarbons, including cyclopenta[c,d]pyrene and 3,4-dihydrocyclopenta[c,d]pyrene,[9] 6-H-benzo[c,d]pyrene-6-one,[10] and benzo[e]pyrene.[11] A complete understanding of its properties on reduction could have broad implications.

III.   REDUCTIVE CLEAVAGES REVISITED

In an extensive review of reductive cleavage reactions of carbon–carbon bonds in radical anions and/or dianions, I stressed the key point that sufficient overlap must occur between the electron-rich orbitals and the bond(s) undergoing cleavage.[12] In this context the following two case studies from the recent literature are of considerable interest.

Hammerich and Saveant recently reported that the C_9–C_{10} bond of 9,9′-bianthryl could be directly cleaved simply by addition of two, three, or four electrons to the π-system.[13] This does not seem reasonable in light of the point made in the preceding paragraph and also in light of the fact that Huber and Müllen could prepare clean, stable solutions of the *tetraanion* of 9,9′-bianthryl (lithium salt) in THF-d_8 at $-20°$C.[14] Insight into the mechanism of this reaction can be gained from consideration of the following observations. Cyclic voltammery of 9,9′-bianthryl in dry DMF containing Bu$_4$NI shows two reversible waves at -1.83 and -2.15 V versus SCE (compare -1.94 and -2.56 V for anthracene) and two irreversible waves at -2.74 and -2.85 V.[13] A repetitive scan voltammogram which included these waves showed the rapid formation of anthracene. Adventitious water, DMF, or Bu$_4$NI were postulated as possible proton sources. Preparative reduction at -2.15 V showed slow cleavage to afford 7% **9**, 30% **10**, and 60% **11 + 12**, whereas electrolysis at -2.80 V afforded 5% **9** and 75% **10**.

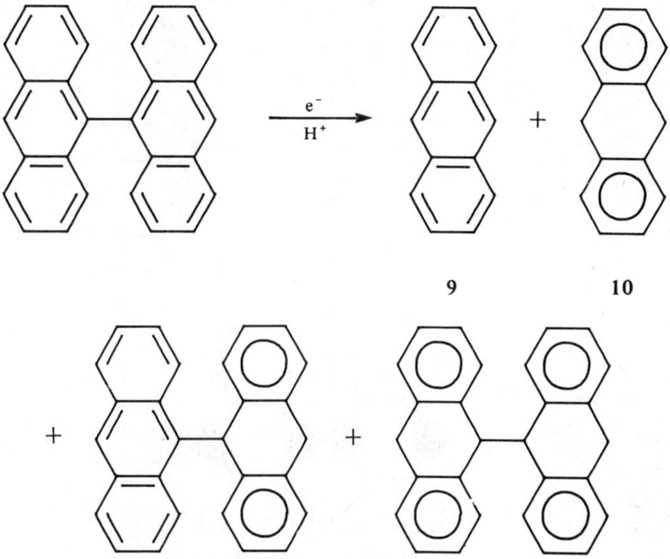

This is undoubtedly a very complex reaction, the course of which is potentially dependent on the concentration of mono-, di-, tri-, and tetraanions present, the concentration of protons, the relative rates (both forward and reverse) of cleavage and protonation, the role of ion pairing, and other factors. However, I believe that the cleavage step must take place by way of species with sp^3 carbons at the C$_9$ and C$_{9'}$ positions (such as **13**). This permits the highest occupied orbital in the precursor (**14**) to overlap with the bond undergoing cleavage.

The ^{1}H NMR spectra of both anthracene and 9,9′-bianthryl display peaks

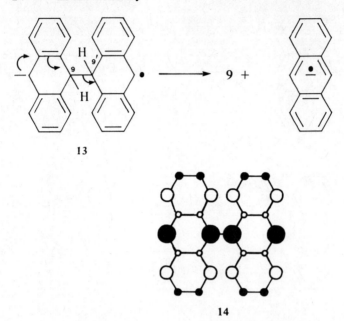

13

14

at almost identical positions with the exception of an upfield shift of two two-proton signals of *ca.* 1 ppm ($H_{1,8}$) and *ca.* 0.5 ppm ($H_{2,7}$) on going from the monomer to the dimer.[14] This was reasonably attributed to shielding by the diamagnetic ring current of the neighboring π-system in a conformation in which the two anthryl groups are approximately perpendicular to each other. In the tetraanion of 9,9'-bianthryl the $H_{1,8}$ signal is assigned to a doublet *ca.* 0.5 ppm downfield from another doublet assigned to $H_{2,7}$, based on a presumed perpendicular conformation and paramagnetic ring currents in the anthryl groups.

Curiously, the doublet assigned to $H_{1,8}$ is shifted only *ca.* 0.1 ppm downfield from the $H_{1,4,5,8}$ multiplet in 9^{2-} whereas the doublet assigned to $H_{4,5}$ is shifted *ca.* 0.4 ppm *upfield* of the latter signal.[14,15] It is possible that these assignments are reversed. One of the two antibonding orbitals that become occupied on reduction of 9,9'-bianthryl to the tetraanion is nonbonding at C_9-C_9', but the other is bonding at this bond (14). Thus one might expect that the anthryl groups in the tetraanion would rotate to a conformation closer to coplanarity than in 9,9'-bianthryl itself. This could cause a small shielding of $H_{1,8}$ compared to 9^{2-} owing to the neighboring π-system but would leave the chemical shift of $H_{2,7}$ relatively unaffected.

A reductive cleavage similar to that just proposed has been recently reported by Edlund and Eliasson.[16] These workers showed that the previously reported reduction of azulene (15) with lithium in THF-d_8 to the corresponding

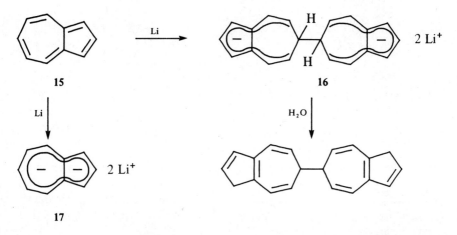

dianion[17] actually affords a dimer of the azulene radical anion (16). Dimer 16 shows no evidence for a paramagnetic ring current, and the C_{sp^3}—C_{sp^3} link is clearly seen as signals at δ 2.54 and 43.8 ($J_{13_{C-H}}$ = 123 Hz) in the ^{1}H and ^{13}C NMR spectra, respectively. Additional contact with lithium in an ultrasonic bath led to reductive cleavage of 16 to afford the azulene dianion (17). The ^{1}H[16] and ^{7}Li[18] NMR spectra of 17 clearly establish it as a paratropic species.

In 1975 Hafner and co-workers reported that treatment of 18 with potassium in THF at −30°C afforded a single product, the 60 MHz ^{1}H NMR spectrum of which displayed a broad singlet at δ 6.61 and a doublet of doublets ($J = 5$, 15 Hz) at δ 7.16 in an area ratio of 3:1, whereas the ^{13}C NMR spectrum displayed five signals.[19] It was claimed that the initial product was 19, which, in the presence of potassium metal at 20°C, slowly suffered reductive cleavage to afford 20.

As in the case of 9,9′-bianthryl, one has a serious problem with a reductive cleavage in which the electron-rich π-system (19 or the corresponding trianion-radical or tetraanion-diradical) is orthogonal to the π-bond undergoing cleavage. The assignment of the *doublet of doublets* ($J = 5$, 15 Hz) at δ 7.16 in the ^{1}H NMR spectrum of 19 also presents a problem. The situation is clarified, however, when it is recognized that the initially observed intermediate is actually the *cis,cis,cis-trans*-cyclononatetraenyl anion (22), which could arise by cleavage

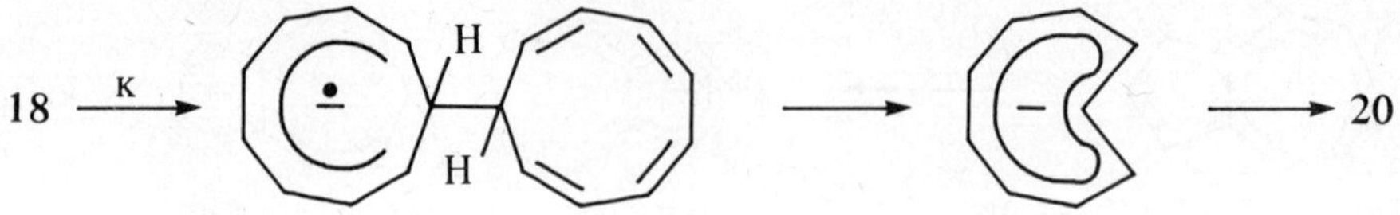

or diradical-dianion

21   22

of **21**. The ^{1}H NMR spectrum of the reduction product of **18** corresponds closely with that reported for **22**[20] with the exception of a triplet ($J = 15$ Hz) at δ -3.52 in **22** which could easily have been overlooked. Anion **22** is known to isomerize to **20** with $\Delta G^{\ddagger}$ between 29.8 and 34.8 kcal/mol in a reaction that is catalyzed by alkali metals.[21]

This interpretation is further supported by the observation that treatment of **18** with potassium *amide* in ammonia-THF-d_8 afforded a deep-red solution whose 90 MHz ^{1}H NMR spectrum displayed an 11-Hz doublet at δ 7.67 and a multiplet centered at δ 6.87 with an area ratio of *ca.* 1:3.[22] This is undoubtedly the previously sought[19] dianion **19**.

Previously, bis-1,3,5,7-cyclooctatetraen-1-yl (**23**) had been reduced to the corresponding dianion[23,24] and tetraanion (**24**).[24,25] The latter species displays an 11-Hz doublet at δ 6.36 (the "ortho" protons) and a broad multiplet at δ 5.3–6.1 with an area ratio of 2:5.[24,25] Paquette et al. concluded that there was no evidence that the two rings in **24** are mutually interactive and that they adopt a perpendicular or near-perpendicular conformation.[25] It was also asserted that the "ortho" protons fall "well within" the deshielding region of the adjacent ring when in this conformation.[25]

I believe that there is insufficient evidence to justify this conclusion. As in the case of the tetraanion of 9,9'-bianthryl, the two highest occupied molecular orbitals in **24** will be either bonding or nonbonding at the inter-ring π-bond in any conformation other than the perpendicular. This would tend to cause a twisting away from the perpendicular and would also lead to a deshielding of the "ortho" protons, as is observed. The same analysis also applies to dianion **19**.[22]

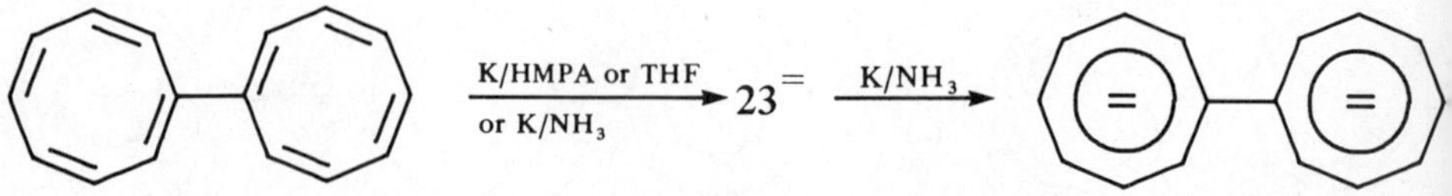

23   24

IV.   THE RISE AND FALL AND RISE OF BISHOMOAROMATICITY

An interesting example of how ideas vary in their degree of acceptance as new facts and new tools are brought to bear is provided by the concept of bishomoaromaticity in delocalized anions.

Nearly two decades ago Brown and Occolowitz[26] reported that 26 underwent allylic H/D exchange in DMSO-d_6/KOt-Bu more than 10^4 times faster than did 25. These observations were explained on the basis of a bishomoaromatic

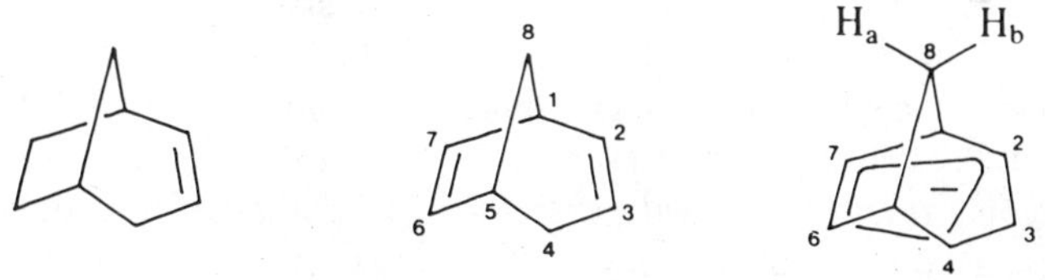

25   26   27

transition state arising from delocalization of negative charge to the $C_6 - C_7$ double bond. This interpretation appeared to receive strong support from the direct observation by ^{1}H NMR spectroscopy of anion 27 by Brown[27] and by Winstein and co-workers.[28] Both groups observed that H_6 and H_7 of 27 (potassium salt in THF-d_8) are shifted upfield by an average of 2.2 ppm relative to the corresponding protons in 26, indicating charge dispersal in the anion. Furthermore, H_{8a} and H_{8b} are shifted upfield by an average of 1.1 ppm on going from 26 to 27, consistent with a ring current effect in the anion.

Nearly 10 years later new results were published which raised questions concerning the NMR criteria for bishomoaromaticity. Trimitsis and Tuncay reported that base-promoted allylic H/D exchange was only 3.3 times faster in 29 than in 28 under conditions similar (but not identical) to those employed for 25 and 26 and concluded that there was essentially no π-participation of the $C_6 - C_7$ double bond in 30 in the transition state.[29] Nevertheless, the ^{1}H NMR spectrum of the *lithium* salt of anion 30 in THF-d_8-hexane showed upfield shifts for $H_{6,7}$ and $H_{8a,8b}$ relative to 29 that were 45% and 60%, respectively,

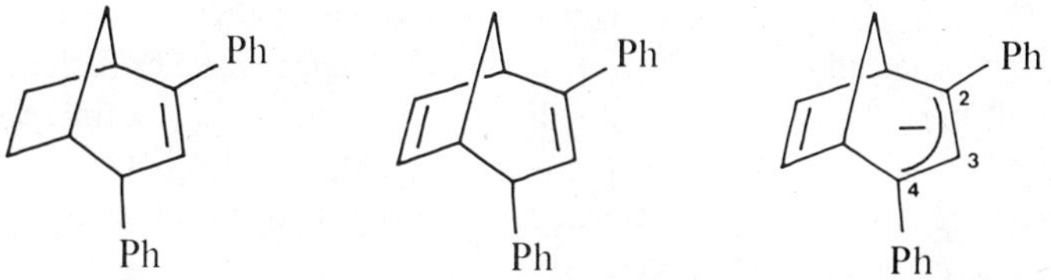

28   29   30

as large as the shifts observed for the corresponding protons on going from **26** to
27.[30] It was concluded that these shifts resulted not from a bishomoaromatic
delocalization of negative charge in **30** but from anisotropic effects due to
charge delocalization at $C_{2,3,4}$.

Despite difficulties of interpretation owing to large anisotropic changes
which might be associated with conformational changes of the phenyl groups,
these results raise some interesting questions. In addition to the interpretation
favored by Trimitsis and Tuncay, namely, that neither the kinetic nor the NMR
results support bishomoaromaticity,[29,30] several other interpretations should be
considered. One possibility is that ring current effects might be observed even
though there is no significant bishomoaromatic stabilization in **30**. Another is
that the transition state for deprotonation of **29** may not reflect the stabilization
resulting from delocalization in **30**.

The foregoing results, along with data for other potentially homoconjugated
anions (such as the cyclohexadienyl anion),[31] as well as results for related
radical[32] and polyenic systems,[33] none of which show definitive evidence for
homoaromatic interactions, led recently to several theoretical investigations of
27 and related anions. Minimal basis set *ab initio* (STO-3G) calculations were
performed at MNDO-[34] or MINDO/3-[35] optimized geometries. Both Kaufmann
et al.[34] and Grutzner and Jorgensen[35] calculated a greater gas phase acidity for
26 than for **25** but concluded that this is *not* due to bishomoaromatic delocalization in the anion derived from the former compound. This conclusion was
based on four lines of argument. (1) No significant geometric changes expected
for bishomoaromaticity (such as a reduction of the C_2-C_7 distance in **27** compared with the corresponding cation) were calculated by MNDO.[34] (2) The
allyl HOMO in **27** was not found to mix significantly with the C_6-C_7 π^*
orbital.[34,35] (3) A total increase of negative charge of 0.05 electrons was calculated for C_6 and C_7 in **27** compared to **26**, but this was attributed to charge
polarization rather than to electron transfer from the allyl portion of **3**.[34] (4) A
greater acidity was calculated for **26** relative to **25** but was attributed to an
inductive effect of the C_6-C_7 double bond in **26**.[34,35] However, the calculated
acidity difference was substantially less than was actually observed.

Brown et al. subsequently presented a number of criticisms of the foregoing
analyses.[36] These authors noted that the semiempirical calculational methods
employed for geometry optimizations may have provided inappropriate geometries. This criticism can be expanded with some additional comments. The
HOMOs, and especially the LUMOs of negatively charged species such as **27**, will
be more diffuse than the corresponding orbitals in their neutral precursors
(e.g., **26**). Thus larger basis sets than provided by MNDO, MINDO/3, or STO-3G
will be required to describe HOMO-LUMO π-interactions and the corresponding
minimum energy geometries adequately.

It was also pointed out that alkenyl inductive effects are not of the mag-

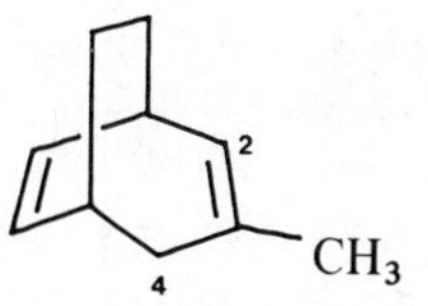

31 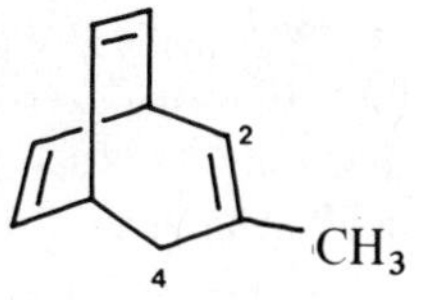   32 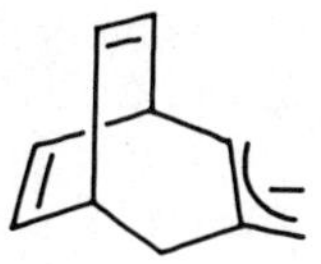   33

nitude to explain adequately the large relative rate effects involved in the H/D exchange reactions.[36] Of particular significance is the observation that **31** and **32** undergo H/D exchange only at C_2 and C_4 in KOt-Bu/DMSO-d_6 and not at the methyl group.[37] Formation of allyl anions with a primary carbon at one end (e.g., **33**) would be expected in the absence of homoconjugative interactions.

Brown et al. performed *ab initio* STO-3G calculations of the interaction of an allyl anion with an ethylene molecule at various angles of approach [0° (planar), 30°, and 90°].[36] It was found that the energy of the allyl HOMO and LUMO decreased and increased, respectively, and that negative charge was transferred to the ethylene moiety as the separation between the two units was decreased to 2.4 Å. However, the significance of these results is unclear since in all cases the *total* energy of the molecular system increases below *ca.* 2.8 Å.

Recently, Christl and co-workers have examined by ^{13}C NMR spectroscopy.[38] The signals for C_6 and C_7 in **27** are shifted upfield by 39.0 and 48.5 ppm, respectively, relative to the corresponding signals in **26**. That these shifts are due to charge delocalization, as in **27**, and not to rapid equilibration between **27a**, **27b**, and **27c**, is based on several lines of evidence. First, substitution of deuteriums at C_2 and at C_2 and C_4 causes small (< 0.5 ppm) *intrinsic* upfield shifts at most of the positions. If **27b** and **27c** were in rapid equilibrium with **27a**, then much larger upfield shifts, particularly for C_6 and C_7, would be expected because the deuteriums at C_2 and C_4 would favor the cyclopropyl sites in **27b** and **27c** over the olefinic sites in **27a**.

Second, there is a total upfield shift of the five carbons of cyclopentadiene of 56 ppm on deprotonation to the cyclopentadienyl anion, but a much larger effect (136 ppm) for C_2, C_3, C_4, C_6, and C_7 in **27** on abstraction of a proton from C_4 of **26**. The additional 80 ppm in the latter reaction is ascribed to partial π-bonding at C_2–C_7 and C_4–C_6 in **27** which causes these carbons to adopt

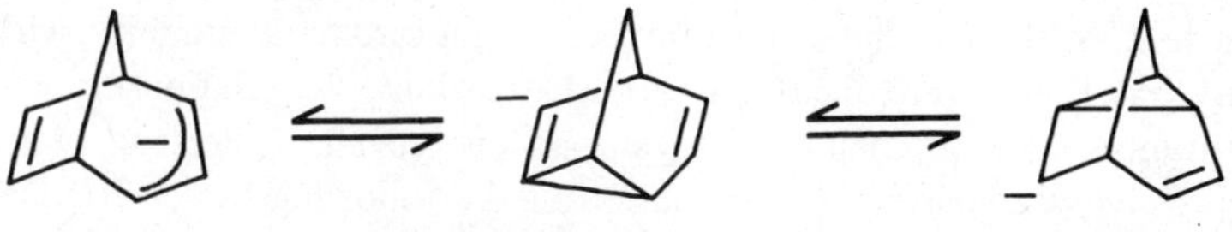

27a   27b   27c

hybridizations intermediate between sp^2 and sp^3. It would be interesting to see how $J_{13_{C-H}}$ changes at these positions in 27 relative to 26.

Recently, Huber et al. have shown that reduction of 34 with lithium in THF-d_8 affords dianion 35.[39] Analysis of the AA'BB' spin systems reveals that the olefinic π-system of 34 ($J_{23} = 11.6\,Hz$, $J_{34} = 7.2\,Hz$) becomes highly delocalized in 35 ($J_{23} = 8.6\,Hz$, $J_{34} = 7.9\,Hz$). Furthermore, the C_{11} methylene protons are shifted upfield by 3.3 ppm owing to the diatropic character of 35. Curiously, the bridgehead methyl protons are also shifted upfield by > 0.7 ppm, whereas the methyl carbons are shifted downfield by 10 ppm. This suggests that some substantial geometry changes may be occurring at the bridgehead positions. Information on energetics is provided by electrochemical studies which show that addition of the second electron to 34 ($E_2^0 \cong -1.8\,V$) is easier than addition of the first ($E_1^0 \cong -2.1\,V$). This is opposite to the electrochemical behavior of two neighboring butadiene moieties.

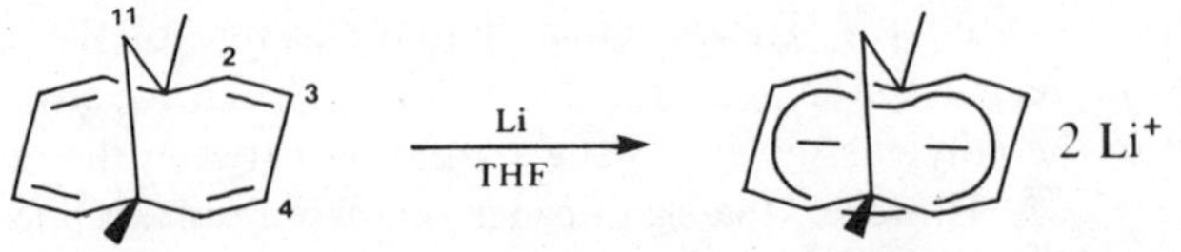

In summary, the current weight of evidence clearly favors the original proposal[26] of bishomoaromatic delocalization in 27 and related anions. Gas phase proton affinities for 27 and analogous systems would be of considerable interest and would put these ideas on a more quantitative footing. Finally, recent work in this field illustrates the dangers of attempting to sort out relatively small effects by performing relatively small basis set calculations on approximate geometries.

V.   ANTIAROMATICITY DISGUISED

A thermodynamic definition of aromaticity is commonly applied, namely, aromatic delocalization confers enhanced π-stability on a cyclic system compared to an isoelectronic open chain reference. Since it is usually difficult to obtain thermodynamic evidence concerning resonance energies, other criteria of aromaticity are often applied, most commonly the presence of a diamagnetic ring current. Although the danger in correlating aromatic stability with diatropicity has been pointed out on various occasions, the relationship between ring currents and aromaticity in bicyclic and polycyclic π-electron systems is not completely understood. Consideration of the following fused $4n$ π-electron ring systems provides some important insights into this relationship.

In 1976 we reported the preparation and direct observation of the 8,8-benzobicyclo[5.2.0]nonatetraenyl anion (**36**).[40] Of particular interest was the observation that the benzo protons were *deshielded* by *ca.* 0.4 ppm relative to the corresponding protons in **37**, despite the greater π-electron density in **36**. Other examples of this phenomenon have been reported, both in neutral hydrocarbons (**38–40**)[41–44] and, more recently, in anions [**41** (and its 8,9-diphenyl analogue) and **42**].[45,46] In all of these compounds (**36, 38–42**), most of which can be considered to be constructed of two fused $4n$ π-electron rings of different size, two important observations should be noted. First, the protons on one of the rings are substantially deshielded relative to appropriate model compounds, (ie., they show evidence for a diamagnetic ring current. In addition to the benzo protons in **36**, this includes the four-membered ring protons in **38, 39**, and **41**, the methyl group in **38**, the five-membered ring protons in **40**, and the ortho phenyl protons in **41** and **42**. Second, the larger of the fused rings can be considered to range from weakly paratropic to moderately diatropic, but is, in all cases, *less diatropic* than the smaller ring.

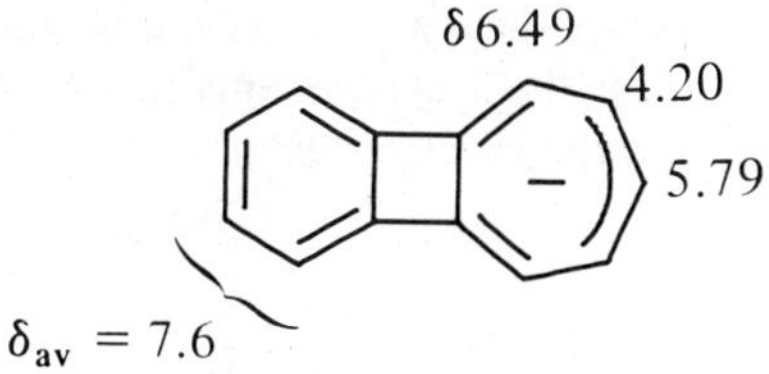

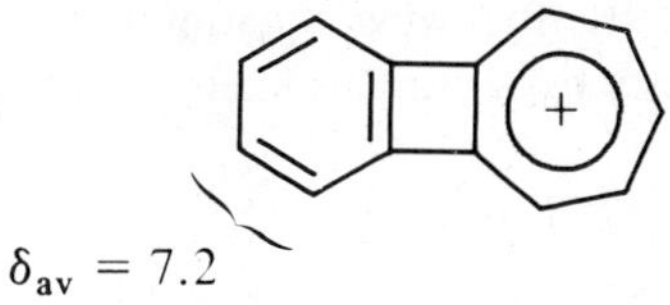

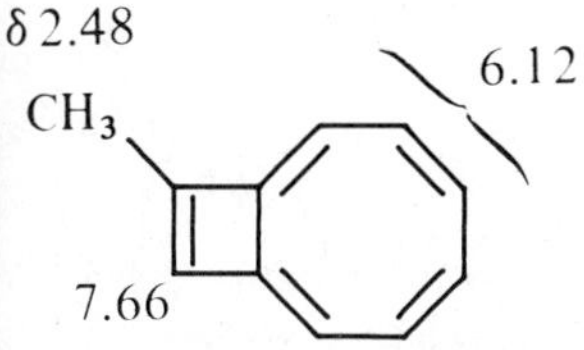

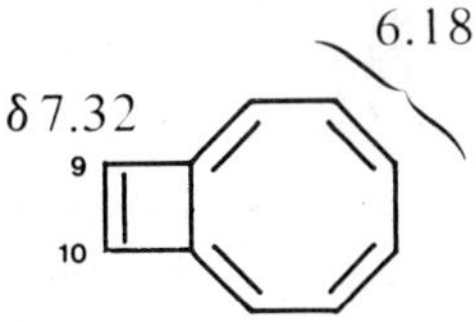

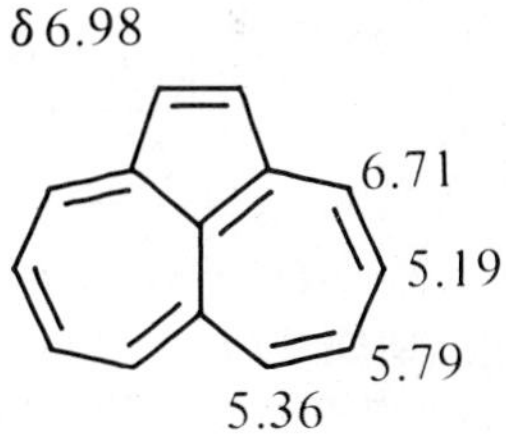

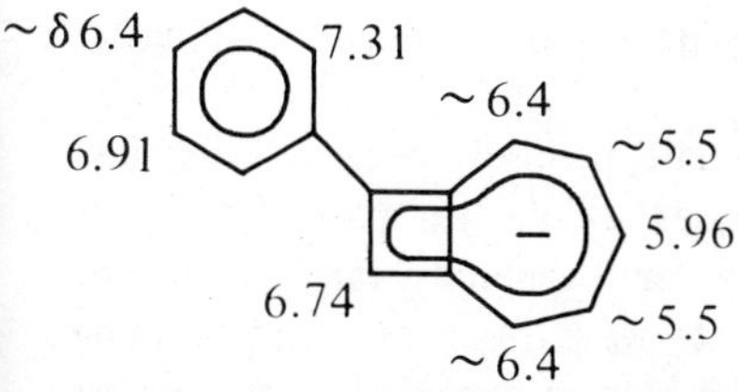

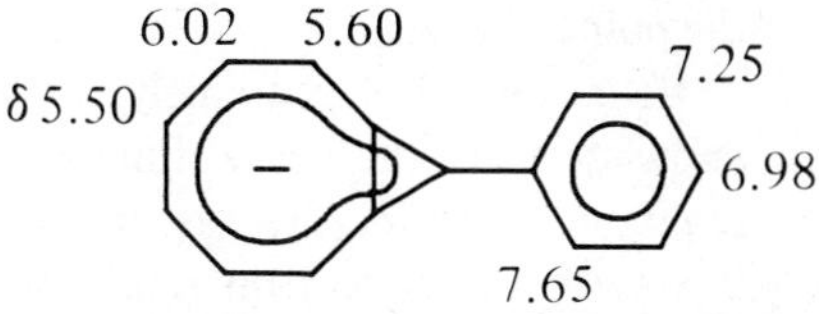

The existence of both diatropic and paratropic rings in the same molecule was first recognized by Jung.[47] Subsequently Schröder recognized the dichotomy in the chemical shift data for **38** and considered the question of the aromaticity of this molecule to be unresolved.[41] Recently Kabuto and Oda have concluded, primarily on the basis of an X-ray analysis of the molecular structure of the 9,10-diphenyl derivative of **39**, that **39** is a 10-π-electron aromatic compound, although its aromaticity is weaker than that of azulene.[43] In contrast, Aihara has concluded from a theoretical study that **39** is diatropic but antiaromatic in nature.[48] This designation of **39** as antiaromatic has been reinforced by Allinger on the basis of molecular mechanics calculations (MMP2), although the calculated molecular structure showed much more π-bond fixation than did the experimental structure.[49]

As of this time, no clear-cut picture has emerged regarding the exact nature of compounds such as **36** and **38–42**. Since the stabilization or destabilization associated with cyclic delocalization in these compounds is likely to be small, it will be difficult to obtain definitive thermodynamic evidence. However, the following physical model should go a long way toward explaining the unusual ^{1}H NMR chemical shifts of these compounds.

In 1950 Ramsey, building upon Van Vleck's theory of molecular magnetism, derived the following equation (given here for the z component) for the shielding tensor σ_{zz} arising from electron motion in the xy plane[50,51]

$$\sigma_{zz} = \sigma_{zz}^d + \sigma_{zz}^p$$

$$= \frac{e^2}{2mc^2} \left\langle 0 \left| \frac{x^2 + y^2}{r^3} \right| 0 \right\rangle - \left(\frac{e\hbar}{2mc} \right)^2 \sum_n \left\{ \frac{\langle 0|L_z|n\rangle \langle n|(2L_z/r^3)|0\rangle}{E_n - E_0} \right.$$

$$\left. + \frac{\langle 0|(2L_z/r^3)|n\rangle \langle n|L_z|0\rangle}{E_n - E_0} \right\} \tag{2.1}$$

where the wave functions for the ground state 0 and the excited state n have energies E_0 and E_n, respectively. The diamagnetic (first-order perturbation) term σ_{zz}^d involves only motion of electrons in the ground state and is positive (shielding) at the point of origin, generally taken to be the center of the ring for cyclic systems. The paramagnetic (second-order perturbation) term σ_{zz}^p involves a sum over all excited states and is generally negative (deshielding) at the point of origin.

Equation 2.1 is extremely difficult to evaluate precisely because a detailed knowledge of the wave functions and energies of all the excited states is required. However, because of its low energy and generally large degree of mixing in the π-electron rings, the excited state wave function corresponding predominantly to a $\pi \rightarrow \pi^*$ transition from the highest occupied to the lowest

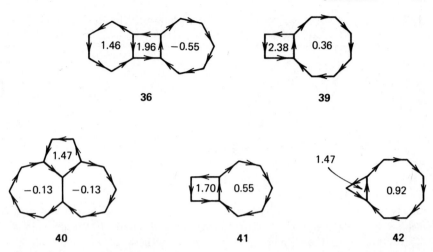

FIGURE 2.1. Directions of the ring currents created by magnetic field-induced mixing of the lowest $\pi \rightarrow \pi^*$ state into the ground state, and magnitudes of total ring currents (sum of first- and second-order terms) in benzene units calculated by the London–McWeeny method.[55] Diamagnetic currents are positive and paramagnetic currents are negative.

unoccupied molecular orbital (HOMO → LUMO) is expected to make a major contribution to the second-order term. The sense of the current created by this transition (the "transition current circulation") is given by the transition length vectors at each bond, which, in turn, can be calculated from the Hückel molecular orbital coefficients.[52] The results of such calculations as well as those of complete London–McWeeny ring current calculations[53–56] are given in Figure 2.1.

What do these calculations tell us? First, in all cases the paramagnetic ring currents flow in opposite directions in the constituent rings. Second, the total ring current is paramagnetic in the larger ring and diamagnetic in the smaller ring, in agreement with the experimental ^{1}H chemical shifts.

How can these results be explained? When a molecule is placed in a magnetic field its electron distribution will be modified to minimize its total energy. This can be achieved by acquisition of orbital angular momentum by some of the electrons through (very slight) mixing of ground and excited state wave functions. This mixing can be visualized as causing a circulation of electrons that creates a magnetic moment oriented parallel to the applied field (a paramagnetic moment), thereby lowering the total energy. However, in some of the orbital mixing, most notably that associated with the HOMO → LUMO transition, the electronic circulation occurs in *opposite* senses in the two rings. The individual moments associated with each ring would therefore be oriented in opposite

TABLE 2.1.   Relative π-electron ring current intensities in **36** calculated by the London–McWeeny method[a]

Ring Current in Ring[b]	Ring Current Contribution from Ring		Total
	A	B	
A	− 0.47	2.85	2.38
B	0.59	− 0.23	0.36

[a]Calculated from Eq. 25 of Reference 55; relative to benzene.
[b]Ring A = four-membered ring; ring B = eight-membered ring.

directions (according to the right-hand rule). Naturally a net energy lowering is achieved by aligning the larger component moment parallel to the applied field. This requires that the smaller component moment be oriented against the field (diamagnetic). Since these moments are proportional to the area of the ring as well as to the size of the reduced π-bond currents,[55] the larger component moment occurs in the larger ring in each of the compounds in Figure 2.1. This means that a diamagnetic ring current is generated in the smaller ring.

The contributions to individual ring currents contributed by each component ring (from London–McWeeny calculations) for **39** are listed in Table 2.1. Note that the major (diamagnetic) contribution to the smaller ring *is due to the larger ring*. This is explained nicely on the basis of the analysis just presented and lends support to our assumption that the HUMO → LUMO transition is a major contributor to, or is at least indicative of, the total paramagnetic term.

Because the diamagnetic ring currents arise primarily from the second-order term in Eq. 2.1 (as opposed to the first-order term, corresponding to the well-known circulation of electrons in the ground state, as in benzene) we term them "second-order diamagnetic ring currents." Ironically, although the latter have sometimes been taken as evidence for aromaticity, they actually arise from the same factor (a small HOMO-LUMO energy gap) which is associated with anti-aromatic destabilization! Finally, it should be noted that somewhat related approaches have been employed by Cornwell,[57] Santry,[58] and Grutzner[59] to explain unusual degrees of shielding in ^{19}F, ^{13}C, and ^{17}O spectra, respectively.

VI.   AROMATICITY AND ANTIAROMATICITY FOILED

The field of carbanions is providing an increasingly rich trove of examples of formally aromatic and antiaromatic molecules. However, sometimes the structural and electronic requirements for aromaticity are in competition with other energetic requirements in a molecule. This is even more apparent in antiaromatic

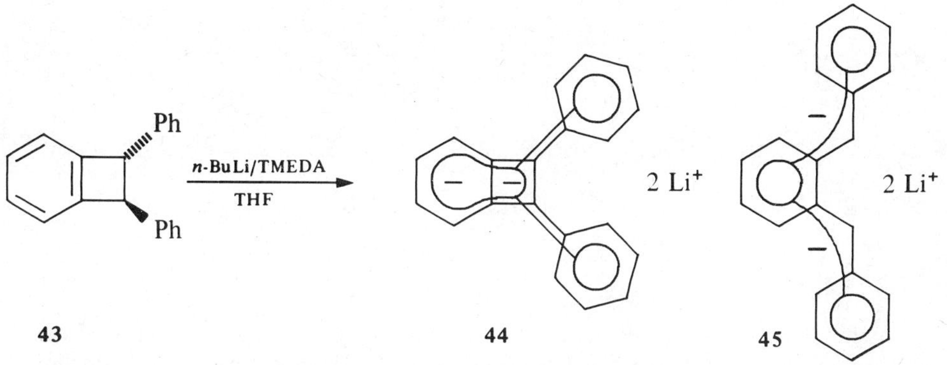

43   44   45

molecules because of the inherent destabilization associated with antiaromaticity. In this section I discuss some interesting structural and electronic
consequences of cyclic π-delocalization in carbanions.

The cyclobutadiene dianion has long been sought as a potential 6π-electron
aromatic compound. Recently Boche and co-workers have generated dianions
44 and **47** by deprotonation of precursors with strong base.[60] Interestingly,
the empirically derived electron density distribution (from the ^{13}C δ values) in
44 is almost the same as in the acyclic analogue **45**. Less than half (*ca.* 48%) of
the negative charge appears in the benzocyclobutene moiety of **44**. Furthermore,
analysis of the AA′XX′ system of the benzo ring shows $^3J_{HH}$ values (π-bond
fixation) which are indicative of an "olefinic" rather than an "aromatic" four-
membered ring. This, along with the relatively low acidity of the monoanion **43**
(which is not deprotonated with *n*-butyllithium), suggests that **44** is not an
aromatic molecule because of unfavorable electron repulsion in the four-
membered ring.

Treatment of **46** with an exceptionally strong base [$(CH_3)_3SiCH_2K$ in THF]
led to formation of dianion **47**.[61] As in the case of **44**, empirical charges determined from the ^{13}C NMR spectrum indicate that only about 36% of the negative
charge is located on the four-membered ring carbons. The MNDO π-bond order
between the four- and six-membered rings is actually slightly larger than that

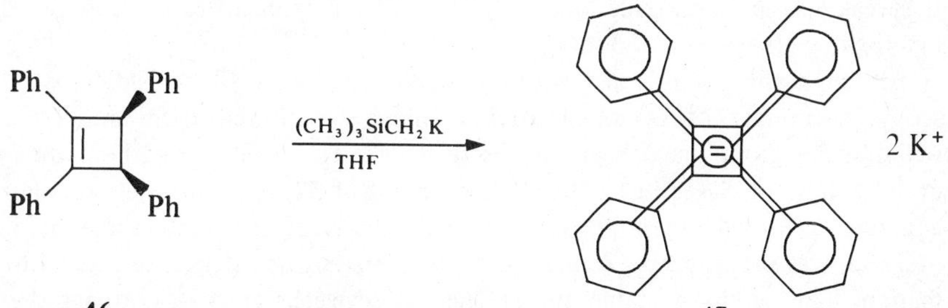

46   47

48	49	50	51	52	53
ΔH_f 209.7	195.6	181.3	15.0	35.9	18.9

(kcal/mol)

within the four-membered ring. These data, along with the low acidity of the monoanion of **46** and the MNDO-calculated enthalpy of formation, do not serve to designate **47** as "aromatic." The conclusion that charge repulsion plays a dominant role was further supported by MNDO calculations which showed that **48** is less stable than isomeric dianions **49** and **50**, whereas the cyclic neutral diboron compound **51** is more stable than isomers **52** and **53**.[62] This calculation also points up the advantage of "Y delocalization" (**50** and **53**).

The cycloheptatrienyl anion is one of the simplest antiaromatic systems. MNDO and MINDO/3 calculations suggest that the planar D_{7h} singlet **54** suffers second-order Jahn–Teller distortions to planar **55** and **56** as essentially isoenergetic ground states.[63] Chemical evidence of the instability of this anion is

54	**55**	**56**

provided by the rapid (possibly two-step) 6 + 2 cycloaddition of the potassium salt to cycloheptatriene in liquid ammonia at $-70°$,[64] and its isomerization at 6×10^{-6} torr in the gas phase to the benzyl anion following exothermic (by 17 kcal/mol) deprotonation of cycloheptatriene by hydroxide ion.[65] Interestingly, the latter isomerization does not occur on deprotonation of cycloheptatriene by the weaker base, methoxide ion, at 10^{-6} torr[66] or upon deprotonation by $^-$OH in a flowing afterglow at 0.5 torr.[67]

Although the parent ion has not been directly observed by NMR spectroscopy, a number of cycloheptatrienyl anions substituted with an electron withdrawing group have been so observed.[68,69] The case of 7-carbomethoxy-cycloheptatriene **57** is instructive.[68] Treatment of **57** with potassium amide in ammonia afforded anion **58** which, on the basis of the small value of J_{34} (4.3 Hz), was proposed by Zwaard and Kloosterziel to exist in a folded conformation. This anion is stable for at least three weeks at $-15°C$ under these

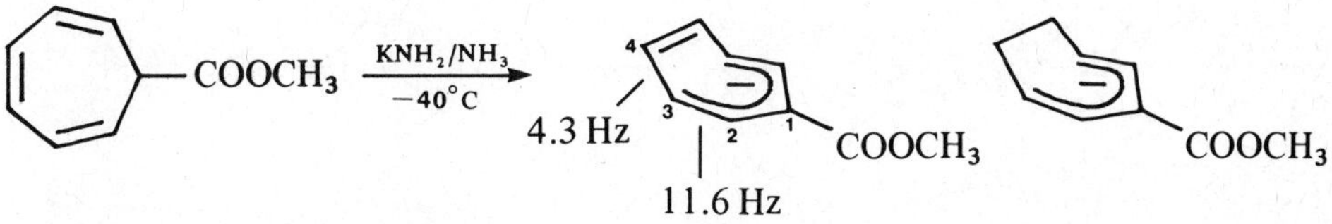

57   58   59

conditions. In fact, this anion could not be quenched by addition of water to the ammonia solution!

When the ^{1}H chemical shifts of 58 are compared with those of dienyl anion 59, it is seen that the ring protons of 58 are shifted upfield by a paramagnetic ring current by *ca.* 2 ppm. If the parent anion (55, 56) is indeed planar, as indicated by semiempirical calculations, then the folding of 58 might be made possible by charge withdrawal by the carbomethoxy group. This, in turn, would suggest that the parent anion remains planar owing to the requirements of charge dispersal.

An indication of the energetic effect of cyclic delocalization in a dibenzo-cycloheptatrienyl anion was obtained by Tolbert and Ali, who determined that 60 is *ca.* 2 pK$_a$ units more acidic than 61.[70] That this is due to a destabilization due to antiaromaticity is suggested by ^{1}H NMR data which indicate that 62 is moderately paratropic.

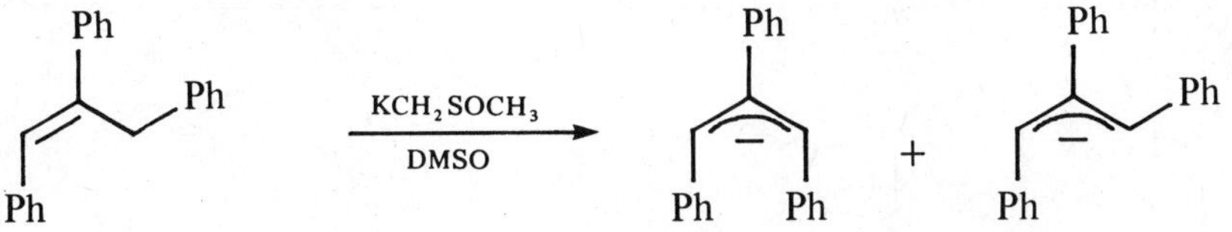

60

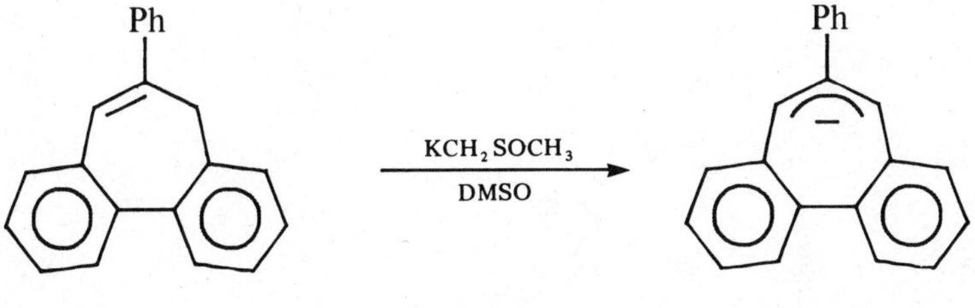

61   62

A related acidity study was reported by Zwaard and Kloosterziel who detected *only* anion 64 by ^{1}H and ^{13}C NMR spectroscopy on treatment of 63 with potassium amide in liquid ammonia.[71] These workers estimate that the

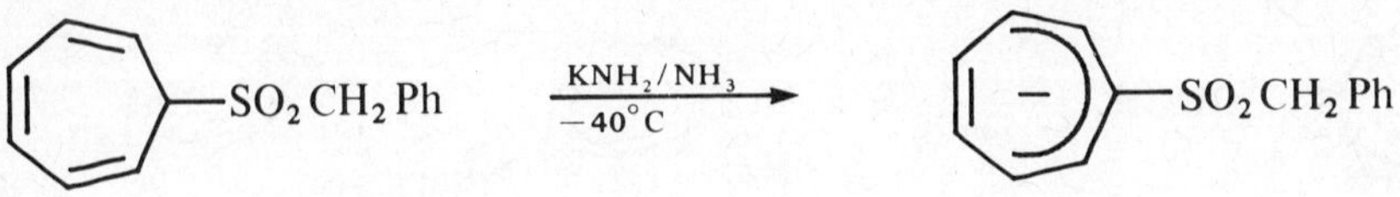

63   **64**

proton of the cycloheptatriene ring is at least 100 times more acidic than that of the benzyl group. This supports earlier conclusions that cycloheptatriene is more acidic than toluene in solution, a relative ranking that also holds in the gas phase.

The cycloheptatrienyl anion ($C_7H_7^-$) has more than a theoretical interest and may ultimately prove to be a useful reactive intermediate in synthesis, playing a role somewhat akin to that of benzyne. For example, $C_7H_7^-$ has been shown to add to C_9 of anthracene in liquid ammonia at $-70°C$ to afford **65** on quenching,[64] and to react further (at $\geqslant 20°C$ in ether) to form **66** and **67**.[72]

Ab initio STO-3G calculations have indicated that benzocyclopropene (**68**) is substantially more acidic than toluene.[73] This appears to be associated with a

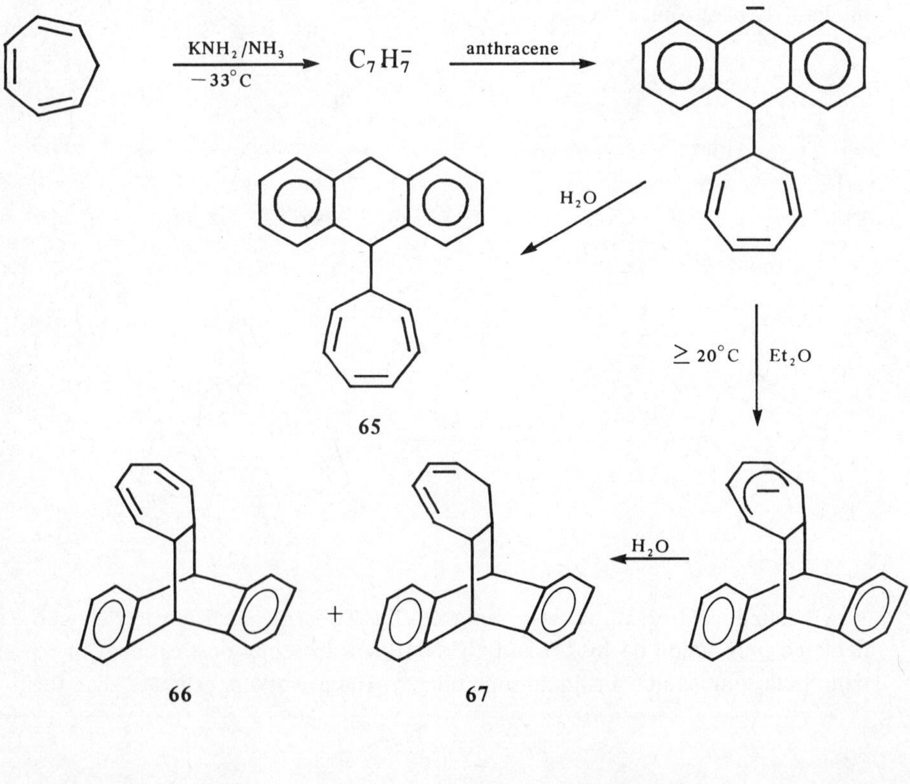

65

66   +   **67**

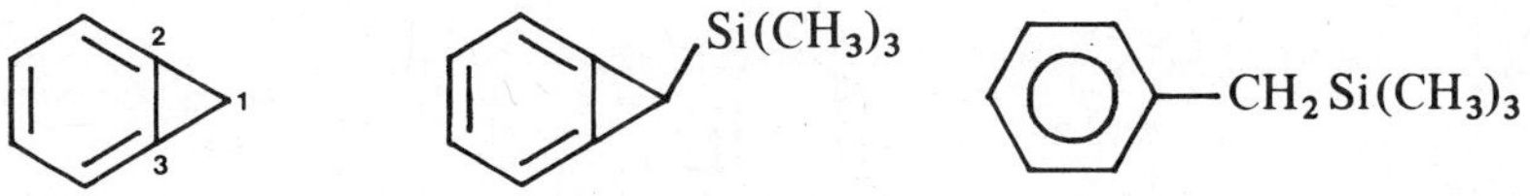

68   **69**   **70**

high polarizability of the C_1-C_2 and C_1-C_3 carbon–carbon bonds. These results are in accord with the greater ease of CH_3ONa/CH_3OH-promoted cleavage of **69** compared with **70**, which suggests pK_a values of *ca.* 36 and 41, respectively.[74]

The cyclooctatetraene ring represents another cyclic 8-π-electron ring system that has received some attention. Most cyclooctatetraene rings are folded in a tub conformation and therefore display little evidence of cyclic delocalization. Anion **72** represents an interesting exception since the eight-membered ring is constrained to be planar or nearly planar. Treatment of **71** with *n*-butyllithium in THF/hexane produced an anion that is formulated as **72**.[75] The argument was

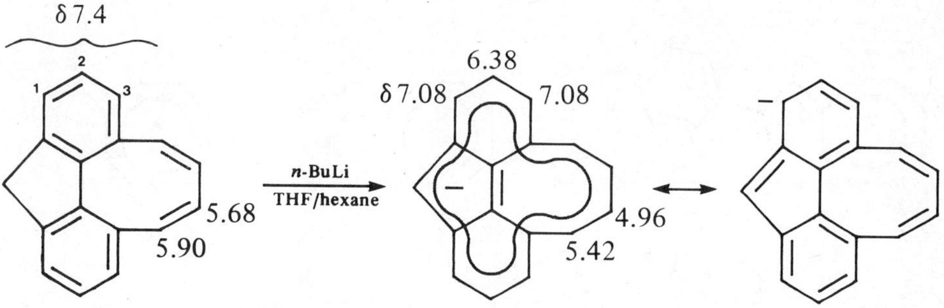

71   **72**   **72a**

presented that the upfield shift of H_2 and H_9, in particular, cannot be explained by a charge shift since H_1 and H_3 should be the positions most shielded by negative charge owing to resonance forms such as **72a**. It was concluded that this is a weakly paratropic peripheral 16-π-electron system (**72**).

Although I agree with the conclusion that **72** is weakly paratropic, I believe that some of the data require a different interpretation. From a theoretical standpoint, it is anticipated that $4n$ π-electron peripheries will suffer distortion or reorganization in some way. One reasonable possibility would be to shift a cyclooctatetraene ring, as in **72a**, toward a dimethylenecyclooctatriene ring, as in **72b**. This would immediately explain the higher field shift of H_2 compared to H_1 and H_3 as a charge-induced shift. This interpretation is supported by the results of self-consistent (ω-technique) HMO calculations which show C_2 to have a greater charge than C_1 or C_3.

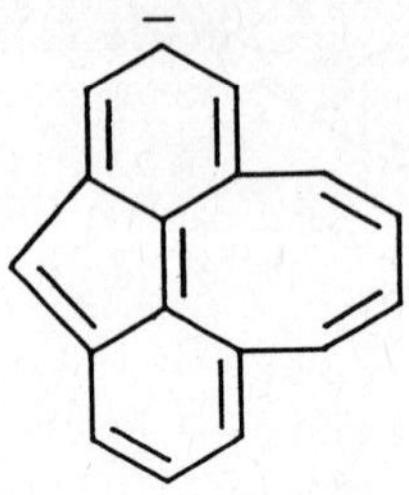

72b

Another interesting example of the behavior of cyclooctatetraene rings concerns the energetics of C_8-C_9 bond rotation in anion **73**. The low energy of this process (14.4 kcal/mol at 4°C) for **73c** is of intrinsic interest.[76] The lack of a dependence of the rate on the concentration of **73c** or of lithium amide, as well as the absence of an observable incorporation of deuterium into **73c** in ammonia-d_3, demonstrates that this bond rotation is an unimolecular process.

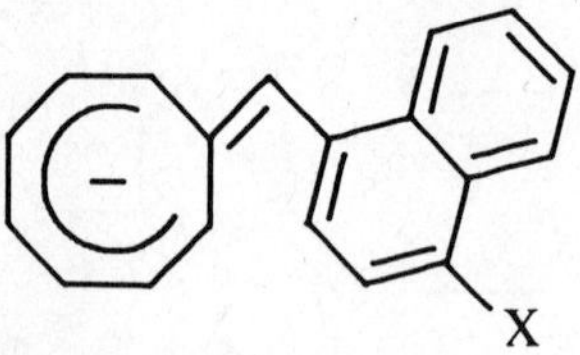

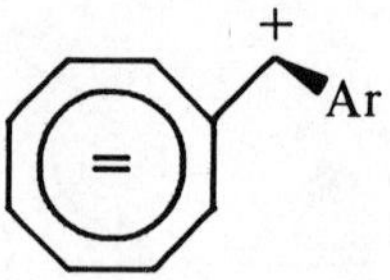

74 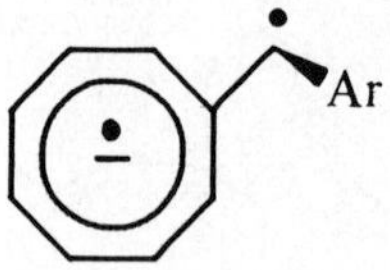   **75**

73a $X = OCH_3$
73b $X = CH_3$
73c $X = H$
73d $X = Cl$

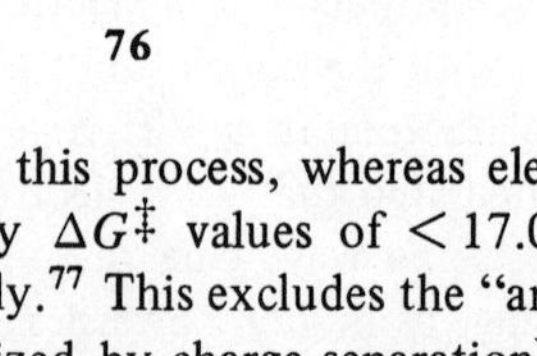

76

Electron-donating groups retard this process, whereas electron-withdrawing groups accelerate it, as shown by $\Delta G^{\ddagger}$ values of < 17.0, 16.4, 14.4, and 12.2 kcal/mol for **73a–d**, respectively.[77] This excludes the "aromatic" transition state **74** (which would be destabilized by charge separation) and radical-anion transition state **75**, but supports transition state **76**. It is proposed that folding of the eight-membered ring in **76** provides a significant driving force for this reaction.

ACKNOWLEDGMENTS

It is a pleasure to acknowledge the National Science Foundation for support of our research in carbanion chemistry. I am also grateful to John Kubicek and Ronald Mueller for performing some of the molecular orbital and ring current calculations cited in this review.

VII.   REFERENCES

1.*   K. Müllen, *Helv. Chim. Acta*, **61**, 2307 (1978).

2.   A. Minsky, J. Klein, and M. Rabinovitz, *J. Am. Chem. Soc.*, **103**, 4586 (1981).

3.*   A. Minsky, A. Y. Meyer, and M. Rabinovitz, *J. Am. Chem. Soc.*, **104**, 2475 (1982).

4.   H. Spiesecke and W. G. Schneider, *Tetrahedron Lett.*, 468 (1961).

5.*   C. Tintel, J. Cornelisse, and J. Lugtenburg, *Recl. Trav. Chim. Pays-Bas*, **102**, 231 (1983).

6.*   a) B. Eliasson, T. Lejon, and U. Edlund, submitted to *J. Chem. Soc., Chem. Commun.* See also b) C. Schnieders, K. Müllen, and W. Huber, submitted for publication.

7.   R. G. Lawler and C. V. Ristagno, *J. Am. Chem. Soc.*, **91**, 1534 (1969).

8.   A. Minsky, A. Y. Meyer, and M. Rabinovitz, *Tetrahedron Lett.*, **23**, 5351 (1982).

9.   C. Tintel, J. Cornelisse, and J. Lugtenburg, *Recl. Trav. Chim. Pays-Bas*, **102**, 14 (1983).

10.   C. Tintel, M. van der Brugge, J. Lugtenburg, and J. Cornelisse, *Recl. Trav. Chim. Pays-Bas*, **102**, 220 (1983).

11.   C. Tintel, J. Lugtenburg, G. A. J. van Amsterdam, C. Erkelens, and J. Cornelisse, *Recl. Trav. Chim. Pays-Bas*, **102**, 228 (1983).

12.   S. W. Staley in B. S. Thyagarajan, Ed., *Selective Organic Transformations*, Vol. 2, Wiley, New York, 1972, p. 309.

13.   O. Hammerich and J. -M. Savéant, *J. Chem. Soc. Chem. Commun.*, 938 (1979).

14.*   W. Huber and K. Müllen, *J. Chem. Soc., Chem. Commun.*, 698 (1980).

15.   K. Müllen, *Helv. Chim. Acta*, **59**, 1357 (1976).

16.*   U. Edlund and B. Eliasson, *J. Chem. Soc., Chem. Commun.*, 950 (1982).

17.   T. Schaefer and W. G. Schneider, *Can. J. Chem.*, **41**, 966 (1963).

18.   R. H. Cox, H. W. Terry, Jr., and L. W. Harrison, *Tetrahedron Lett.*, 4815 (1971).

19.   K. Hafner, S. Braun, T. Nakazawa, and H. Tappe, *Tetrahedron Lett.*, 3507 (1975).

20.*   G. Boche, H. Weber, D. Martens, and A. Bieberbach, *Chem. Ber.*, **111**, 2480 (1978).

21.   G. Boche and A. Bieberbach, *Chem. Ber.*, **111**, 2850 (1978).

22.   S. W. Staley, B. A. Bucklin, C. K. Dustman, and G. E. Linkowski, to be published.

23.   G. R. Stevenson and J. C. Concepción, *J. Am. Chem. Soc.*, **95**, 5692 (1973).

24.   S. W. Staley, C. K. Dustman, K. L. Facchine, and G. E. Linkowski, submitted to *J. Am. Chem. Soc.;* G. E. Linkowski, Ph.D. Dissertation, University of Maryland, 1975.

25.   L. A. Paquette, G. D. Ewing, and S. G. Traynor, *J. Am. Chem. Soc.*, **98**, 279 (1976).

26.*   J. M. Brown and J. L. Occolowitz, *Chem. Commun.*, 376 (1965); *J. Chem. Soc. B*, 411 (1968).

27.   J. M. Brown, *Chem. Commun.*, 638 (1967).

28.   S. Winstein, M. Ogliaruso, M. Sakai, and J. M. Nicholson, *J. Am. Chem. Soc.*, **89**, 3656 (1967).

29.*   G. B. Trimitsis and A. Tuncay, *J. Am. Chem. Soc.*, **97**, 7193 (1975).

30.*   G. B. Trimitsis and A. Tuncay, *J. Am. Chem. Soc.*, **98**, 1997 (1976).

31.   G. A. Olah, G. Asensio, H. Mayr, and P. v. R. Schleyer, *J. Am. Chem. Soc.*, **100**, 4347 (1978).

32.   (a) R. Sustmann and R. W. Gellert, *Chem. Ber.*, **111**, 42 (1978): (b) T. Kawamura, Y. Takeichi, S. Hayashida, M. Sakamoto, and T. Yonezawa, *Bull. Chem. Soc. Jpn.*, **51**, 3069 (1978).

33.   K. N. Houk, R. W. Gandour, R. W. Strozier, N. G. Rondan, and L. A. Paquette, *J. Am. Chem. Soc.*, **101**, 6797 (1979).

34.   E. Kaufmann, H. Mayr, J. Chandrasekhar, and P. v. R. Schleyer, *J. Am. Chem. Soc.*, **103**, 1375 (1981).

35.   J. B. Grutzner and W. L. Jorgensen, *J. Am. Chem. Soc.*, **103**, 1372 (1981).

36.*   J. M. Brown, R. J. Elliott, and W. G. Richards, *J. Chem. Soc., Perkin Trans. II*, 485 (1982).

37.   S. W. Staley and D. W. Reichard, *J. Am. Chem. Soc.*, **91**, 3998 (1969).

38.*   M. Christl, H. Leininger, and D. Brückner, *J. Am. Chem. Soc.*, **105**, 4843 (1983).

39.   W. Huber, K. Müllen, R. Busch, W. Grimme, and J. Heinze, *Angew. Chem., Int. Ed. Engl.*, **21**, 301 (1982).

40.   S. W. Staley, F. Heinrich, and A. W. Orvedal, *J. Am. Chem. Soc.*, **98**, 2681 (1976).

41.   G. Schröder, S. R. Ramadas, and P. Nikoloff, *Chem. Ber.*, **105**, 1072 (1972).

42.   M. Oda and H. Oikawa, *Tetrahedron Lett.*, **21**, 107 (1980).

43.   C. Kabuto and M. Oda, *Tetrahedron Lett.*, **21**, 103 (1980).

44.   A. Minsky, A. Y. Meyer, K. Hafner, and M. Rabinovitz, *J. Am. Chem. Soc.*, **105**, 3975 (1983).

45.   M. Oda, T. Watabe, and T. Kawase, *Tetrahedron Lett.*, **22**, 4249 (1981).

46.   T. Kawase and M. Oda, *Tetrahedron Lett.*, **23**, 2677 (1982).

47.*   D. E. Jung, *Tetrahedron*, **25**, 129 (1969).

48.   J. Aihara, *J. Am. Chem. Soc.*, **103**, 5704 (1981).

49.*   N. L. Allinger and Y. H. Yuh, *Pure Appl. Chem.*, **55**, 191 (1983).

50.   N. F. Ramsey, *Phys. Rev.*, **78**, 699 (1950).

51.   A. Carrington and A. D. McLachlan, *Introduction to Magnetic Resonance*, Harper and Row, New York, 1967, p. 57

52.   S. M. Warnick and J. Michl, *J. Am. Chem. Soc.*, **96**, 6280 (1974).

53.   F. London, *J. Phys. Radium*, **8**, 397 (1937).

54.   R. McWeeny, *Mol. Phys.*, **1**, 311 (1958).

55.   R. B. Mallion, *Mol. Phys.*, **25**, 1415 (1973).

56.   R. C. Haddon, *Aust. J. Chem.*, **30**, 1 (1977).

57.*   C. D. Cornwell, *J. Chem. Phys.*, **44**, 874 (1966).

58.   D. P. Santry, cited by J. Mason, *Adv. Inorg. Chem. Radiochem.*, **18**, 197 (1976).

59.   M. Bawendi and J. B. Grutzner, *Abstracts of Papers*, 185th ACS National Meeting, Seattle, WA, March, 1983, ORGN 192; J. B. Grutzner, personal communication.

60.*   G. Boche, H. Etzrodt, M. Marsch, and W. Thiel, *Angew. Chem., Int. Ed. Engl.*, **21**, 132 (1982).

61.   G. Boche, H. Etzrodt, M. Marsch, and W. Thiel, *Angew. Chem., Int. Ed. Engl.*, **21**, 133 (1983).

62.   T. Clark, D. Wilhelm, and P. v. R. Schleyer, *Tetrahedron Lett.*, **23**, 3547 (1982).

63.   A. W. Zwaard, A. M. Brouwer, and J. J. C. Mulder, *Recl. Trav. Chim. Pays-Bas*, **101**, 137 (1982).

64.   S. W. Staley and A. W. Orvedal, *J. Am. Chem. Soc.*, **96**, 1618 (1974).

65.*   R. L. White, C. L. Wilkins, J. J. Heitkamp, and S. W. Staley, *J. Am. Chem. Soc.*, **105**, 4868 (1983).

66.   C. A. Wight and J. L. Beauchamp, *J. Am. Chem. Soc.*, **103**, 6499 (1981).

67.*   C. H. DePuy and V. M. Bierbaum, *Acc. Chem. Res.*, **14**, 146 (1981).

68.*   A. W. Zwaard and H. Kloosterziel, *Recl. Trav. Chim. Pays-Bas*, **100**, 126 (1981).

69.   K. M. Rapp, T. Burgemeister, and J. Daub, *Tetrahedron Lett.*, 2685 (1978).

70.*   L. M. Tolbert and M. Z. Ali, *J. Org. Chem.*, **47**, 4793 (1982).

71.   A. W. Zwaard and H. Kloosterziel, *J. Chem. Soc., Chem. Commun.*, 391 (1982).

72.   G. Kaupp, H. -W. Grüter, and E. Teufel, *Chem. Ber.*, **115**, 3208 (1982).

73.   C. Eaborn and J. G. Stamper, *J. Organometal. Chem.*, **192**, 155 (1980).

74.   C. Eaborn, R. Eidenschink, S. J. Harris, and D. R. M. Walton, *J. Organometal. Chem.*, **124**, C27 (1977).

75.   I. Willner and M. Rabinovitz, *J. Org. Chem.*, **45**, 1628 (1980).

76.   S. W. Staley, C. K. Dustman, and G. E. Linkowski, *J. Am. Chem. Soc.*, **103**, 1069 (1981).

77.*   S. W. Staley and C. K. Dustman, *J. Am. Chem. Soc.*, **103**, 4297 (1981).

3

CARBENES

ROBERT A. MOSS

Department of Chemistry, Rutgers University, New Brunswick, New Jersey 08903

MAITLAND JONES, JR.

Department of Chemistry, Princeton University, Princeton, New Jersey 08544

I. INTRODUCTION

In the three years between Volume 2 and this latest edition of *Reactive Intermediates* only a few reviews of carbene chemistry have appeared. There is an Account of the chemistry of unsaturated carbenes by Peter Stang[1] and a review in *Tetrahedron* by K. G. Taylor on the chemistry of carbenes flanked by heteroatoms.[2] Taylor's review does not explore the theoretical treatments of carbenes with "unorthodox" neighbors such as lithium and beryllium. These we discuss in this chapter. The more physical aspects of carbene chemistry and, in particular, singlet-triplet energy separations are dealt with in a chapter by E. R. Davidson in *Diradicals*, edited by W. T. Borden.[3] Finally, Lionel Salem has a few pages on carbene structures in *Electrons in Chemical Reactions*.[4] Although we discuss material from these reviews from time to time in this chapter, we neither summarize them nor deal extensively with the subjects they already describe very well.

II. PHYSICAL MEASUREMENTS, CALCULATIONS, AND STRUCTURES

A.   Singlet-Triplet Splitting in Methylene

In the first two volumes in this series we mentioned the experiment of Lineberger and co-workers[5] in which a difference in energy of 19.5 ± 0.7 kcal/mol was determined experimentally for the singlet and triplet states of methylene. Although this number still stands unrefuted — that is, there is no published explanation of why it is incorrect — there seems to be general agreement now that the real value must be about 10 kcal/mol (see discussion in Volume 2).

Indeed the Lineberger paper seems to be sinking without a trace. It is rarely cited, although an exception is the discussion by Davidson mentioned before.[3] It would certainly be nice to know what is wrong with this provocatively long-lived experiment. Note added in proof: Now we do. See: M. Okumura, L. I.-C. Yeh, D. Normand, J. J. H. van den Bresen, S. W. Bustamente, and Y. T. Lee, *Tetrahedron,* In Press (1984).

B.   Singlet-Triplet Splittings in Other Carbenes

The chapter by Davidson[3] gives a nice summary of the effect of substitution on singlet-triplet energy gaps. It has been argued that a change in ground state from triplet to singlet as one changes from CH_2 to CF_2 is a result of the greater electronegativity of fluorine. The electron-withdrawing property of fluorine stabilizes the nonbonding electrons in the σ orbital, which is doubly occupied in the singlet and only singly occupied in the triplet. Thus an electronegativity effect preferentially stabilizes the singlet.[6,7] More electropositive elements such as lithium or boron would favor the ground state triplet.[7] Dilithiocarbene itself has been the subject of detailed calculations,[8] and not only is the triplet found to be the ground state but there are no less than three triplets lower in energy than the first singlet. The energy separation between the lowest triplet and the singlet is estimated at 21.3 kcal/mol. The singlet is linear (see below) and rises in energy as the Li-C-Li angle decreases.

A counterargument runs along more traditional lines and focuses not on electronegativity but upon the π-donating ability of elements such as fluorine. The earlier suggestion that such effects dominate[9] has been supported by the computational evidence of Feller, Borden, and Davidson[10] that along the series CX_2, $(X = NH_2, OH, F, FH+)$ the π-donation effects dominate electronegativity. The opposite effects of π donors and π acceptors have also been emphasized by Houk, Liebman, and their co-workers.[11] For the moment it would seem that traditional π-donation effects are sufficient to explain the change in ground state.

The role of electronegativity and π donation has also been discussed with regard to the shapes of singlet carbenes. The intriguing suggestion has been made that there may be linear singlets.[12,13] Groups such as Li, BeH, and BH_2 all have available empty p-orbitals and may preferentially stabilize an sp hybridized singlet. The more conventional bent structure with carbon roughly sp^2 hybridized will be preferred if the p-orbitals on the flanking atoms are either unavailable, as in methyl, or filled as in nitrogen, oxygen, and the halogens. Whether one describes these linear singlets in conventional terms involving doubly bonded resonance forms[13] or in terms of two-electron three-center bonds[12] seems to us largely a matter of taste.

In Volume 2 there is a discussion of the possibility that some of the chemistry of CBr_2 might be interpreted in terms of a triplet state. There is a subsequent calculation[14] that places the singlet approximately 8 kcal/mol lower than the triplet and thus argues against our suggestions.

Carboxycarbene (H–C–COOH) has been calculated to have a triplet ground state.[15] The initial value for the singlet-triplet gap produced by DZ SCF calculations of 30 kcal/mol was refined by reference to calculations for methylene, addition of d functions on the carbene carbon, and still other ways to give a smaller value of 7.6 kcal/mol.

III.   GENERATION OF CARBENES

A.   Excited Diazo Compounds and/or Excited Carbenes

Although no real progress has been made in this area since Volume 2, at least two new examples of unusual behavior have appeared. Each instance simply shows that direct photolysis of diazo compounds leads to an intermediate different from that in either the thermal or photosensitized decomposition. This kind of behavior was first noted more than 20 years ago by Frey and his group,[16] and a number of instances are noted in Volume 2.

The Shechter group has found that carbene **1** generated from either the diazo compound or diazirine undergoes intramolecular migrations of methyl and phenyl more indiscriminately than it does when generated thermally or by photosensitization.[17] In similar work, Nickon and Zurer have found a much more selective formation of olefins from the spiro diazo compound **2** in the thermal Bamford–Stevens reaction than in the photochemical.[18]

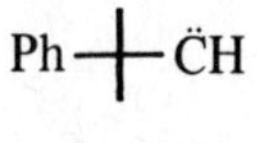

$$Ph {-\!\!\!+\!\!\!-} \ddot{C}H$$

1

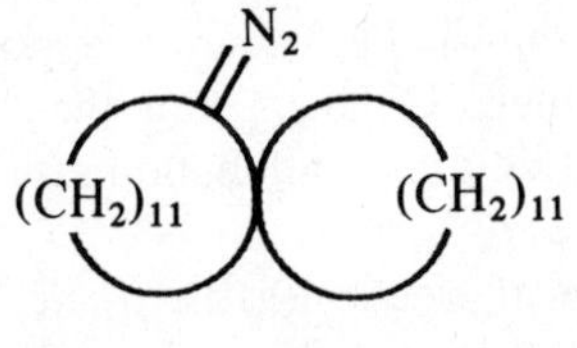

$$(CH_2)_{11} \quad (CH_2)_{11}$$

2

Once again it is not yet possible to distinguish reactions of an excited carbene from those of an excited diazo compound but, as noted in Volume 2, it is certain that something odd is going on.

Excited triplet diphenylcarbene has been observed spectroscopically, along with the excited singlet,[19] and trapped using methanol or isoprene.[20] In the first of these papers[19] the ground state triplet diphenylcarbene is found to be produced in two ways: the major route passes through the lowest singlet diphenylcarbene and the minor one through an excited triplet diphenylcarbene. As the ground state appears within 15 psec, carbon–nitrogen bond breaking and intersystem crossing must each occur in less than this time. Excited triplet carbene has a lifetime of 4 nsec and thus is not a major source of the ground state triplet which appears in 110 psec. The singlet-triplet gap in diphenylcarbene is estimated at 3.9 kcal/mol, which seems in good agreement with previous data.[21]

In the second paper,[20] which must set some sort of a record for misspelled authors' names in the references, the excited triplet is generated in two ways and its decay in methanol or isoprene measured. From these data bimolecular reaction rate constants can be determined. These are $2.1 \pm 0.3 \times 10^9$ L/mol (sec) for isoprene and $3.1 \pm 0.4 \times 10^8$ L/mol (sec) for methanol. The excited triplet reacts 10^4 times faster with isoprene than does the ground state triplet.[22] The trapping experiments were done in acetonitrile, and thus reaction with methanol and isoprene must be faster than trapping of the carbene by acetonitrile. In the absence of a good intuitive feeling about what the reactions of excited carbenes should be like, it is hard to do more than make that observation.

Why might an excited diazo compound produce the kind of changes attributed to carbenes? An answer was suggested by Professor W. Kirmse.[23] We conventionally write diazo compounds in which the carbon bears a partial negative charge and the nitrogen to which it is attached, a partial positive charge. This will not be the case in the photoexcited state. Here the charges will be reversed, and the carbon will bear a partial positive charge. In this case, migration of a methyl group with loss of nitrogen will produce 2-methyl-2-butene from *tert*-butyl diazomethane just as is observed.[24] Kirmse has provided a

$$(CH_3)_3C-\overset{-}{C}H-\overset{+}{N_2} \xrightarrow{\ h\nu\ } (CH_3)_3C-\overset{+}{C}H-\overset{-}{N_2} \xrightarrow{\ CH_3\sim\ } (CH_3)_2C=CHCH_3 + N_2$$

compelling example of this kind of "umpolung" of an excited state in a series of photodecompositions of ketenes.[25] Direct photolysis of ketene **3** in pentane leads to the carbene product **4**. Photolysis at $-60°C$ in methanol, however,

| 3 | 4 | 6 | 5 |

produces an equal amount of the rearranged product **5**. The carbon of the carbonyl group carries a partial negative charge in the excited state and is protonated in methanol to give the cation **6**, which undergoes typical Wagner–Meerwein rearrangement to give, ultimately, **5**. This paper also provides nice examples of the relatively little-used ketene photolysis route to carbenes.

B.   Metal-Mediated Transfers of Carbenes

Two interesting modifications of the Simmons–Smith procedure have been reported. In the first, ultrasound was used to clean continually the surface of mossy zinc during the reaction, thus avoiding the need to prepare a zinc–copper couple or other active zinc reagent (e.g., $R_2 Zn$). With mossy zinc and $CH_2 I_2$ in refluxing dimethoxyethane, good isolated yields of cyclopropanation products were obtained from methyl oleate (91%), α-pinene (67%), and camphene (74%).[26]

The second development employed OH-directed cyclopropanation of allylic β-hydroxysulfoximine diastereomers as the key step in the synthesis of optically active cyclopropanes.[27] For example, racemic ketone **7** reacted with lithium (+)-(S)-N,S-dimethyl-S-phenylsulfoximine to give the diastereomeric β-hydroxysulfoximines, **8**. Chromatographic separation afforded pure diastereomers,

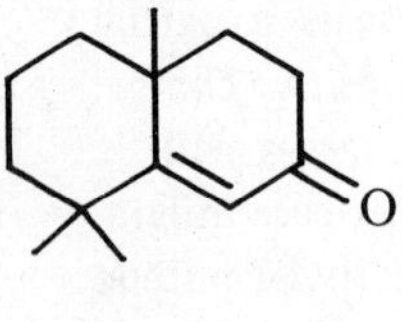

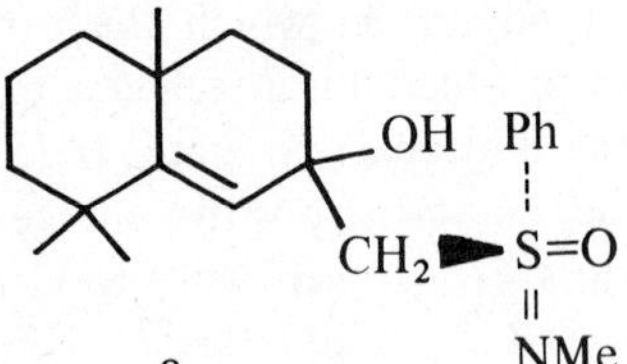

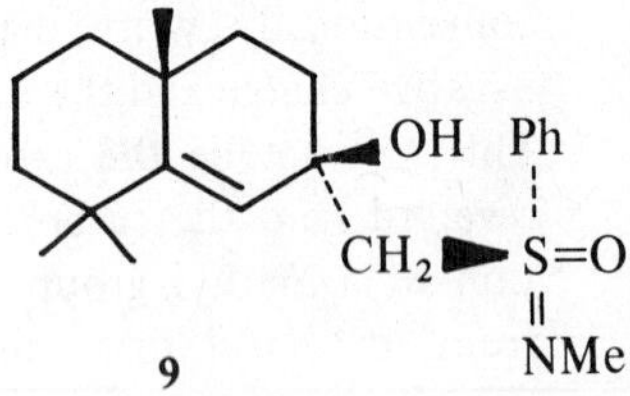

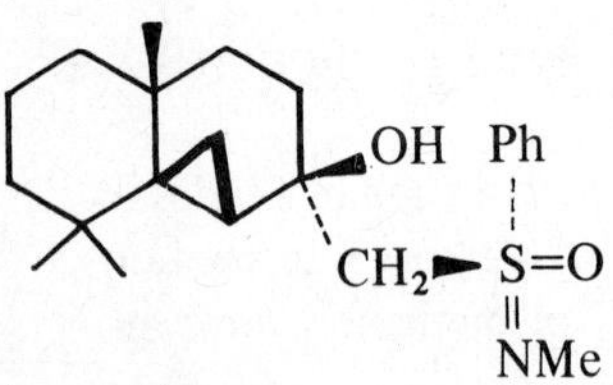

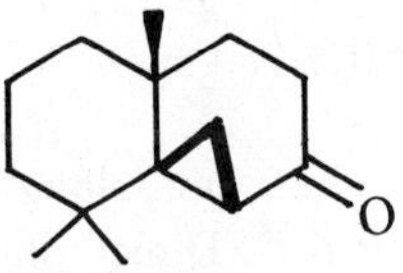

for example, **9**, which underwent stereospecific cyclopropanation [$CH_2 I_2$, Zn(Ag)] to **10** in 80% yield. Thermolysis of **10** at 100°C reversed the sulfoximine addition, affording (−)-dihydromayurone, **11** (92%), which could be converted into (+)-thujopsene, **12**, in 86% yield and 98% optical purity. From the other isolated diastereomer of **8** the enantiomers of **11** and **12** could be obtained similarly. Conversion of **7** to **11** (and its enantiomer) constitutes a formal resolution and directed cyclopropanation of a racemic ketone. The method was successfully applied to several other enones.[27]

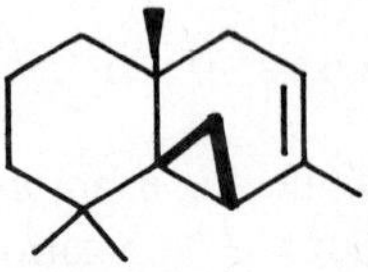

The Simmons–Smith reagent normally behaves as an electrophile. Its reactivity toward terminal methylene groups is low.[28] In contrast, the reagent generated from CH_2N_2 and $Pd(OAc)_2$ (ether, $5°C$) transfers "CH_2" in good yield. Cyclopropanations of terminal double bonds were readily achieved: n-decene (89%), 4-vinylcyclohexene (77%), and limonene (82%).[29] These reactions may proceed by way of prior coordination of the alkene to the Pd, with the ultimate intermediate being a metal-carbene-alkene complex (compare Reference 38).

Metal carbene complexes are discussed in detail in Chapter 4, but we mention here several reagents that transfer alkyl- and arylcarbenic fragments to olefins. Successful iron-based methylene transfer[30] has led to iron reagents designed to transfer more complicated carbenic ligands. For example, the benzylidene iron complex 13 transfers "phenylcarbene" to a variety of alkenes under mild conditions ($-78°C$, CH_2Cl_2, 1–2 h) in good to excellent yields.[31] Alkenes ranging from ethylene (47% isolated yield) to tetramethylethylene (50%) could be

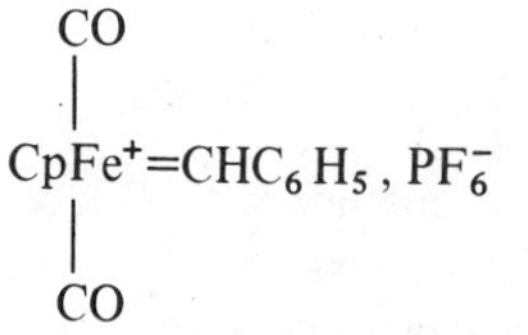

$$CpFe^+{=}CHC_6H_5, PF_6^-$$

with CO ligands above and below (13)

$$(CO)_5W{=}CHC_6H_5$$

14

cyclopropanated, and even the usually unreactive *trans*-stilbene reacted in 96% yield. Additions to *cis*- (57%) and *trans*-butene (52%) were stereospecific. Very high *syn*-phenyl stereoselectivities were observed; some syn/anti ratios at $0°C$ were propene (7.8), styrene or *cis*-butene (≥ 100), cyclopentene (≥ 200), and trimethylethylene (≥ 50). These properties of 13 recall similar reactions of the tungsten complex 14,[32] and a similar mechanistic rationale was offered for cyclopropanations with 13.[31] For attack on a terminal alkene, transition state 15 was proposed.[31,32] Here, R lies preferentially trans to $Cp(CO)_2Fe$, and the partial positive charge developed at the β alkenic carbon is stabilized by ipso interaction with the phenyl ring. Cyclopropane closure occurs by electrophilic attack of C_β on the C–Fe bond, leading to a syn arrangement of R and phenyl in the product.

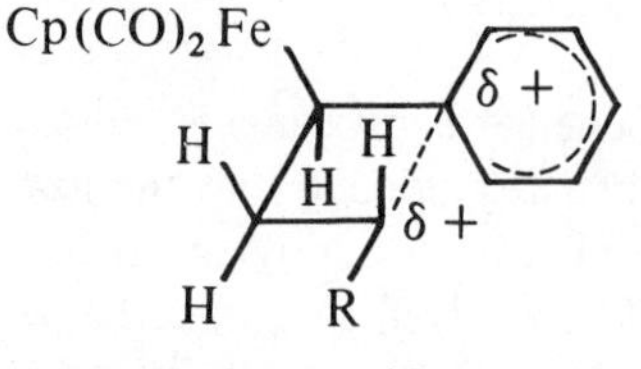

15

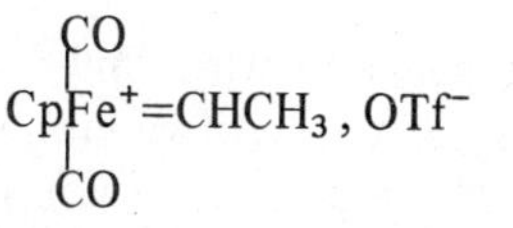

$$CpFe^+{=}CHCH_3, OTf^-$$

16

$$CpFe{-}CHOCH_3$$

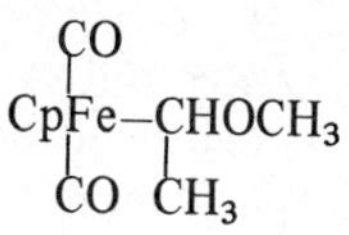

17

Iron carbene complexes also transfer the methylcarbene ligand.[33,34] The unstable complex **16**, prepared by the action of trimethylsilyl triflate on **17** in CH_2Cl_2 solution at $-78°C$ transferred "CH_3CH" to styrene (isolated yield, 75%), cyclooctene (60%), and methylenecyclohexane (86%). Syn stereoselectivities (4–5:1) were again observed.[33] Similarly, unstable complex **18**, prepared *in situ* by methylation of **19** with methyl fluorosulfonate, transferred "CH_3CH" to alkenes in CH_2Cl_2 solution at $25°C$.[34] Some cyclopropanation conversions (determined by gas chromatography) were cyclooctene (70%), 1-decene (70%), *cis*-5-decene (44%), and styrene (67%). Syn stereoselectivities

$$
\begin{array}{cc}
\text{CO} \quad \text{CH}_3 \\
| \qquad | \\
\text{CpFeCHSC}_6\text{H}_5, \text{FSO}_3^- \\
| \quad | \quad + \\
\text{CO} \quad \text{CH}_3
\end{array}
\qquad\qquad
\begin{array}{c}
\text{CO} \\
| \\
\text{CpFeCHSC}_6\text{H}_5 \\
| \quad | \\
\text{CO} \quad \text{CH}_3
\end{array}
$$

18	**19**

were observed with cyclooctene and styrene, but not with 1-decene or cyclododecene. No reactions occurred with *trans*-5-decene or *trans*-stilbene. These CH_3CH transfer reagents are somewhat inconvenient because of their instability and the several step sequences needed to prepare them, but they do represent promising approaches to the ethylidenation of alkenes.

Less successful (thus far) have been Fe complexes designed to transfer dimethylcarbene ligands. Two groups generated **20** by HBF_4 protonation of **21** and observed the cyclopropanation of 1-decene, isobutene, and styrene in 10—40% yields.[35,36] However, products were not obtained from *cis*-5-decene, 4-methylcyclohexane, or cyclooctene.

$$
\begin{array}{c}
\text{CO} \\
| \\
\text{CpFe}^+=\text{CMe}_2, \text{BF}_4^- \\
| \\
\text{CO}
\end{array}
\qquad\qquad
\begin{array}{c}
\text{CO} \quad \text{CH}_2 \\
| \qquad \| \\
\text{CpFe—CCH}_3 \\
| \\
\text{CO}
\end{array}
$$

20	**21**

Successful cyclopropanation with dimethylcarbene (or equivalent) is notoriously difficult because of competitive intramolecular reaction, so that the low yields observed with **20** are perhaps unsurprising. What is pleasantly surprising, however, is the rather good yield (46% based on unrecovered alkene) of hexamethylcyclopropane obtained by reaction of 2,2-dibromopropane with ethereal

n-BuLi in tetramethylethylene at $-70°C$.[37] Presumably, this is a carbenoid reaction of Me_2CBrLi. Cyclopropanation of styrene also proceeded in reasonable yield (44%), but reactions with cyclohexene (27%), 1-decene (9%), and cyclooctene (16%) were less successful. Norbornene did not react. Still, given the simplicity of the procedure, this is a reaction worth remembering.

There has been strong interest in transition metal catalysts, particularly dirhodium tetraacetate, $Rh_2(OAc)_4$, for reactions of ketocarbenes and carboalkoxycarbenes. $Rh_2(OAc)_4$ is a good catalyst for the diazoester/carboalkoxycarbenoid cyclopropanation of alkenes.[38,39] The intermediate carbenoid is electrophilic (dihydropyran reacts ~ 3 times faster than cyclohexene at $22°C$), but subject to steric effects (tetramethylethylene reacts ~ 0.3 times as rapidly as cyclohexene).[38] $Rh_2(OAc)_4$ appears to be more efficient than previously employed Cu catalysts. Good yields of cyclopropanes could be obtained from 1:1 alkene/diazoester in ether at $25°C$ using diazoester/catalyst ratios of 100–200, although it is necessary to add the diazoester very slowly.[39]

With 1- or 2-substituted 1,3-butadienes, $Rh_2(OAc)_4$ gave better overall yields than $Rh_6(CO)_{16}$, $CuCl \cdot P(O-i-C_3H_7)_3$, or $PdCl_2 \cdot 2C_6H_5CN$.[40] Regioselectivities were marked and depended on R in substrates 22 and 23. Using $EtOOCCHN_2/Rh_2(OAc)_4$ at $25°C$, 1-substituted dienes 22 were preferentially attacked at the terminal double bond, whereas 2-substituted dienes 23 mainly

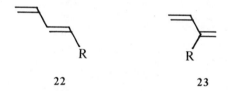

22   23

added at the disubstituted double bond.[28,40] Very large regioselectivities were observed with 22, R = Cl (> 100) and 23, R = OCH_3 (> 200).[40] The regioselectivities observed with the several catalysts and given dienes were proportional, so that "metal carbene regioselectivity" indices could be defined. Relative to $Rh_2(OAc)_4$, $CuCl \cdot P(O-i-C_3H_7)_3$ and $Rh_6(CO)_{16}$ provided more regioselective intermediates (factors of 1.44 and 1.10), whereas the relative regioselectivity was lower with the Pd catalyst (0.48).[40]

$Rh_2(OAc)_4$ was studied as a catalyst for ylide generation by way of reactions of diethyl diazomalonate and ethyl diazoacetate with allyl halides, sulfides, and amines.[41] The same catalyst has also been used in an efficient synthesis of the tricyclo[2.1.0.0^{2,5}]pentanone system.[42] Ethyl or $tert$-butyl diazoacetate was catalytically decomposed in 1,4-diacetoxy-2-butyne to give cyclopropene 24, which, after conversion to diazoketone 25, afforded tricyclopentanone 26 in 25–30% isolated yield by $Rh_2(OAc)_4$ catalyzed intramolecular addition. The

Rh-catalyzed reaction is much superior to earlier tricyclopentanone syntheses based on Cu ketocarbenoid reactions.

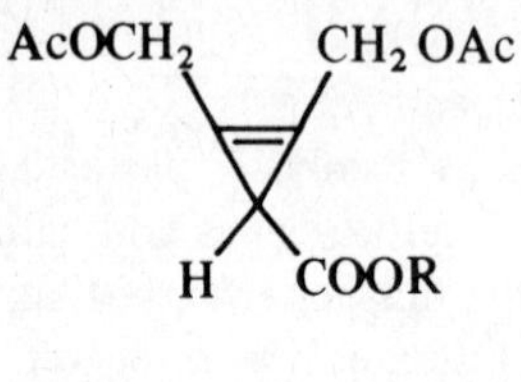

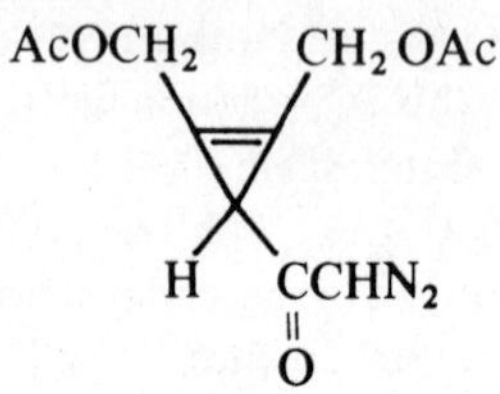

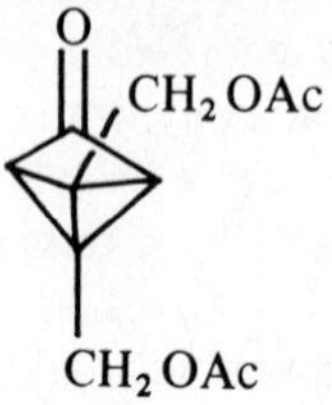

| 24 | 25 | 26 |

Callot used iodorhodium (III) mesotetraaryl porphyrin, **27**, as a catalyst for the carboethoxycarbenoid cyclopropanation of alkenes, demonstrating "unnatural" syn stereoselectivities.[43] With Ar = mesityl, **27** catalyzed the addition of "CHCOOEt" to cyclohexene (60°C in 1,2-dichloroethane) to yield 77% of the 7-carboethoxynorcaranes with syn/anti = 1.17. The isomer ratio

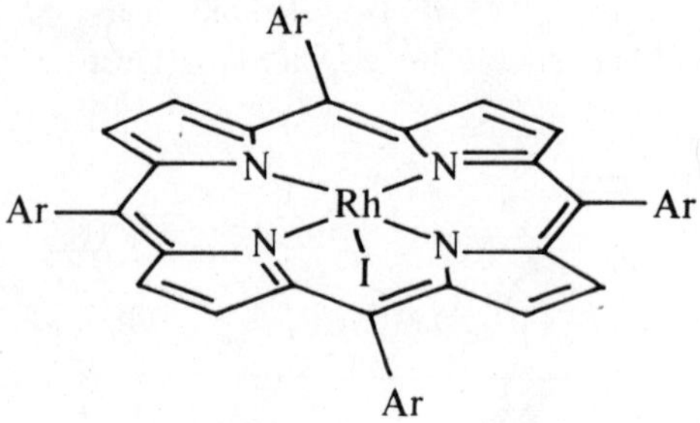

27

was 0.74 with **27**, Ar = phenyl, 0.32 with Rh(II) pivalate, and only 0.12 with CuCl as catalyst. The tendency toward synthesis of the more hindered cyclopropane using **27** was accentuated with more hindered substrates. *cis*-4-Methyl-pentene-2 gave 82% of the CHCOOEt adduct, syn/anti = 6.52 using **27**, Ar = mesityl, but the isomer ratio decreased to 4.9 with **27**, Ar = phenyl, 2.2 with Rh pivalate, and 0.56 with CuCl. A Rh carbene complex, **28**, was suggested as the key intermediate.[43] Because of steric congestion about Rh in **28** (particularly when R = CH_3), the π orbital of a *cis*-alkene can productively approach the carbenic carbon most readily when the alkene's vinyl protons are directed downward, toward the complex, and its alkyl substituents R' are directed away. This orientation leads to the *syn*-carboethoxycyclopropane isomer. The consistent reduction in syn/anti ratios observed when R in **28** is changed from

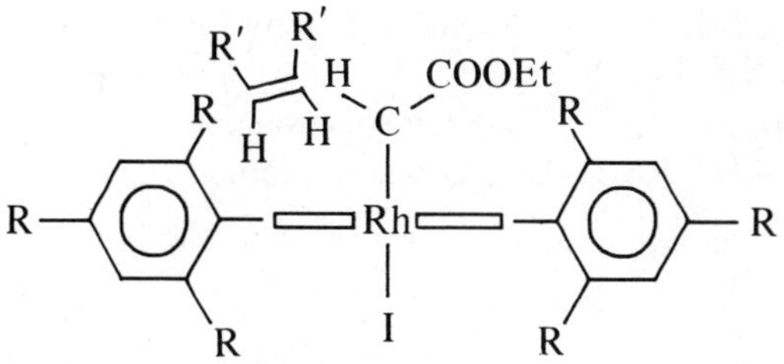

28

CH_3 to H is supportive of this steric explanation, as is the preferential formation of anti isomers when simple Rh catalysts are used.[43]

Rhodium complexes (**27**, Ar = mesityl or phenyl) also catalyze insertion reactions, for example, the insertion of "CHCOOEt" into C—H bonds of *n*-alkanes.[44] Yields were moderate (17–50%), but selectivity on a per-bond basis was sensitive to Ar in **27**. More terminal C—H insertion occurred with the sterically more demanding catalyst. For example, with *n*-decane, the primary/secondary ratio was 0.14 with **27**, Ar = phenyl, but 0.67 when Ar = mesityl. Also, the selectivity increased as the substrate became larger; for example, from 0.44 with *n*-hexane to 0.83 with *n*-dodecane (both with **27**, Ar = mesityl). Parallel increases were not observed when Ar = phenyl.

These phenomena could be explained with a model similar to **28**, in which an alkane C—H bond replaces the alkene's π bond. Attack of a primary C—H bond on the encumbered carbenoid carbon would be easier than attack of a more sterically hindered secondary C—H bond. The mesityl catalyst produces a carbenoid from ethyl diazoacetate that is approximately 30 times more selective for attack at the substrate's terminal CH_3 than is the carbenoid generated by Rh pivalate. This "abnormal" selectivity is opposed to the commonly observed tert > sec > prim insertion sequence displayed by carbenoids. We must note, however that the products of **27**-CHCOOEt with decane C—H bonds are sufficiently distributed over $C_1–C_5$ sites[44] to render the reaction of little synthetic utility at present, despite the unusual per bond selectivity.

$Rh_2(OAc)_4$ catalyzes both intermolecular[45,46] and intramolecular[47] carbenic insertions. Good yields of esters were obtained from the catalyzed decomposition of ethyl diazoacetate in $C_5–C_8$ cycloalkanes.[45] A preference for insertion at C_2 was detected with *n*-alkanes, and a "lipophilicity effect" favored insertion with larger *n*-alkanes. For example, $Rh_2(OAc)_4$-mediated MeOOCCH competitive insertion into cyclooctane C—H bonds was 7.7 times faster than insertion into cyclopentane C—H bonds. This preference disappeared when the carbene was thermally or photochemically generated from the diazoester.[45] "Insertions" of CHCOOR into the O—H bonds of saturated alcohols were catalyzed by

$Rh_2(OAc)_4$. A substantial preference persisted for O–H insertion over C=C addition.[46]

Intramolecular insertion affords cyclopentanone **30** from diazoester **29** in 68% isolated yield.[47] $Rh_2(OAc)_4$ catalysis promotes this reaction under very

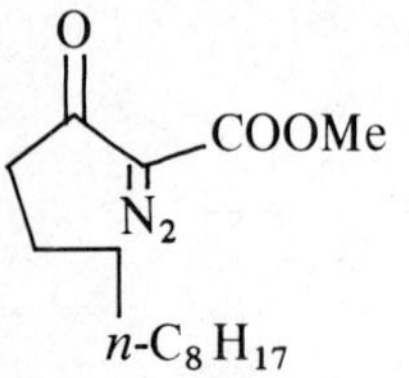

mild conditions (25°C, 30 min in CH_2Cl_2). Related examples indicate utility for the preparation of key intermediates for the synthesis of cyclopentane-containing natural products.[47] Interestingly, with **31**, 48% of the desired allyl-cyclopentanone is formed by C–H insertion. Only 10% of intramolecular C=C addition occurs.

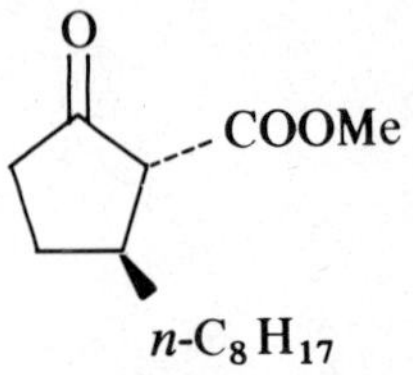

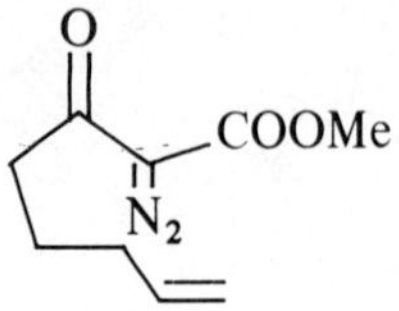

Finally, both $Rh_2(OAc)_4$ and **27** (Ar = phenyl) catalyze the formation of stilbenes from aryldiazomethanes. Yields are good, and there is a marked preference for formation of the Z isomer.[48]

C.   Unsaturated Carbenes

a.   *Alkylidenecarbenes (Methylene Carbenes)*

Gilbert offered an extension of his *in situ* diazoalkene preparation (Eq. 3.1).[49] Condensation of acetone with the anion of dimethyl diazomethylphosphonate presumably affords diazoethene **32**, from which isopropylidenecarbene (dimethylmethylene carbene) **33** arises by nitrogen loss. Substitution of aldehydes

$$(MeO)_2P(O)CHN_2 + Me_2C=O \xrightarrow[\text{THF, }-78°C]{\text{KO-}t\text{-Bu}} (MeO)_2P(O)OK + [Me_2C=C=N_2]$$

32

$$\xrightarrow[-78°C]{16 \text{ hrs}} [Me_2C=C:]$$

(3.1)

33

or aryl ketones for acetone in Eq. 3.1 affords fair to good yields of alkynes. These products are simply rationalized by intramolecular rearrangements of derived carbenes. For example, 2-furfuraldehyde afforded 50% of 2-ethynyl-furan, presumably by way of a 1,2 shift in carbene **34**. Similarly, from acetophenone, 67% of 1-phenyl-1-propyne was formed, perhaps by way of **35**. The intramolecular rearrangement to alkynes is too rapid to divert where the inter-

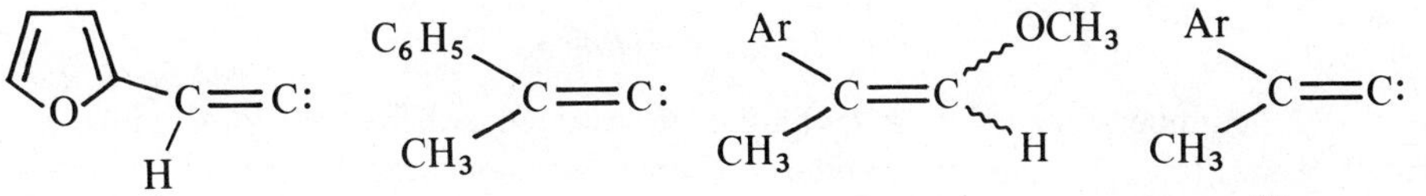

34 **35** **36** **37**

mediate RR'C=C: has R or R'=H, phenyl or a phenyl group with an electron-donating group. With *p*-nitroacetophenone as the substrate, however, 32% of the intermediate can be diverted to **36** by reaction with methanol, while 55% of the alkyne is concomitantly formed. Apparently, when Ar = *p*-nitrophenyl, carbene **37** rearranges competitively with solvent trapping by methanol. The electron-withdrawing *p*-nitrophenyl group would slow the intramolecular re-arrangement of **37** (relative to that of **35**) whether aryl or methyl migration was responsible for alkyne formation.

The diazoalkene method of Eq. 3.1 may prove to be quite useful. Diazo-alkene **32** can be intercepted by 1,3-dipolar addition to 3,3-dimethylcyclo-propene. Simultaneously, 16% of nitrogen-free product **39** is formed, perhaps by capture of carbene **33**, followed by fragmentation of intermediate bicyclo-butane **38** (Eq. 3.2).[50] In an intramolecular variant of this reaction, the

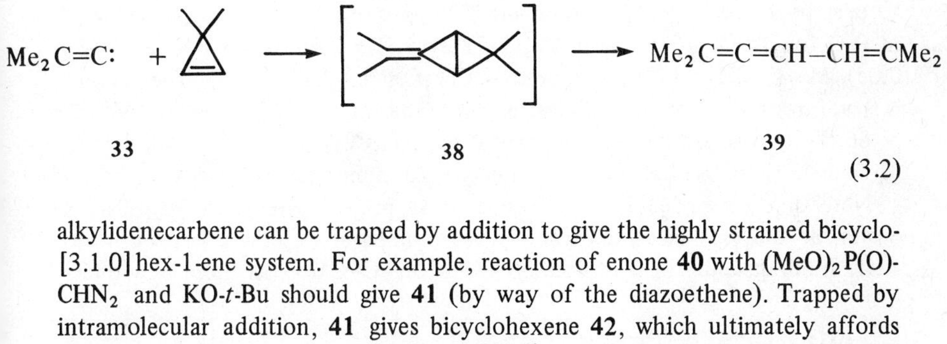

$$Me_2C=C: \ + \ \bigtriangleup \ \longrightarrow \ \left[\ \right] \ \longrightarrow \ Me_2C=C=CH-CH=CMe_2$$

33 **38** **39**

(3.2)

alkylidenecarbene can be trapped by addition to give the highly strained bicyclo-[3.1.0]hex-1-ene system. For example, reaction of enone **40** with $(MeO)_2P(O)CHN_2$ and KO-*t*-Bu should give **41** (by way of the diazoethene). Trapped by intramolecular addition, **41** gives bicyclohexene **42**, which ultimately affords isomeric dimers in 45% and 7% yields.[51] Note, however, that although

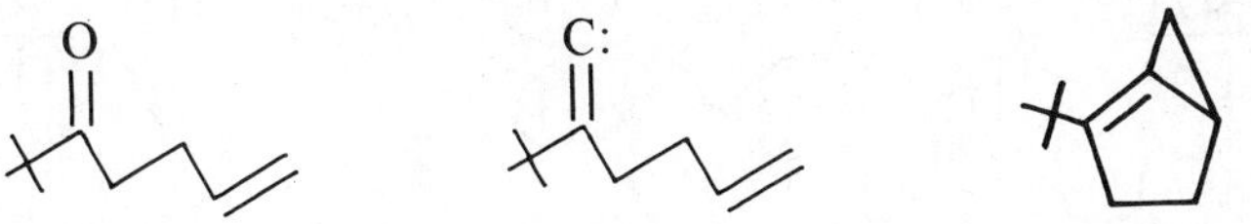

40 **41** **42**

alkylidenecarbenes have been cited as intermediates in these reactions, there is no conclusive evidence in their favor. Other mechanisms, which contain nitrogen-bearing intermediates and exclude free carbenes, are conceivable.[49]

b.   *Cumulenylidenes*

A "cumulenylidene" is an "extended unsaturated carbene."[1,52,53] Alkylidene-carbenes (e.g., **33**) and alkenylidenecarbenes (vinylidenecarbenes, **43**) have been discussed previously and in *Reactive Intermediates,* Volume 2 (pp. 69–74). Recent effort has been devoted to the generation of higher members of the series.[1,53] Using the nomenclature recommended by Stang,[1] the parent carbene,

$$R_2C{=}C{=}C{:} \qquad R_2C{=}C{=}C{=}C{:} \qquad R_2C{=}\overset{\overset{\displaystyle OTf}{|}}{C}{-}C{\equiv}CH \qquad R_2C{=}C{=}C{=}C\diagup\diagdown$$

43   **44**   **45**   **46**

$R_2C{=}C{:}$ is an *alkylidenecarbene*, and the name for each succeeding member of the series reflects the addition of another double bond: $R_2C{=}C{=}C{:}$ (alkenyli-denecarbene); $R_2C{=}C{=}C{=}C{:}$ (alkadienylidenecarbene); $R_2C{=}C{=}C{=}C{=}C{:}$ (alka-trienylidenecarbene), and so on.

Alkadienylidenecarbenes (**44**, R=H, CH_3, C_6H_5) have been generated from triflates **45** by base catalyzed α-elimination with KO-*t*-Bu in dimethoxyethane at 0°C.[54,55] Carbenes **44** (R=CH_3, C_6H_5) could be trapped by addition to $Me_2C{=}CMe_2$, affording alkenes **46**. Products of addition of **44** (R=H, CH_3) to cyclohexene were isolated as dimers.[54] Studies of relative rates of additions[28] at 0°C for **44** (R=CH_3) gave: $Me_2C{=}CMe_2$ (5.55 ± 0.4), $Me_2C{=}CHMe$ (2.55 ± 0.05), $Me_2C{=}CH_2$ (1.00), and *trans*-MeCH=CHMe (0.39 ± 0.01). The carbene is therefore an electrophile with a selectivity index $m = 0.77$[54] (compare, Vol. 2, p. 86 ff). It is somewhat less discriminating than CCl_2 ($m = 1.00$). Alkylidene-carbene **43** (R=CH_3) has m approximately 0.70, similar to that of **44**.

Note that although the alkyl groups of **44** project parallel to the vacant p orbital at the (sp) divalent carbon of **44** (compare, **44′**) there is no obvious steric hindrance to the addition of **44** to $Me_2C{=}CMe_2$, as there is with **33**

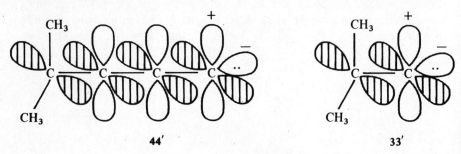

44′   **33′**

(compare, **33′**).[56] Obviously, the greater separation of the *gem*-dimethyl groups from the reaction center in **44** negates serious steric interaction with the substrate's alkyl substituents. Carbenes **44** (R=CH₃, C₆H₅) could also be trapped by M—H insertion into R_3SiH, R_3GeH, or R_3SnH, affording allenes **47**.[55]

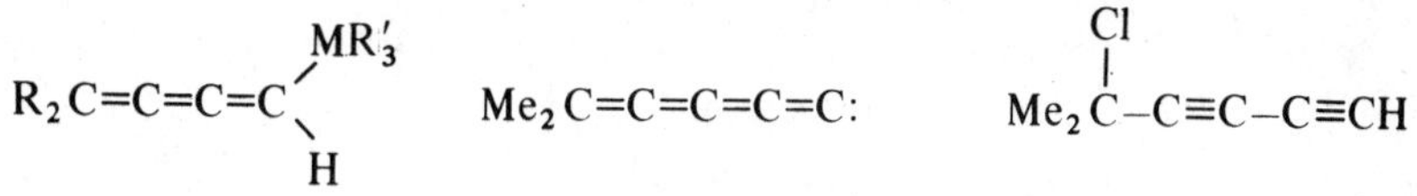

$$R_2C=C=C=C\begin{matrix} \nearrow MR_3' \\ \searrow H \end{matrix} \qquad Me_2C=C=C=C=C: \qquad \begin{matrix} Cl \\ | \\ Me_2C-C\equiv C-C\equiv CH \end{matrix}$$

47   48   49

The alkatrienylidenecarbene **48** (or its complex with base) was generated by the action of *tert*-butoxide on diynyl chloride **49** (glyme, −30°C).[52] Addition to $Me_2C=CMe_2$ afforded 15% of a dimer of the expected addition product. The next higher member of the series, alkatetraenylidenecarbene **50** was generated from triflate **51** by elimination with *tert*-butoxide/glyme at 20°C.[57] Addition reactions were observed with $Me_2C=CMe_2$ and cyclohexene, although the

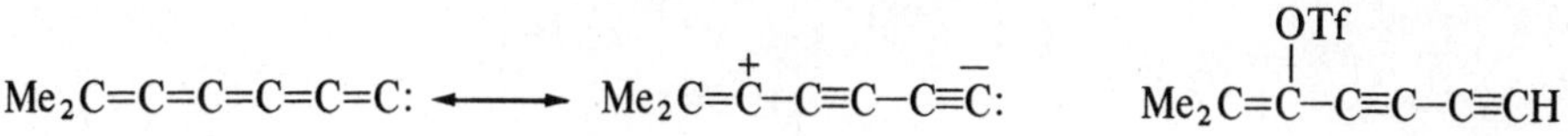

$$Me_2C=C=C=C=C=C: \longleftrightarrow Me_2C=\overset{+}{C}-C\equiv C-C\equiv\overset{-}{C}: \qquad \begin{matrix} OTf \\ | \\ Me_2C=C-C\equiv C-C\equiv CH \end{matrix}$$

50   50′   51

products were unstable. The expected product from $Me_2C=CMe_2$ was **52**, which was presumably converted to the isolated **53** by base-catalyzed isomerization. This type of isomerization was cleverly prevented when adamantyl analogue **54** was added to $Me_2C=CMe_2$. The anticipated adduct formed in 53% yield and resisted isomerization because of the difficulty of forming double bonds at the bridgehead position of adamantane.

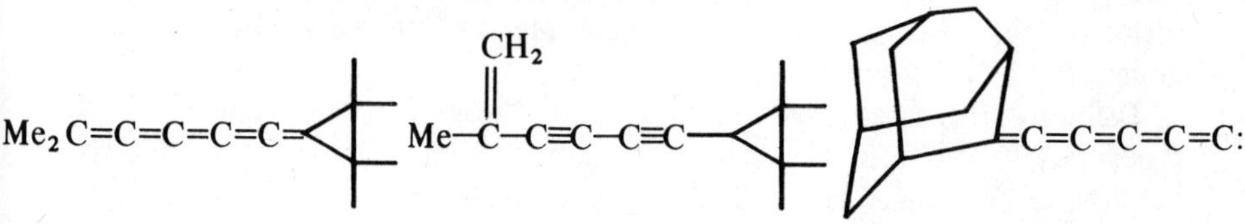

52   53   54

Carbene **50** also inserted into the M-H bonds of Et_3SiH, Et_3GeH, and Et_3SnH. With the silane and germane, cumulenes analogous to **47** were formed, but with Et_3SnH the unexpected **55** was produced in 10.6% yield. The origin of this product is unclear, but could be related to the high acidity and low bond

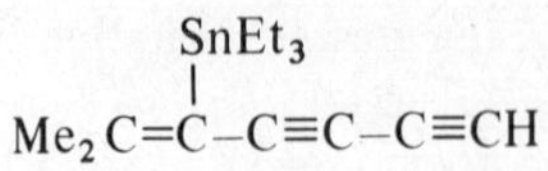

$$\underset{55}{Me_2C=\overset{\overset{\displaystyle SnEt_3}{\displaystyle |}}{C}-C\equiv C-C\equiv CH}$$

energy of Sn–H (relative to Si–H or Ge–H), together with charge-separated ionic character in **50** (compare resonance contributor **50′**).

D.   Dihalocarbenes

Two new precursors for difluorocarbene have been discovered.[58,59] Bis(trifluoromethyl)cadmium glyme complex **56** reacted with acetyl bromide or chloride in $Me_2C=CMe_2$ at $-27°C$ for 30 min, affording approximately 70% of the CF_2 adduct along with 96% of acetyl fluoride.[58] Additions to *cis*- and

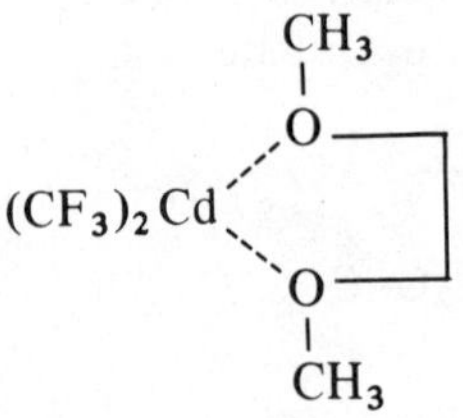

$$(CF_3)_2Cd\underset{\displaystyle O}{\overset{\displaystyle O}{\big<}}\quad \begin{array}{c} CH_3 \\ | \\ \\ \\ | \\ CH_3 \end{array} \qquad \mathbf{56}$$

trans-butene were stereospecific, but cyclopropane yields were low. The mechanism of CF_2 generation is still unclear, but the addition reactions proceed under mild conditions.

In a second new method, CBr_2F_2 was reduced with metallic lead to generate CF_2 which was captured by several alkenes.[59] Typical reactions were carried out over several hours in warm CH_2Cl_2 with $Bu_4N^+Br^-$ added to complex product $PbBr_2$. Lead is the limiting reagent in these reactions (excess CBr_2F_2), and yields of CF_2 adducts (based on Pb) were: $Me_2C=CMe_2$ (80–90%), α-methylstyrene (55%), and styrene (~ 17%).[59]

Dichlorocarbene has been directly generated from chloroform and powdered NaOH at 30–40°C in an ultrasound bath.[60] With a low-powered sonifier (45 kHz, 35 W), small-scale dichlorocyclopropanations of alkenes could be accomplished in good yields without the need for either phase transfer catalysts or expensive bases. Typical isolated yields from alkenes were: $Me_2C=CMe_2$ (74%), α-methylstyrene (97%), cyclooctene (99%), and 1-octene (81%). *Both* mechanical stirring and ultrasonic irradiation were essential to the high yields, and the selectivity of the CCl_2 resembled that of CCl_2 generated from chloroform and KO-*t*-Bu.

The major property of superoxide ion ($O_2{}^{\cdot-}$) toward chloroform is basicity, that is, CCl_2 is generated by α-elimination. Potassium superoxide, solubilized in benzene with a quaternary ammonium chloride salt, reacted with chloroform and cyclohexene to yield dichloronorcarane (20%).[61] In a polar solvent such as DMF the reaction takes a different course, and C–Cl bond cleavages occur.

Additions of CCl_2 ($CHCl_3$, 50% aq. NaOH, triethylbenzylammonium chloride) to allylic alcohols **57–59** proceeded in good yields. In each case, the *syn*-hydroxyl cyclopropane isomer predominated.[62] Syn/anti ratios were as high as 13:1 (**57**) and 15:1 (**58**). Considerably less syn selectivity was observed with

59, and adducts could not be isolated from cyclopentenol itself. The stereoselectivity is reminiscent of that observed in the Simmons–Smith reaction and suggests that CCl_2 (or CCl_3^-?) can be coordinated to the allylic oxygen (hydrogen bonded to OH?) and then delivered to the double bond. In contrast, CCl_2 thermally generated from $PhHgCCl_2Br$ added only anti (to OH) with cycloheptenol-cyclononenol.[63] The reversal from syn to anti cyclopropanation stereoselectivity as the 3-cycloalkenol ring sizes increases also finds analogy in Simmons–Smith methylenation (although here the crossover occurs between cycloheptenol and cyclooctenol).[64] Stereoselectivity reversals are probably associated with OH conformational preferences that are a function of ring size. In medium-ring cycloalkenols, CCl_2 may be coordinated to pseudoequatorial OH groups and not deliverable to the "syn face" of the double bond. Without relative reactivity studies to detect accelerated cyclopropanation, it is unclear if the anti dichlorocyclopropanation observed with molecules such as 3-cycloheptenol reflects a stereoselective OH-mediated CCl_2 addition or simple anti cyclopropanation due to steric hindrance to syn cyclopropanation originating at the OH group.

The regioselectivity of the Reimer–Tiemann reaction of phenol with CCl_2 is strongly affected by added α-cyclodextrin. The normal p/o substitution ratio changes from 0.71 to 4.65 in $0.15\,M$ aqueous cyclodextrin.[65] Unfortunately, the percent-conversion drastically decreases over this range from 12% to 0.95%. NMR binding studies indicate that $CHCl_3$ effectively competes with phenol (phenoxide ion) for the cyclodextrin binding sites. It is reasonably suggested that CCl_2 may be (reversibly) generated within the cyclodextrin toroid and forced to react at the para position of a phenoxide ion which "backs into" the cyclodextrin cavity.[65] Although the yields in this reaction are too low for immediate synthetic application, the principles are suggestive.

IV.   CARBENOIDS

Continuing studies of Li carbenoids (see, Volume 2, pp. 81–83), Seebach reported ^{13}C NMR chemical shifts for a wide variety of ^{13}C labeled α-halolithium compounds in THF solution at $-100°C$ to $-110°C$.[66] With species such as **60** and **61** (X=Cl, Br, or I) or **62**, the observation of $^{13}C-^7Li$ coupling indicated slow exchange of carbenoid ligands on the NMR time scale, and pointed to

$$H_2CLiX \qquad X_3CLi \qquad CH_3CBr_2Li$$

60   **61**   **62**   **63**

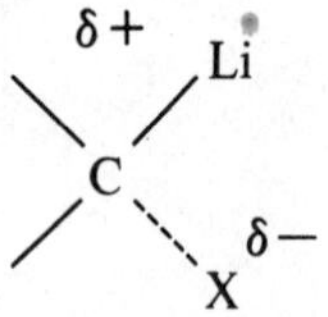

intact C–Li bonds. The replacement of H or X by Li to produce carbenoids such as **60–62** leads to significant NMR *deshielding* of the carbenoid carbon.[67] This apparent decrease of electron density is consistent with weakening of the carbenoid's C–X bonds, and apparently corresponds to an electrophilic lithium carbenoid (sp^2-like carbon, compare, **63**), in agreement with experimental results.

However, values of $^{13}C-^{13}C$ and $^1H-^{13}C$ coupling constants between substituent atoms and the ^{13}C carbenoid center were much smaller than in analogous non-Li compounds.[66] Because of the proportionality between coupling constant and s character, this finding suggests that the substituent-carbenoid carbon bonds have low s (high p) character, a conclusion not in keeping with inferences based on the chemical shift results. It must also be remembered that the stable carbenoids observed in THF at $-100°C$, whatever their structure, may not be representative of the species that react electrophilically with alkenes at higher temperatures. Moreover, the presence of strongly solvating THF may be important to carbenoid stability. Continuing work in this area would be very welcome.

V.   REACTIVITY OF CARBENES

A.   1,4-Additions

Little has been added in this area over the three years. The paper we cited in Volume 2 as "in press" by Klumpp and Kwantes has appeared,[68] as have two theoretical studies.[69, 70] In all three cases the idea that 1,4-addition in general, or to norbornadienes in particular, is dependent in any way on the nucleophilicity of the carbene is discounted. Klumpp and Kwantes show that the 1,4-addition *and* endo 1,2-addition respond similarly to changes in the electronic nature of a substituent at C7. As the 1,2-addition is surely not nucleophilic in character, it is no longer reasonable to suppose that nucleophilic character plays an important role in the 1,4-addition. The two theoretical works respond similarly: σ approach is more important in 1,2-addition than it is in 1,4. That is, nucleophilicity is of *less* value in 1,4-addition than it is in 1,2.[69] Klumpp and Kwantes suggest that the transition state (64) for the usual endo 1,2-addition will be

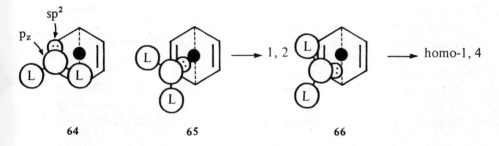

64   65   66

destabilized by interaction of a group attached to the carbene with the remote norbornadiene double bond. Other transition states with the groups on the carbene pointing outside may become important. One of these (65) leads to endo 1,2-addition, the other (66) to 1,4. More work, both experimental and theoretical, is surely needed before this difficult question is resolved.

B.   Quantitative Characterization of Carbenic Selectivity

a.   HOMO-LUMO Treatments

The application of frontier molecular orbital (FMO) theory to carbene/alkene cycloaddition was discussed in some detail in Volume 2 of *Reactive Intermediates* (pp. 86–90). This reaction can be formulated in terms of simultaneous orbital interactions between (1) the carbene's lowest unoccupied orbital (LU, usually taken as p) and the alkene's highest occupied orbital (HO = π), and (2) the alkene's LU (π^*) and the carbene's HO (σ).[71, 72]

Rondan, Houk, and Moss calculated LU and HO energies for 12 carbenes by *ab initio* (STO-3G/4-31G) methods.[73] The carbene LU energies E_{CXY}^{LU} were linearly correlated with m_{CXY}, the empirical carbene/alkene selectivity index.[74] Electron-donating substituents on the carbene raised E_{CXY}^{LU} and increased carbenic selectivity, whereas electron-withdrawing substituents induced opposite effects. Both m_{CXY} and E_{CXY}^{LU} correlated with ΔE_{stab}, the carbene stabilization energy, defined as the negative of the 4-31G energy of the isodesmic reaction

(Eq. 3.3). The most stable carbenes (highest ΔE_{stab}) reacted least exothermically with alkenes and exhibited the greatest selectivity (highest m_{CXY}).[73]

$$CH_2 + CH_3X + CH_3Y \rightarrow CXY + 2CH_4 - \Delta E_{stab} \qquad (3.3)$$

From calculated CXY LU and HO energies and measured alkene orbital energies, one could estimate the *differential* orbital energies ($\epsilon_{CXY}^{LU} - \epsilon_{alkene}^{HO}$), the "electrophilic" interaction, and ($\epsilon_{alkene}^{LU} - \epsilon_{CXY}^{HO}$), the "nucleophilic" interaction, for many carbene/alkene additions.[74] This admittedly primitive procedure correctly reflected the dominant, experimentally observed electrophilicity of CCl_2 and CF_2, the nucleophilicity of $C(OCH_3)_2$, and the ambiphilicity of CH_3OCCl. Neglecting overlap, carbenic philicity was determined by the smaller of the differential orbital energies (i.e., the stronger of the two interactions). Ambiphilic behavior was observed when neither interaction was dominant, as with CH_3OCCl.

Similarly, Schoeller and Brinker formulated the reaction of a singlet carbene and an alkene in terms of the interactions of their respective FMO's.[75] With one-electron perturbation theory, the energies of these interactions are given by (Eq. 3.4a and 3.4b), where σ and p are the carbene HO and LU, and π and π^*

$$(a) \quad \delta E_I = \frac{\langle \pi / H / p \rangle^2}{E_p - E_\pi} \qquad (b) \quad \delta E_{II} = \frac{\langle \sigma / H / \pi^* \rangle^2}{E_\sigma - E_{\pi^*}} \qquad (3.4)$$

are the alkene HO and LU. When $|\delta E_I| > |\delta E_{II}|$, the carbene was predicted to behave as an electrophile toward alkenes; when $|\delta E_{II}| > |\delta E_I|$, as a nucleophile. Ambiphilic behavior is anticipated when $\delta E_I \sim \delta E_{II}$.[75] For CXY, E_σ rises as the electronegativities of X and Y decrease, so that the nucleophilicity of the carbene should increase (electrophilicity decrease) in the order X, Y = $NMe_2 >$ OMe > F. Electrophilic behavior of CXY should be most strongly expressed when X and Y are most electronegative.

Schoeller calculated the energies of the molecular orbitals of various carbenes by STO-3G methods.[76] In increasing energy, the orderings were

$$LU: \quad CCl_2 > CF_2 < C(SMe)_2 < C(OMe)_2 < C(NH_2)_2$$

$$HO: \quad CCl_2 < CF_2 < C(OMe)_2 < C(NH_2)_2 < C(SMe)_2$$

The absolute values of these orbital energies are not very meaningful, but the orderings are of interest. Specifically, from the HO ordering $C(SMe)_2$ would appear to have the "most nucleophilic σ orbital."[76] However, the prediction that $E_{CF_2}^{HO}$ lies above $E_{CCl_2}^{HO}$ is an artifact of the calculations. At the 4–31G level the situation is reversed, and σ_{CF_2} is lower than σ_{CCl_2} by approximately

2 eV.[73] Indeed, the measured ionization protentials of CCl_2 and CF_2 are 9.76[77] and 11.7[78] eV, respectively. This is quite important, for the very low-lying σ orbital of CF_2 is the principal reason for the exclusively electrophilic behavior of CF_2 thus far observed with alkenes.

Schoeller defined a carbene selectivity index S, as in Eq. 3.5, where the δE

$$S = \frac{\partial}{\partial \Delta}(\delta E_{\pi \to \phi_3} + \delta E_{\phi_1 \to \pi*}) \tag{3.5}$$

terms represent the usual FMO interactions (compare, Eq. 3.4a,b) and Δ represents a change in the π or $\pi*$ alkene orbitals induced by olefinic substituent variation.[79] S values (at $\Delta = 0$) were calculated for a number of carbenes and correlated with the experimental m_{CXY} selectivity parameters.[74] Good correlations were observed for the dihalocarbenes, $CH_3 CCl$, and CH_2, but not for the phenylhalocarbenes or (it can be shown) for monohalocarbenes. Species such as $C(NH_2)_2$, $C(OH)_2$, and cyclopropenylidene had $2.5 < S < 2.9$, indicative of nucleophilic carbenes.[79] The correlation between m_{CXY} and 4–31G carbene LU energies[73] appears to be better than that between m_{CXY} and S.

More recently, Schoeller[80] examined the correlation between $\log k_{rel}^{obsd}$ for electrophilic carbenes and the ionization potentials of Moss' standard set of alkene substrates.[74] A two-parameter equation

$$[\log k_{rel} = a\,IP + b\,Coeff + C]$$

gave good results. Here, IP is the alkene ionization potential (i.e., E_π) and "Coeff" is the larger olefinic carbon atomic coefficient in the π or HOMO. The ratio a_{CXY}/a_{CCl_2} was then used to define a selectivity factor S_{CXY}, which correlated very well with m_{CXY}. This result is not surprising because m_{CXY} and S_{CXY} both correlate the relative reactivities of the same set of alkenes toward electrophilic carbenes. Both are experimental measures, with m directly related to differential activation parameters and S related to the alkene HO's that help to determine these parameters. Extended discussions of these matters appear in Reference 73.

Using FMO theory, Schoeller also examined the obedience of carbene/alkene cycloadditions to the reactivity/selectivity principle (RSP). Carbenic series such as CF_2, CBr_2, and CCl_2 exhibit normal RSP behavior (selectivity and reactivity inversely correlated), whereas a pair such as $C(OMe)_2$ and $C(NMe)_2$ is predicted to show an inverse RSP (selectivity and reactivity directly correlated).[81] These trends are traced to the influence of various carbenic substituents on the energies of the carbenes FMO's, particularly the LU.

b. *Carbenic Philicity*

Carbenes can be experimentally classified as electrophiles, nucleophiles, or ambiphiles depending on their selectivity toward alkenes.[74] The empirical olefinic selectivity of a carbene m_{CXY} can be represented relative to that of CCl_2 as the slope of a log-log correlation of the two carbenes' relative addition rates with a common series of olefinic substrates. It was also found that (1) m_{CXY} values are generally well correlated by (Eq. 3.6), in which $\Sigma_{X,Y}$ represents the sum of the appropriate σ constants for the substituents of CXY, and

$$m_{CXY} = -1.10\,\Sigma_{X,Y}\sigma_R^+ + 0.53\,\Sigma_{X,Y}\sigma_I - 0.31 \qquad (3.6)$$

(2) m_{CXY} values are linearly related to calculated carbenic LU energies[73,81] and stabilization energies.[73] The quantity m_{CXY}, either measured or calculated from Eq. 3.6, is therefore a convenient descriptor of carbenic selectivity. It has been observed that carbenes for which $m_{CXY} \lesssim 1.50$ behave as electrophiles toward alkenes, those with $m_{CXY} \gtrsim 2.2$ behave as nucleophiles, and those with $1.5 < m_{CXY} < 2.2$ behave as ambiphiles.[74]

To probe the lower end of the m_{CXY} scale, CF_3CCl was photolytically generated from 3-(trifluoromethyl)-3-chlorodiazirine and added to various alkenes.[82] Additions to the 2-butenes were stereospecific, and the carbene did not insert into allylic C–H bonds. Relative reactivity data, however, showed the carbene to be quite unselective, with $m_{CF_3CCl}^{obsd}$ approximately 0.19. This value is significantly below $m_{CH_3CCl}^{obsd}$ (0.50^{74}) and illustrates the loss of selectivity attending the CF_3 for CH_3 substitution in XCCl. Decreased selectivity is probably associated with increased carbenic reactivity (normal RSP behavior) because 4–31G studies of CF_3CCl give $\Delta E_{stab}^{CF_3\,CCl} = 1.4\,kcal/mol$ (see, Eq. 3.3), an extremely low value, much below $\Delta E_{stab}^{CH_3\,CCl}$ ($29.3\,kcal/mol^{73}$). A more probable value for $\Delta E_{stab}^{CF_3\,CCl}$ is approximately $18\,kcal/mol$,[82] still substantially below the CH_3CCl value, indicating that CF_3CCl is intrinsically more reactive than CH_3CCl.

The quantity $m_{CF_3CCl}^{obsd}$ was not particularly well reproduced by calculation from Eq. 3.6. The expanded correlation (Eq. 3.7) was offered, which included a term in σ^- to cope with the strongly electron-withdrawing CF_3 group. Eq. 3.7

$$m_{CXY} = -1.70\,\Sigma_{X,Y}\sigma_R^+ + 0.76\,\Sigma_{X,Y}\sigma_I + 0.64\,\Sigma_{X,Y}\sigma_R^- - 0.66 \qquad (3.7)$$

nicely correlated m_{CXY}^{obsd} of nine other carbenes and gave $m_{CF_3\,CCl}^{calcd} = 0.25$ in reasonable agreement with $m_{CF_3CCl}^{obsd}$. CF_3CCl was thus characterized as a highly reactive, unselective carbene. Its behavior reflects the destabilizing influence of the CF_3 substituent. CF_3CCl is the most unselective carbene for which m_{CXY}

has been determined and serves as a good model for poorly stabilized, singlet, disubstituted carbenes.[82]

In principle, an empirical carbenic selectivity scale could be based on comparative σ values determined by competitive additions to styrene substrates. Such a scale might parallel m_{CXY} for which simple alkenes ($Me_2C=Me_2$, $Me_2C=CHMe$, $Me_2C=CH_2$, *cis-* and *trans-*MeCH=CHMe) serve as a standard substrate set. Dürr has embarked on such a project.[83] Hammett ρ values were determined at various temperatures for PhCCl generated from KO-*t*-Bu/18-crown-6 and benzal chloride. Additions to a set of substituted styrenes gave $\rho_{PhCCl}^{280K} = -0.33$, versus σ^+[84], which should be compared to the more selective CCl_2, $\rho_{CCl_2}^{273K} = -0.69$, versus σ^+.[85] These results parallel the selectivities on the m_{CXY} scale ($m_{PhCCl}^{298K} = 0.83$, $m_{CCl_2}^{298K} = 1.00$).[74]

CH_3OCCl ($m_{CXY}^{calcd} = 1.59$) behaved as an *ambiphile* toward the alkene set $Me_2C=CMe_2$, $Me_2C=CH_2$, *trans-*MeCH=CHMe, $CH_2=CHCOOMe$, $CH_2=CHCN$.[86] In contrast to the electrophiles CCl_2 or CH_3CCl, where relative addition rates continuously decreased along the set (decreasing reactivity with decreasing olefinic π-electron availability), CH_3OCCl manifested a parabolic dependence of relative addition rate on alkene π-electronic character. It was reactive toward either electron-rich or electron-poor alkenes, but minimally reactive toward alkenes of intermediate nature. This behavior could be rationalized by FMO analysis.[74]

Further investigations of CH_3OCCl ambiphilicity were carried out with styrene[87] and 6,6-dimethylfulvene substrates.[88] The ambiphilicity of CH_3OCCl persisted toward $X-C_6H_4CH=CH_2$ with relative reactivities (25°C) of 1.50 ∓ 0.03 ($X = p-OCH_3$), 1.07 ∓ 0.01 ($p-CH_3$), 1.00 (H), 1.04 ∓ 0.01 ($m-Cl$), and 1.27 ∓ 0.02 ($m-NO_2$).[87] CCl_2 and CF_2 behaved as electrophiles toward these substrates, with decreasing relative reactivities along the series and ρ approximately -0.6. Estimates of $(\epsilon_{CH_3OCCl}^{LU} - \epsilon_{X-styrene}^{HO})$ and $(\epsilon_{X-styrene}^{LU} - \epsilon_{CH_3OCCl}^{HO})$ (see above) were in accord with observations. The former (electrophilic term) was smaller and dominant when $X = p-OCH_3$ or $p-CH_3$, but the latter (nucleophilic term) was dominant when $X = m-Cl$ or $m-NO_2$. A similar analysis revealed constant dominance of the electrophilic term for CCl_2 or CF_2 across the entire substrate set.[87]

In additions to **67**, electrophilic CCl_2 exclusively attacked an endocyclic double bond,[89] whereas nucleophilic $C(OMe)_2$ exclusively cyclopropanated the exocyclic bond.[88] In this case CH_3OCCl also functioned as a nucleophile,

67

attacking the fulvene exocyclic double bond.[88] Estimates of the differential frontier orbital energy terms, coupled with examination of the atomic coefficients in the HO and LU of **67**, provided a satisfactory rationale for the behavior of the three carbenes. For CH_3OCCl in particular, the energy terms were approximately equal (ambiphilicity), and the exocyclic regioselectivity of addition was attributed to overlap control by the very high coefficient (0.65) at C_6 in the LU of **67**. This makes the carbene HO/fulvene LU interaction dominant and mandates the nucleophilic CH_3OCCl addition.[88]

As shown by the data in Table 3.1, phenoxychlorocarbene (generated by diazirine thermolysis at $25°C$) is also an ambiphile.[90]

Like CH_3OCCl, but in contrast to the electrophilic behavior of CH_3CCl or CCl_2 the relative reactivities of PhOCCl describe a parabolic relation with the set of alkenes ordered by decreasing π-electron availability. (The depression of the $PhOCCl/Me_2C=CMe_2$ relative reactivity was attributed to steric factors operative with the larger PhOCCl but not with MeOCCl.[90])

On the m_{CXY} scale, m_{PhOCCl} calculated from (eq. 3.6) is 1.49, essentially equal to m_{CF_2} (1.48[74]) and significantly less than m_{CH_3OCCl} (1.59). Although the relative reactivities of CF_2 toward methyl acrylate and acrylonitrile have yet to be determined, in all alkenic additions of which we are aware, CF_2 behaves as an electrophile.[28, 74] This includes additions to substituted styrenes.[91] where MeOCCl behaves as an ambiphile.[87] Thus the "border" of electrophilicity and ambiphilicity now appears to be experimentally located at $m_{CXY} \approx 1.48 - 1.49$, with CF_2 and the electrophiles on the lower side and PhOCCl and MeOCCl on the higher.

TABLE 3.1.   Relative reactivities of XCCl toward olefins

| Olefin | X in XCCl | | | |
	PhO^a	MeO^b	Me^c	Cl^d
$Me_2C=CMe_2$	3.0	12.6	7.44	78.4
$Me_2C=CH_2$	7.3	5.43	1.92	4.89
$t-MeCH=CHEt^e$	1.00	1.00^f	1.00^f	1.00^f
$CH_2=CH-n-C_4H_9^e$	0.36			
$CH_2=CHCOOMe^e$	3.7	29.7	0.078	0.060
$CH_2=CHCN^e$	5.5	54.6	0.074	0.047

^{a}Reference 90, 25°C.
^{b}Reference 86a, 25°C.
^{c}Reference 86b, 25 C.
^{d}Reference 86b, 80°C.
^{e}The overall k_{rel} is the sum of both *syn*-Cl and *anti*-Cl additions of XCCl (except for $X=Cl$) to this olefin.
^{f}The standard olefin is *trans*-butene instead of *trans*-pentene.

The ambiphilicity of PhOCCl toward the alkenes of Table 3.1 is in keeping with FMO considerations.[90] This was anticipated because of the similarity in STO-3G/4–31G calculated HO and LU energies from PhOCCl (-10.78 and $2.02\,\mathrm{eV}$)[90] and MeOCCl (-10.82 and 2.46).[73] One might anticipate that this parallelism of PhOCCl and MeOCCl behavior should also extend to styrene substrates, where MeOCCl is an ambiphile.[87] However, the PhOCCl generated from α,α-dichloroanisole (50% aq. NaOH, phase transfer catalysis) added as a *nucleophile* to X–C_6H_4CH=CH$_2$ (X = p–CH$_3$O, p–CH$_3$, p–Cl, m–Br), with ρ(vs. σ) = 1.11 (7°C), 0.40 (27°C), or 0.23 (47°C).[92] On the other hand, when PhOCCl was generated by thermolysis (25°C) of phenoxychlorodiazirine, ambiphilic cyclopropanation of X–C_6H_4CH=CH$_2$ was observed with relative reactivities: 1.95 (p–CH$_3$O), 1.32 (p–CH$_3$), 1.00 (H), 1.17 (p–Cl), 1.42 (p–CF$_3$), and 1.33 (m–NO$_2$).[93] Perhaps the nucleophilic species generated by α-elimination from PhOCHCl$_2$[92] and trapped by styrenes is actually PhOCCl$_2^-$, with cyclopropanes subsequently formed by cyclization of **68**. If anion addition were the slow step in the sequence, a positive Hammett ρ would be expected.

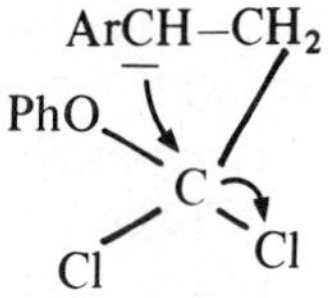

68

In this scenario, the species generated from the diazirine is the carbene, and its ambiphilicity agrees with expectations based on FMO considerations.[93] Further experiments are desirable to resolve the apparent conflict.

Two important communications have appeared concerning the philicity of siloxycarbenes.[94,95] Previously, it had been reported that photolysis of acylsilane **69** afforded Si-to-O rearrangement product **70**, a nucleophilic siloxycarbene which could be intercepted by addition to diethyl fumarate, but not to cyclohexene, tetramethylethylene, or ketene dimethylacetal.[96] Additionally, there was product-based evidence for reactions between **70** and carbonyl groups which was interpreted in terms of a nucleophilic carbene.[97]

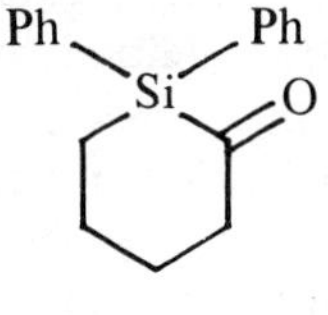

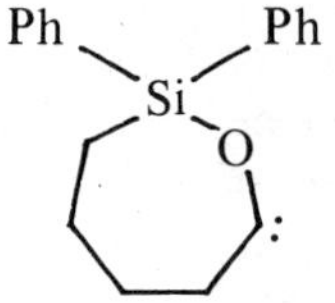

69   **70**

Dalton et al. now report photolytic studies of ketone **71** indicating that irradiation (366 nm) in 2-propanol gives acetal **73** by trapping intermediate siloxycarbene **72**,[94] but that irradiation in diethyl fumarate gives cyclopropane **74** by direct interception of the excited singlet and triplet states of ketone **71**. Carbene **72** is not involved.[95] Acetophenone photosensitization and diene-quenching experiments implicate the triplet excited state of **71** as a precursor

$$Me_3Si\overset{\overset{O}{\|}}{C}CH_3 \qquad Me_3SiO\overset{..}{C}CH_3 \qquad \underset{\underset{OCHMe_2}{|}}{Me_3SiOCHCH_3}$$

$$\begin{array}{c} COOEt \\ EtOOC \diagdown \bigtimes \diagup \\ CH_3 \quad OSiMe_3 \end{array}$$

$$\text{\textbf{71}} \qquad\qquad \text{\textbf{72}} \qquad\qquad \text{\textbf{73}}$$

74

of **73**.[94] The lifetime of triplet **71**, measured by Stern–Volmer quenching with cyclohexadiene in acetone, is 1.29×10^{-8} sec and is independent of added 2-propanol up to $3.9 M$. The failure of 2-propanol to quench triplet **71** implies that the triplet reacts not directly with the alcohol but with a second intermediate derived from the triplet, presumably **72**. Further experiments were interpreted to suggest competition in **72** between reversion to **71** and capture by 2-propanol. A competitive rate constant ratio of approximately 0.23 indicated that high alcohol concentrations are needed for efficient trapping. The data do not reveal the multiplicity of **72**, but carbene/OH "insertions" are generally assigned to singlet carbenes. One plausible pathway for **72** → **73** (there are others), would be photoexcitation of **71** to a singlet state, intersystem crossing to triplet **71**, α-cleavage to radical pair **75**, and spin inversion/collapse of **75** to singlet **72**.

$$Me_3Si\cdot \quad \cdot\overset{\overset{O}{\|}}{C}CH_3 \qquad \longleftrightarrow \qquad Me_3Si\cdot \quad \overset{\overset{\cdot O}{|}}{\underset{..}{C}}CH_3$$

75

In contrast to the alcohol-trapping experiments, the formation of **74** from **71** does not appear to go through carbene **72**.[95] The quantum yield for cyclopropane formation from **71** is unaffected by addition of $10 M$ 2-propanol, which should compete with fumarate for the carbene. Moreover, fumarate strongly quenches the fluorescence of excited singlet **71**, although the quantum yield of **74** increases with increasing [fumarate], suggesting that **74** arises by reaction of singlet **71** with fumarate. Additionally, benzophenone photosensitization generates **74** from **71** and fumarate, indicating that triplet **71** can also be intercepted by fumarate. (The triplet portion of the addition reaction can be competitively quenched by 1,3-cyclohexadiene.)

The mechanism in (Eq. 3.8) was offered to account for the formation of **74**.[95] Addition of singlet or triplet **71** to fumarate generates short-lived 1,4-biradical **76**, which affords 1,3-biradical **77** by $C \rightarrow O$ migration of $SiMe_3$.

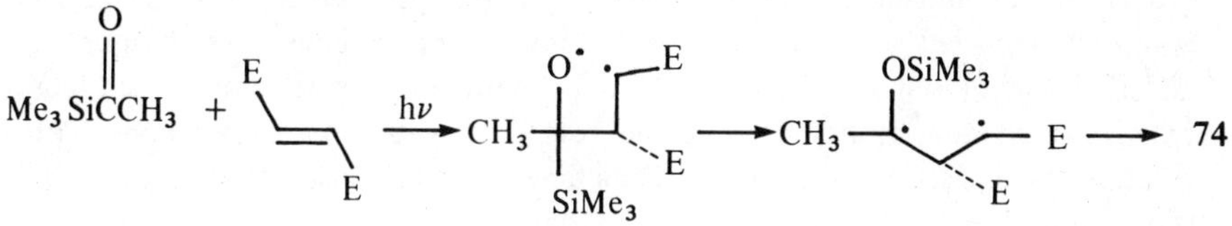

(3.8)

Closure of **77** gives **74**. Such a mechanism explains why cyclopropanes are not photochemically obtained from **69** and electron-rich alkenes. Carbonyl excited states prefer reaction with electron-poor alkenes.[95] The generality of the results obtained with **71** has not yet been established, but the formation of cyclopropanes such as **74**, or the product from **69** and fumarate, do not necessarily require the intermediacy of a siloxycarbene. The philicity of this species must still be regarded as uncertain.

c.   *Absolute Rate Constants of Carbene Reactions*

The application of nanosecond and picosecond laser-flash photolysis to reaction of carbenes offers major opportunities. Absolute rate constants and Arrhenius parameters will now be available for reactions previously studied only by relative reactivity methods and differential activation parameters. The new data may well require significant revision of our conception of carbene/alkene cycloaddition.

Studies of the reactions of singlet and triplet fluorenylidene with alkenes are discussed in Section V.C. Fluorenylidene also reacts with aliphatic ketones to give carbonyl ylides that can cyclize to oxiranes or be quenched by oxygen, or diethyl fumarate.[98] Laser-flash photolytic generation of the carbene from 9-diazofluorene in acetone gave transient uv spectra of triplet fluorenylidene, of ylide **78**, derived by addition to acetone, and of 9-fluorenyl radical, formed

78

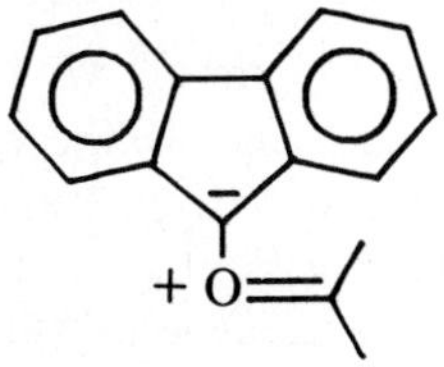

by H abstraction. Direct observation of the triplet carbene in a mixture of acetone and CD_3CN gave $k = 1 \times 10^7 M^{-1} sec^{-1}$ as the bimolecular rate constant for ylide formation.[98]

Absolute rate constants were also reported for reactions of PhCCl and p-$CH_3OPhCCl$ with methanol and *tert*-butanol.[99] The dependence of k_{obs} on [ROH] in acetonitrile for these apparently singlet OH "insertions" was complex and best analyzed as competitive reactions of the carbene with monomeric and (H-bonded) oligometric alcohols. With methanol, the oligomers were more reactive than the monomer, but the reverse was true for *tert*-butanol. For reactions with MeOH oligomers in isooctane the rate constants were 2.9×10^9 (PhCCl) and $4.3 \times 10^9 M^{-1} sec^{-1}$ ($CH_3OPhCCl$) per methanol unit; that is, close to diffusion controlled. Rate constants for reactions with monomeric methanol could be obtained by extrapolation to [MeOH]=0; for example, $2 \times 10^7 M^{-1} sec^{-1}$ for p-$CH_3OPhCCl$. The analogous reaction with *tert*-butanol was slower [$2.5 \times 10^6 M^{-1} sec^{-1}$], presumably for steric reasons.[99]

Photolysis of phenylchlorodiazirine in aerated isooctane at 25°C gave a transient absorption for singlet PhCCl at λ_{max} approximately 295 nm, similar to the absorption seen in a 3-methylpentane matrix at 77 K.[100] Although transient decay was second order in isooctane, pseudo-first-order quenching kinetics were observed in the presence of alkenes, and concomitant cyclopropane formation was demonstrated. Absolute carbene/alkene addition rate constants were determined from correlation of the rate constant for transient decay and [alkene]. Ratios of the absolute rate constants for pairs of alkenes were similar to ratios of relative rate constants determined by conventional competition experiments based on relative product yields, indicating that the kinetics of the transient were associated with product formation.

The absolute rate constants for PhCCl/alkene additions vary somewhat with the experimental apparatus.[100-102] The most recently measured set[101] includes: $Me_2C=CMe_2$ (2.8×10^8), $Me_2C=CHMe$ (1.3×10^8), *trans*-MeCH=CHEt (5.5×10^6), $CH_2=CH-n-C_4H_9$ (2.2×10^6), all in L/mol (sec) at 23°C with reproducibilities of $\mp$ 10%. The rate constants are significantly lower than the diffusion-controlled limit, cover a range of approximately 130, and display the increasing reactivity as a function of increasing olefinic alkylation typical of electrophilic carbenic additions.[28, 74]

These studies were extended to a series of p-substituted phenylchlorocarbenes ($p-X = CH_3O$, CH_3, H, Cl, CF_3).[101] Each carbene was allowed to react with the four alkenes cited above, thus generating a 5-carbene $\times$ 4-alkene rate constant matrix. Significant trends were (1) With any carbene the rate constants decreased regularly with decreasing olefinic alkylation. A factor of approximately 100 was observed in passing from $Me_2C=CMe_2$ to $CH_2=CH-n-C_4H_9$. (2) With any alkene p-$CH_3OPhCCl$ was the least reactive and p-CF_3PhCCl the most reactive carbene. In the carbene substituent order $CH_3O < CH_3 < H <$

$Cl < CF_3$, k_{obs} increased regularly, with the total variation again approximately two orders of magnitude. (3) The most rapid reaction (p–$CF_3PhCCl +$ $Me_2C=CMe_2$) had $k = 1.5 \times 10^9$ L/mol (sec) and the slowest reaction (p–$CH_3OPhCCl + CH_2 = CH$–n–C_4H_9) had $k = 1.3 \times 10^5$ L/mol (sec), so that the total rate constant variation was approximately four orders of magnitude. (4) For all four alkene substrates the Hammett ρ (vs. σ_p^+) for addition of X–PhCCl was essentially constant and approximately $+ 1.5$.

Qualitatively, $p > 0$ might be interpreted to mean that electron-donor (ED) substituted arylchlorocarbenes are more stabilized than ones substituted with electron-withdrawing (EW) groups[73] and therefore react more slowly with a given alkene. Recent evidence that some carbene/alkene additions may proceed by way of kinetically significant intermediates (e.g., 79 in Eq. 3.9, a charge transfer complex) complicates the matter,[102] but rationalization of the ρ values can still be made by focusing on the intermediate. When X is EW, passage of 79 to product over "electrophilic" transition state 80 is favored because X stabilizes enhanced negative charge imposed on the carbenic carbon. Dissociation of 79 to carbene and alkene is disfavored, however, because EW substituents destabilize carbenes.[73] When X is ED, these considerations reverse. Now 80 is destabilized, and the dissociation of 79 to carbene and alkene is favored. Thus EW groups enhance product formation from 79, but ED groups favor reversion of 79 to ArCCl, and ρ based upon the disappearance of ArCCl should be positive.[101] The absence of carbenic selectivity change ($\sim$ constant ρ) as the alkene is altered is

$$\text{(3.9)}$$

unanticipated and still unexplained. It was suggested that for a particular carbene/alkene system where complex formation is involved in the rate-determining step, a very fast reaction (high reactivity) may still be selective if the factors that determine selectivity control the passage of the intermediate to product.[101]

The postulation of intermediate 79 in the addition of ArCCl to alkenes stems from a study of the temperature dependence of k_{obs} for the addition of PhCCl to alkenes.[102] When log k_{obs} was plotted against $1/T$ in the normal way, additions to *trans*-pentene and 1-hexene were characterized by normal, albeit very low, activation energies (~ 1 kcal/mol) and preexponential factors in the range of 10^7-10^8. However, additions to $Me_2C=CMe_2$ and $Me_2C=CHMe$ displayed *negative* activation energies of -1.7 and -0.77 kcal/mol, respectively.[102]

One way to rationalize these kinetic results is to postulate a reversibly formed dissociable intermediate (e.g., 79) in the addition reaction (see eqs. 3.10 and 3.11), where C is the intermediate and A is the alkene. Under these conditions k_{obs} would be related to the rate constants in Eqs. 3.10 and 3.11 by

$$A + PhCCl \underset{k_{-1}}{\overset{k_1}{\rightleftharpoons}} C \qquad (3.10)$$

$$C \xrightarrow{k_2} \text{cyclopropane} \qquad (3.11)$$

$$\frac{k_{-1}}{k_2} = \frac{k_1 - k_{obsd}}{k_{obsd}} \qquad (3.12)$$

(Eq. 3.12).[102] Taking $k_1 \sim k_{diffusion}$ (at temperature T in isooctane), the ratio k_{-1}/k_2 could be calculated from k_{obsd}. Plots of log (k_{-1}/k_2) versus $1/T$ were now linear for all four alkenes, and yielded the *differential* activation parameters in Table 3.2 for the competition between completion of cycloaddition versus dissociation of complex C.

The additions of PhCCl appear to be entropy dominated. However, the activation entropies are comparable for the alkenes, "so that the observed rate constants depend upon the relative enthalpies for cyclopropanation vs. dissociation of the complexes. This *de facto* enthalpic control [translated to $\Delta\Delta G^{\ddagger}$] is expressed in larger k_{obsd} values for cyclopropanations of the more highly substitued alkenes ... in accord with well-known features of carbene/ alkene addition reactions."[102]

The importance of entropy, manifested in Table 3.2, recalls the suggestion of Skell[103] that cycloadditions of reactive carbenes might proceed by way of entropy-dominated transition states, whereas additions of more stabilized

TABLE 3.2.   Differential activation parameters for PhCCl + alkenes

Alkene	$\Delta\Delta G^{\ddagger}$ (kcal/mol)	$\Delta\Delta H^{\ddagger}$ (kcal/mol)	$\Delta\Delta S^{\ddagger}$ (eu)
$Me_2C=CMe_2$	2.2	-4.4	-23
$Me_2C=CHMe$	2.7	-3.7	-22
$t-MeCH=CHEt$	4.3	-1.6	-20
$CH_2=CH-n-C_4H_9$	4.5	-1.5	-21

carbenes might occur by way of enthalpy-dominated transition states. With the availability of absolute rate constants, these ideas will soon be subject to direct tests.

d.   Temperature Dependence of Carbenic Selectivity

Giese examined the relative reactivities of CF_2, CCl_2, and CBr_2 toward $Me_2C=CMe_2$, $Me_2C=CHMe$, and *trans*-MeCH=CHMe (all relative to $Me_2C=CH_2$) at various temperatures from 270–420 K.[104] For each pair of alkenes the carbenic selectivities, determined by the relative reactivities, underwent reversal at higher temperatures. For example, whereas the selectivity order based on $\log (k_{Me_2C=CMe_2}/k_{Me_2C=CH_2})$ at 270 K was $CF_2 > CCl_2 > CBr_2$, the order at 420 K was $CBr_2 > CCl_2 > CF_2$. The former result (at $-3°C$) is the "normal" selectivity sequence encountered at $25°C$;[74] the higher temperature reversal requires comment.

The study was extended to CFCl and CBrCl using the $Me_2C=CHMe/Me_2C=CH_2$ substrate pair.[105] Correlations of $\log (k_{Me_2C=CHMe}/k_{Me_2C=CH_2})$ versus $1/T$ for CXY again showed a temperature-dependent order; see Figure 3.1. As temperature *increased*, the selectivities of CF_2 and CFCl *decreased*, those of CClBr and CBr_2 *increased*, and that of CCl_2 remained nearly constant. At the

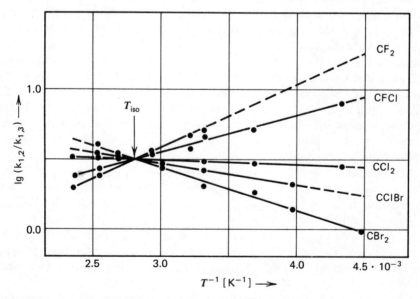

FIGURE 3.1.   Selectivity $[\log(k_{Me_2C=CHMe}/k_{Me_2C=CH_2}]$ of CXY versus 1/T. T_{iso} is the isoselective temperature, and the temperature limits are $2.0 \times 10^{-3}/K^{-1}$ $(227°C)$ and $4.5 \times 10^{-3}/K^{-1}$ $(-50.8°C)$. Reprinted with permission from reference 105.

common intersection point of the correlation lines ($\sim 360\,K$, $87°C$), the *iso-selective temperature*, all five carbenes had comparable selectivities. Below T_{iso}, the normal selectivity order[74] $CF_2 > CFCl > CCl_2 > CBrCl > CBr_2$ obtained; above it, the order was inverted.[105] Previous work showed that the same iso-selective temperature held for CF_2, CBr_2, and CCl_2 with $Me_2C=CMe_2/Me_2C=CH_2$ and *trans*-$MeCH=CHMe/Me_2C=CH_2$ substrate pairs.[104]

The existence of an isoselective temperature for the family of CXY + alkene competitions suggests that this reaction set obeys a differential form of the isokinetic relationship, or compensation law (Eq. 3.13),[106, 107] where δ_{CXY} represents a substituent change in CXY, $\Delta\Delta H^{\ddagger}$ and $\Delta\Delta S^{\ddagger}$ are the differential activation parameters for additions of CXY to the two competing alkenes,

$$\delta_{CXY}\Delta\Delta H^{\ddagger} = T_{iso}\delta_{CXY}\Delta\Delta S^{\ddagger} \tag{3.13}$$

and the proportionality constant is the isoselective temperature T_{iso}, which can be related to an isokinetic temperature.[107]

The existence of T_{iso} is also a strong indication that all of the reactions proceed by an identical mechanism.[106] By (Eq. 3.13), we expect variations of $\Delta\Delta H^{\ddagger}$ and $\Delta\Delta S^{\ddagger}$ to compensate each other partially in their effects on $\Delta\Delta G^{\ddagger}$.[106a] Indeed, for the five competitive carbene additions of Fig. 3.1, there was a proportionality of $(\Delta H^{\ddagger}_{Me_2C=CH_2} - \Delta H^{\ddagger}_{Me_2C=CHMe})$ and $(\Delta S^{\ddagger}_{Me_2C=CH_2} - \Delta S^{\ddagger}_{Me_2C=CHMe})$. A detailed analysis of the dependencies of differential enthalpic and entropic terms on substrate structure and carbenic substituent led to the conclusion that entropic effects dominate the selectivities of CBr_2 and $CBrCl$, whereas electronic (enthalpic) factors determine the selec-tivities of CF_2 and $CFCl$.[105] This conclusion would seem to be in harmony with Skell's suggestion[103] (see above). However, Exner has warned that "when the isokinetic relationship holds, it is useless [for reasons based on expected experi-mental errors] to discuss separately the values of ΔH and ΔS in addition to ΔG."[106b]

The carbenic selectivity reversal observed above T_{iso} is concisely reflected in an inversion of the m_{CXY}^{obsd} sequence.[105, 108] "Two-point" m_{CXY} values were determined from log $(k_{Me_2C=CHMe}/k_{Me_2C=CH_2})$ at 20 and $120°C$ for the car-benes of Figure 3.1. At $20°C$ a normal[74] ordering of m_{CXY} was observed, but at $120°C$ (above T_{iso}, $87°C$) the order was inverted: CBr_2 (1.20), $CClBr$ (1.05), CCl_2 (1.00), $CFCl$ (0.85), CF_2 (0.75).[105] A similar m_{CXY} inversion was observed with four-point m_{CXY} correlations of CF_2, CCl_2, and CBr_2 at 20, 60, and $100°C$.

Fitting the 60 and $100°C$ data to the form of correlation (Eq. 3.6) that was derived from $25°C$ data led to (Eqs. 3.14 and 3.15).[108] Comparison of Eqs. 3.6, 3.14, and 3.15 suggests that the resonance contribution $\Sigma\sigma_R^+$ of the carbenic substituents to selectivity is relatively temperature independent, whereas their inductive contribution $\Sigma \sigma_I$ is strongly temperature dependent, reverses sign between 25 and 60°C, and continues to decrease with increasing temperature. The analysis[108] concludes that as temperature increases, strong inductively withdrawing carbenic substituents (e.g., F, $\sigma_I = +0.50$) should cause a con-

$$m_{CXY}^{60°} = -1.1\Sigma_{X,Y} \sigma_R^+ - 2.6\Sigma_{X,Y} \sigma_I + 2.8 \qquad (3.14)$$

$$m_{CXY}^{100°} = -1.0\Sigma_{X,Y} \sigma_R^+ - 7.4\Sigma_{X,Y} \sigma_I + 7.1 \qquad (3.15)$$

tinuing decrease in m_{CXY}. However, the "constant" term in Eqs. 3.6, 3.14, and 3.15 shows a strong positive correlation with temperature, so that carbenes with weaker inductively withdrawing substituents (e.g., Br, $\sigma_I +0.44$) will display a net increase in m_{CXY} with increasing temperature because of the influence of the "constant."

This analysis[108] is disturbing because of its reliance on the temperature-dependent behavior of the "constant" term. This parameter tends to "accumulate" experimental errors from the data and, it has been warned, can lessen the significance of dual substituent parameter equations.[109]

A second rationalization of the temperature-dependent behavior of m_{CXY} was based on the intermediate complex formulation of carbenic addition (Eqs. 3.10, 3.11),[102, 108] and invoked to explain temperature-dependent relative reactivities of CBr_2 and CCl_2 additions to CH_3O-, Ph-, and alkyl-substituted alkenes. The principal points were: (1) the effects of CH_3O and Ph substituents on alkene reactivity operate mainly on k_2 of (Eq. 3.11), rather than on the complex formation equilibrium (Eq. 3.10); (2) dissection of $\Delta\Delta G^{\ddagger}$ into $\Delta\Delta H^{\ddagger}$ and $\Delta\Delta S^{\ddagger}$ indicates that the enhanced reactivity of CH_3-, CH_3O-, or Ph-substituted alkenes toward CCl_2 or CBr_2 is largely governed by $\Delta\Delta S^{\ddagger}$.[110] The conclusions are in keeping with the data, but we repeat Skell's caution that "this activation parameter-reactivity correlation . . . is contingent on the ability of the Arrhenius equation to separate effectively $\Delta H^{\ddagger}$ and $\Delta S^{\ddagger}$ terms . . . of small magnitude . . ."[103] Recalling the apparent operation of enthalpy/entropy compensation in these carbenic additions (see above), we take note of these dissections,[110] but conservatively concentrate on the behavior of $\Delta\Delta G^{\ddagger}$ as a function of CXY substituents.

Similar conclusions (and perhaps a similar caution) attend a study of CCl_2 additions to a series of p-substituted α-methylstyrenes and $R(CH_3)C=CH_2$ (R=Et, i−Pr, tert-Bu).[111] Relative addition reactivities with the styrenes were temperature independent between 253 and 353 K, whereas relative reactivities increase at higher temperature with $R(CH_3)C=CH_2$. It was suggested that the apparently normal electronic effects in the styrene series (electron donor-substituted styrenes react more rapidly) are largely expressed through $\Delta\Delta S^{\ddagger}$, whereas, as shown by dissection of $\Delta\Delta G^{\ddagger}$, the steric inhibition of CCl_2 addition found with $R(CH_3)C{=}CH_2$ is generated by changes in both $\Delta\Delta H^{\ddagger}$ and $\Delta\Delta S^{\ddagger}$.[111]

Notwithstanding our cautious attitude toward dissection of $\Delta\Delta G^{\ddagger}$ into enthalpic and entropic contributions, and associated specific mechanistic analyses, we cannot ignore the importance of entropic factors in carbenic selectivity. The situation may be further clarified through absolute kinetic studies, where one can work directly with $\Delta H^{\ddagger}$ and $\Delta S^{\ddagger}$, rather than with relative rate data and differential activation parameters. Note also that correlation equations such as (Eq. 3.6) not only adequately reproduce experimental carbenic selectivities, but also are directly related to *ab initio* calculated properties of CXY that do not include entropy. Clearly, the origins of carbenic selectivity are not yet completely understood. Further study is needed, particularly to determine the generality of intermediate complexes in the cycloaddition process.

e.   *Allylic and Homoallylic Carbenes*

An example of the Skattebøl rearrangement appears in (eq. 3.16).[112,113] An appropriate mechanistic formulation involves the ionic rearrangement of

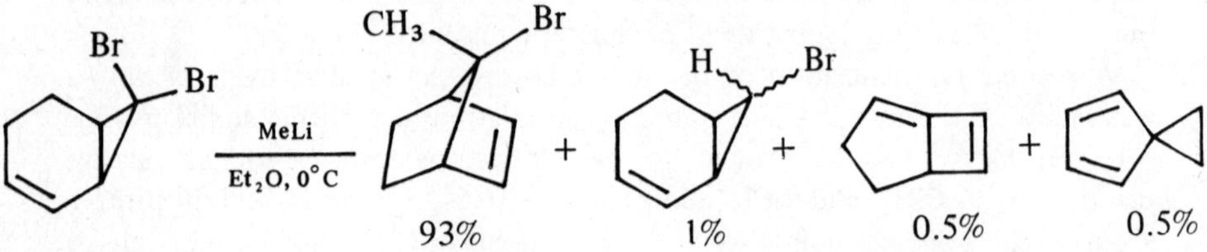

$$\tag{3.16}$$

carbenoid **81** to ion pair **82**, which is converted to the principal product by sequential reactions with MeLi and RBr.[113] Formulation of the rearrangement in terms of the corresponding free carbenes, **83** and **84**, is untenable because of the stereospecificity of formation of the 7-substituted norbornene in (Eq. 3.16), and also because the major products expected from free **84** are the two dienes shown in (Eq. 3.16) but actually formed there in only trace amounts.[114]

In a significant experimental advance, Brinker and Ritzer prepared a glass

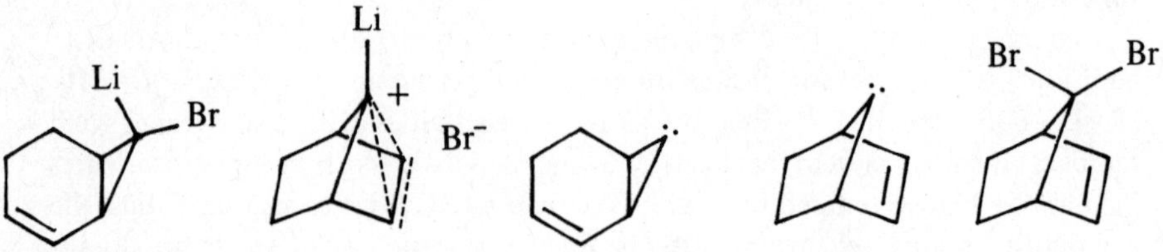

81   82   83   84   85

tube filled with glass turnings coated with methyllithium.[112] When the reaction (Eq. 3.16) was carried out by passing the dibromonorcarene through the tube (*in vacuo*, 24–140°C), the yields of dienes increased to 25.1 and 4.8% respectively, whereas 7-methyl-7-bromonorbornene decreased to 14.7% (however, the yield of 7-bromonorcarene increased to 10.5%). The outcome of this "gas phase" methyllithium-tube reaction rather closely resembles the chemistry of **84** generated by the Bamford–Stevens pyrolysis of 7-norbornenone tosyl-hydrazone salts.[112,114] Indeed, similar results were obtained when 7,7-dibromo-norbornene was passed through the MeLi tube (diene yields, 47.5 and 12.6%), although treatment of the dibromide with ethereal MeLi in solution at 0°C gave a product distribution similar to that of (Eq. 3.16); that is, only traces of the dienes.[112]

The MeLi tube method appears to have important synthetic and mechanistic applications; both carbenoids and species much closer to free carbenes can now be prepared from the same halide precursors.

Bicyclo[3.2.1]octa-2,6-dien-4-ylidene, **86**, has been the subject of three separate reports.[115–117] Of interest has been the importance of delocalization (**86′**),[117] in analogy to a suggestion made for the related carbene 87 (87′).[118]

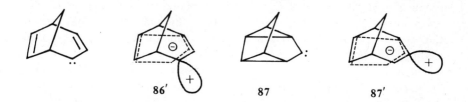

$$\text{86'} \qquad \text{87} \qquad : \qquad \text{87'}$$

Carbene **86** may have been encountered serendipitously during the reaction of dichloride **88** with the lithium radical anion of *p,p′*-di-*tert*-butylbiphenyl (Eq. 3.17).[115] When the reaction mixture was quenched with D_2O, the product was 86.5% d_2 with approximately 0.9 D at each of the indicated positions. The suggested mechanism involved formation of carbenoid **89** and rearrangement to **86** (presumably as a LiCl carbenoid) by way of carbenoid **90**. Two one-electron reductions of **86** by (Ar)Li, followed by protonation, would complete

$$\underset{\textbf{88}}{\text{Cl} \quad \text{Cl}} \quad \xrightarrow{\text{ArLi}} \quad \xrightarrow{H_2O} \quad \underset{\underset{\textbf{68\%}}{(D)}}{\overset{(D)}{}} \qquad\qquad (3.17)$$

the reaction. In fact, the reductions could also occur at other stages of the mechanism, bypassing **86** or its carbenoid. The rearranging species could be what was termed[115] a "carbene radical anionoid."

In an extended study **86** and its monoene analogue **91** were generated either by photolysis of the corresponding diazo compounds or by pyrolysis of the tosylhydrazone sodium salts.[116] Most interestingly, when photolytically generated at $0°C$, both **86** and **91** behaved as nucleophiles toward a series of styrenes.

Care was taken to exclude 1,3-dipolar additions of the diazoalkenes followed by nitrogen loss from intermediate pyrazolines. This process required 9 h for completion in the dark at $0°C$, whereas the photolyses were over in 10 min. The photolytic ρ (vs. σ) was $+0.25 \pm 0.03$ for **86** and $+0.68 \pm 0.05$ for **91**. Corresponding ρ values for $0°C$ dark, dipolar addition reactions were $+1.26$ and $+1.35$, respectively.

The nucleophilic behavior of **86** and **91** suggests that both carbenes are best formulated as singlet[119] σ^2 nucleophiles, ruling out a significant contribution of the (p^2) delocalized **86'** structure to **86**. Calculations support these conclusions.[116] Thus HO and LU orbital energies were calculated for **86** and **91** at the 4-31G level, based on STO-3G optimized structures. The values obtained (in electron volts) were: **86**, 1.70 (LU) and -8.48 (HO); **91**, 2.16 (LU) and -8.56 (HO). With the use of these carbene frontier orbital energies[116] and normalized styrene π and π^* energies,[87] estimates of $(\epsilon^{LU}_{carbene} - \epsilon^{HO}_{X\text{-styrene}})$ and $(\epsilon^{LU}_{X\text{-styrene}} - \epsilon^{HO}_{carbene})$ indicate the latter differential energies to be *smaller* than the former for both **86** and **91**, consistent with nucleophilic additions to the styrenes. The philicities of **86** and **91** appear to be a consequence of their relatively high-lying HO's.

Additionally, STO-3G calculations were made of σ and p approach trajectories for the reaction of **91** with styrene. A large (12.1 kcal/mol) "exchange repulsion" component was observed for the p (electrophilic approach) relative to the σ (nucleophilic approach). Overall, the σ approach was favored by approximately 5 kcal/mol despite charge transfer and electrostatic contributions that modestly favored the p approach.[116]

The properties of **86**, generated by pyrolysis of tosylhydrazone salts, are not

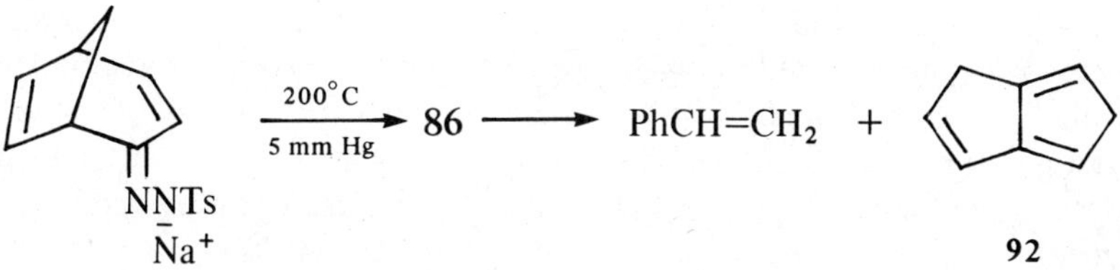

$$\xrightarrow[\text{5 mm Hg}]{200°C} \quad 86 \longrightarrow \quad PhCH=CH_2 \quad +$$

92

(3.18)

as clear. Murahasi et al. report the formation of styrene (56%) and triene **92** (4.7%) (Eq. 3.18).[116] On the other hand, Freeman and Swenson pyrolyzed the analogous Li salt (120–198°C, nominal 0.1 mm Hg) and observed very low yields (1.5–3%) of **93–98** in the indicated distribution.[117] Product ratios, but not product identities, were altered when different methods of Li salt

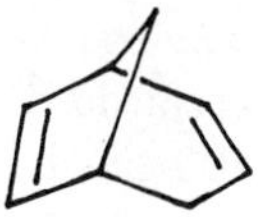

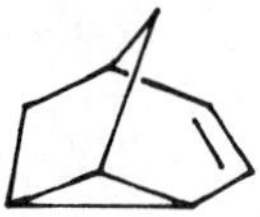

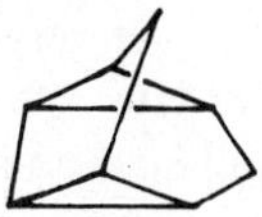

93   94   95
44%   7%   10%

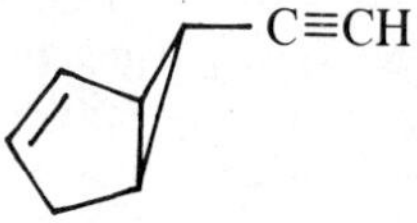

96 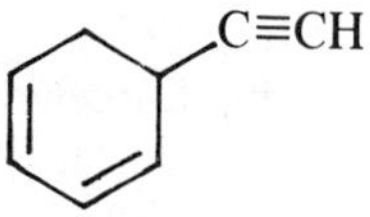   97   98
24%   9%

decomposition were used. It is difficult to explain the disparities between the two sets of products, and mechanistic speculation[116,117] must be treated with skepticism.

C.   Fluorenylidene

Among the most fascinating developments in carbene chemistry over the last three years surely must be ranked the rise and fall of the "new properties of fluorenylidene" (**99**). It is a long, complicated story and all one really need

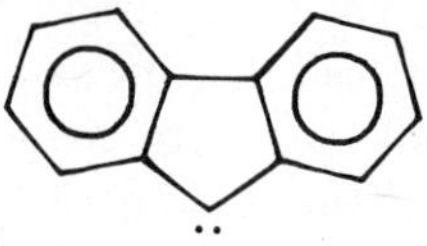

99

do to catch up is to read the summary communication of Griller, Montgomery, Scaiano, Platz, and Hadel[120] which appears back-to-back with a reevaluation of earlier work by Schuster.[121] We briefly summarize here for it makes an interesting, if ultimately disappointing story. It begins in late 1980 and early 1981[122] with a report from the Schuster group that laser photolysis of diazo-fluorene in acetonitrile led to a transient absorbing at 470 nm which decayed to a new material with absorptions at 400 and 500 nm. On various bases the first transient was assigned the structure of singlet fluorenylidene and the second species the triplet. This work was exciting not only because it looked exactly like the kind of modern carbene chemistry that should be done, but also because it seemed to be yielding rich dividends immediately. For instance, the 470-nm transient would have represented the first direct observation of a singlet carbene in solution. Moreover, it was clear from the very first that if these assignments were correct a great deal of quantitative information would be available on carbene reactions for the very first time. Indeed, numbers were rapidly forth-coming. However, in late 1981 changes began to be made,[123] and in a paper by Wong, Griller, and Scaiano it was noted that the 500-nm absorption belonged not to triplet fluorenylidene as previously assigned, but rather to the 9-fluorenyl radical. As we show shortly, this seemingly minor adjustment was more impor-tant than it looked at first. Earlier in 1981, the Schuster group had begun to report on the chemical properties of fluorenylidene.[124] These properties grew from the merely surprising and interesting to the truly astonishing. For instance, the assignment of the 500-nm band to the fluorenyl radical meant that if the original assignment of the 470-nm band to singlet fluorenylidene was correct, this singlet carbene was abstracting hydrogen. This is scarcely a normal property for a singlet. Even more amazing was the report[122] that singlet fluorenylidene reacted with 2-pentene in a nonstereospecific fashion. There is another more subtle problem: if this work is correct to this point, we have two fluorenylidenes to deal with. There is a singlet that reacts nonstereospecifically with olefins and abstracts hydrogen and a triplet with similar chemical, if not physical, properties. But yet another fluorenylidene lurks in the older literature.[125] In 1965 it was shown that butadiene could be used to scavenge selectively triplet fluorenylidene leaving behind an intermediate that reacted *stereospecifically* with *cis*-4-methyl-2-pentene. This would make a total of three fluorenylidenes, and it is difficult, to say the least, to see what the "extra" intermediate might be. A careful analysis of the situation would lead to the inevitable conclusion that something was awry, and we think it not unreasonable that a preliminary finger might have been pointed at the unusual solvent used for all of this chemistry, acetonitrile. A naive observer might have assumed that the very first thing any of the inter-mediates would do in acetonitrile would be to react, and this is exactly what happens. It turns out that the 470-nm band decays to the 400-nm band *only* in the presence of acetonitrile and not in other solvents. All is resolved if the

470-nm band is reassigned to the triplet with the 400-nm band belonging to the product of reaction with the acetonitrile, presumably the ylide. The 500-nm band remains attached to the 9-fluorenyl radical. It is important to note that the decay of the triplet (470-nm band) to the ylide (400-nm band) does not require that the triplet be reacting with acetonitrile. The triplet could be in equilibrium with another intermediate (the singlet, for instance) that reacts with acetonitrile.

Now one can be much more comfortable, if less excited, by the situation: triplet fluorenylidene is formed, presumably from a singlet unobservable, even at picosecond time scales.[120,121] In acetonitrile it forms the ylide which absorbs at 400 nm. It also abstracts hydrogen from solvent to give the 9-fluorenyl radical. Thus the triplet exhibits typical properties and both abstracts hydrogen and adds nonstereospecifically to olefins. The ylide, previously assigned the triplet carbene structure, is trapped by a variety of agents. In passing one can say that the various rates assigned to this "triplet" were unusual in some cases. Styrene and isoprene were less reactive than diethylfumarate, for instance. Now that the reactive intermediate appears to be an ylide, the rates are no longer surprising.

In summary, the older 1965 work[125] seems to be largely vindicated, but not entirely.[126] Although the broad outlines are apparently correct, there are details that need revision. In particular, it is now clear[120,126] that the hexafluorobenzene used in the early work as an "inert" medium is not quite inert. Some 15 or 20% of the product is formed from the reaction of fluorenylidene with hexafluorobenzene. The structure of this product or products is not yet known. Also unknown are the structures of the presumed ylide and the compounds formed by its reaction with olefins. One's feeling that the ylide structure is correct and that products are produced by olefin trapping of this ylide is reinforced by a recent paper in which appropriate products were isolated from a different carbene (Eq. 3.19).[127]

Nitrile ylides have been detected when 1-naphthyldiazomethane is decomposed by laser flash photolysis in nitriles. These ylides have lifetimes measured in hundreds of microseconds.[128]

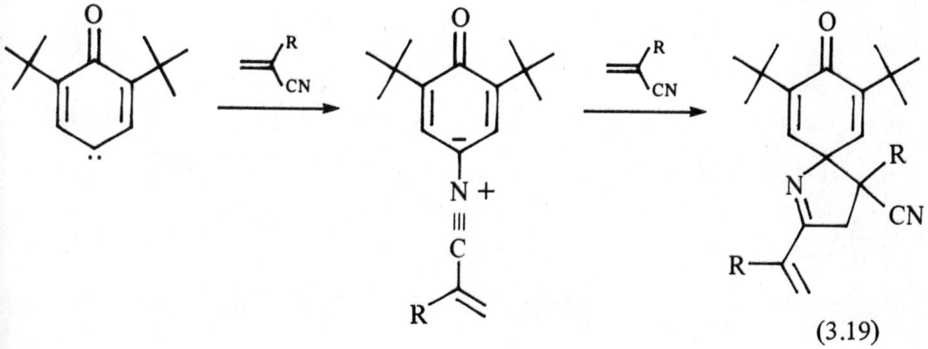

(3.19)

An important question is whether singlet and triplet fluorenylidene equilibrate, as is the case for diphenylcarbene,[129] or whether the decay from the initially formed singlet to the triplet is a one-way street. The early work of Jones and Rettig[125] showed that the addition to cis olefins became increasingly less stereospecific as hexafluorobenzene was added to the reaction mixture. The interpretation was that the supposedly inert fluorocarbon decreased the rate of bimolecular addition, allowing intersystem crossing to the triplet to yield nonstereospecific addition. As mentioned earlier, addition of butadiene resulted in scavenging of the triplet and isolation of the stereospecific reactions of the singlet. It has now been suggested that singlet and triplet fluorenylidene are fully equilibrated and that the role of hexafluorobenzene is not as previously claimed.[126] The sensitized decomposition of diazofluorene in methanol leads to singlet product, 9-methoxyfluorene.[126] An analysis of the rates of reaction, as measured by the disappearance of the 470-nm band, shows that equilibration is fast relative to addition to olefins or reactions with methanol, but slow compared to addition to butadiene. Thus the old explanation[125] for the butadiene effect can stand, but a new explanation may have to be found for the decrease in stereospecificity of the reaction. Schuster and his group[126] suggest that the energy difference between singlet and triplet, calculated by them at 1.1 kcal/mol, is increased in the presence of hexafluorobenzene and that the stereochemical change is a result of increased amount of triplet at equilibrium.

Gaspar and his group have reached similar conclusions.[130a] Triplet scavengers such as butadiene or styrene increased the stereospecificity of cyclopropanation in the reaction of fluorenylidene 99 with cis-dichloroethylene, thus indicating an increased percentage of cyclopropane from the singlet. In addition, the absolute and relative yield of 9-(2,2-dichloroethylidene)fluorene 100, a product of reaction of triplet 99 with olefin, decreased. Methanol, a singlet trap, was without effect on the specificity of cycloaddition. This requires a rather delicate balancing of rates: triplet 99 must react with butadiene more quickly than it undergoes intersystem crossing, at least in the "uphill" direction. Singlet 99 must react with methanol more slowly than intersystem crossing in both directions.

100

Added hexafluorobenzene, but *not* other diluents including both fluoro-carbons and benzene, induces a decrease in stereospecificity of cyclopropanation and yield of **100**. Again a "special" role for hexafluorobenzene is implied. The authors suggest formation of a triplet-mimicking carbenoid.

Some support for this idea comes from an esr experiment in which a "not easily interpretable" radical signal is obtained from irradiation of diazofluorene in hexafluorobenzene at $-127°C$.[130c]

In contrast, perfluorocyclobutane was found to decrease the stereospecificity of addition of indenylidene.[130b] Although a variety of special roles can be imagined for hexafluorobenzene, ranging from pi complex formation to produc-tion of carbene-mimicking diradicals, it is not clear how a saturated fluorocarbon could act in a similar fashion. This may be the first real example of collisional deactivation of a carbene in solution.

What lessons are to be learned from this episode? This is an important ques-tion because every effort should be made to avoid its repetition. There is inevi-tably unfortunate fallout. The original assignment of the 400-nm band to triplet fluorenylidene led to the conclusion that cyclopropanation is preceded by complex formation.[131] The reassignment of this absorption to the ylide renders this conclusion invalid (see however Section V.B.c).[102] This reviewer has already seen in three research proposals references to the "unusual" properties of singlet fluorenylidene and generalizations to the behavior of other species. References are beginning to creep into the literature. The fact that fluorenylidene became the focus for mistakes surely does not lessen one's confidence that one wave of the future is exactly this kind of combination of physical and organic chemistry. It is our opinion that this remains an extraordinarily exciting area that is very likely to be most productive. What should be improved in the future, one suspects, is the collaboration between those using these exciting modern methods and the more traditional practitioners in the field.

D.   Ylides

a.   *Carbonyl Ylides*

The intensive study lavished upon the reaction of carbenes with carbon—carbon double bonds has not been matched in the area of carbon—oxygen double bonds. This has been a long-recognized deficiency, but the reaction of divalent carbon with carbonyl groups has nonetheless remained something of a poor relation in the family of carbene chemistry. Although many recog-nized the existence of the problem, no one did very much about it. Happily the last few years has seen a change for the better as a number of papers has appeared on the subject.

The reaction is a really old one, as it dates at least to the work of Dieckmann in 1910,[132a] who was really just revising earlier work of Buchner in 1885.[132b]

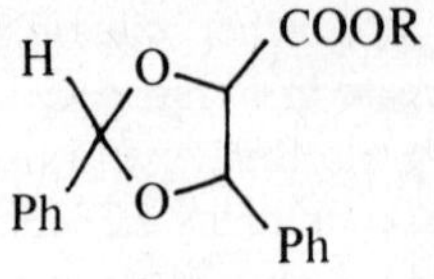

Dieckmann established the structure of the major product of the reaction of a carbene with an aldehyde as the 1,3-dioxolane **101**. Similarly, Bradley and Ledwith[133] found three products including a 1,3-dioxolane from the reaction of methylene with acetone. In two papers de March and Huisgen have now described the related additions of dicarbomethoxycarbene to benzaldehyde (Eq. 3.20).[134] The major products are the dioxolanes **102** and **103**, but the oxirane **104** is formed as well. Although **104** is not an intermediate in the formation of **102** and **103**, it can be converted to them by heating to higher temperatures in benzaldehyde. A kinetic analysis reveals a common intermediate in the formation of **102–104**, and it is logically assumed to be the carbonyl ylide **105**, which is trapped by more aldehyde to give the major products.[134] Confirmation comes from trapping of the ylide with electrophilic olefins as in (Eq. 3.21).[134]

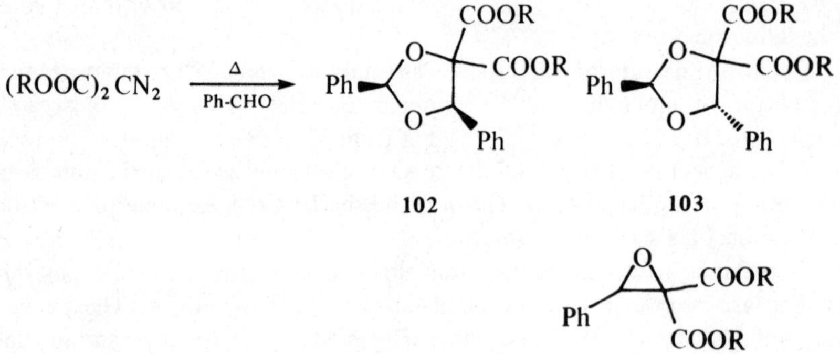

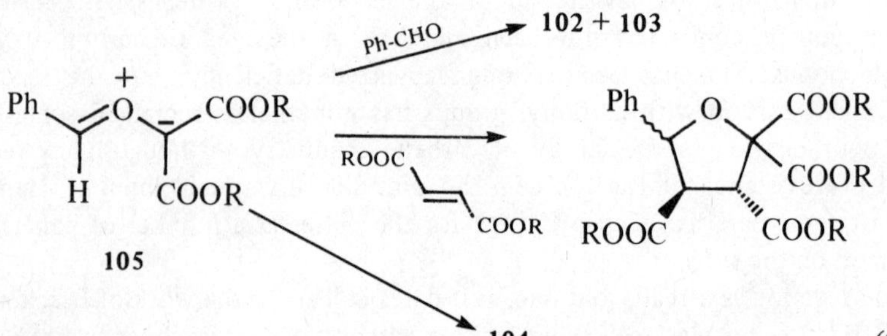

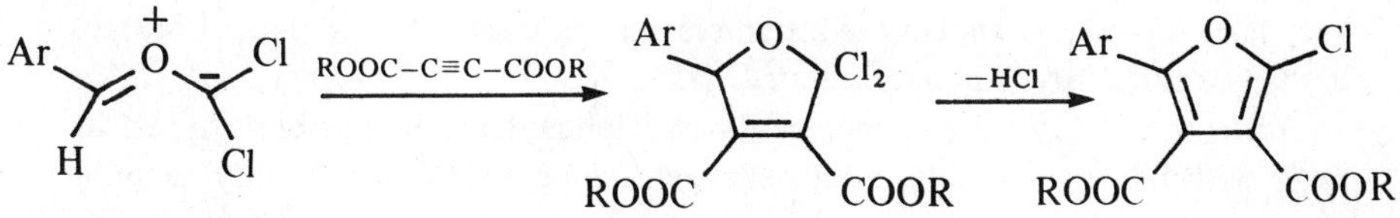

$$(3.22)$$

A similar trapping of an ylide has been reported by Gil and Landgrebe (Eq. 3.22).[135]

Is carbonyl ylide formation reversible? The answer appears to be yes. In two papers examining carbonyl ylides formed not by carbene addition but by decomposition of heterocycles, strong evidence for the fragmentation of the ylide into a carbonyl compound and a carbene has been presented.[136] In one of them, pyrolysis of **106** in carbon tetrachloride (surely a curious choice of solvent) led to the products shown in Eq. 3.23). Dicyclopropylketone clearly points to an ylide fragmentation, although the other partner, carbene **107** may be partly lost to radical reactions. Biacetyl is presumably derived from **107**. Fragmentation in the other sense would give the observed acetic anhydride and the products of dicyclopropylcarbene. Compound **108** may be the end product of a reaction involving dicyclopropylcarbene and carbon tetrachloride, but it is a trifle disconcerting not to find 1-cyclopropylcyclobutene mentioned. Although reaction of singlet dicyclopropylcarbene with carbon tetrachloride could be

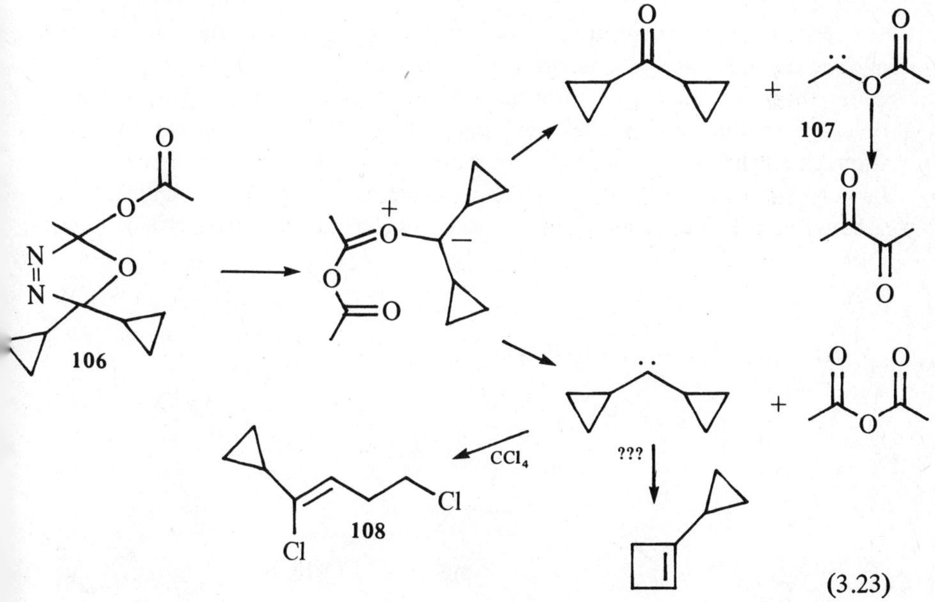

$$(3.23)$$

very fast, one would expect the intramolecular expansion to compete favorably. Perhaps it is destroyed by radicals.[137]

Arylcarbenes also give carbonyl ylides. Diphenylcarbene has been shown to react with benzophenone at liquid nitrogen temperature in a *tert*-butyl alcohol glass.[138] More recently Wong, Griller, and Scaiano have examined the reaction of fluorenylidene with a variety of ketones (see Section V.B.c for quantitative data).[98] They are able to observe the carbonyl ylide along with the 9-fluorenyl radical created by abstraction of hydrogen by the triplet. Addition may be due to triplet fluorenylidene, although the authors are careful to point out that addition could be due to a small amount of singlet in equilibrium with the triplet. We think this is probably the case, but radicals do add to carbonyls at the oxygen, so it is difficult to be certain (Eq. 3.24). 1-Napthylcarbene has also been shown to add to ketones to give ylides. Here two mechanisms can be

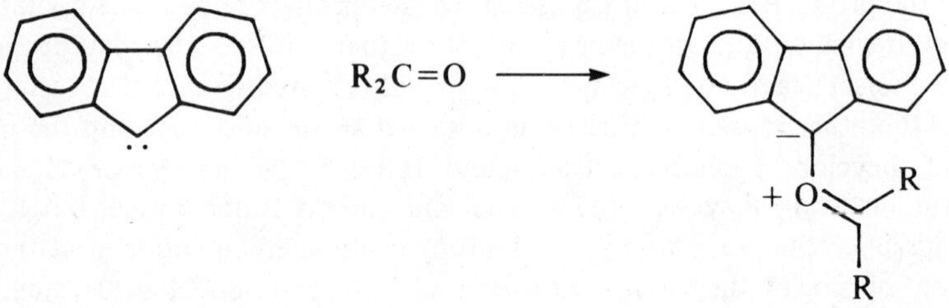

$$(3.24)$$

identified as the ylide spectrum is formed by an initial "instantaneous" jump, followed by a slower 60-nsec growth. It is reasoned that the rapid reaction occurs through the singlet carbene and the slower from the ground state triplet.[128] In this case it does not appear that ylide formation is reversible. Although all the details are surely not worked out, the picture one derives from these recent papers and their earlier predecessors is reasonably simple. The carbonyl ylide is formed and then either decays intramolecularly to give oxirane,

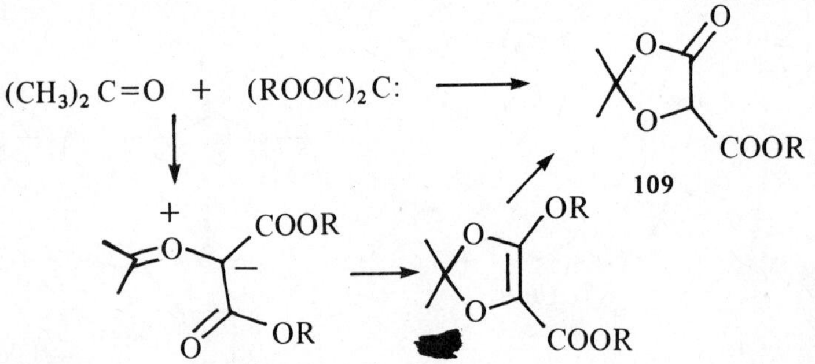

$$(3.25)$$

or is trapped in intermolecular fashion to give the 1,3-dioxolane. But, once again, all is not really so simple. If one makes a very simple perturbation on Huisgen's[134] system and replaces benzaldehyde with acetone, the entire nature of the reaction changes. No longer is the product a 1,3-dioxolane, but instead the dioxolan-4-one **109**.[139] This reaction is by no means fully understood yet, and not too much is known beyond the structure of the major product. For some reason the carbonyl ylide is trapped intramolecularly in this case (although not in the examples of de March and Huisgen), and a methyl group is lost, either by hydrolysis of an intermediate enol ether or some other route (Eq. 3.25).

A very closely related reaction has been reported by Alonso and Chitty (Eq. 3.26).[140] Methyl α-diazoacetoacetate adds to a variety of aldehydes to give the 1,3 dioxole-4-carboxylates. Although the authors describe the reaction as a 1,3 dipolar addition of the carbene to the aldehyde, and although is catalyzed by copper salts, it can be described as an addition of the carbene or

$$\text{Ph-CHO} + \quad \overset{\text{O}\quad\text{O}}{\underset{\text{N}_2}{\|\quad\|}}\text{OR} \longrightarrow \quad \overset{\text{Ph}}{\underset{\text{H}}{}}\overset{+}{\diagdown}\text{O}\diagdown\text{COOR} \longrightarrow \text{Ph}-\overset{\text{O}}{\underset{\text{O}}{\diagup}}\diagdown\overset{\text{COOR}}{}$$

$$(3.26)$$

carbenoid to the aldehyde to give an ylide which is then captured intramolecularly as in the previous unpublished work.[139] Why the capture occurs intramolecularly and not intermolecularly is not easy to see. Capture is by the more reactive keto end of the ylide, and perhaps the increased reactivity of the keto end over the ester end is responsible. A simple test of this primitive surmise is obviously available.

It seems likely that the flurry of renewed activity in the carbonyl ylide area is the beginning of an intense and thorough exploration of this neglected reaction.

b. Ylides from Single-Bonded Heteroatoms

Although the reaction of carbenes with the OH bonds of alcohols is common and often used as a diagnostic for the singlet state, it has been remarked quite

$$\text{R}_2\text{C:} + \text{R'OH} \quad \xrightarrow[\text{direct}]{\text{ylide}} \quad \overset{-}{\text{R}_2\text{C}}-\overset{+}{\text{O}}\diagup\overset{\text{H}}{\underset{\text{R'}}{}} \longrightarrow \text{R}_2\text{CH}-\text{OR'}$$

$$\text{proton-}\atop\text{ation} \quad \overset{+}{\text{R}_2\text{CH}} + \overset{-}{\text{OR'}}$$

sensibly by Kirmse that very little information exists on the mechanism of this reaction.[141] He points out that in principle three mechanisms exist: a direct insertion, formation of an ylide followed by proton transfer, and protonation of the carbene to give a carbocation followed by collapse to give the product. He and his co-workers provide us with examples of the second two possibilities. Reaction of cyclopentadienylidene 110 with CH_3OD gives 111 and 112, but no 113 as determined by analysis of Diels–Alder adducts. Thus a direct insertion is eliminated because the reaction was run at $-70°C$, below the temperature at which hydrogen or deuterium shifts could be facile.

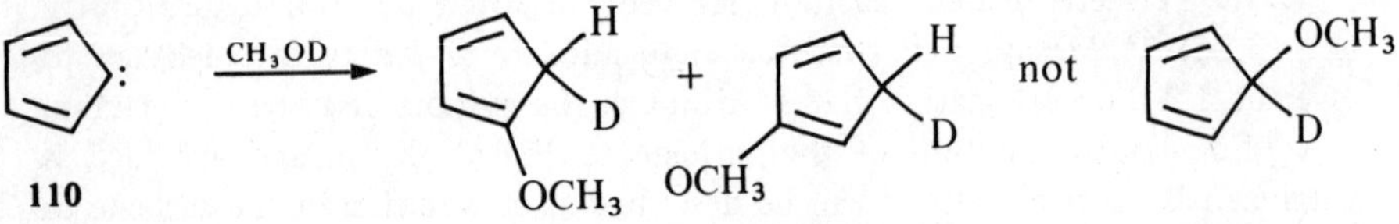

By contrast, cycloheptatrienylidene 114 reacts with EtOD to give 7-ethoxy-tropilidene 115 with the deuterium randomly positioned (Eq. 3.27). Although this is consistent with protonation of the carbene to give the tropilium ion 116 followed by reaction with ethoxide, it is necessary to eliminate both protonation of the diazo compound to give a diazonium ion 117 and intervention of the cyclic allene 118. The former is eliminated by use of a mixture of chloro-tropylidenes as carbene source, and the latter possibility seems remote, as strained allenes are known to react to give vinyl rather than allyl ethers.[141] It remains to be seen whether examples of the first mechanism, direct OH inser-tion, will one day appear.

That extraordinarily complicated situations can be encountered in reactions with alcohols is shown by the work of Griller, Liu, and Scaiano[99] in which they show that, at least for chlorophenylcarbenes, reaction may take place with aggregated alcohols. For details see Section V.B.c.

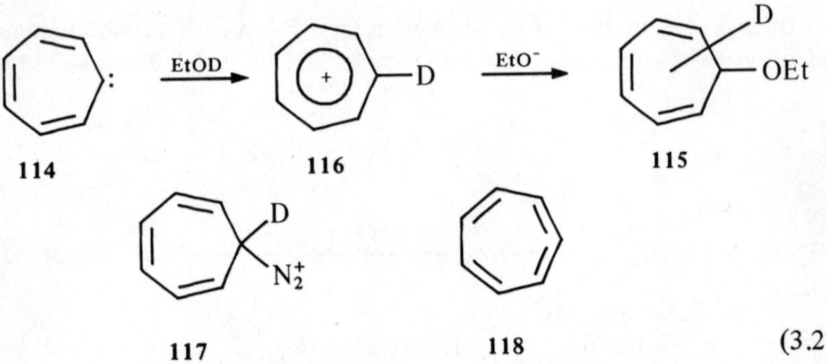

(3.27)

E.   Phenylcarbenes, Cycloheptatrienylidenes, and Their Interconversion

Theory, spectroscopy, and experiment have all been brought to bear on the question of the structure of the material formed when either phenylcarbene or cycloheptatrienylidene is generated.

At the MNDO level, Waali[142] finds that cycloheptatetraene 118 is more stable than the carbene 114 by 23 kcal/mol. If the barrier to interconversion of 118 and 114 were high, one still might be able to see the chemistry derived from 114. Waali finds no barrier, and indeed the carbene becomes the transition state for the interconversion of two twisted, cyclic allenes, 118 and 118'.

118   114   118'

Another set of calculations, this time at the *ab initio* level, agrees that the best description of the intermediate is the cyclic, nonplanar allene.[143]

The same conclusions have been reached experimentally.[144] Harris and W. M. Jones took advantage of the potential chirality of the cyclic allene 118 and the necessarily achiral nature of the carbene 114. Two kinds of products are observed (Eq. 3.28). Generation in the presence of diphenylisobenzofuran gives the trapped product 119, which certainly looks like it comes from the allene. In the presence of styrene, however, the spirononatriene 120 is formed, and this is most simply explained as the product of a conventional carbene addition. However, both products can be rationalized as coming from *either* 114 or 118. When the intermediate is generated from a chiral precursor 121, 119 is formed optically active. This shows that indeed the cyclic, twisted allene

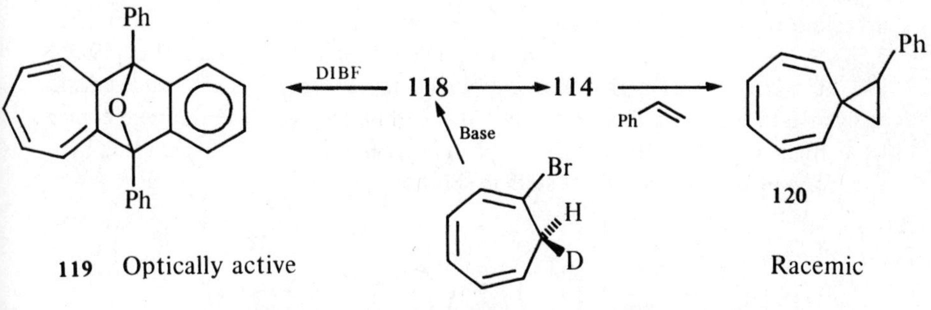

119   Optically active

121   Optically active

120   Racemic

(3.28)

is involved in its formation. However, **120** was not optically active. Despite the calculations alluded to before, this fact implicates the carbene and suggests a connection between **118** and **114** occupying an energy minimum. Recall the work of Kirmse[141] in which the intermediate reacted with ethanol to give the 7-ethoxycycloheptatriene. This is not the product expected of the cyclic allene but can be rationalized by protonation of the carbene. Perhaps the calculations need to be refined further.

Spectroscopic evidence has also appeared.[145] Chapman and his co-workers showed that the material formed from either irradiation of triplet phenyl-carbene, irradiation of the diazirine **122**, or pyrolysis of phenyldiazomethane at 500°C was probably the cyclic allene (Eq. 3.29). The evidence that the intermediate is not the cyclopropene **123** is compelling, and the infrared spectrum

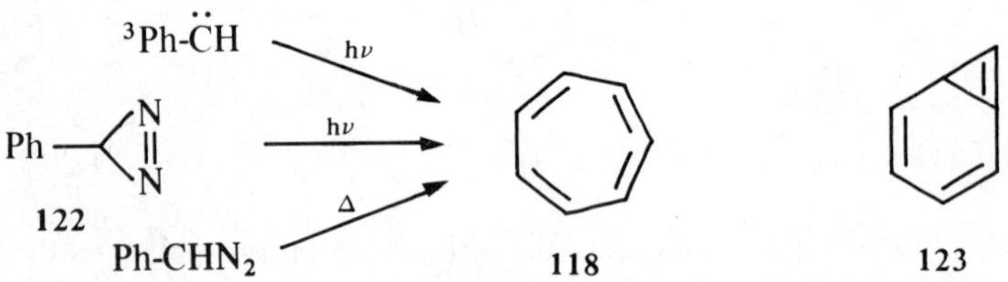

$$(3.29)$$

published seems appropriate for **118**. This work does not conflict with the experiments of W. M. Jones, and, in particular, it does not mean that his apparent trapping of the carbene **114** is incorrect or misleading. The spectroscopic experiment is a kind of "thermodynamic" experiment, and the carbene certainly could be accessible from the lower energy allene under W. M. Jones' conditions. If the carbene **114** were sufficiently reactive, **120** could be formed from it, even if the allene **118** were more stable.

Ordinarily the phenylcarbene rearrangement does not occur in solution, at least for simple phenylcarbenes. Thus irradiation of diphenyldiazomethane in solution produces no fluorene,[146] whereas gas phase pyrolysis does (Eq. 3.30).[147] However, decomposition of diphenyldiazomethane using the 249-nm line of a laser revealed fluorene and several dimeric products.[148] Mechanistic analysis is difficult at this point, but the authors suggest that fluorene comes from further photolysis of singlet diphenylcarbene. Perhaps we are seeing here a reaction of the excited singlet state in solution.

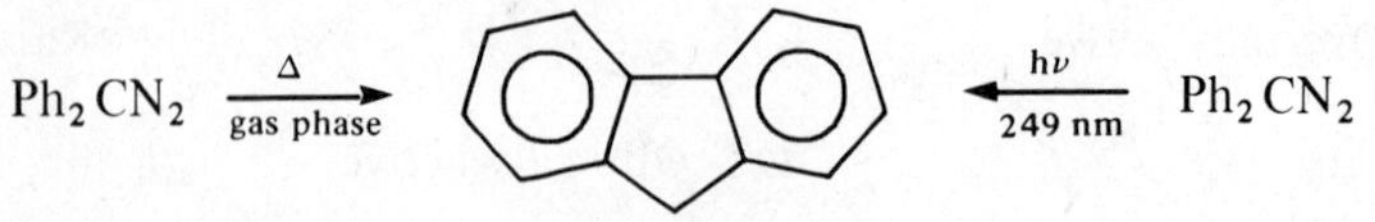

F.   Reactions at Low Temperature and in Matrices

The mechanism of decomposition of triplet diphenylcarbene in a matrix is a hydrogen abstraction process that occurs by quantum mechanical tunneling of the hydrogen atom.[149] The triplet is established as the reactive species by the observation of a large isotope effect when deuterium is substituted for hydrogen at the 2 position of isopropanol. When the substitution is made at the OH site, no isotope effect is noticed. A plot of log [diphenylcarbene] versus time is nonlinear.[150] This is to be expected if the carbene is lodged in several different sites in the matrix. The decay of the carbene in these different sites will take place at different rates. A plot of concentration versus time$^{1/2}$ is linear.[149,150] This effect should not be observed in single crystals and indeed clean, first-order kinetics were observed under such conditions.[151] Arrhenius parameters are reported for hydrogen abstraction and are thought diagnostic of hydrogen atom transfer by tunneling. This has important consequences for the previously observed chemistry in matrices in which hydrogen abstraction dominates addition to olefins, the reaction generally prevalent at higher temperatures. The new mechanism appears as both the alternative processes, addition to double bonds or abstraction of hydrogen are severely slowed at low temperature. Indeed, it is calculated in this paper[149] that in the absence of tunneling triplet diphenylcarbene would be stable at 77 K.

It has also been observed by Tomioka[152] that in the reaction of arylcarbenes with alcohols the C−H bonds become relatively more reactive compared to the O−H bonds as the temperature decreases (Eq. 3.31). Presumably this simply reflects the increased competitive advantage of the lower energy triplet at low temperatures. Although at room temperature the O−H/C−H insertion selectivity increased with the electron-donating ability of Z, at low temperature both electron-releasing and electron-withdrawing groups increased O−H insertion. It has not yet been possible to explain this effect.

$$^3 Z-Ph-\overset{..}{C}H \quad \underset{R_2CHOH}{\longleftrightarrow} \quad {}^1 Z-Ph-\overset{..}{C}H \qquad (3.31)$$

$$Z-Ph-CH_2-CR_2 \qquad\qquad Z-Ph-CH_2-O-CHR_2$$
$$\underset{OH}{|}$$

Two papers have examined intramolecular migrations taking place in carbenes isolated in matrices. In the first of these[153] carbene **124** was found to give **125**, **126**, and **127** in solution (Eq. 3.32). Lowering the temperature simply increased the amount of triplet product, **127**. The triplet is increasingly favored

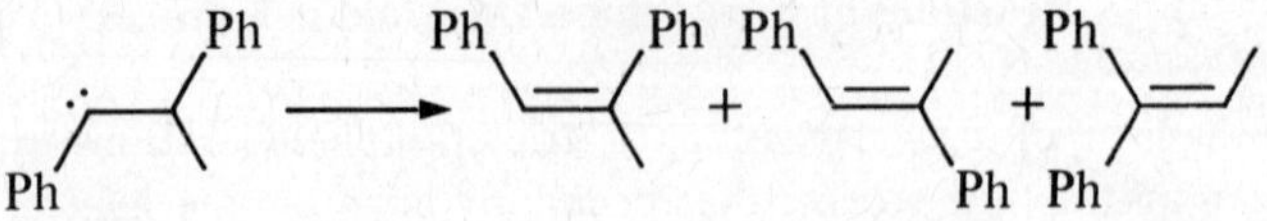

$$(3.32)$$

at low temperature, and therefore the triplet reaction, phenyl migration, increases. However, when the temperature dropped beyond the freezing point **127** was substantially suppressed. It was assumed that the relatively large motion needed to migrate the large phenyl allowed the smaller hydrogen to move to whatever singlet carbene remained at low temperature. This paper has been reexamined in more recent work[154] and intermolecular reactions of both spin states with a matrix uncovered. The matrix clearly has an enormous effect upon reactions of the triplet. It is worth quoting the penultimate sentence of the body of this paper, "It would appear that ESR kinetic studies of unimolecular reactions of triplets will reveal more about environmental factors than about the desired reaction."[154]

Matrix effects apparently also operate on olefin cyclopropanation reactions. Changes in the ratio of cis to trans product as phenylcarbene adds to styrenes occur as the temperature is lowered. A complete analysis is not yet possible beyond the general statement that as the temperature goes down, reactions of triplets will be maximized, and it is not surprising to see the triplet produce an increase in trans cyclopropane. The presence of acetonitrile (see Section V.C) in some of the solid-phase studies complicates the issue as it may very well be reactive.[155,156]

We should also mention that the low-temperature chemistry of some nonaromatic carbenes may be emerging. Thus cyclopentadienylidene **110** has been allowed to dimerize, react with carbon monoxide to give ketene **128**, and photolyze to an unknown acetylene (Eq. 3.33).[157]

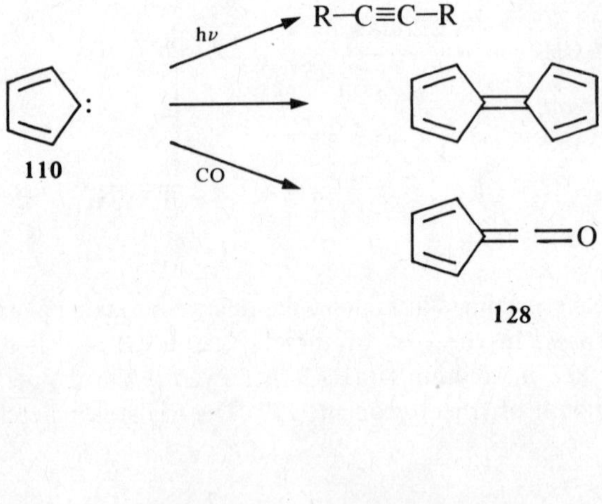

$$(3.33)$$

G.   Wolff Rearrangement

It has long been known[158] that irradiation of methyldiazoacetate in alcohols leads to the following set of products (Eq. 3.34):

$$N_2CHCOOMe \xrightarrow[\text{i-PrOH}]{h\nu} \text{i-PrOCH}_2COOMe + Me_2C(OH)CH_2COOMe$$

$$\qquad\qquad\qquad\qquad\qquad\quad 129 \qquad\qquad\qquad\qquad 130$$

$$\text{(3.34)}$$

$$+ \text{ MeOCH}_2COO\text{-}i\text{-Pr} \qquad \text{i-PrOCH}_2COO\text{-}i\text{-Pr} + CH_3COOMe$$

$$\qquad 131 \qquad\qquad\qquad\qquad 132 \qquad\qquad\qquad 133$$

When the reaction is photosensitized, product **133** greatly increases, **131** and **132** are eliminated, and some **129** and **130** are still formed. It is logical to presume that **133** comes from the triplet carbene by double hydrogen abstraction and that **131** and **132** are singlet products of some kind. Compounds **129** and **130** could be products of either singlet or triplet, although the fact that they are greatly reduced on photosensitization would suggest a singlet source (Eq. 3.35).[159]

$$N_2CHCOOMe \xrightarrow[\text{sens.}]{h\nu,\ i\text{-PrOH}} \quad 133 \quad + 129 + 130$$

$$\qquad\qquad\qquad\qquad\qquad \text{(increase)}$$

$$\text{(3.35)}$$

$$N_2CHCOOMe \xrightarrow{h\nu,\ i\text{-PrOH}} \quad 129\text{--}132 \quad \text{not} \quad 133$$

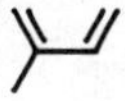

Triplet quenching eliminates compound **133**, but **129–132** are unaffected (Eq. 3.35). This reinforces the view suggested above that **129–132** are singlet products. Why are not **129** and **130** eliminated in the sensitized reaction as are **131** and **132**? This is presumably because **131** and **132** are products of the singlet diazo compound whereas **129** and **130** come from the singlet carbene. The sensitized reaction eliminates **131** and **132** because it bypasses the singlet diazo compound, but it does not eliminate **129** and **130** because these are formed by singlet carbene in equilibrium with the triplet carbene. This leads to the mechanistic scheme in Eq. 3.36:

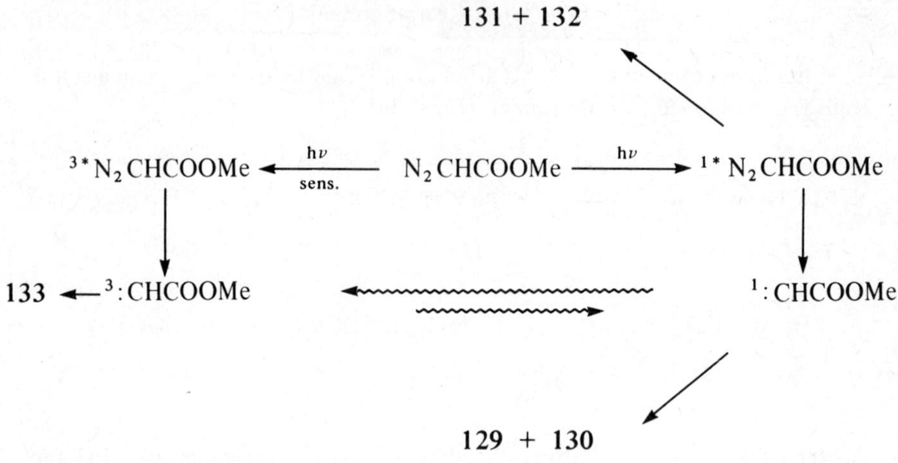

$$131 + 132$$

$$^{3*}N_2CHCOOMe \xleftarrow[\text{sens.}]{h\nu} N_2CHCOOMe \xrightarrow{h\nu} {}^{1*}N_2CHCOOMe$$

$$133 \leftarrow {}^{3}:CHCOOMe \qquad\qquad\qquad {}^{1}:CHCOOMe$$

$$129 + 130$$

$$(3.36)$$

Note that in this mechanism the singlet diazo compound both loses nitrogen to give the singlet carbene and undergoes rearrangement. It has earlier been shown[160] that a more simple relative of diazoacetic ester, diazoacetone, undergoes only Wolff rearrangement. The singlet carbene does not appear to be formed. This is rationalized by invoking the old suggestion of Kaplan and Meloy[161] that Wolff rearrangement occurs from the s–Z form 134 whereas the s–E form 135 leads to carbene. Diazoacetone is largely in the s–Z arrangement and so gives Wolff rearrangement, not carbene. Diazoacetic ester is approximately half in each form and so does both reactions.

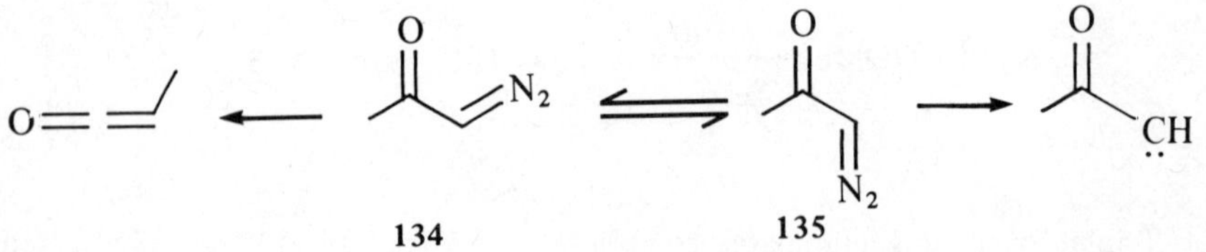

$$\text{134} \qquad\qquad \text{135}$$

This overall scheme is used to rationalize behavior of other diazo compounds, such as 136. This compound is known to exist preferentially in the s–Z conformation, and it is no surprise that direct irradiation gives Wolff rearrangement almost exclusively without the formation of products of O–H or C–H insertion, which would be diagnostic of carbenes (Eq. 3.37). Photosensitized decomposition reduces — but does not eliminate — Wolff rearrangement. Products of

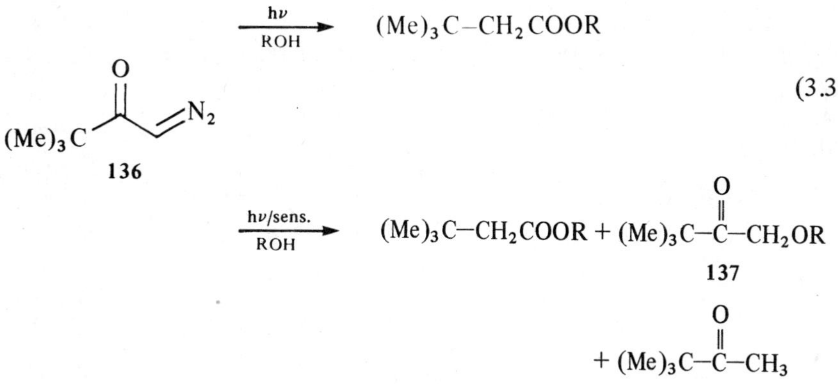

$$\xrightarrow[\text{ROH}]{h\nu} \quad (Me)_3 C-CH_2 COOR$$

(3.37)

136

$$\xrightarrow[\text{ROH}]{h\nu/\text{sens.}} \quad (Me)_3 C-CH_2 COOR + (Me)_3 C-\overset{\overset{\displaystyle O}{\|}}{C}-CH_2 OR$$

137

$$+ (Me)_3 C-\overset{\overset{\displaystyle O}{\|}}{C}-CH_3$$

138

O–H insertion (**137**) and double abstraction (**138**) appear. It seems that triplet carbene is produced which both gives **138** and undergoes intersystem crossing to the singlet carbene that gives **137**. The persistence of the Wolff rearrangement product, even under photosensitized conditions, implies that in this case the singlet carbene can also undergo Wolff rearrangement. It is suggested that the difference in the behavior of the two reactions shown in Eq. 3.37 has to do with the relative migratory aptitudes of alkyl versus alkoxy.[159]

It has been pointed out that the existence of ketene formation directly from the diazo compound complicates earlier searches for an interconversion of ketocarbenes.[162] The authors eliminate diazo compound involvement by photosensitizing the decomposition of the diazo compounds. They find that typical ketocarbenes (**139**) undergo oxygen shift rapidly compared to intermolecular reaction with alcohol or, in some cases, intramolecular hydrogen shift (Eq. 3.38).

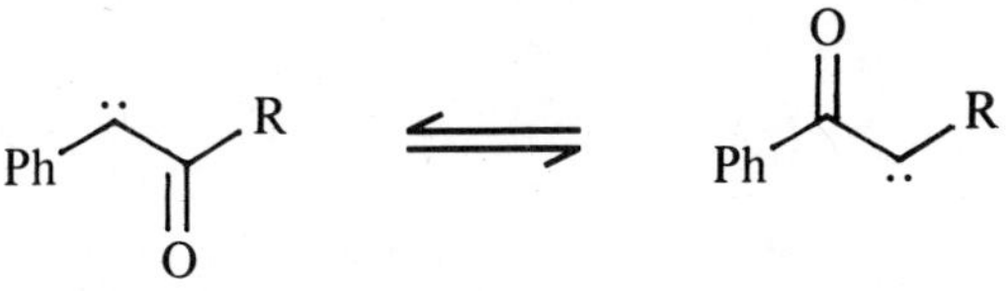

139

(3.38)

Failure to take account of a previously reported Wolff rearrangement apparently led to the incorrect assignment of many products of a direct photolysis of diazo compound **140**.[163] It had been earlier shown that although the sensitized irradiation of **140** in olefins led to cyclopropanes, the direct irradiation did not. Singlet **141** proceeded to fragments through a trappable ketene formed by Wolff rearrangement.[164] The claim[163] that irradiation at wavelengths shorter than those used previously led to good yields of cyclopropanes has been retracted.[165] The products thought to be spirocyclopropanes are, alas, cyclobutanones formed by unexceptional cycloaddition of the ketene 142 (Eq. 3.39).

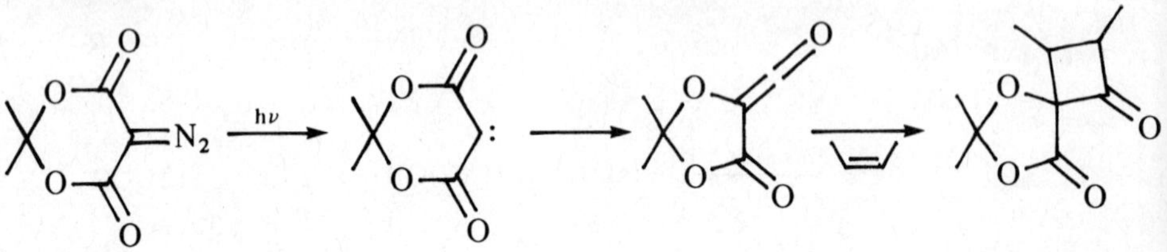

$$(3.39)$$

H.   Insertion Reactions

After the long treatment of this subject in Volume 2, the discussion in this section can be relatively brief.

The high barriers to triplet migrations have been confirmed by calculation,[166] as has the existence of a very low barrier for rearrangement of singlet vinylidene.[167] The intriguing suggestion has been made that vinylidene might be accessible from twisted ethylene and that ethylidene might give acetylene by a hydrogen loss (Eq. 3.40).[168] Rearrangements of a number of substituted vinylidenes to the corresponding acetylenes have been described.[49]

$$CH_3 - \ddot{C}H \longrightarrow H - C \equiv C - H \qquad (3.40)$$

The question of whether a migrating phenyl group would prefer arrangement 143 or 144 has been addressed experimentally.[169] In both 145 and 146 hydrogen rearrangement is preferred to phenyl by a factor of 10^3. The advantage of

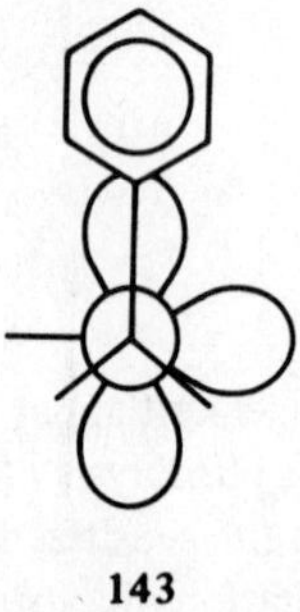

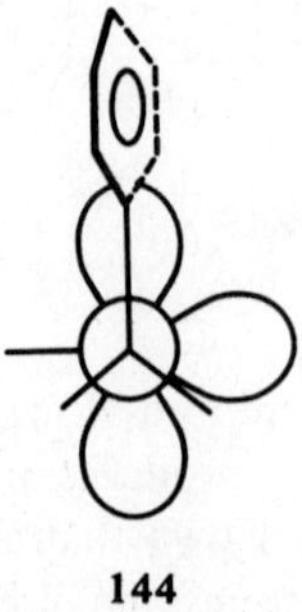

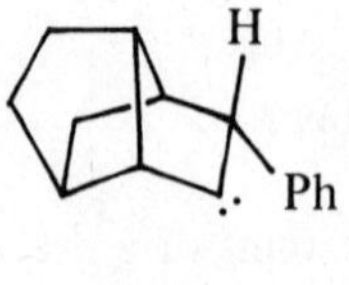

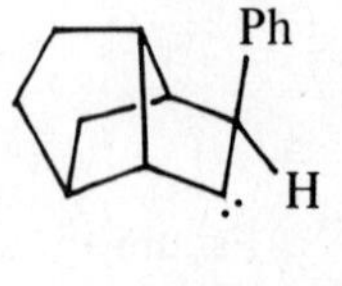

hydrogen in 145 probably has to do with the better overlap with an empty p orbital, but 146 must be a different case. The authors suggest that phenyl

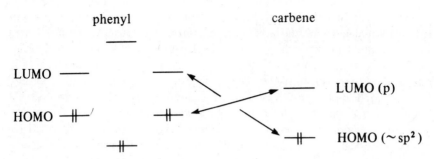

phenyl   carbene

migration is retarded because of the inability to achieve orientation **143** for steric reasons. This seems a reasonable explanation to us, as does the intuitive idea that **143** should be preferred to **144**. Orientation **143** corresponds to a phenyl group HOMO/carbene LUMO interaction and **144** to a phenyl group LUMO/carbene HOMO interaction. As can be seen from an energy-level diagram, in general one would assume the former to be the more important.

I.   Interesting Miscellanea

Although cyclopropenylcarbene **147** was shown to fragment to acetylenes even in a matrix at 10 K, (Eq. 3.41)[170] similar carbenes have been more cooperative.[171] Eisenbarth and Regitz showed that irradiation of **148a** and **148b** led

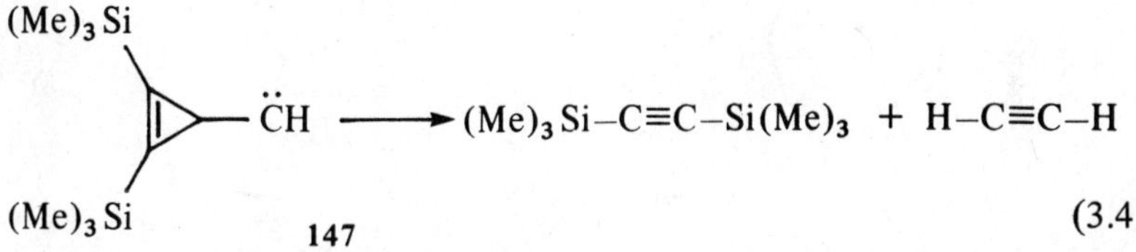

$(Me)_3Si$

$(Me)_3Si$ **147** $\ddot{C}H \longrightarrow (Me)_3Si-C\equiv C-Si(Me)_3 + H-C\equiv C-H$

$$(3.41)$$

to cyclobutadiene **149** and acetylenes. The direct irradiation of **148a** produced the products in the ratio 70–30; the sensitized irradiation in the ratio 20–70, along with 10% of an unknown compound (Eq. 3.42).

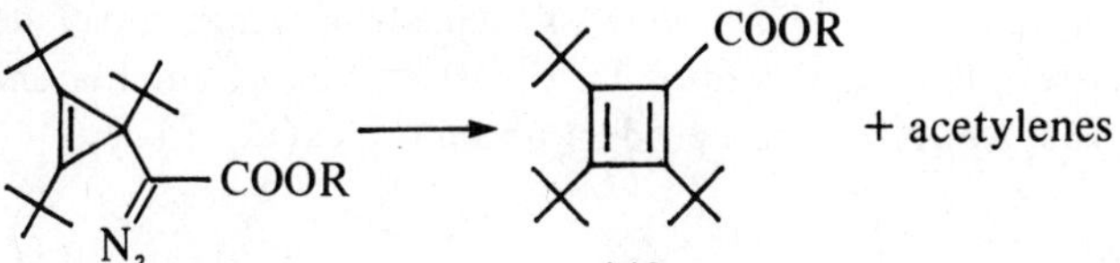

COOR

COOR   + acetylenes

N_2 **148**   **149**

$$(3.42)$$

a, R = *tert*-Bu   direct hν 70   30
b, R = Me   sens. hν 20   70

This change in ratio is explained by postulating that singlet **150** rearranges to cyclobutadiene and the acetylenes are formed through the triplet state. There are problems with the proffered explanation. It is known that the ring expansion and fragmentation reactions of cyclopropylcarbenes proceed with retention of stereochemistry.[172] Thus a singlet carbene, not a triplet, is very strongly implicated in both reactions of cyclopropylcarbenes. The data surely show, however, that a triplet is involved in acetylene formation in the photosensitized reaction of cyclopropenylcarbenes. Perhaps intermediate **151** is formed (although no analogous intermediate is apparently produced for cyclopropanylcarbenes) and then fragments after intersystem crossing (Eq. 3.43).

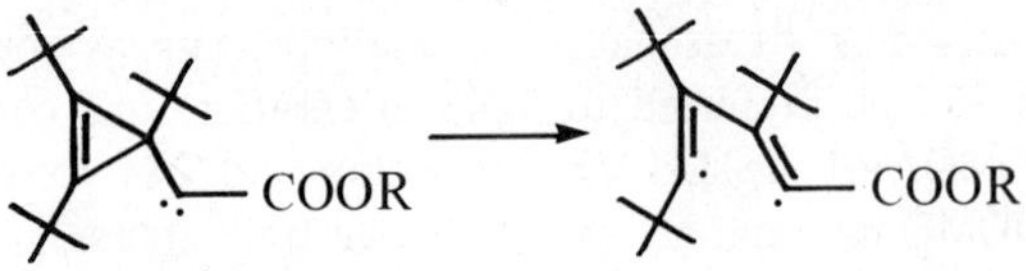

$$\text{(3.43)}$$

150   **151**

The real cyclononatetraenylidene has finally been generated.[173] It is produced from the diazo compound which is accessible through a modification of the older, incorrect procedure.[174] The ultimate product is dimer **152**, which is thought to arise by hydrogen abstraction by triplet carbene and dimerization (Eq. 3.44). There are many other possible routes, however, and we do not think the case for triplet carbene is strong.

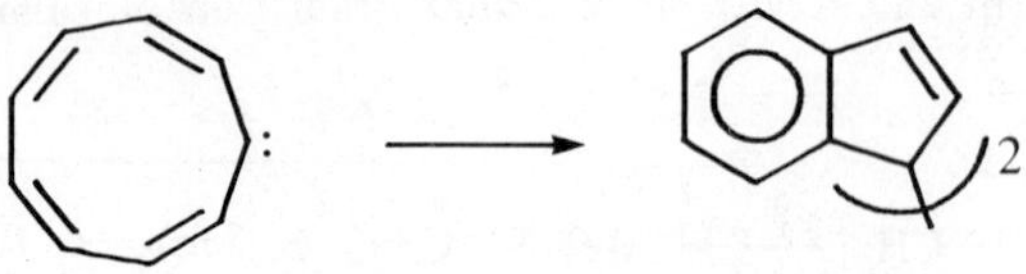

$$\text{(3.44)}$$

152

Although the following report does not strictly fit in our chapter, we cannot resist mentioning that when atomic carbon is allowed to react with ammonia or ammonia/water at $-196°$C, amino acids are produced on workup.[175]

The first direct observation of a triplet dialkylcarbene has been reported. Di-*tert*-butylcarbene is not linear, as an analysis of the E/D ratio yields a bond angle of approximately 143°. Products obtained on warming include **153** and **154**, which are presumably formed by singlet carbene in equilibrium with the triplet, and **155**, which may be a product of the triplet (Eq. 3.45).[176]

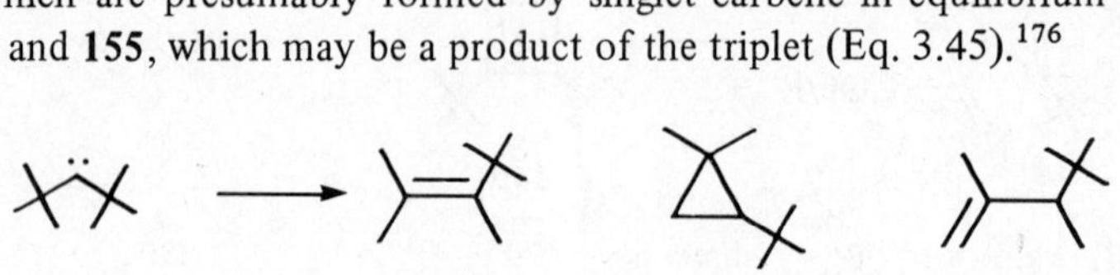

$$\text{(3.45)}$$

153   **154**   **155**

The intermolecular reactions of adamantanylidene (156) have also been described. Here the "dialkyl" carbene is protected from intramolecular rearrangement by its geometry.[177] That difficult intramolecular rearrangements will occur provided that intermolecular escape is blocked is shown by the formation of 157 from diadamantylcarbene (Eq. 3.46).[178]

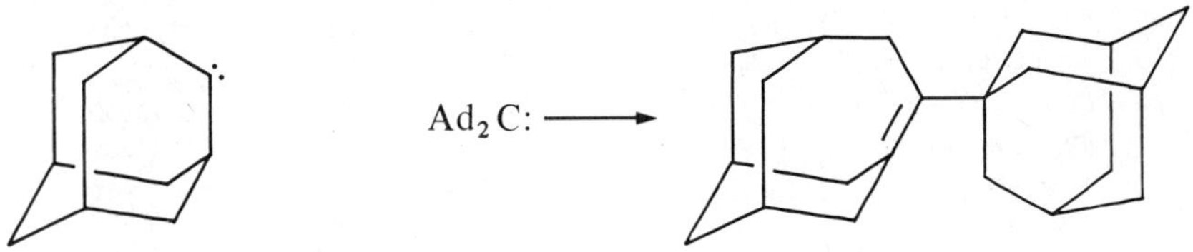

$$Ad_2C: \longrightarrow$$

156   157

(3.46)

Over the years carbenes have produced an enormous number of highly strained and sometimes otherwise inaccessible molecules. One might think that the end of the list had been reached, but not so. Thus 159 is the reported product of carbene 158 (Eq. 3.47).[179] Hydrazone 160 is also formed. Other propellanes not quite so spectacularly strained have been produced by similar methods.[180,181] In this case formation of 159 seems surprising to us. Why, for instance, was not 161 the only product? Its formation is not mentioned in the article at all. In earlier cases[180,181] products of carbon–hydrogen insertion either were observed[181] or were rendered impossible by the geometry of the system.[180] Could it be that this is not a carbene addition at all, but rather an example of a thermally produced pyrazoline 162 which isomerizes to 160 and loses nitrogen to give the diradical precursor of 159 (Eq. 3.47)?

158   159   not   161

160   162   (3.47)

VI.   REFERENCES

1.   P. S. Stang, *Acc. Chem. Res.*, **15**, 348 (1981).

2.   K. G. Taylor, *Tetrahedron*, **38**, 2751 (1981).

3.*   W. T. Borden, *Diradicals*, Wiley, New York, 1982.

4.   L. Salem, *Electrons in Chemical Reactions*, Wiley, New York, 1982.

5.   P. F. Zittel, G. B. Ellison, S. V. ONeill, E. Herbst, W. C. Lineberger, and W. P. Reinhardt, *J. Am. Chem. Soc.*, **98**, 3731 (1976).

6.   C. W. Bauschlicher, Jr., H. F. Schaefer, III, and P. S. Bagus, *J. Am. Chem. Soc.*, **99**, 7106 (1977).

7.   J. F. Harrison, R. C. Liedtke, and J. F. Liebman, *J. Am. Chem. Soc.*, **101**, 7162 (1979).

8.   A. Mavridis and J. F. Harrison, *J. Am. Chem. Soc.*, **104**, 3827 (1982).

9.   N. C. Baird and K. F. Taylor, *J. Am. Chem. Soc.*, **100**, 1333 (1978).

10.* D. Feller, W. T. Borden, and E. R. Davidson, *Chem. Phys. Lett.*, **71**, 22 (1980).

11.* P. H. Mueller, N. G. Rondan, K. N. Houk, J. F. Harrison, D. Hooper, B. H. Willen, and J. F. Liebman, *J. Am. Chem. Soc.*, **103**, 5049 (1981).

12.   W. W. Schoeller, *J. Chem. Soc., Chem. Commun.*, 124 (1980).

13.   L. Pauling, *J. Chem. Soc., Chem. Commun.*, 688 (1980).

14.   C. W. Bauschlicher, *J. Am. Chem. Soc.*, **102**, 5492 (1980).

15.   K. S. Kim and H. F. Schaefer, III, *J. Am. Chem. Soc.*, **102**, 5389 (1980).

16.   H. M. Frey and I. D. R. Stevens, *J. Am. Chem. Soc.*, **84**, 2647 (1962); H. M. Frey, *J. Chem. Soc.*, 2293 (1962); H. M. Frey and I. D. R. Stevens, *J. Chem. Soc.*, 3514 (1963); *J. Chem. Soc.*, 4700 (1964); *J. Chem. Soc.*, 3101 (1965); H. M. Frey, *Pure Appl. Chem.*, **9**, 527 (1964); A. M. Mansoor and I. D. R. Stevens, *Tetrahedron Lett.*, 1733 (1966).

17.* A. L. Kraska, K. -T. Chang, S. -J. Chang, C. G. Moseley, and H. Shechter, *Tetrahedron Lett.*, **23**, 1627 (1982).

18.   A. Nickon and P. St. J. Zurer, *J. Org. Chem.*, **46**, 4685 (1981).

19.   C. Dupuy, G. M. Korenowski, M. McAuliffe, W. Hetherington, and K. B. Eisenthal, *Chem. Phys. Lett.*, **72**, 272 (1981).

20.   Y. Wang, E. V. Sitzmann, F. Novak, C. Dupuy, and K. B. Eisenthal, *J. Am. Chem. Soc.*, **104**, 3238 (1982).

21.   G. L. Closs, in R. A. Moss and M. Jones, Jr., Eds., Carbenes, Vol. 2, Wiley, New York, 1975, chapter 4.

22.   K. B. Eisenthal, N. J. Turro, M. Aikawa, J. A. Butcher, Jr., C. Dupuy, G. Hefferon, W. Hetherington, G. M. Korenowski, and M. J. McAuliffe, *J. Am. Chem. Soc.*, **102**, 6563 (1980).

23.   W. Kirmse to M. Jones, Jr., Bochum, West Germany, March 1982, personal communication.

24.   For a large number of examples of thermal, photochemical, and catalyzed decompositions of diazo compounds see W. Kirmse and K. Horn, *Berichte*, **100**, 2698 (1967).

25.   W. Kirmse and W. Spaleck, *Angew. Chem. Int. Ed. Engl.*, **20**, 776 (1981).

26.   O. Repič and S. Vogt, *Tetrahedron Lett.*, **23**, 2729 (1982).

27.* C. R. Johnson and M. R. Barbachyn, *J. Am. Chem. Soc.*, **104**, 4290 (1982).

28.   R. A. Moss in M. Jones, Jr. and R. A. Moss, Eds., *Carbenes*, Vol. 1, Wiley, New York, 1973.

29.   M. Suda, *Synthesis*, 714 (1981).

30.   S. Brandt and P. Helquist, *J. Am. Chem. Soc.*, **101**, 6473 (1979).

31.   M. Brookhart, M. B. Humphrey, H. J. Kratzer, and G. O. Nelson, *J. Am. Chem. Soc.*, **102**, 7802 (1980).

32.   C. P. Casey, S. W. Polichnowski, A. J. Shusterman, and C. R. Jones, *J. Am. Chem. Soc.*, **101**, 7283 (1979); see the discussions in M. Jones, Jr. and R. A. Moss, Eds., *Reactive Intermediates*, Vol. 2, Wiley-Interscience, 1981, pp. 67–69, 140–142.

33.* M. Brookhart, J. R. Tucker, and G. R. Husk, *J. Am. Chem. Soc.*, **103**, 979 (1981).

34.* K. A. M. Kremer, P. Helquist, and R. C. Kerber, *J. Am. Chem. Soc.*, **103**, 1862 (1981).

35.   C. P. Casey, W. H. Miles, H. Takuda, and J. M. O'Connor, *J. Am. Chem. Soc.*, **104**, 3761 (1982).

36.   K. A. M. Kremer, G. H. Kuo, E. J. O'Connor, P. Helquist, and R. C. Kerber, *J. Am. Chem. Soc.*, **104**, 6119 (1982).

37.   P. Fischer and G. Schaefer, *Angew. Chem. Int. Ed. Engl.*, **20**, 862 (1981).

38.* A. J. Anciaux, A. J. Hubert, A. F. Noels, N. Petiniot, and P. Teyssié, *J. Org. Chem.*, **45**, 695 (1980).

39.   M. P. Doyle, D. V. Leusen, and W. H. Tamblyn, *Synthesis*, 787 (1981).

40.* M. P. Doyle, R. L. Dorow, W. H. Tamblyn, and W. E. Buhro, *Tetrahedron Lett.*, **23**, 2261 (1982).

41.   M. P. Doyle, W. H. Tamblyn, and V. Bagheri, *J. Org. Chem.*, **46**, 5094 (1981).

42.   H. Irngartinger, A. Goldmann, R. Schappert, P. Garner, and P. Dowd, *J. Chem. Soc., Chem. Commun.*, 455 (1981); P. Dowd, P. Garner, R. Schappert, H. Irngartinger, and A. Goldmann, *J. Org. Chem.*, **47**, 4240 (1982).

43.* H. J. Callot and C. Piechocki, *Tetrahedron Lett.*, **21**, 3489 (1980); H. J. Callot, F. Mertz, and C. Piechocki, *Tetrahedron*, **38**, 2365 (1982).

44.* H. J. Callot and F. Mertz, *Tetrahedron Lett.*, **23**, 4321 (1982).

45.   A. Demonceau, A. F. Noels, A. J. Hubert, and P. Teyssié, *J. Chem. Soc., Chem. Commun.*, 688 (1981).

46.   A. F. Noels, A. Demonceau, N. Petiniot, A. J. Hubert, and P. Teyssié, *Tetrahedron*, **38**, 2733 (1982).

47.   D. F. Taber and E. H. Petty, *J. Org. Chem.*, **47**, 4808 (1982).

48.   B. K. Ravi Shankar and H. Shechter, *Tetrahedron Lett.*, **23**, 2277 (1982).

49.* J. C. Gilbert and U. Weerasooriya, *J. Org. Chem.*, **47**, 1837 (1982); J. C. Gilbert, U. Weerasooriya, and D. Giamalva, *Tetrahedron Lett.*, 4619 (1979).

50.   P. M. Lahti and J. A. Berson, *J. Am. Chem. Soc.*, **103**, 7011 (1981).

51.   R. F. Salinaro and J. A. Berson, *Tetrahedron Lett.*, **23**, 1447 (1982); **23**, 1451 (1982).

52.   W. J. le Noble, S. Basak, and S. Srivastava, *J. Am. Chem. Soc.*, **103**, 4639 (1981).

53.   P. J. Stang, *Israel J. Chem.*, **21**, 119 (1981).

54.* P. J. Stang and T. E. Fisk, *J. Am. Chem. Soc.*, **102**, 6813 (1980); **101**, 4772 (1979).

55.   P. J. Stang and M. R. White, *J. Am. Chem. Soc.*, **103**, 5429 (1981).

56.   M. S. Newman and T. B. Patrick, *J. Am. Chem. Soc.*, **91**, 6461 (1969).

57.*   P. J. Stang and M. Ladika, *J. Am. Chem. Soc.*, **103**, 6437 (1981); **102**, 5406 (1980).

58.   L. J. Krause and J. A. Morrison, *J. Chem. Soc., Chem. Commun.*, 671 (1980); *J. Am. Chem. Soc.*, **103**, 2995 (1981).

59.   H. P. Fritz and W. Kornrumpf, *Z. Naturforsch.*, **36b**, 1375 (1981).

60.*   S. L. Regen and A. Singh, *J. Org. Chem.*, **47**, 1587 (1982).

61.   S. T. Purrington and G. B. Kenion, *J. Chem. Soc., Chem. Commun.*, 731 (1982).

62.   R. H. Ellison, *J. Org. Chem.*, **45**, 2509 (1980).

63.   D. Seyferth and V. A. Mai, *J. Am. Chem. Soc.*, **92**, 7412 (1970).

64.   C. D. Poulter, E. C. Friedrich, and S. Winstein, *J. Am. Chem. Soc.*, **91**, 6892, (1969).

65.   M. Komiyama and H. Hirai, *Bull. Chem. Soc. Jpn.*, **54**, 2053 (1981).

66.*   D. Seebach, H. Siegel, J. Gabriel, and R. Hässig, *Helv. Chim. Acta*, **63**, 2046 (1980).

67.   H. Siegel, K. Hiltbrunner, and D. Seebach, *Angew. Chem. Int. Ed. Eng.*, **18**, 785 (1979); D. Seebach, H. Siegel, K. Muller, and K. Hiltbrunner, *Angew. Chem. Int. Ed. Engl.*, **18**, 784 (1979).

68.   G. W. Klumpp and P. M. Kwantes, *Tetrahedron Lett.*, **22**, 831 (1981).

69.   N. L. Bauld and D. Wirth, *J. Comput. Chem.*, **2**, 1 (1981).

70.   W. W. Schoeller and N. Aktekin, *J. Chem. Soc., Chem. Commun.*, 20 (1982).

71.   C. Mayor and C. Wentrup, *J. Am. Chem. Soc.*, **97**, 7467 (1975); C. Wentrup, C. Mayor, and R. Gleiter, *Helv. Chim. Acta*, **55**, 2628 (1972).

72.   W. M. Jones, R. A. LaBar, U. H. Brinker, and P. H. Gebert, *J. Am. Chem. Soc.*, **99**, 6379 (1977); note 27.

73.*   N. G. Rondan, K. N. Houk, and R. A. Moss, *J. Am. Chem. Soc.*, **102**, 1770 (1980).

74.*   R. A. Moss, *Acc. Chem. Res.*, **13**, 58 (1980); see also R. A. Moss, C. B. Mallon, and C. -T. Ho, *J. Am. Chem. Soc.*, **99**, 4105 (1977).

75.   W. W. Schoeller and U. H. Brinker, *Z. Naturforsch.*, **35b**, 475 (1980).

76.*   W. W. Schoeller, *Tetrahedron Lett.*, **21**, 1505 (1980).

77.   J. A. Shapiro and F. P. Lossing, *J. Phys. Chem.*, **72**, 1552 (1968).

78.   D. E. Milligan and M. E. Jacox, *J. Chem. Phys.*, **47**, 703 (1967).

79.*   W. W. Schoeller, *Tetrahedron Lett.*, **21**, 1509 (1980).

80.   W. W. Schoeller, N. Aktekin, and H. Friege, *Angew. Chem. Int. Ed. Engl.*, **21**, 932 (1982).

81.   W. W. Schoeller, *Angew. Chem. Int. Ed. Engl.*, **20**, 698 (1981).

82.   R. A. Moss, W. Guo, D. Z. Denney, K. N. Houk, and N. G. Rondan, *J. Am. Chem. Soc.*, **103**, 6164 (1981).

83.   W. Brück and H. Dürr, *Tetrahedron Lett.*, **23**, 2175 (1982).

84.   This value is calculated from the data in Reference 83, $\rho_{\text{PhCCl}}^{280\text{K}} = -0.58$ versus σ.

85.   R. R. Kostikov, A. P. Molchanov, G. V. Golovanova, and I. G. Zenkevich, *J. Org. Chem. USSR*, **13**, 1846 (1977).

86.   (a) R. A. Moss, M. Fedorynski, and W. -C. Shieh, *J. Am. Chem. Soc.*, **101**, 4736 (1979). (b) R. A. Moss and R. C. Munjal, *Tetrahedron Lett.*, 4721 (1979).

87.*   R. A. Moss, W. Guo, and K. Krogh-Jespersen, *Tetrahedron Lett.*, **23**, 15 (1982).

88*. R. A. Moss, C. M. Young, L. A. Perez, and K. Krogh-Jespersen, *J. Am. Chem. Soc.*, **103**, 2413 (1981).

89.   A. Amaro and K. Grohmann, *J. Am. Chem. Soc.*, **97**, 3880 (1975); H. Hart, R. L. Holloway, C. Landry, and T. Tabata, *Tetrahedron Lett.*, 4933 (1969).

90.* R. A. Moss, L. A. Perez, J. Włoskowska, W. Guo, and K. Krogh-Jespersen, *J. Org. Chem.*, **47**, 4177 (1982).

91.   R. A. Moss and C. B. Mallon, *J. Am. Chem. Soc.*, **97**, 344 (1975).

92.   W. Brück and H. Dürr, *Angew. Chem. Int. Ed. Engl.*, **21**, 916 (1982).

93.   R. A. Moss and L. A. Perez, *Tetrahedron Lett.*, **24**, 2719 (1983).

94.* J. C. Dalton and R. A. Bourque, *J. Am. Chem. Soc.*, **103**, 697 (1981).

95.* R. A. Bourque, P. D. Davis, and J. C. Dalton, *J. Am. Chem. Soc.*, **103**, 699 (1981).

96.   A. G. Brook, H. W. Kucera, and R. Pearce, *Can. J. Chem.*, **49**, 618 (1971).

97.   A. G. Brook, R. Pearce, and J. B. Pierce, *Can. J. Chem.*, **49**, 1622 (1971).

98.   P. C. Wong, D. Griller, and J. C. Scaiano, *J. Am. Chem. Soc.*, **104**, 6631 (1982).

99.* D. Griller, M. T. H. Liu, and J. C. Scaiano, *J. Am. Chem. Soc.*, **104**, 5549 (1982).

100.* N. J. Turro, J. A. Butcher, Jr., R. A. Moss, W. Guo, R. C. Munjal, and M. Fedorynski, *J. Am. Chem. Soc.*, **102**, 7576 (1980). Note that quantitative data in this reference is revised later.[102]

101. R. A. Moss, L. A. Perez, N. J. Turro, I. R. Gould, and N. P. Hacker, *Tetrahedron Lett.*, **24**, 685 (1983).

102.* N. J. Turro, G. F. Lehr, J. A. Butcher, Jr., R. A. Moss, and W. Guo, *J. Am. Chem. Soc.*, **104**, 1754 (1982).

103. P. S. Skell and M. S. Cholod, *J. Am. Chem. Soc.*, **91**, 7131 (1969).

104. B. Giese and W. -B. Lee, *Angew. Chem. Int. Ed. Engl.*, **19**, 835 (1980).

105.* B. Giese, W. -B. Lee, and J. Meister, *Annalen*, 725 (1980).

106. Reviews (a) G. W. Klumpp, *Reactivity in Organic Chemistry*, Wiley-Interscience, New York, 1982, pp. 274–282, 352–358; (b) O. Exner, *Prog. Phys. Org. Chem.*, **10**, 411 (1973); (c) J. E. Leffler and E. Grunwald, *Rates and Equilibria of Organic Reactions*, Wiley, New York, 1963, pp. 315ff.

107. B. Giese and W. -B. Lee, *Tetrahedron Lett.*, **23**, 3561 (1982).

108.* B. Giese and W. -B. Lee, *Berichte*, **114**, 3306 (1981).

109. S. Ehrenson, R. T. C. Brownlee, and R. W. Taft, *Prog. Phys. Org. Chem.*, **10**, (1973), especially pp. 38–39.

110. B. Giese, W. -B. Lee, and C. Neumann, *Angew. Chem. Int. Ed. Engl.*, **21**, 310 (1982).

111. B. Giese and C. Neumann, *Tetrahedron Lett.*, **23**, 3557 (1982).

112.* U. H. Brinker and J. Ritzer, *J. Am. Chem. Soc.*, **103**, 2116 (1981); L. Skattebøl, *Tetrahedron*, **23**, 1107 (1967).

113. For a discussion of the rearrangement see M. Jones, Jr. and R. A. Moss, Eds., *Reactive Intermediates*, Vol. 2, Wiley-Interscience, New York, 1981, pp. 113–121.

114. For a discussion of norbornenylidene and related carbenes, see M. Jones, Jr. and R. A. Moss, Eds., *Reactive Intermediates*, Vol. 2, Wiley-Interscience, New York, 1981, pp. 111–112.

115.   J. Stapersma, P. Kuipers, and G. W. Klumpp, *Rec. Trav. Chim. Pays-Bays*, **101**, 213 (1982).

116.   S. -I. Murahashi, K. Okumura, T. Naota, and S. Nagase, *J. Am. Chem. Soc.*, **104**, 2466 (1982).

117.   P. K. Freeman and K. E. Swenson, *J. Org. Chem.*, **47**, 2033 (1982).

118.   R. G. Bergman and V. J. Rajadhyaksha, *J. Am. Chem. Soc.*, **92**, 2163 (1970).

119.   Both **86** and **91** added stereospecifically to *cis*- and *trans*-butenes when generated photolytically at low temperature.[116]

120.*   D. Griller, C. R. Montgomery, J. C. Scaiano, M. S. Platz, and L. Hadel, *J. Am. Chem. Soc.*, **104**, 6813 (1982). Be sure also to see D. Griller, L. Hadel, A. Nazran, M. S. Platz, P. C. Wong, T. G. Savino and J. C. Scaiano, *J. Am. Chem. Soc.*, **106**, 2227 (1984).

121.*   B. -E. Brauer, P. B. Grasse, K. J. Kaufmann, and G. B. Schuster, *J. Am. Chem. Soc.*, **104**, 6814 (1982).

122.   J. J. Zupancic and G. B. Schuster, *J. Am. Chem. Soc.*, **102**, 5958 (1980); J. J. Zupancic, P. B. Grasse, and G. B. Schuster, *J. Am. Chem. Soc.*, **103**, 2423 (1981).

123.   P. C. Wong, D. Griller, and J. Scaiano, *J. Am. Chem. Soc.*, **103**, 5334 (1981).

124.   J. J. Zupancic and G. B. Schuster, *J. Am. Chem. Soc.*, **103**, 944 (1981).

125.   M. Jones, Jr. and K. R. Rettig, *J. Am. Chem. Soc.*, **87**, 4013, 4015 (1965).

126.*   P. B. Grasse, B. -E. Brauer, J. J. Zupancic, K. J. Kaufmann and G. B. Schuster, *J. Am. Chem. Soc.*, **105**, 6833 (1983).

127.   A. S. Kende, P. Hebeisen, P. J. Sanfilippo, and B. H. Toder, *J. Am. Chem. Soc.*, **104**, 4244 (1982). For an earlier suggestion see W. L. Magee and H. Shechter, *Tetrahedron Lett.*, 4697 (1979).

128.   L. M. Hadel, M. S. Platz, and J. C. Scaiano, *J. Am. Chem. Soc.*, in press, 1983.

129.   G. L. Closs and B. E. Rabinow, *J. Am. Chem. Soc.*, **98**, 8190 (1976).

130.*   (a) P. P. Gaspar, C. -T. Lin, B. L. Whitsel Dunbar, and P. Balasubramanian, *J. Am. Chem. Soc.*, **106**, 2128 (1984). (b) R. A. Moss, D. P. Mack, and C. M. Young, *J. Am. Chem. Soc.*, **105**, 5859 (1983). (c) P. P. Gaspar, unpublished work.

131.   P. C. Wong, D. Griller, and J. C. Scaiano, *Chem. Phys. Lett.*, **83**, 69 (1981).

132.   (a) W. Dieckmann, *Ber. Deutsch. Chem. Ges.*, **43**, 1024 (1910). (b) E. Buchner and Th. Curtius, *Ber. Deutsch. Chem. Ges.*, **18**, 2371 (1885).

133.   J. N. Bradley and A. Ledwith, *J. Chem. Soc.*, 3480 (1983).

134.*   P. de March and R. Huisgen, *J. Am. Chem. Soc.*, **104**, 4952, 4953 (1982).

135.   H. S. Gill and J. A. Landgrebe, *Tetrahedron Lett.*, **23**, 5099 (1982).

136.   M. Békhaze and J. Warkentin, *J. Org. Chem.*, **47**, 4870 (1982), *J. Am. Chem. Soc.*, **103**, 2473 (1981). For earlier precedents see G. W. Griffin, *Angew. Chem. Int. Ed. Engl.*, **10**, 537 (1971); P. D. Bartlett and N. Shimizu, *J. Am. Chem. Soc.*, **100**, 4268 (1978).

137.   H. M. Ensslin and M. Hanack, *Angew. Chem. Int. Ed. Engl.*, **6**, 702 (1967); S. Nishida, I. Moritani, E. Tsuda, and T. Teraji, *J. Chem. Soc., Chem. Commun.*, 781 (1969).

138.   H. Tomioka, T. Miwa, S. Suzuki, and Y. Izawa, *Bull. Chem. Soc. Jpn.*, **53**, 753 (1980). For an intramolecular example see M. Hamaguchi and T. Ibata, *Tetrahedron Lett.*, 4475 (1974).

139.   M. Jones, Jr., T. M. Ford, M. C. Boraas, and P. R. Roberts, unpublished.

140.   M. E. Alonzo and A. W. Chitty, *Tetrahedron Lett.*, **22**, 4181 (1981).

141.*   W. Kirmse, K. Loosin, and H. -D. Sluma, *J. Am. Chem. Soc.*, **103**, 5935 (1981).

142.   E. Waali, *J. Am. Chem. Soc.*, **103**, 3604 (1981).

143.   L. Radom, H. E. Schaefer III, and M. A. Vincent, *Nouv. J. Chim.*, **4**, 411 (1980). For earlier calculation see R. L. Tyner, W. M. Jones, Y. Öhrn, and J. R. Sabin, *J. Am. Chem. Soc.*, **96**, 3765 (1974); M. J. S. Dewar and D. Landman, *J. Am. Chem. Soc.*, **99**, 6179 (1977).

144.*   J. W. Harris and W. M. Jones, *J. Am. Chem. Soc.*, **104**, 7329 (1982).

145.   P. R. West, O. L. Chapman, and J.-P. LeRoux, *J. Am. Chem. Soc.*, **104**, 1779 (1982).

146.   H. Tomioka, G. W. Griffin, and K. Nishiyama, *J. Am. Chem. Soc.*, **101**, 6009 (1979).

147.   H. Staudinger and R. Endle, *Ber. Deutsch. Chem. Ges.*, **46**, 1437 (1913).

148.   N. J. Turro, M. Aikawa, J. A. Butcher, Jr., and G. W. Griffin, *J. Am. Chem. Soc.*, **102**, 5128 (1980).

149.*   M. S. Platz, V. P. Senthilnathan, B. B. Wright, and C. W. McCurdy, Jr., *J. Am. Chem. Soc.*, **104**, 6494 (1982).

150.   C.-T. Lin and P. P. Gaspar, *Tetrahedron Lett.*, **21**, 3533 (1980).

151.   D. C. Doetschman and C. A. Hutchison, *J. Chem. Phys.*, **56**, 3964 (1972).

152.   H. Tomioka, S. Suzuki, and Y. Izawa, *J. Am. Chem. Soc.*, **104**, 3156 (1982).

153.*   H. Tomioka, H. Ueda, S. Kondo, and Y. Izawa, *J. Am. Chem. Soc.*, **102**, 7817 (1980).

154.*   H. Tomioka, N. Hayashi, Y. Izawa, V. P. Senthilnathan, and M. S. Platz, *J. Am. Chem. Soc.*, **105**, 5053 (1983).

155.   H. Tomioka, Y. Ozaki, Y. Koyabu, and Y. Izawa, *Tetrahedron Lett.*, **23**, 1917 (1982).

156.   For related work on diazoamide reactions see H. Tomioka, M. Kondo, and Y. Izawa, *J. Org. Chem.*, **46**, 1090 (1981).

157.   M. S. Baird, I. R. Dunkin, N. Hacker, M. Poliakoff, and J. J. Turner, *J. Am. Chem. Soc.*, **103**, 5190 (1981).

158.   T. DoMinh, O. P. Strausz, and H. E. Gunning, *J. Am. Chem. Soc.*, **91**, 1261 (1969).

159.*   H. Tomioka, H. Okuno, and Y. Izawa, *J. Org. Chem.*, **45**, 5278 (1980).

160.   H. D. Roth and M. L. Manion, *J. Am. Chem. Soc.*, **98**, 3392 (1976). H. D. Roth, *Acc. Chem. Res.*, **10**, 85 (1977).

161.   F. Kaplan and G. R. Meloy, *J. Am. Chem. Soc.*, **88**, 950 (1968).

162.   H. Tomioka, H. Okuno, S. Kondo, and Y. Izawa, *J. Am. Chem. Soc.*, **102**, 7123 (1980).

163.   T. Livinghouse and R. V. Stevens, *J. Am. Chem. Soc.*, **100**, 6479 (1978). R. V. Stevens, *Pure Appl. Chem.*, **51**, 1317 (1979).

164.   M. Jones, Jr., W. Ando, M. E. Hendrick, A. Kulczycki, Jr., P. M. Howley, K. F. Hummel, and D. S. Malament, *J. Am. Chem. Soc.*, **94**, 7469 (1972). S. L. Kammula, H. L. Tracer, and M. Jones, Jr., *J. Org. Chem.*, **42**, 2931 (1977).

165.   R. V. Stevens, G. S. Bisacchi, L. Goldsmith, and G. E. Strouse, *J. Org. Chem.*, **45**, 2708 (1980).

166.   L. B. Harding, *J. Am. Chem. Soc.*, **103**, 7469 (1981).

167. Y. Osamura, H. F. Schaefer, III, S. K. Gray, and W. H. Miller, *J. Am. Chem. Soc.*, **103**, 1904 (1981).

168. E. M. Evlethard and A. Sevin, *J. Am. Chem. Soc.*, **103**, 7414 (1981).

169. A. Nickon and J. K. Bronfenbrenner, *J. Am. Chem. Soc.*, **104**, 2022 (1982).

170. G. Maier, H. Hoppe, H. P. Reisenauer, and C. Krüger, *Angew. Chem. Int. Ed. Engl.*, **21**, 437 (1982).

171. P. Eisenbarth and M. Regitz, *Angew. Chem. Int. Ed. Engl.*, **21**, 913 (1982). Acetylenes were missed in earlier work: S. Masamune, N. Nakamura, M. Suda, and H. Ona, *J. Am. Chem. Soc.*, **95**, 8481 (1973).

172. R. R. Gallucci and M. Jones, Jr., *J. Am. Chem. Soc.*, **98**, 7704 (1976).

173. E. E. Waali and C. W. Wright, *J. Org. Chem.*, **46**, 2201 (1981).

174. D. Lloyd and N. W. Preston, *Chem. and Ind.*, 1039 (1966). See also E. E. Waali, J. L. Taylor, and N. T. Allison, *Tetrahedron Lett.*, 3873 (1977).

175. P. B. Shevlin, D. W. McPherson, and P. Melius, *J. Am. Chem. Soc.*, **105**, 488 (1983).

176.* J. Gano, R. H. Wettach, M. S. Platz, and V. P. Senthilnathan, *J. Am. Chem. Soc.*, **104**, 2326 (1982).

177. R. A. Moss and M. J. Chang, *Tetrahedron Lett.*, **22**, 3747 (1981).

178. S. F. Sellers, T. C. Klebach, F. Hollowood, M. Jones, Jr., and P. v. R. Schleyer, *J. Am. Chem. Soc.*, **104**, 5492 (1982).

179. V. Vinković and Z. Majerski, *J. Am. Chem. Soc.*, **104**, 4027 (1982).

180. K. Mlivarić-Majerski and Z. Majerski, *J. Am. Chem. Soc.*, **102**, 1418 (1980).

181. D. P. G. Hamon and V. C. Trenerry, *J. Am. Chem. Soc.*, **103**, 4962 (1981).

4

METAL – CARBENE COMPLEXES

CHARLES P. CASEY

Department of Chemistry, University of Wisconsin, Madison, Wisconsin 53706

I.   INTRODUCTION

This chapter covers reports on the syntheses and reactions of metal–carbene complexes for the years 1980–1982. There has been an explosion of publications in this area, and this review is by no means comprehensive. It concentrates on new types of reactions observed in the last several years. An excellent review by Herrmann[1] on the chemistry of bridging methylene complexes appeared in 1982; consequently, this topic is not covered exhaustively. A review by Brown of the use of metal–carbene complexes in synthesis appeared in 1980[2] and an annual survey by Hegedus appeared in 1982.[3] Highlights of their groups' recent research on metal–carbene complexes were published by Schrock,[4,5] Mansuy,[6] and Dolgoplosk.[7]

II.   CYCLOPROPANES FROM METAL–CARBENE COMPLEXES

In 1966 Pettit and Jolly reported that the methoxymethyl iron compound 1 reacted rapidly with HBF_4 and that in the presence of cyclohexene, norcarane was formed.[8] An iron methylene complex was suggested as the reactive intermediate responsible for the cyclopropanation. Subsequently, Brookhart was able to spectroscopically observe the related, more stable, phosphine-substituted carbene complex 2.[9] Helquist has developed 3 as a convenient precursor of the iron methylene complex for use as a cyclopropanating agent.[10]

Extension of this chemistry to the synthesis of benzylidene iron complexes as reagents for the synthesis of phenyl cyclopropanes was relatively straight-

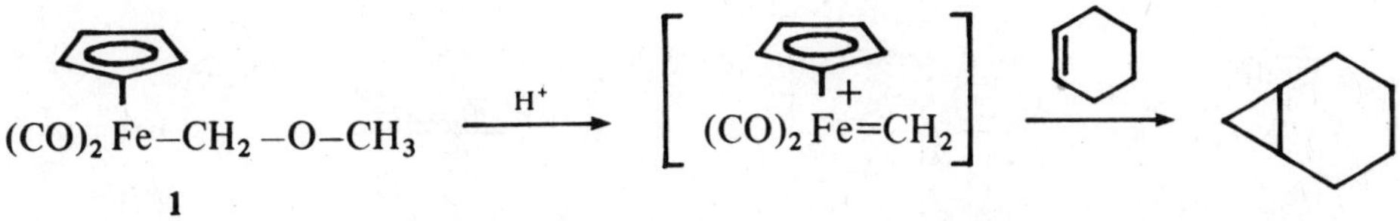

$$(CO)_2 Fe-CH_2-O-CH_3 \xrightarrow{H^+} \left[(CO)_2 Fe=CH_2 \right]^+$$

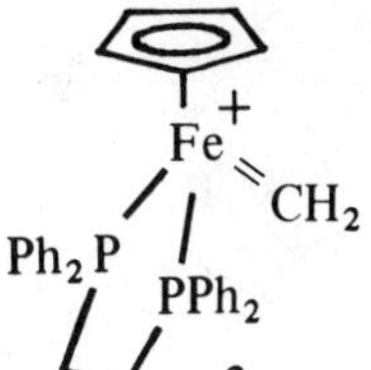

1

$$\underset{Ph_2P \quad PPh_2}{Fe^+ \diagdown CH_2}$$

2

$$(CO)_2 Fe-CH_2-S^+ \diagup^{CH_3}_{CH_3}$$

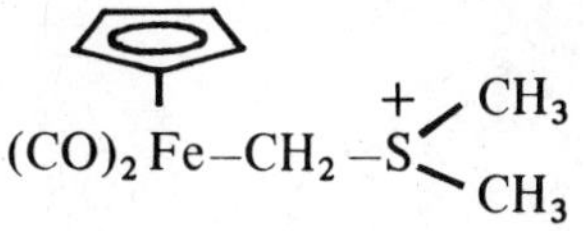

3

forward. Reaction of $(C_5H_5)(CO)_2FeCH(OCH_3)C_6H_5$ with $(C_6H_5)_3C^+$ results in abstraction of methoxide and generation of the isolable iron benzylidene complex **4**, which reacts with alkenes to give high yields of *syn*-cyclopropanes[11,12] The alkene is proposed to react directly with the electrophilic carbene ligand without prior complexation to iron. The syn selectivity of the reaction is explained in terms of transition state **5**, in which the alkyl substituent on the alkene and the large $CpFe(CO)_2$ unit are trans to one another, and the positive charge developing on one of the alkene carbon atoms is stabilized by electron donation from the phenyl ring.

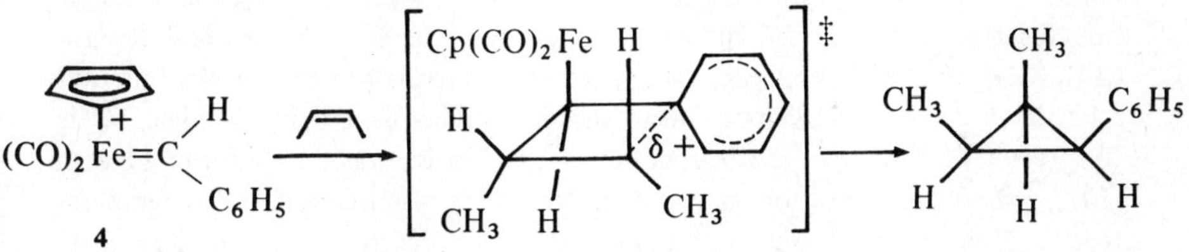

$$(CO)_2 Fe=C \diagup^H_{C_6H_5}$$

4

$$\left[\begin{array}{c} Cp(CO)_2 Fe \quad H \\ H, \\ CH_3 \quad H \quad \quad \delta + \\ CH_3 \end{array} \right]^{\ddagger}$$

5

$$\underset{H \quad H \quad H}{CH_3 \diagdown CH_3 \diagdown C_6H_5}$$

6

Extension of this chemistry to alkyl-substituted carbene complexes is more difficult because side reactions are common in the synthesis of the required precursors and also because the alkyl-substituted carbene complexes exhibit a strong tendency to rearrange to alkene complexes. Nevertheless, several successful routes to ethylidene transfer reagents have been developed. Reaction of $(C_5H_5)(CO)_2 Fe=C(OCH_3)CH_3^+$, **6**, with either BH_4^- or $HBEt_3^-$ gave the ether **7**;[13,14] no side reaction involving deprotonation of **6** to give $(C_5H_5)(CO)_2 FeC(OCH_3)=CH_2$ was noted. Reaction of **7** with $Me_3SiOSO_2CF_3$ in the presence of alkenes led to high yields of cyclopropanes.[13] For example, cyclooctene reacted with **7** to give a 87% yield of a 4.5:1 mixture of *syn:anti* methylcyclopropanes. The stable phosphine substituted ethylidene complex $(C_5H_5)(CO)$

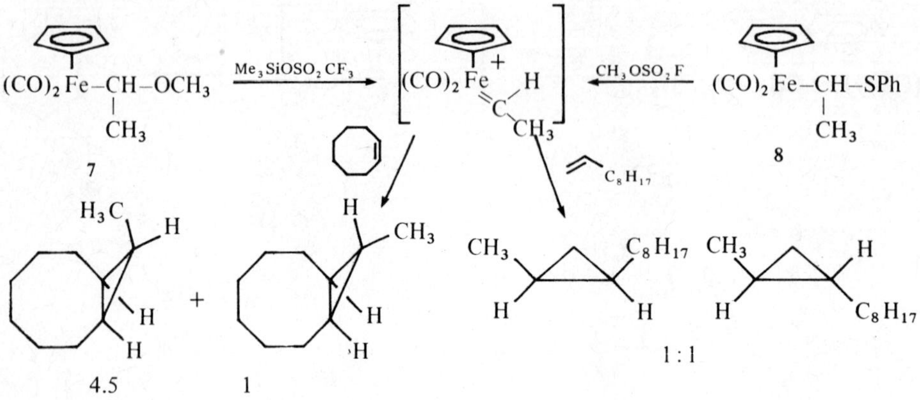

(Ph$_3$P)Fe=CHC$_2$H$_5^+$ was observed by NMR. Helquist has developed a related synthesis of this ethylidene transfer reagent. Reaction of the iron-substituted phenylsulfide, 8, with CH$_3$OSO$_2$F in the presence of alkenes led to good yields of methylcyclopropanes.[15] For example, even the relatively unreactive 1-decene reacted to give a 1:1 mixture of *cis*- and *trans*-2-methyl-1-octylcyclopropane.

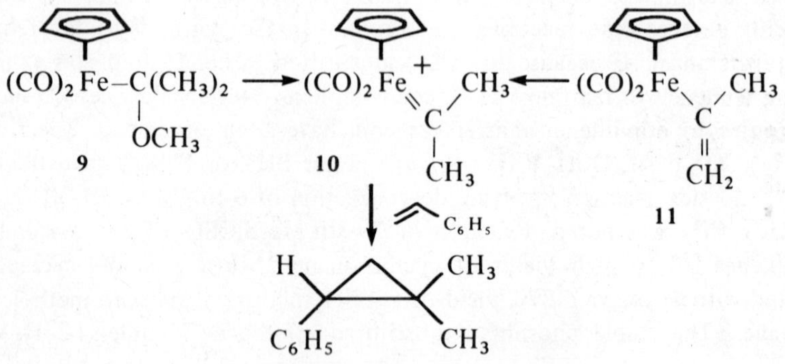

In attempting to develop a synthesis of the methyl ether precursor 9 of the dimethylcarbene iron complex 10, Casey and Miles studied the reactions of methylorganometallics with methylmethoxycarbene complex 6.[16] Deprotonation to the vinyl iron complex C$_5$H$_5$(CO)$_2$FeC(OCH$_3$)=CH$_2$ was a major side reaction with CH$_3$Li; demethyloxylation occurred with CH$_3$MgI to give C$_5$H$_5$(CO)$_2$FeCOCH$_3$, but LiCuMe$_2$ gave moderate yields of the desired methyl ether 9. A more convenient route to the dimethylcarbene iron complex, 10, is by protonation of the vinyl iron complex 11. Treatment of either 9 or 11 with HBF$_4$ led to the formation of complex 10, which was isolated at low temperature.[16] This dimethylcarbene complex underwent rearrangement to a propene complex upon warming to $-10°$C, and consequently formed cyclopropanes only with alkenes such as styrene and isobutylene, which are highly reactive toward electrophiles. Helquist observed that **10** reacted with alkenes as unreactive as 1-octene to give low yields of cyclopropanes.[17] The development of a better dimethylcarbene transfer reagent is clearly needed; the molybdenum and tungsten carbene complexes recently synthesized by Brookhart should be investigated in this regard.[18]

Helquist has begun studies of intramolecular cyclopropanation, making use of iron carbene complexes.[17]

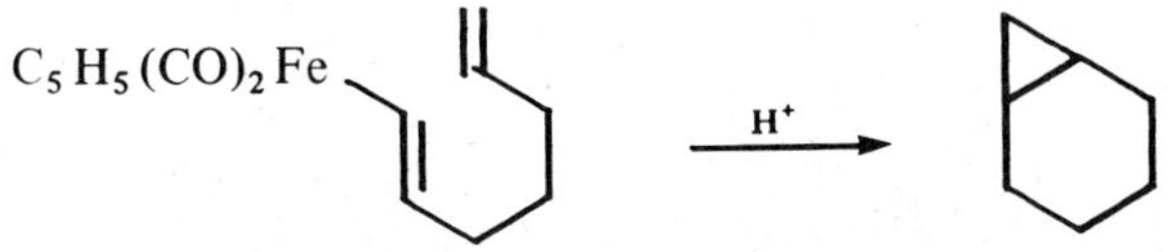

Although protonation of vinyl organometallics is a convenient route to metal–carbene complexes, side reactions involving attack of the carbene complex on the vinyl organometallic precursor are sometimes observed. Cutler has reported that protonation of **12** leads to formation of up to 78% of the β-metallocarbonium ion complex **13**.[14]

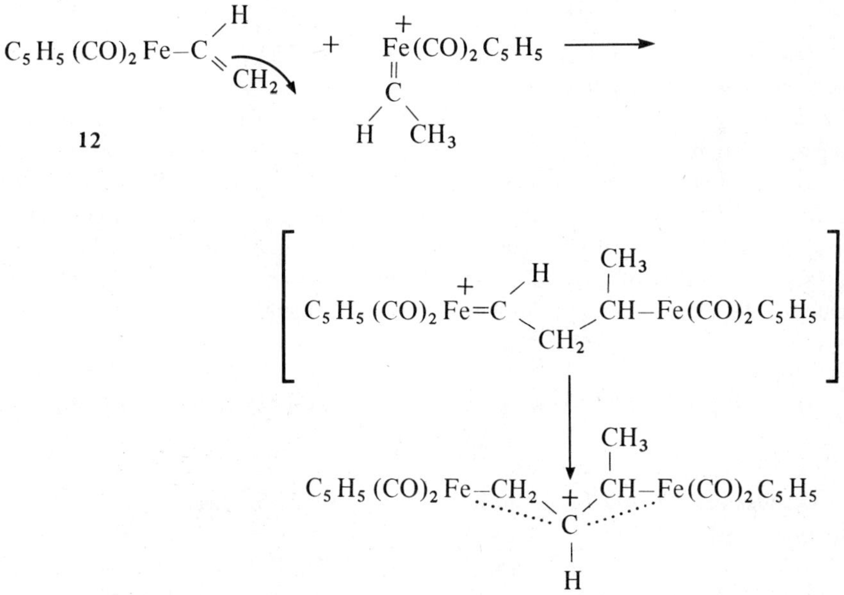

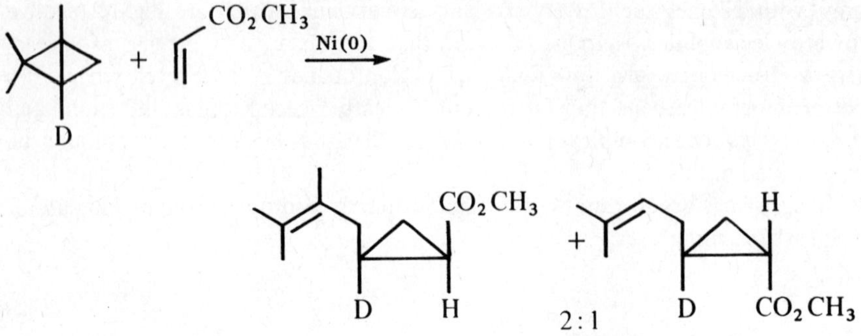

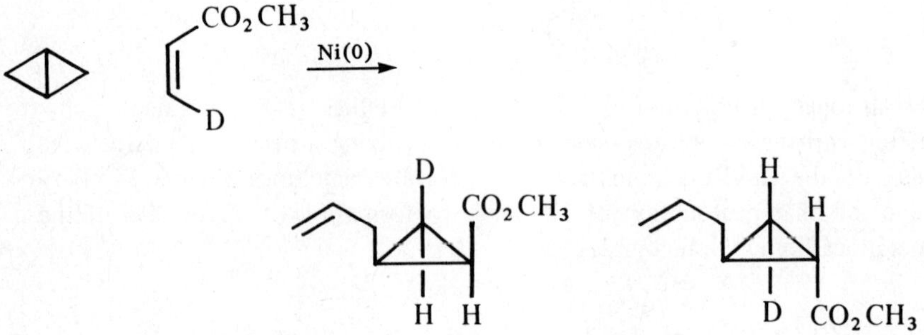

Transition metals catalyze the ring opening of bicyclo[1.1.0]butanes, and one proposed pathway involves an intermediate allylcarbene complex.[19] Noyori has reported the use of bis(1,5-cyclooctadiene)nickel(0) as a catalyst for the reaction of bicyclobutanes with electron-deficient alkenes such as methyl acrylate and acrylonitrile to give allylcyclopropane derivatives.[20] Stereochemical studies indicate that the reaction proceeds by cleavages of the central bond and a peripheral bond of the bicyclobutane to produce an allylcarbene nickel complex. The resulting nucleophilic carbene complex reacts stereospecifically with electron-deficient alkenes to produce allyl cyclopropanes.

III.   SYNTHESIS AND REACTIONS OF ELECTROPHILIC CARBENE COMPLEXES

Gladysz has published a definitive paper describing his elegant studies of the stereochemistry of formation of a rhenium benzylidene complex.[21] Reaction of the chiral rhenium benzyl complex, 14, with $(C_6H_5)_3C^+$ results in stereospecific abstraction of one of the two diastereotopic benzylic hydrogen atoms. The kinetic product of this hydride abstraction is the benzylidene complex 15, in which the nitrosyl group is in the plane of the benzylidene ligand, and the phenyl group is syn to the cyclopentadiene ring. Molecular orbital calculations indicate that when the nitrosyl ligand is in the plane of the benzylidene ligand the same d orbital is not required to backbond to both ligands. The initially formed benzylidene complex **15-k** slowly rearranges at 30°C to the more stable isomer **15-t** by rotation of the benzylidene ligand. The structure of the thermodynamically favored (> 99:1) isomer **15-t** was established by X-ray crystallography. Low-temperature photolysis establishes a 55:45 photostationary state of the two isomers.

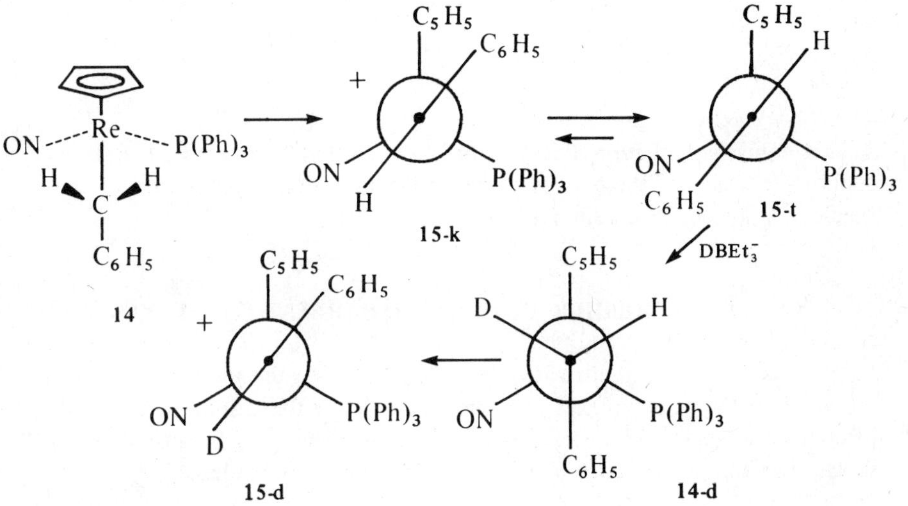

Nucleophiles attack **15** stereospecifically to give diastereomerically pure adducts. For example, Li^+ $DBEt_3^-$ reacts with **15-t** stereospecifically to form the single diastereomer of **14** shown above.[21] Complex **14-d** reacts with $C(C_6H_5)_3^+$ stereospecifically with loss of H and formation of the monodeutero benzylidene complex **15-d**. Nucleophilic attack occurs from the side of the benzylidene group opposite the bulky $P(Ph)_3$ ligand, as rigorously established by X-ray crystallography of the product of attack of $C_6H_5CH_2MgCl$ on **15-t**, which gave $(C_5H_5)(NO)(PPh_3)Re–CH(CH_2C_6H_5)C_6H_5$.

Metal-carbene complexes have been proposed as intermediates in the acid-catalyzed disproportionation of metal formyl compounds.[22,23] For example, Gladysz has found that formyl compound **16** disproportionates to a 1:2 mixture of methyl compound **17** and carbonyl compound **18** upon treatment with CF_3CO_2H.[22] The reaction is thought to proceed by hydride transfer from formyl compound **16** to a molecule of protonated **16** to give one equivalent of **18** and hydroxymethyl metal compound **19**. Protonation of **19** would generate an intermediate methylene complex, **20**, which could be reduced by a second equivalent of formyl compound **16** to methyl compound **17**. Excellent support for each of these steps comes from independent generation of intermediates and from model studies.

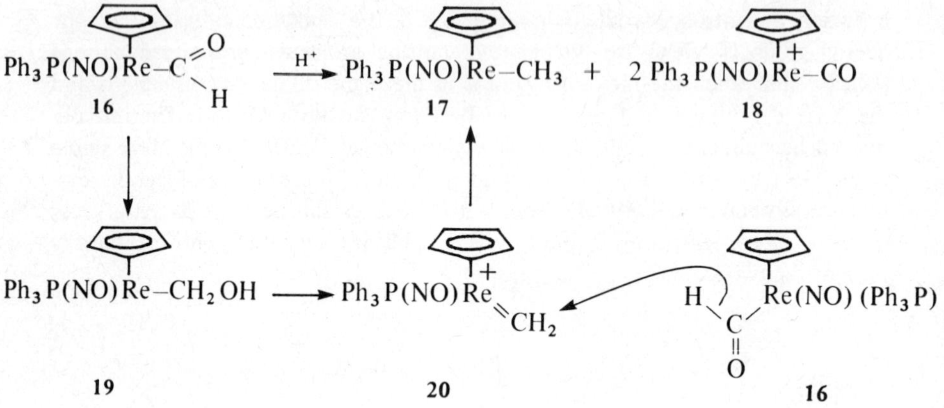

IV.   ELECTROPHILIC CUMULATED CARBENE COMPLEXES

One route to cumulated carbene complexes involves the addition of terminal acetylenes to metal complexes, as exemplified by the formation of **21**.[24,25] The reactions probably involve intermediate acetylene complexes, because Berke has shown that **22** can be transformed to vinylidene carbene complex **23**.[26]

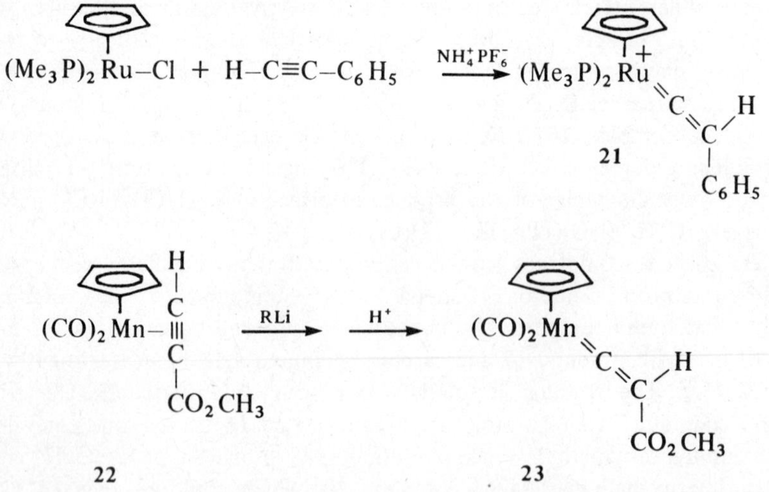

Another route to vinylidene carbene complexes is illustrated by the dehydration of the acyl iron compound **24** by $(CF_3SO_2)_2O$.[27] Hughes has reported the cycloaddition of a vinylidene carbene complex to an iron acetylide to produce the interesting delocalized carbene complex **25**.[28]

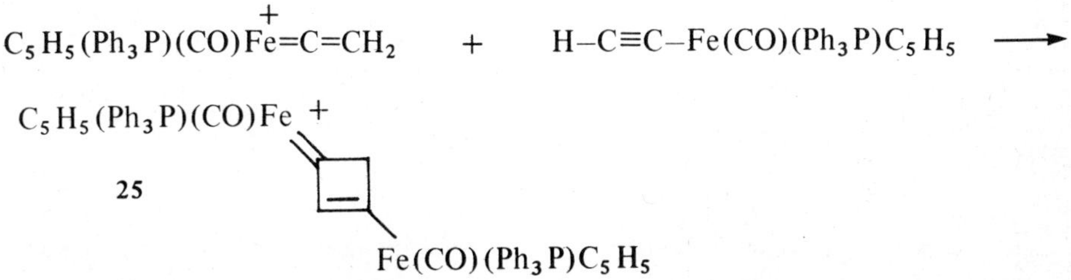

$$C_5H_5(Ph_3P)(CO)\overset{+}{Fe}=C=CH_2 \quad + \quad H-C{\equiv}C-Fe(CO)(Ph_3P)C_5H_5 \longrightarrow$$

Berke[29–31] and Selegue[32] have used propargyl alcohols to synthesize more highly cumulated allenylidene complexes. For example, the *t*-butyl-substituted allenylidene chromium complex, **26**, was synthesized by reaction of $Cr(CO)_6$ with the dianion of a propagyl alcohol, followed by photochemical decarbonylation and treatment with phosgene.[29] A related attempt to prepare a less sterically hindered methyl-substituted allenylidene complex by a similar route led instead to the formation of the dimeric compound **27** by a complex mechanism.[30] Berke has also synthesized the highly cumulated C_3O chromium complex **28**.[33]

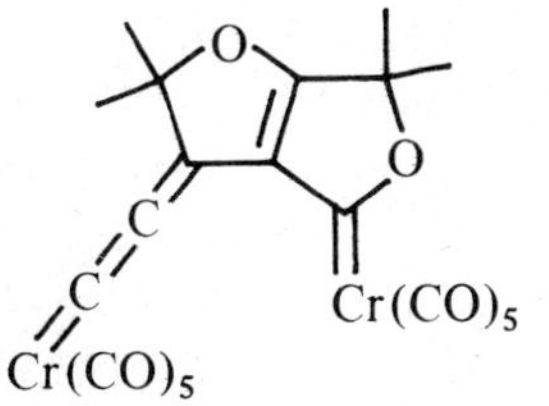

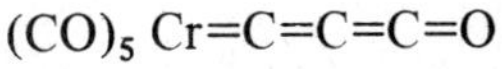

V. INTERCONVERSION OF METAL ALKYLS AND METAL ALKYLIDENE HYDRIDES BY HYDROGEN MIGRATION

The first good evidence for the reversible shift of a hydrogen from the α-carbon of a transition metal alkyl ligand to the transition metal to form a metal–carbene–hydride system came from Green and Cooper's study of $[Cp_2 WLCH_3^+]$.[34] Green has now obtained further evidence for such 1,2 hydrogen shift equilibria in the system where $L = PMe_3$.[35] Equilibration of $Cp_2 W$ $(CH_3)(PMe_3)^+$ and $Cp_2 W(CH_2 PMe_3)H^+$ occurs at 70°C in acetone; the key step in the equilibration is proposed to be insertion of tungsten into an α–CH bond of the methyl tungsten intermediate **29**.

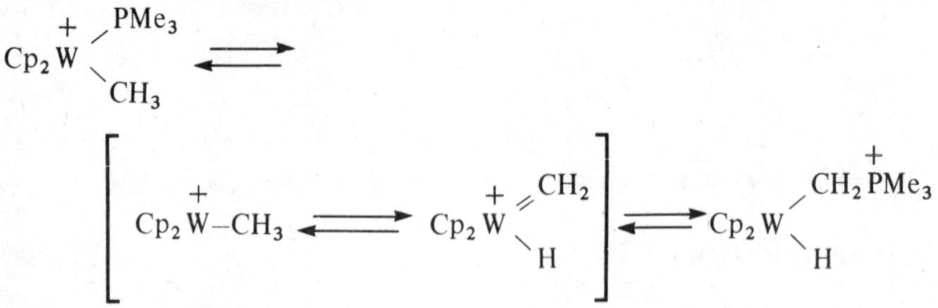

29

Insertions into α–CH bonds can occur when the metal has less than 16e in its outer shell, and when the metal has a vacant orbital of suitable energy and direction. Even when complete transfer of the hydrogen to the metal center does not occur, interaction of the metal center with an α–CH bond can occur. For example, in $Ti(Me_2 PCH_2 CH_2 PMe_2)(CH_3)Cl_3$ there is a three-center 2-electron interaction between Ti and an α–CH bond such that the Ti–C–H angle is 70° and the Ti – H distance is 2.03 Å.[36] It is clear that such interactions can lead to insertion into α–CH bonds.

Schrock has found that reduction of the tantalum(V) alkyl compound **30** leads to insertion of the metal into the α–CH bond and the formation of the tantalum alkylidene hydride complex **31**.[37] There is evidence that the α–CH bond of the alkylidene ligand interacts with the tantalum center by way of a three-center 2-electron Ta–H–C interaction; the low infrared stretch of the CH bond at $2440 \, cm^{-1}$ and the low NMR coupling constant of $J_{CH} = 72 \, Hz$ are characteristic of this interaction. Schrock has reported many structures such as $[Ta(CHCMe_3)Cl_3(PMe_3)]_2$ in which the M=C–H angle is small (84°) and the M=C–R angle is large (161°).[38] This unusual deformation, involving a rotation of the entire alkylidene ligand and weakening of the CH bond, has been attributed to an intramolecular electrophilic interaction of acceptor orbitals of the metal with the carbene lone pair.[39]

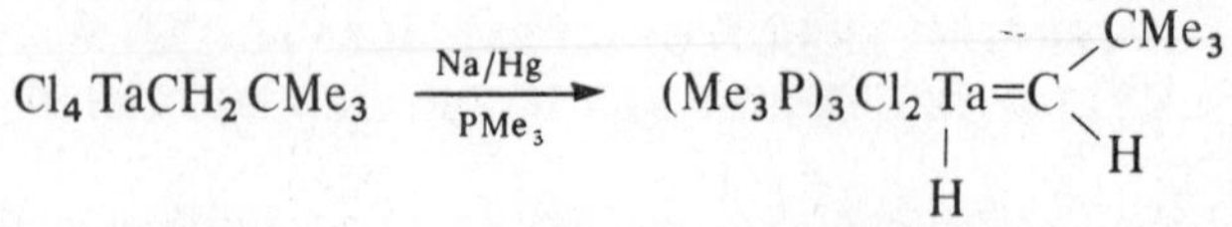

30   31

Schrock has discovered an intriguing system in which α-hydride and β-hydride eliminations are both rapid and result in equilibration of the tantalum neopentyl ethylene complex **32** and the tantalum neopentylidene ethyl complex **33**.[40] Magnetization transfer experiments established that rapid equilibration was occurring at $-30°C$. The small carbon–hydrogen coupling constants seen for the CH_2 unit of the neopentyl group of **32** ($J_{CH} = 98\,Hz$) and for the Ta=CH unit of the neopentylidene group of **33** ($J_{CH} = 80\,Hz$) indicated that Ta–H–C three-center 2-electron interactions were important for both of these compounds.

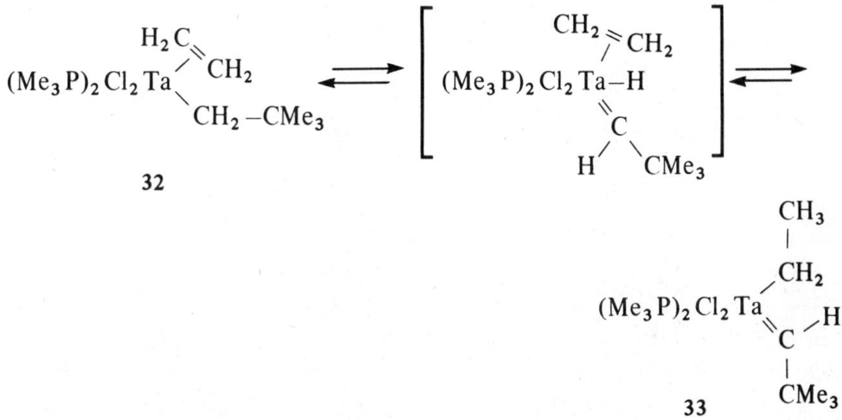

Bercaw found that the zirconoxycarbene complex of niobium hydride, **34**, underwent rapid reversible migration of hydrogen from niobium to the carbene ligand.[41] Spin saturation transfer experiments established that the rate of hydrogen migration was $31\,s^{-1}$ at $32°C$. This was 10^6 faster than the analogous methyl migration.

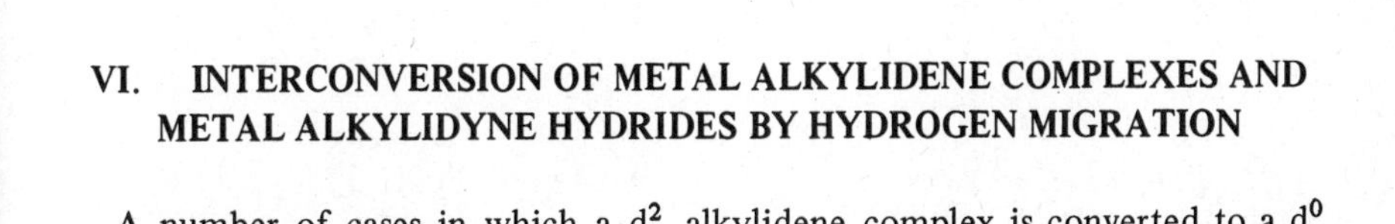

VI.   INTERCONVERSION OF METAL ALKYLIDENE COMPLEXES AND METAL ALKYLIDYNE HYDRIDES BY HYDROGEN MIGRATION

A number of cases in which a d^2 alkylidene complex is converted to a d^0 alkylidyne hydride complex have been observed. This transformation is accompanied by a formal oxidation of the metal complex by two units and by an increase in the electron count at the metal site. Reduction of $(PMe_3)_2Cl_3Ta=CHCMe_3$ in the presence of PMe_3 gives the stable d^2 alkylidene complex **35**.[37] However, reduction of the related complex, $(C_5Me_5)Br_2Ta=CHCMe_3$, leads to formation of an alkylidyne hydride complex, **36**, presumably by way of a hydride migration reaction of an intermediate alkylidene complex. These types of complexes are readily distinguished spectroscopically. The α-proton of the alkylidene ligand of **35** interacts strongly with tantalum; it has an infrared band at $2200\,cm^{-1}$, a ^1H NMR resonance at $\delta -7.9$, and couples to the alkylidene carbon at $\delta 209$ with $J_{CH} = 69\,Hz$. In contrast, the alkylidyne carbon of **36** appears at $\delta 306$ and is not coupled to the hydride ligand.

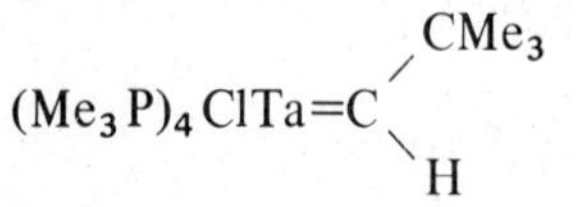

$$(Me_3P)_4ClTa=C\underset{\displaystyle H}{\overset{\displaystyle CMe_3}{<}}$$

$$(C_5Me_5)(Me_3P)_2Ta\equiv C-CMe_3 \atop \qquad\qquad\qquad\quad |\ \ H$$

35 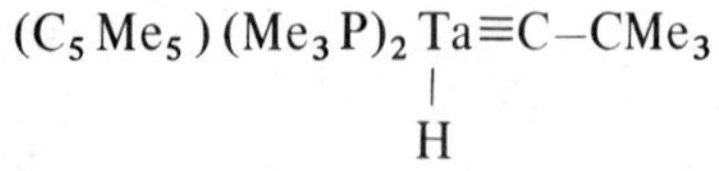   **36**

The position of the equilibrium between a metal alkylidene complex and a metal alkylidyne hydride complex is shifted toward the hydride when the metal center is made more electron deficient.[42] Thus the alkylidene complex **37** is converted to the alkylidyne hydride complex **38** upon treatment with $AlMe_3$. Interaction of $AlMe_3$ with the chloride ligand makes the tantalum center more electropositive and gives rise to the hydride migration. When the alkylidene chloride complex **37** is converted to the corresponding, more ionic iodide complex **39** by treatment with Me_3SiI, an equilibrium mixture of **39** and the rearranged alkylidyne hydride complex **40** is obtained.[42]

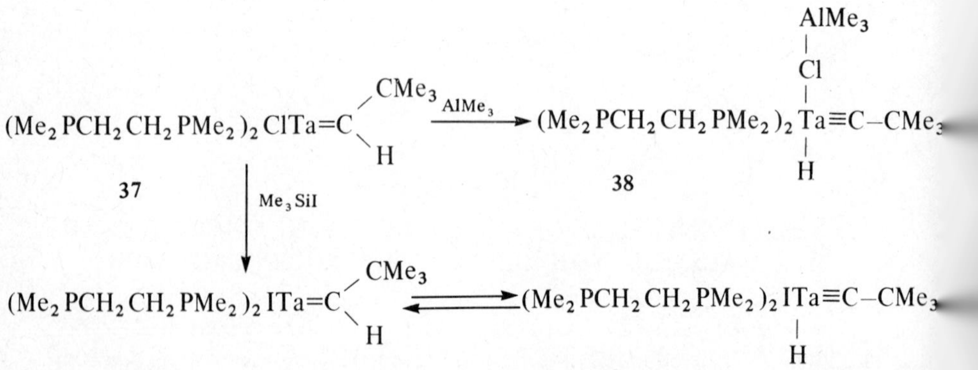

$$(Me_2PCH_2CH_2PMe_2)_2ClTa=C\underset{\displaystyle H}{\overset{\displaystyle CMe_3}{<}} \xrightarrow{\ AlMe_3\ } (Me_2PCH_2CH_2PMe_2)_2Ta\equiv C-CMe_3$$

with $AlMe_3$—Cl on Ta and H:

$$(Me_2PCH_2CH_2PMe_2)_2Ta\equiv C-CMe_3$$

37   **38**

$\Big\downarrow Me_3SiI$

$$(Me_2PCH_2CH_2PMe_2)_2ITa=C\underset{\displaystyle H}{\overset{\displaystyle CMe_3}{<}} \ \rightleftharpoons\ (Me_2PCH_2CH_2PMe_2)_2ITa\equiv C-CMe_3$$

39   **40**

The first terminal methylidyne complex **41** was recently prepared by Schrock, and its protonation reactions have been studied.[43] Protonation of **41** occurs at the methylidyne carbon to produce the unsymmetric methylene complex **42**. In the ^{1}H NMR spectrum of **42** at $-100°$C, the hydrogen bridging between tungsten and carbon appears at $\delta - 7.97$, and the noninteracting methylene hydrogen appears at $\delta 7.05$. Magnetization transfer experiments carried out at low temperature indicate rapid equilibration between the two sites. At room temperature, the methylene protons appear as a single resonance at $\delta - 0.16$. The fact that both methylene protons are bound to carbon in **42** is established by the large coupling ($J_{CH} = 119$ Hz) to the methylene carbon at $\delta 220$ in the ^{13}C NMR spectrum. A minor change in the ligand system from PMe$_3$ to dmpe results in complex **43** which now undergoes protonation at tungsten to give **44**. The methylidyne carbon of **44** appears as a doublet ($J_{CH} = 69$ Hz) at $\delta 263$ at $-30°$C; at room temperature, rapid reversible hydrogen migration from tungsten to carbon results in a triplet ($J_{CH} = 69$ Hz) at $\delta 263$.[43]

$$(\text{Me}_3\text{P})_4\text{ClW}\equiv\text{C–H} \xrightarrow{\text{H}^+} (\text{Me}_3\text{P})_4\text{ClW}{=}\overset{+}{\text{C}}\underset{\text{H}}{\overset{\cdots\text{H}}{\diagup}}$$

41   **42**

$$(\text{Me}_2\text{PCH}_2\text{CH}_2\text{PMe}_2)_2\text{ClW}\equiv\text{C–H} \xrightarrow{\text{H}^+} (\text{Me}_2\text{PCH}_2\text{CH}_2\text{PMe}_2)_2\text{ClW}\equiv\overset{\overset{\displaystyle\text{H}}{|+}}{\text{C–H}}$$

43   **44**

In view of these rapid reversible migrations of hydrogen from methylene ligands to electron-deficient transition metal centers, the nature of the CrCH$_2^+$ and CoCH$_2^+$ ions seen by mass spectroscopy and the metal–carbene bond strengths derived from these measurements should be viewed with caution.[44,45]

VII. INTERCONVERSION OF METAL ALKYLS AND METAL ALKYLIDENE ALKYLS BY CARBON MIGRATION

Pettit and Brady[46,47] demonstrated that the same mixture of hydrocarbons is obtained from heterogeneous Fischer–Tropsch catalysts when either CO and H$_2$ or CH$_2{=}$N$_2$ and H$_2$ are employed. They suggested that surface bound methylene species are key intermediates in Fischer–Tropsch chemistry and that carbon–carbon bond formation occurs by coupling of surface bound alkyl and methylene units. Consequently there has been much interest in the migration of alkyl groups from a metal to a methylene ligand in the same complex. A number of examples of this type of reaction has now been observed.

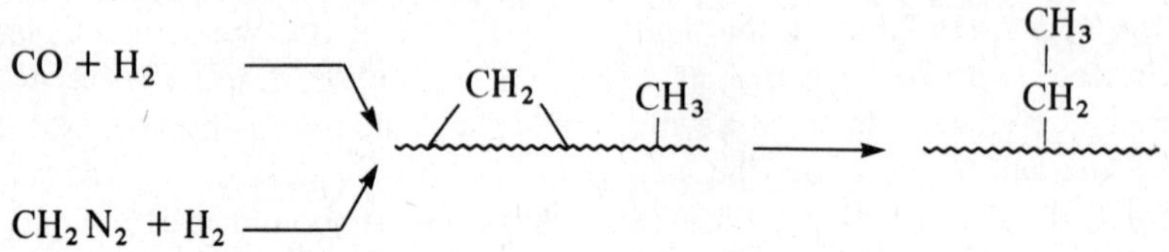

$$CO + H_2 \longrightarrow$$

$$CH_2N_2 + H_2 \longrightarrow$$

Schrock and Sharp[48] reported that $(C_5H_5)_2(CH_3)Ta=CHCH_3$ rearranged to the propylene hydride complex **45** at 75°C. The reaction was suggested to proceed by methyl migration to the ethylidene ligand followed by β-hydride elimination.

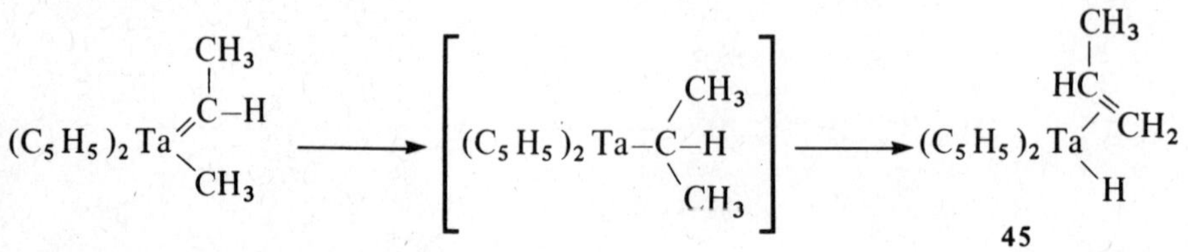

Cooper has found that $(C_6H_5)_3C\cdot$ abstracts a hydrogen atom from the paramagnetic dimethyltungsten compound **46** to produce the ethylene hydride complex **47**.[49] The reaction is proposed to proceed by formation of methylene complex **48**, migration of methyl to the methylene ligand to produce ethyl complex **49**, and β-hydride elimination. Both **48** and **49** can be trapped by added phosphines. When the corresponding neutral diamagnetic complex $Cp_2W(CH_3)_2$ is treated with $(C_6H_5)_3C^+$, **46** and $(C_6H_5)_3C\cdot$ are initially generated; this indicates that hydride abstraction reactions can proceed by an electron transfer mechanism.[50]

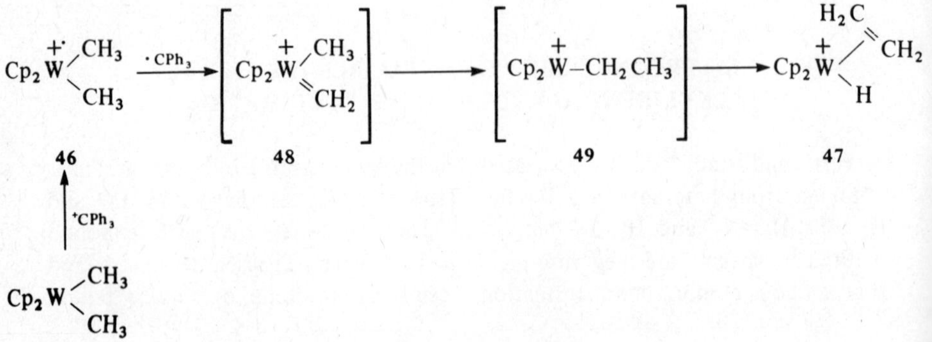

Alkyl migration to a coordinated alkylidene ligand is one proposed step in the complex rearrangement that occurs when tungsten radical cation **50** is treated with $(C_6H_5)_3C\cdot$.[51,52]

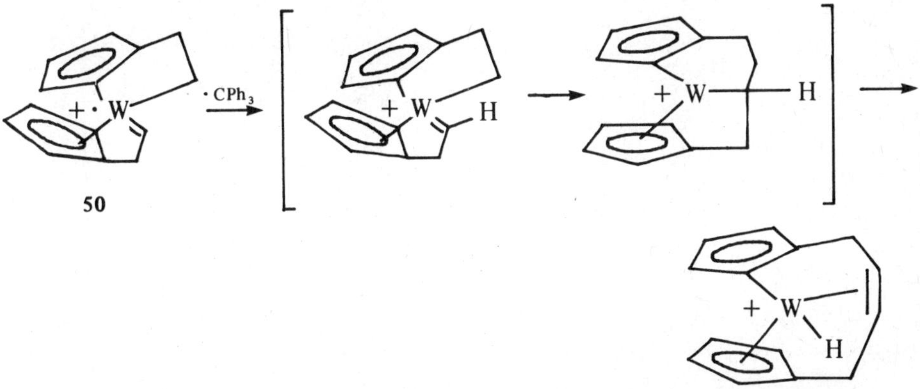

In the case of the reaction of the corresponding methylethyltungsten compound, **51**, a choice between α-hydrogen and β-hydrogen abstraction by $(C_6H_5)_3C^+$ is possible. The reaction proceeds cleanly by electron transfer and α-hydrogen atom abstraction from the methyl group.[50] Abstraction of a hydrogen atom from the site α to the odd electron metal center can lead directly to a M=C double bond. Cooper suggests that α-hydrogen atom abstraction in reactions of $(C_6H_5)_3C^+$ is characteristic of prior electron transfer followed by hydrogen atom abstraction by $(C_6H_5)_3C\cdot$, and that β-hydrogen atom abstraction is the result of attack by $(C_6H_5)_3C^+$ without prior electron transfer. Gladysz has also observed a similar selective α-hydrogen atom abstraction from **52** by $(C_6H_5)_3C^+$ and has suggested that these reactions may proceed by prior electron transfer.[53]

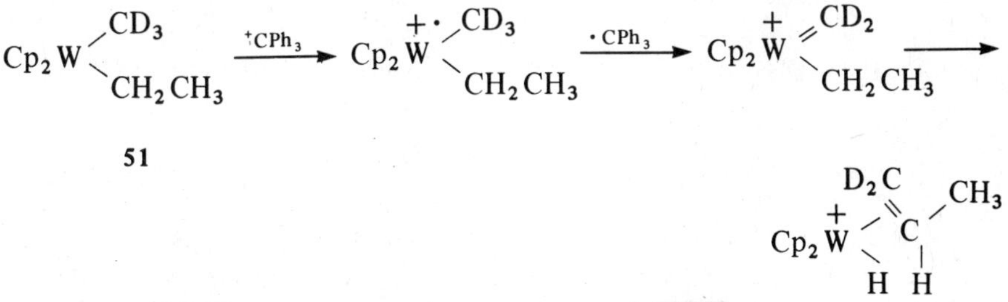

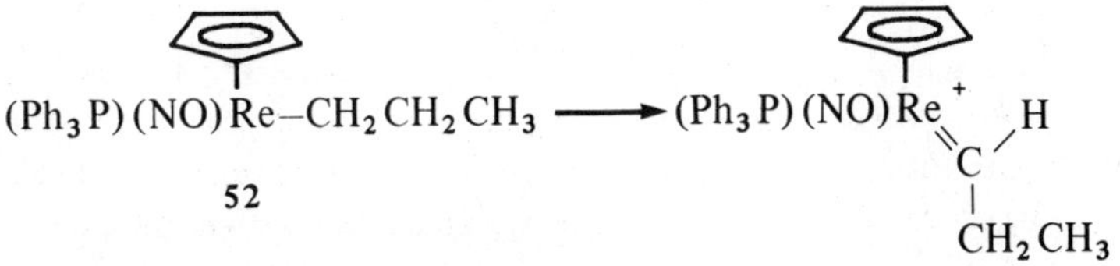

Bercaw found that the rate of methyl migration from niobium to a carbene ligand of **53** was about 10^6 slower than the corresponding hydrogen migration reaction.[41] The intermediate alkyl niobium complex could not be directly observed because of a rapid β-hydride elimination.

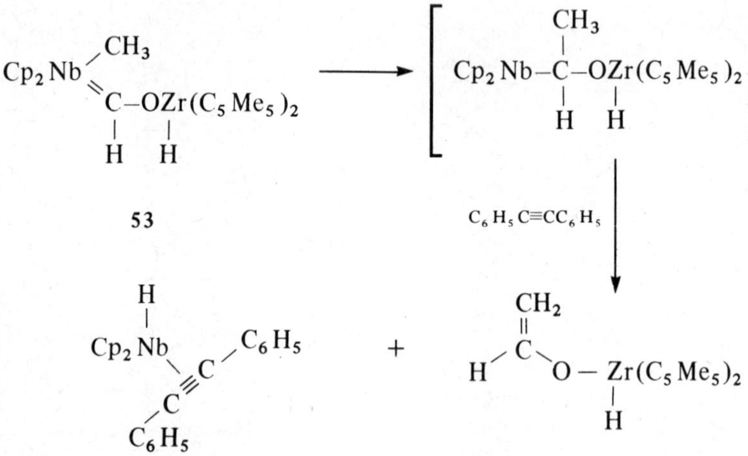

Thorn and Tulip have pointed out that although a large number of stable alkyl alkylidene complexes of early transition metals are known,[54] they do not undergo alkyl migration to the coordinated alkylidene, presumably because of the nucleophilicity of the alkylidene unit. In their own work they generated an iridium compound **54** which possesses a methyl group cis to an electrophilic carbene precursor.[54] Generation of a transient methyl methylene complex led to methyl migration products.

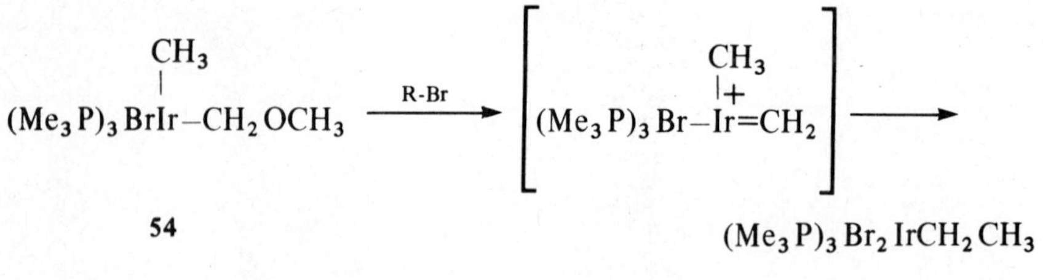

VIII.   METAL–CARBENE COMPLEXES AND OLEFIN METATHESIS

The major achievement in metal–carbene complex chemistry in the last three years has been the isolation of metal–carbene complexes that are active olefin metathesis catalysts. These isolated species all contain an oxo or alkoxy ligand which appears to be essential. This may explain why some metathesis catalysts are active only if traces of oxygen or water are added. Goddard has carried out molecular orbital calculations that indicate that a metal oxo ligand can facilitate olefin metathesis by acting as a variable electron donor.[55] When a metal–alkene–carbene complex is converted to a metallacyclobutane, the electron count at the metal drops by two. This can be compensated by increasing electron donation from the oxo ligand, which involves a change from a metal–oxygen double bond to metal–oxygen triple bond.

$$
\begin{array}{ccc}
\overset{\displaystyle O}{\underset{\displaystyle CH_2=CH_2}{\overset{\parallel}{\underset{|}{M}}=CH_2}} & \rightleftharpoons & \overset{\displaystyle O}{\underset{\displaystyle H_2C-CH_2}{\overset{\mathrm{III}}{\underset{|\;\;|}{M}-CH_2}}}
\end{array}
$$

The importance of oxygen ligands was established by comparison of the reactions of $Cl_3(PMe_3)_2Ta=CHC(CH_3)_3$, **55**, and $(Me_3CO)_2(PMe_3)ClTa=CHC(CH_3)_3$, **56**. Complex **55** reacts with ethylene to give products derived from decomposition of an intermediate metallacyclobutane by β-hydride elimination; no olefin metathesis products were seen.[56,57] In contrast, **56** reacts with alkenes to give metathesis products in good yield.

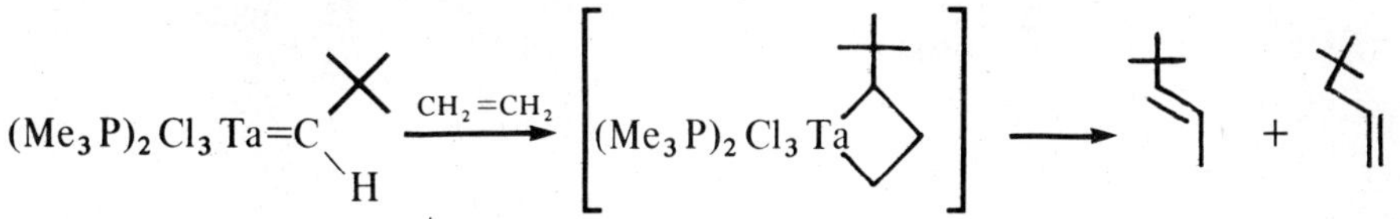

55

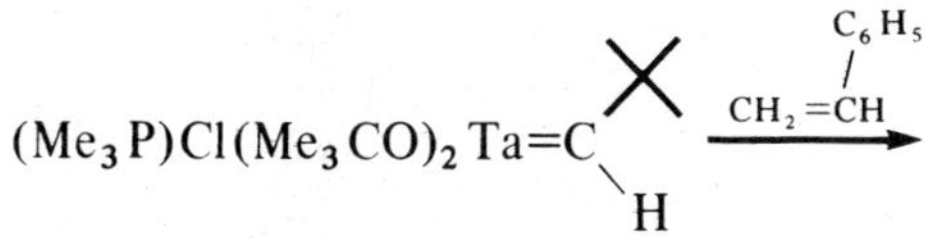

56

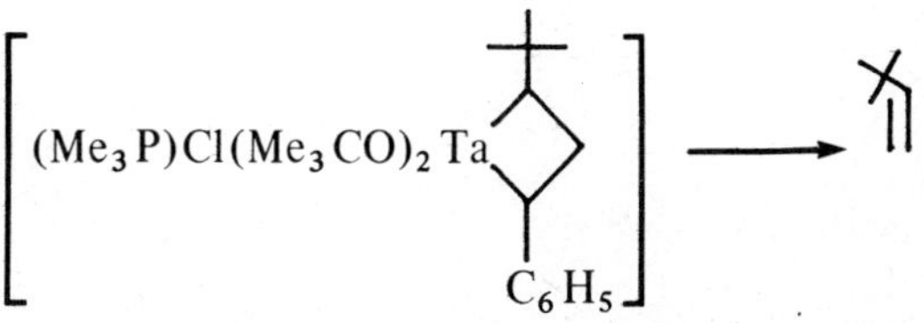

In an attempt to make a tungsten alkylidene complex, Schrock discovered a fantastic reaction in which nearly all ligands are interchanged between a tantalum and a tungsten complex.[58] The reaction produced the interesting tungsten oxoalkylidene complex, **57**. The X-ray crystal structure of this complex indicates that the neopentylidene and oxo ligands are cis and co-planar.[59] This arrangement allows the use of different d orbitals for backbonding to the alkylidene and oxo ligands. The W=CH–CMe$_3$ angle is 140°, indicating that the alkylidene ligand is not distorted; apparently, electron donation from the oxo ligand reduces the need for an electrophilic interaction between tungsten and the α–CH bond of the alkylidene ligand.

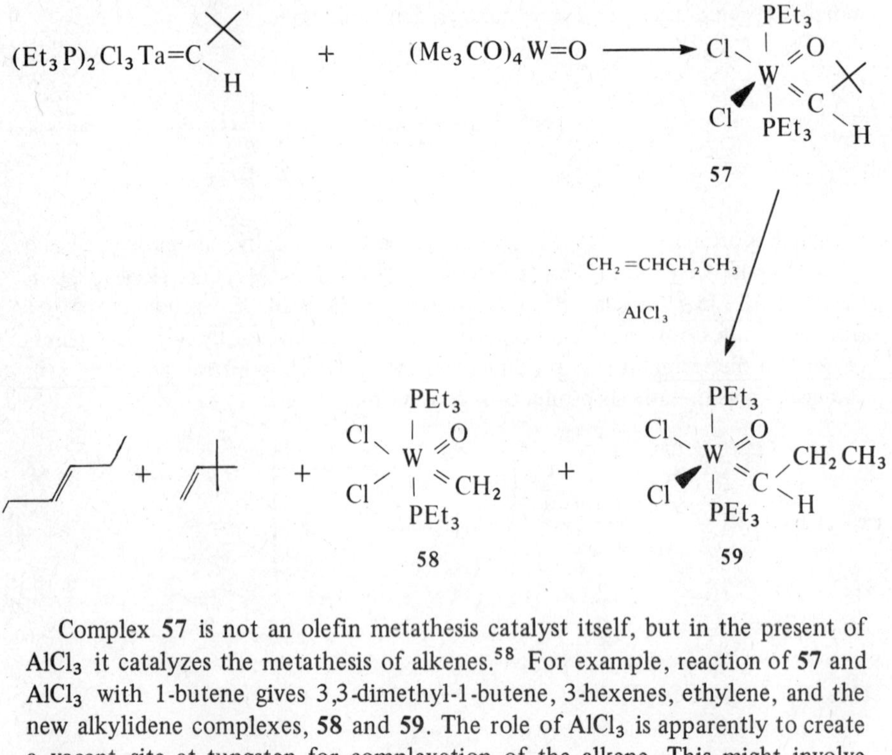

Complex **57** is not an olefin metathesis catalyst itself, but in the present of AlCl$_3$ it catalyzes the metathesis of alkenes.[58] For example, reaction of **57** and AlCl$_3$ with 1-butene gives 3,3-dimethyl-1-butene, 3-hexenes, ethylene, and the new alkylidene complexes, **58** and **59**. The role of AlCl$_3$ is apparently to create a vacant site at tungsten for complexation of the alkene. This might involve either loss of chloride or phosphine since both processes occur readily.

Reaction of **57** with $(C_6H_5CN)_2PdCl_2$ leads to removal of one PEt$_3$ ligand and formation of the stable five-coordinate complex, **60**, which has been characterized by X-ray crystallography.[58,60] Complex **57** is an active catalyst for the metathesis of terminal and internal olefins even in the *absence* of added Lewis acids.

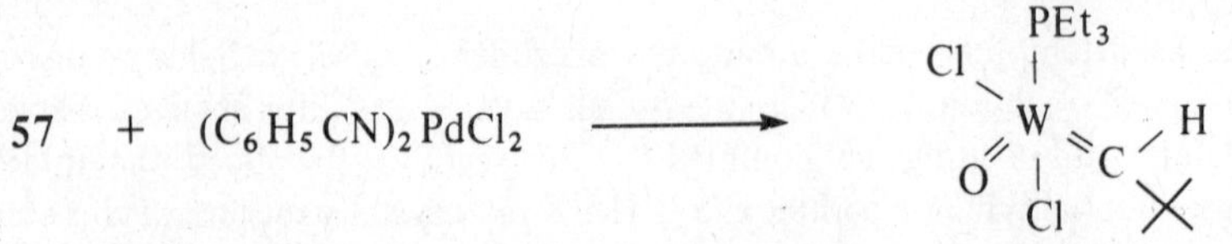

Osborn has also succeeded in isolating active metathesis catalysts. $(Me_3CCH_2O)_2(Me_3CCH_2)_2W=O$ reacts with Lewis acids such as $AlBr_3$ to form 1:1 adducts.[61] These adducts slowly lose neopentane and form the stable isolable metal—carbene complex 61. This carbene complex does not react with cis-2-pentene by itself, but upon addition of $AlBr_3$, a very active catalyst is produced that converts 2-pentene into 2-butenes and 3-hexenes.[62] Low-temperature NMR indicates that the 1:1 adduct 62 is formed. Upon addition of 2-pentene, two new metal—carbene complexes with $W=CHCH_3$ and $W=CHCH_2-CH_3$ groups are seen by NMR.

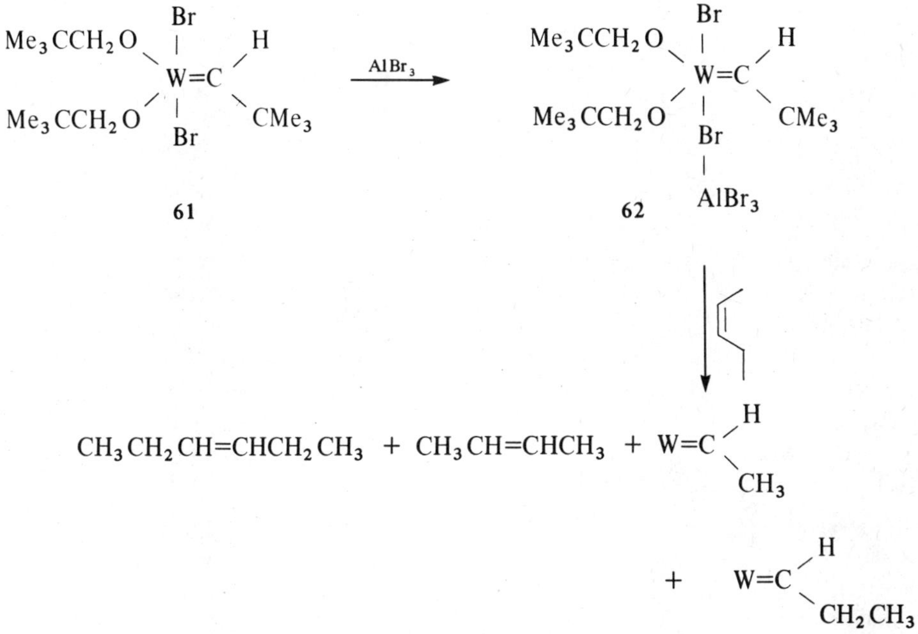

IX.   METALLACYCLOBUTANES IN OLEFIN METATHESIS

In 1979 Tebbe and Parshall showed that methylene titanium complex $Cp_2\overline{TiCH_2AlClMe_2}$, 63, catalyzes the degenerate metathesis of terminal alkenes.[63] In the last three years, Grubbs has investigated this system in great detail and has isolated metallacyclobutane intermediates from the system that are active metathesis catalysts. Reaction of 63 with amines and terminal alkenes such as 3,3-dimethyl-1-butene led to the formation of the stable isolable titanacyclobutane 64.[64] X-ray crystal structures of 64 and related compounds indicate that these titanacyclobutanes are planar and symmetric.[65] The planarity of these titanacycles is significant since several earlier stereochemical models for the moderate stereospecificity of the metathesis reaction presumed a puckered metallacyclobutane. It would be interesting to know whether titanacycles with substituents on the α-carbon atom would also have planar rings. Both the X-ray structure and deuterium isotope effects on carbon chemical shifts support a symmetric structure for the titanacycle. This is significant since molecular orbital calculations of Eisenstein and Hoffmann had suggested that the metalla-cycle would undergo a distortion toward a metal—carbene—alkene complex.[66]

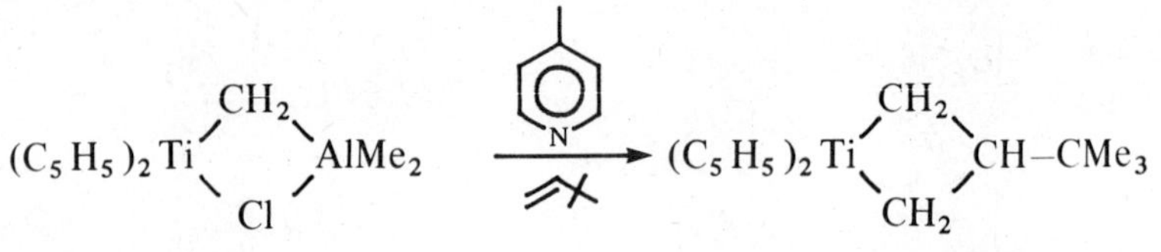

63   64

Metallacyclobutane **64** reacts with Me_2AlCl at $-40°C$ to regenerate **63**.[67] The reaction proceeds with second-order kinetics, first order in **64** and in Me_2AlCl. When this reaction was carried out with **trans-64-d**, isomerization occurred much more rapidly than conversion to **63**. Grubbs explained these results as shown in the following scheme.

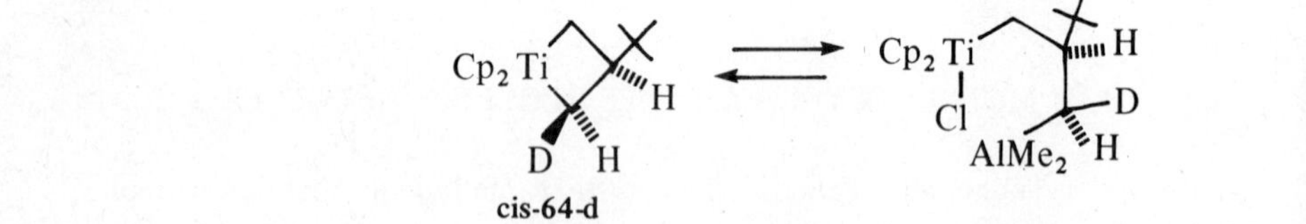

The metallacycle **64** is an active catalyst for the degenerate metathesis of terminal alkenes.[68] Compound **64** also reacts irreversibly with diphenylacetylene to give the titanacyclobutene **65**. The rate of formation of **65** is first order in **64** and independent of acetylene concentration. The rate of cis—trans isomerization of **trans-64-d$_1$** is half the rate of reaction with the acetylene. Both of these rates are consistent with a rate-limiting ring opening of the titanacyclo-butane to a metal–methylene intermediate and free alkene. It should be noted that the resting state of this titanium metathesis system is a metallacycle, whereas the resting states of the well-defined Ta and W metathesis catalysts isolated by Schrock[57,58] are metal–alkylidene complexes.

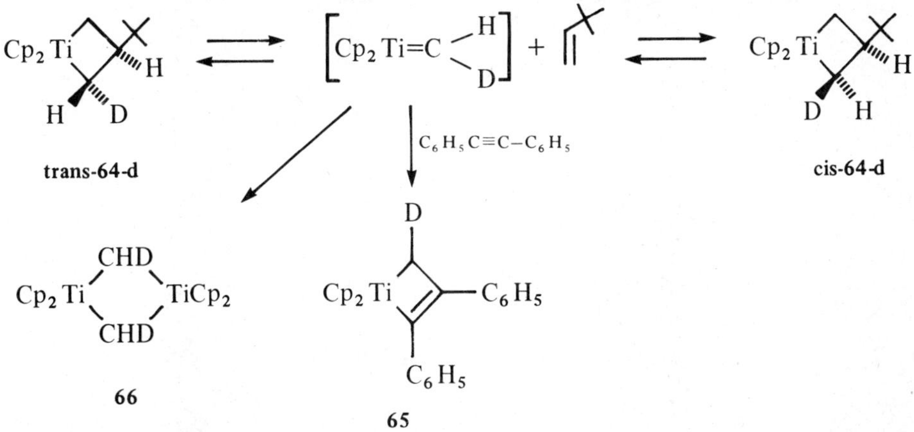

trans-64-d

cis-64-d

$C_6H_5C \equiv C-C_6H_5$

66

65

In the absence of alkenes 64 decomposes to the 1,3-dimetallacyclobutane, 66, which is inactive in metathesis reactions. This is an example of a chain termination reaction of an olefin metathesis catalyst.[69] Evans and Grubbs have used 63 as a reagent for converting esters into vinyl ethers.[70]

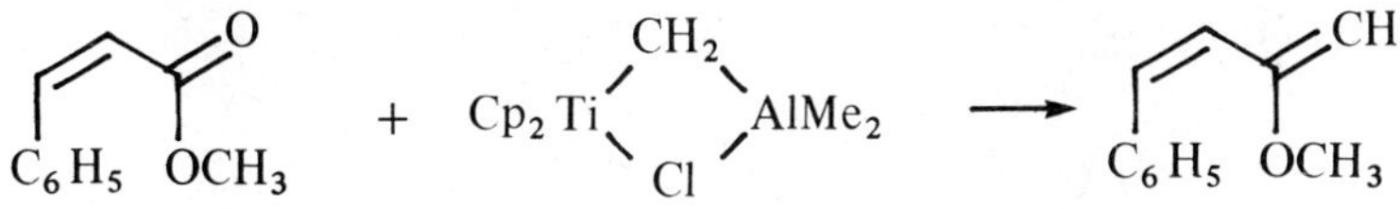

63

X.   METAL–CARBENE–ALKENE COMPLEXES

Metal–carbene–alkene complexes are considered to be key intermediates in the olefin metathesis reaction and in the cyclopropanation of alkenes. Stable complexes of this type have now been observed and isolated. Their intermediacy in cyclopropanation has been demonstrated, but olefin metathesis-like reactions of metal–carbene–alkene complexes have not yet been observed.

In the thermolysis of 67, Casey and Shusterman[71] observed an unstable intermediate metal–carbene–alkene complex, **68**, which underwent decomposition to cyclopropane **69**. The reaction was found to be autocatalytic, and a chain mechanism involving $W(CO)_4$ was proposed.

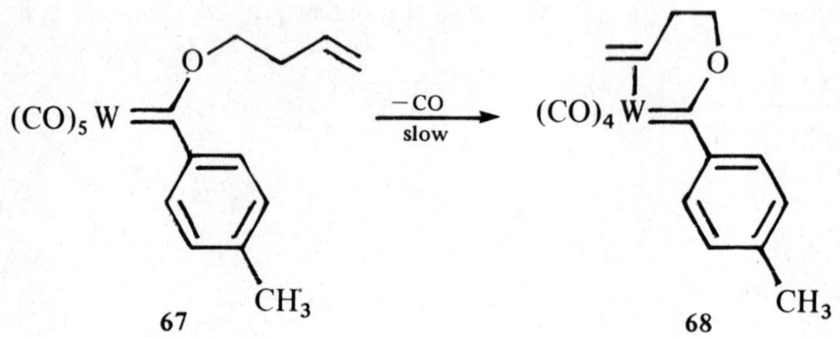

The more stable, amino-substituted metal–carbene–alkene complex, **70**, was isolated. X-ray crystallography showed that the carbene and alkene ligands in **70** were nearly perpendicular to one another.[72] For olefin metathesis, a parallel conformation would be required. The chelate ring size with two atoms linking the alkene to the carbene carbon appears to be particularly favorable since rearrangement of longer-chain alkenes to this chelate size was observed in the synthesis of **71**[72] and **72**.[73]

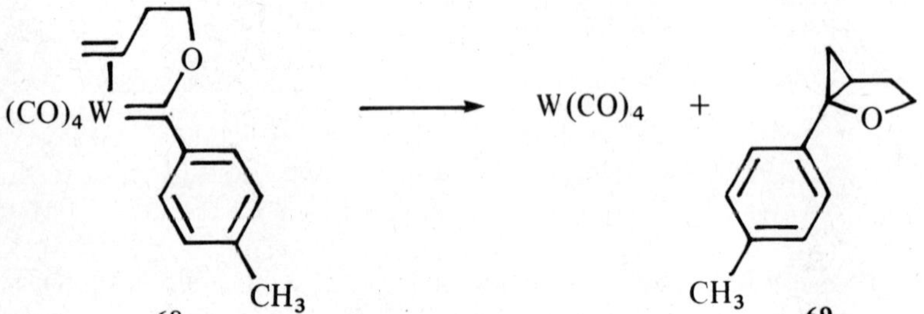

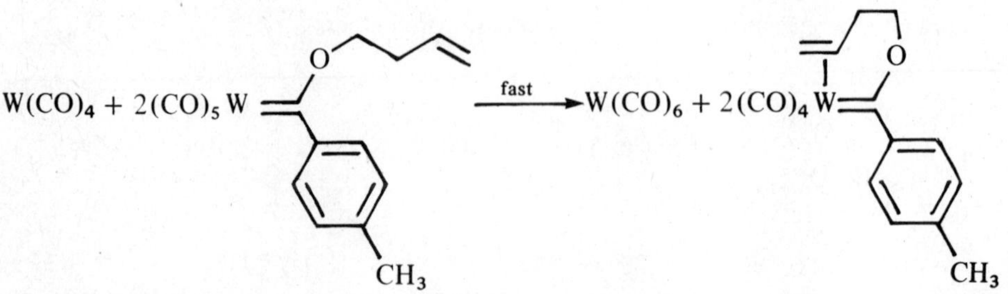

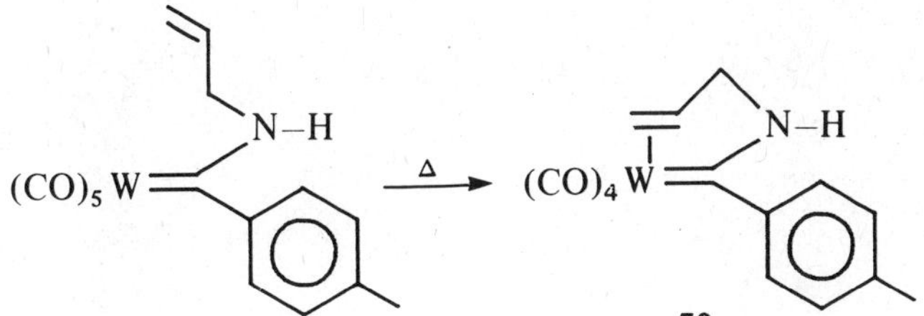

70

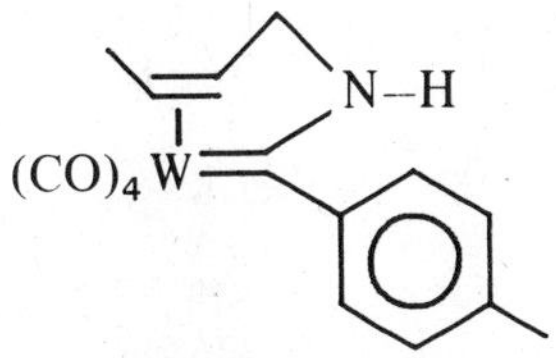

71 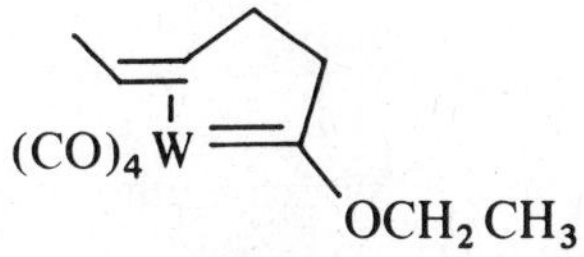   **72**

Casey and Vollendorf were able to block this isomerization with methyl substituents and succeeded in isolating metal–carbene–alkene complex **73** and characterizing it by X-ray crystallography.[74] The two independent molecules of **73** in the unit cell have different conformations, one with the carbene and alkene ligands perpendicular and one with them parallel. On thermolysis, **73** is converted to cyclopropane **74**.

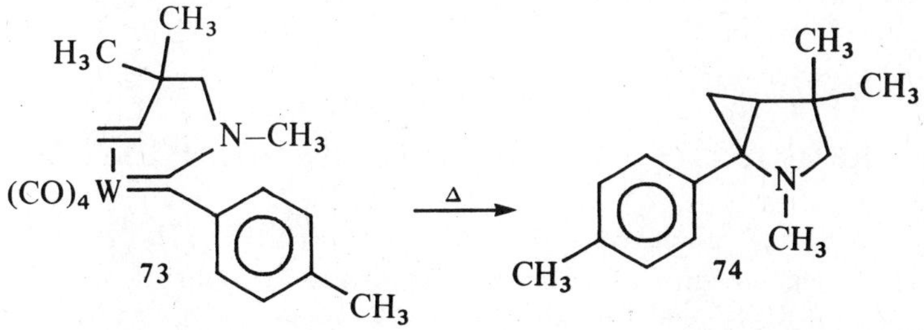

73   **74**

Thorn has found that the methyoxymethyl(ethylene) iridium complex, **75**, reacts readily with $Me_3SiOSO_2CF_3$ to form the hydride allyl compound **76**.[75] The reaction is proposed to proceed by way of an iridium–methylene–ethylene complex, which undergoes cyclization to a metallacyclobutane, followed by β-hydride elimination to form **76**.

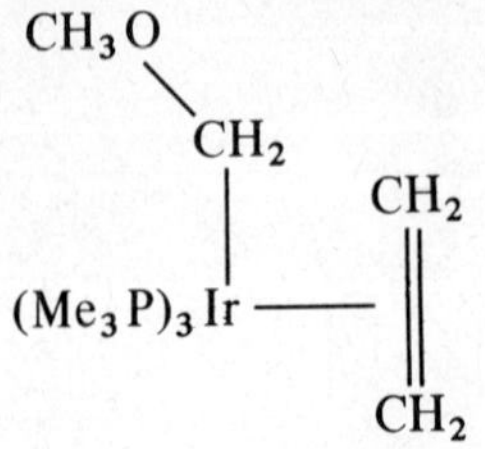

CH$_3$O
  \
   CH$_2$
    | CH$_2$
    | ‖
(Me$_3$P)$_3$Ir —— ‖
                  CH$_2$

75

$\xrightarrow{E^+}$ $\left[\begin{array}{c} \text{CH}_2 \ \text{CH}_2 \\ \| + \ \| \\ (\text{Me}_3\text{P})_3\text{Ir} — \\ \text{CH}_2 \end{array} \right.$ $\longrightarrow$ $(\text{Me}_3\text{P})_3\overset{+}{\text{Ir}} \square \left. \right]$ $\longrightarrow$

$\overset{\overset{\displaystyle H}{+|}}{(\text{Me}_3\text{P})_3\text{Ir}} —— \triangle$

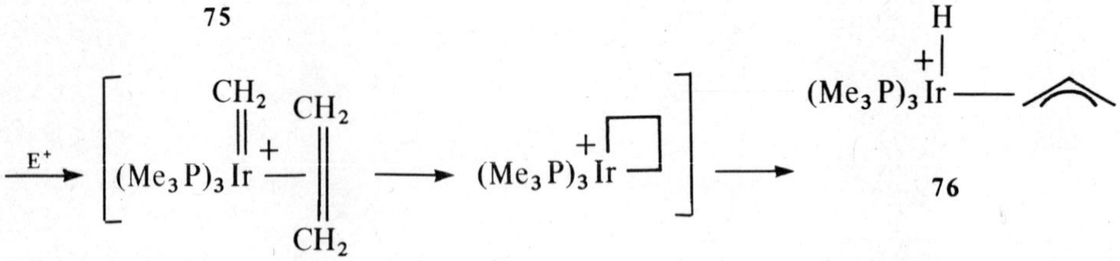

76

Crystal structures of other metal–carbene–alkene complexes have recently been reported, including those of **72**,[73] **77**,[76] and **78**.[77] Molecular orbital calculations dealing with the preferred conformation of metal–carbene–alkene complexes and with their conversion to metallacyclobutanes were reported by Eisenstein and Hoffmann.[66]

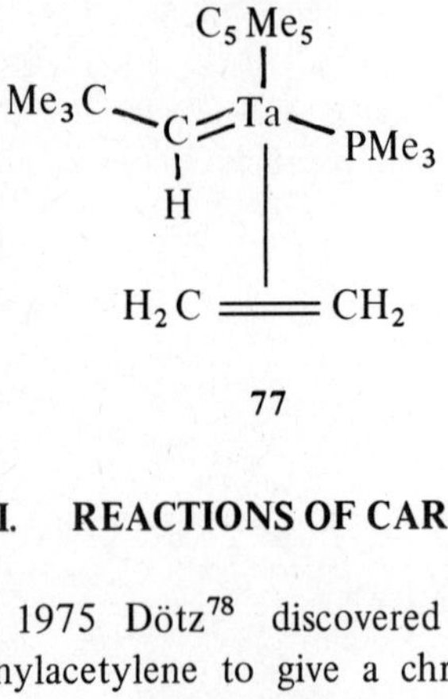

C$_5$Me$_5$
        |
Me$_3$C   Ta
      \ C=   \ PMe$_3$
      / |
     H |
         |
  H$_2$C === CH$_2$

77

MeO$_2$C —   — CO$_2$Me
       \ | /
(CO)$_3$Fe ===
              OMe

(with H at top)

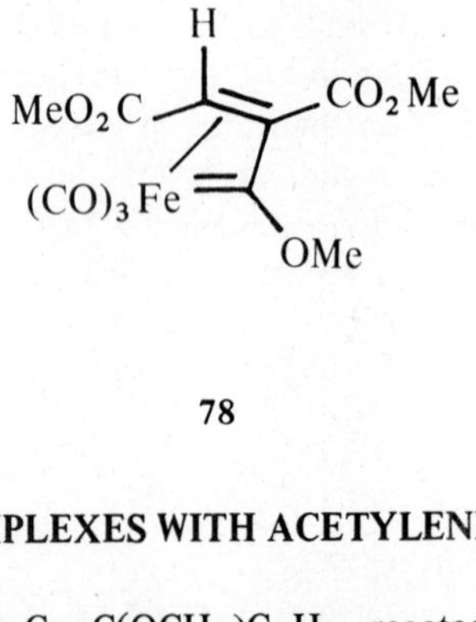

78

XI. REACTIONS OF CARBENE COMPLEXES WITH ACETYLENES

In 1975 Dötz[78] discovered that $(CO)_5Cr=C(OCH_3)C_6H_5$ reacted with diphenylacetylene to give a chromium complex of a naphthol, which incorporates the carbene ligand, the acetylene, and carbon monoxide.

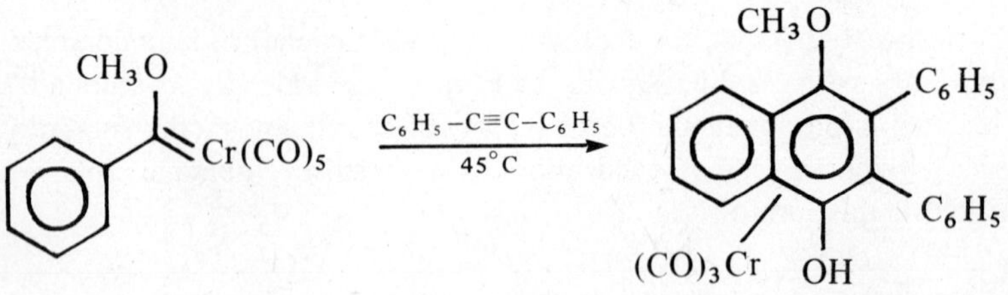

CH$_3$O
    \
     C Cr(CO)$_5$
     ‖
(phenyl ring)

$\xrightarrow[\text{45°C}]{C_6H_5-C\equiv C-C_6H_5}$

(naphthol structure)
CH$_3$O
          C$_6$H$_5$
          C$_6$H$_5$
(CO)$_3$Cr OH

The mechanism of the reaction is undoubtedly complex because side products such as vinylcarbene complexes, indenes, cyclobutenones, and vinylketenes are sometimes obtained. Recent kinetic studies show that the reaction proceeds by reversible loss of CO, followed by coordination of an acetylene and rate-determining reaction of the resulting complex.[79] This probably occurs by ring closure to a metallacyclobutene and ring opening to a vinylcarbene complex. Several possible routes to the naphthol complex from the vinylcarbene complex are shown in the schemes that follow. One involves a cyclization to metallacyclohexadiene **79** as a key step; another involves vinylketene complex **80** as a key intermediate. Recently, Dötz has isolated a vinylketene complex from the reaction of $(CO)_5 Cr=C(OCH_3)C_6H_5$ with $Me_3SiC\equiv CSiMe_3$.[80]

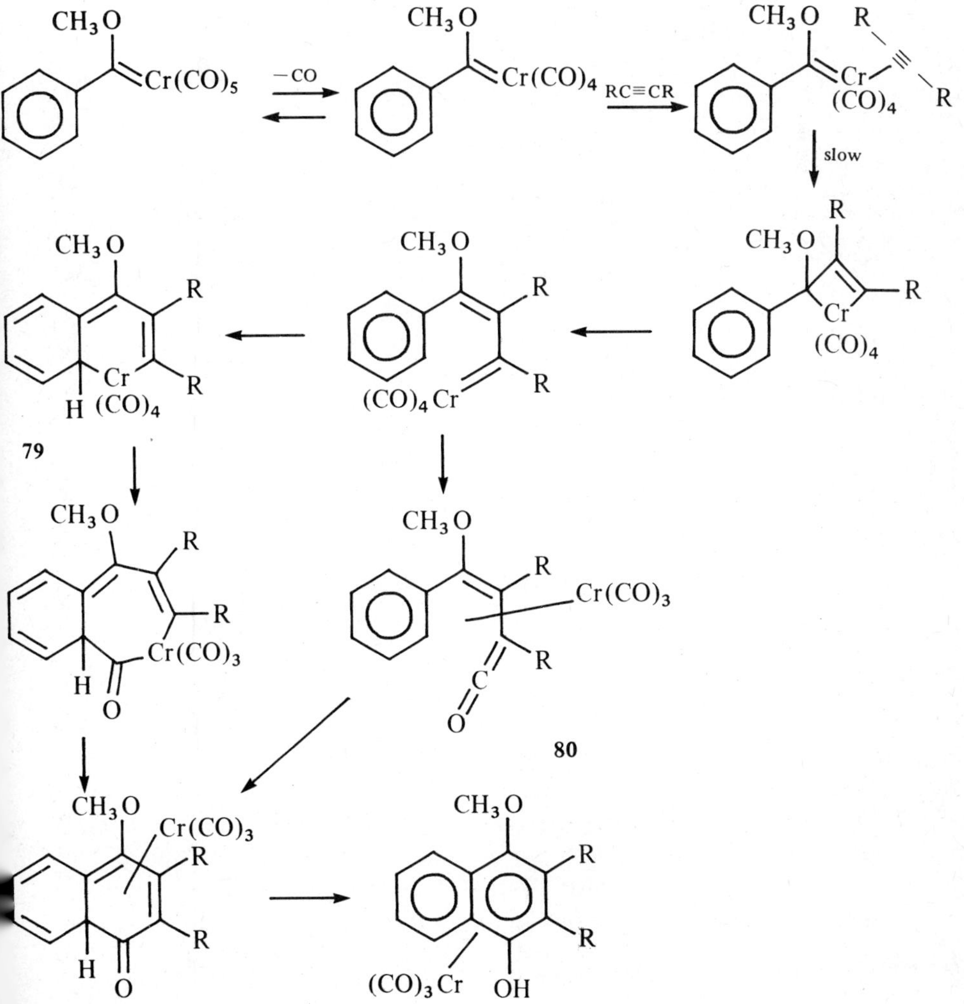

79

80

Because this reaction appears to be an efficient entry into the naphthol ring system common to many natural products, the regiochemistry of the reaction has been examined in detail by synthetic chemists. The reaction usually gives mixtures of uncomplexed and complexed naphthols with chromium bonded to either of the two rings. Therefore, the reaction mixtures are normally oxidized *in situ*, and the resulting uncomplexed naphthoquinones are isolated.

Wulff has examined the regiospecificity of the reactions of the *o*-methoxyphenylcarbene complex, **81**, with alkynes because the *o*-methoxyphenyl unit is a key substituent in anthracyclines.[81] The reaction of **81** with 1-pentyne was highly regioselective, and gave exclusively isomer **82**, which corresponds to incorporation of the unsubstituted end of the acetylene α to the carbene carbon. In contrast, the reaction of 2-pentyne showed little regioselectivity and gave a 1.5:1 mixture of isomers **83** and **84**.

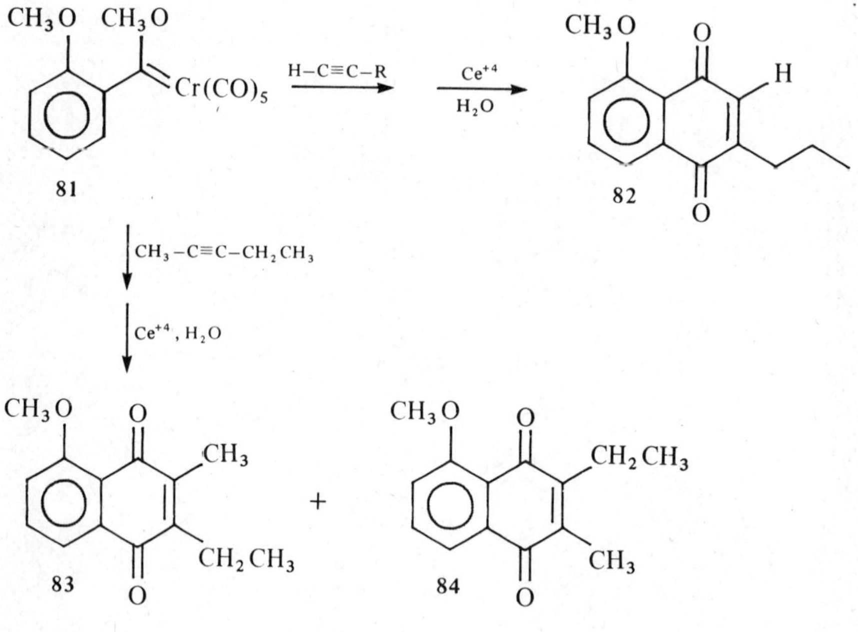

In the case of symmetrically substituted naphthoquinones, the lack of regiospecificity presents no problem. For example, Dötz obtained a mixture of two isomers from the reaction of $(CO)_5Cr=C(OCH_3)C_6H_5$ with allylacetylene **85**, but both isomers were converted by Ag_2O oxidation to Vitamin $K_1(20)$. The stereochemistry of the double bond in the side chain was preserved in this synthesis.[82]

In his synthesis of deoxyfrenolicin, Semmelhack overcame the problem of low regioselectivity of alkyne reactions with chromium carbene complexes by setting up an intramolecular cyclization.[83] In doing so, he developed an excellent

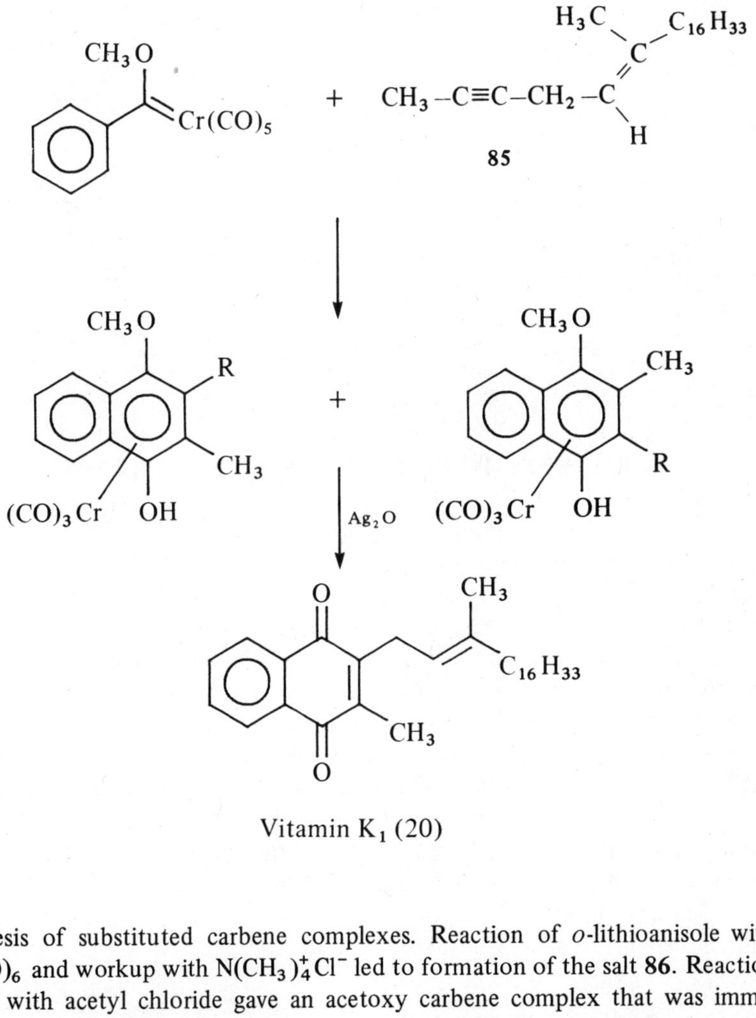

Vitamin K$_1$ (20)

synthesis of substituted carbene complexes. Reaction of *o*-lithioanisole with Cr(CO)$_6$ and workup with N(CH$_3$)$_4^+$Cl$^-$ led to formation of the salt **86**. Reaction of **86** with acetyl chloride gave an acetoxy carbene complex that was immediately reacted with the requisite alcohol to give **87**. This intramolecular cyclization of **87** was completely regioselective and was the key step in this natural product synthesis.

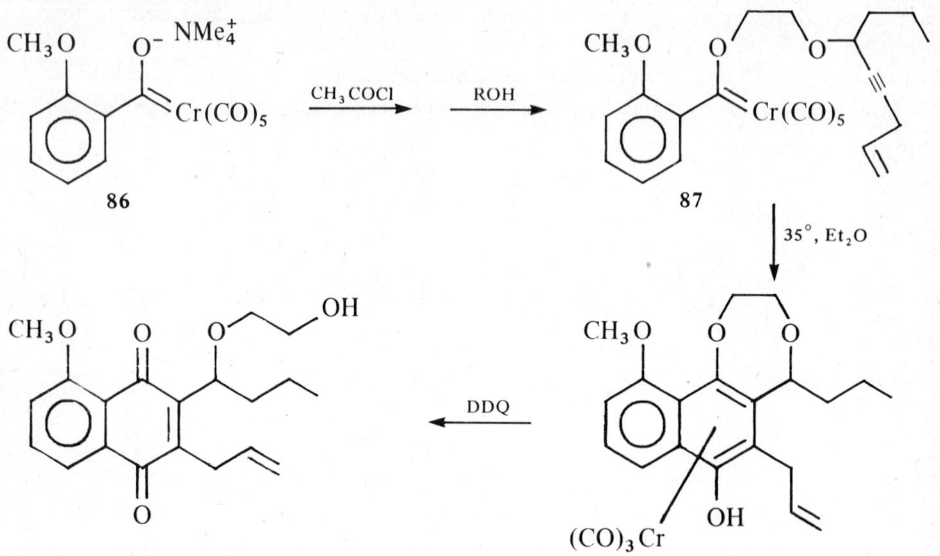

XII.   IRON–PORPHYRIN–CARBENE COMPLEXES

The evidence for formation of cytochrome P 450–Fe(II)–carbene complexes during reductive metabolism of various polyhalogenated compounds has been reviewed by Mansuy.[6] The metabolic reduction of CCl_4, the widely used anaesthetic halothane, $CF_3CHClBr$, and the insecticide DDT, $(p–ClC_6H_4)_2CHCCl_3$, all involve such chemistry. Iron(II)tetraphenylporphyrin carbene complexes have been extensively studied as models for the cytochrome compounds.

Fe(II)TPP (TPP = tetraphenylporphyrin) reacts with CCl_4 in the presence of excess reducing agent to give the very stable dichlorocarbene complex 88.[84] σ-Alkyliron(III)complexes have been suggested as intermediates in the formation of these carbene complexes. The reaction of halothane with Fe(II)TPP in the presence of a reducing agent gives the σ-alkyliron(III)complex 89, the first such complex with a halogen substituent on the carbon bound to the iron.[85] Ordinarily, elimination of this halogen takes place to give an iron–carbene complex, as in the case of CF_3CCl_3 which leads to complex 90.

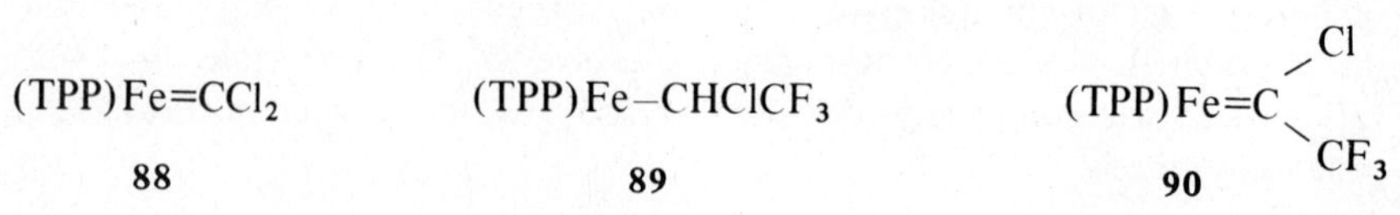

$(TPP)Fe=CCl_2$	$(TPP)Fe–CHClCF_3$	$(TPP)Fe=C{<}^{Cl}_{CF_3}$
88	**89**	**90**

The reaction of Fe(TPP) with DDT (p–ClC$_6$H$_4$)$_2$CH–CCl$_3$ does not stop at the stage of the Fe=CClR complex but proceeds further by elimination of HCl to form the very stable vinylidene carbene complex 91.[86] One electron oxidation of 91 is reversible and leads to the isolation of a species that is spectrally analogous to catalase and horseradish peroxidase compounds.[87] Balch[88] and Goff[89] have provided evidence that the oxidation is accompanied by migration of a porphyrin nitrogen to the carbene atom to form 92.

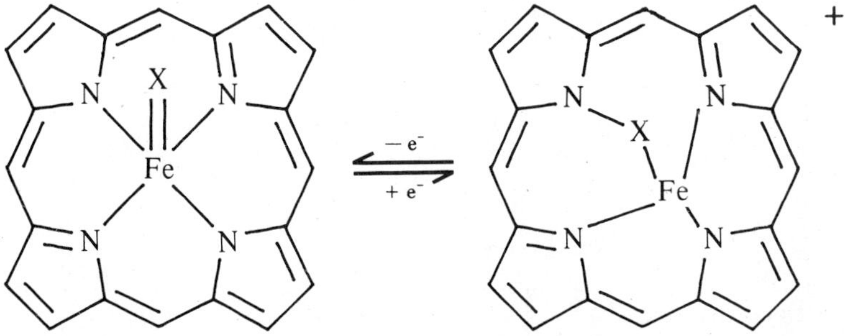

$$X = (p\text{-ClC}_6H_4)_2 C{=}C:$$

91   92

XIII.   BRIDGING CARBENE COMPLEXES

Since Hermann has recently published an excellent review on bridging carbene complexes,[1] only a few comments on these compounds are made here. The three most common routes to bridging carbene complexes are (1) the addition of diazo compounds to metal–metal double bonds, (2) the reaction of gem-diiodides with metal anions, and (3) the reaction of coordinatively unsaturated metal complexes with terminal carbene metal complexes.

The reaction of the cobalt–cobalt double-bonded compound 93 with diazomethane gave the methylene bridged compound 94. Thermolysis of 94 led to loss of CO and regeneration of a Co=Co double bond in 95. Reaction of 95 with diazoethane gave 96, which has both a bridging methylene and a bridging ethylidene ligand.[90] In cases where the required diazo compound was not readily available it could be generated in situ by MnO$_2$ oxidation of the corresponding hydrazone; this method was used for the synthesis of 97.[91]

The second general method for the synthesis of bridging carbene complexes is illustrated by the reaction of 1,1-diiodoethane with $Fe_2(CO)_8^{-2}$, which gives the bridging ethylidene complex **98**.[92] The third method is exemplified by the synthesis of **99** from $(CO)_5Cr=C(OCH_3)C_6H_5$, $Pt(CH_2CH_2)_3$, and PMe_3.[93, 94]

There is a strong tendency for carbenes without electron donor heteroatom substituents to assume a bridging rather than a terminal position in a bimetallic complex. Knox proposed that the cis-trans isomerization of **100** occurs by means of rotation about the metal–metal bond of an intermediate with a terminal carbene ligand.[95] The cis-trans isomerization of methylene complex **101** is much slower ($t_{1/2} = 10$ min at $36°C$) than that of $[C_5H_5(CO)_2Fe]_2$

(fast on NMR time scale at $-35°C$); the slow isomerization of **101** is due to the instability of an intermediate terminal methylene complex.[96] Whereas the monocarbene complex **102** has a terminal carbene ligand, probably due to the reluctance to form a carbonyl bridge, the bis carbene complex **103** has two bridging carbene ligands.[97]

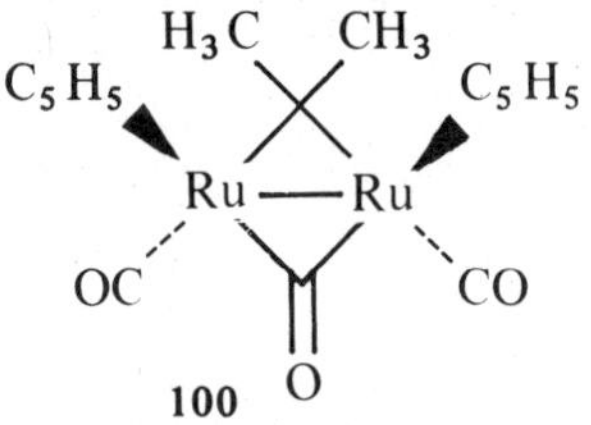

100

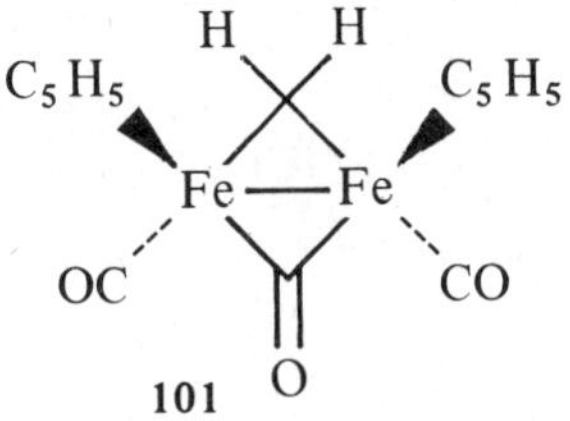

$(CO)_5 Re-Re(CO)_4$

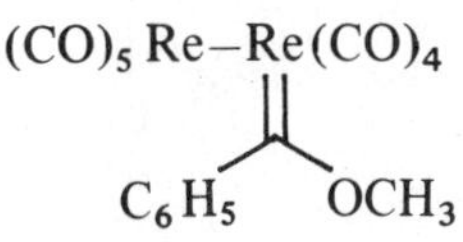

102

101

$(CO)_4 Re -\!\!-\!\!- Re(CO)_4$

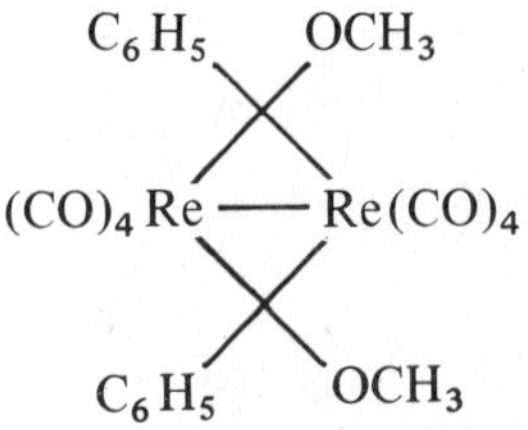

103

X-ray photoelectron spectroscopy studies by Lichtenberger[98] and by Jolly[99] have established that the bridging methylene carbon is highly negatively charged. This may explain why bridging methylene complexes can undergo protonation at the bridging carbon.[96] The resulting bridging methyl compounds, such as **104**, have been shown by NMR and X-ray crystallography to have a three-center 2-electron M — H — C interaction.[100] In contrast, Hermann has shown that protonation of a methylene-bridged rhodium complex occurs selectively at the metal—metal bond to give **105**.[101]

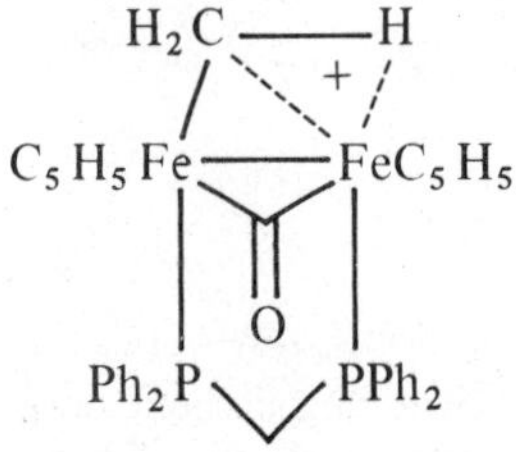

104

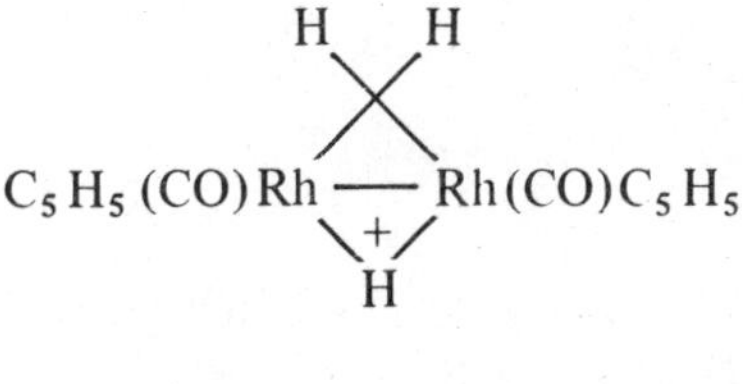

105

Casey and Fagan have shown that the bridging methylene compound **106** undergoes hydride abstraction by $(C_6H_5)_3C^+$ to produce the first bridging methyne complex, **107**, which undergoes an unprecedented addition of its bridging C–H bond to alkenes.[96, 102]

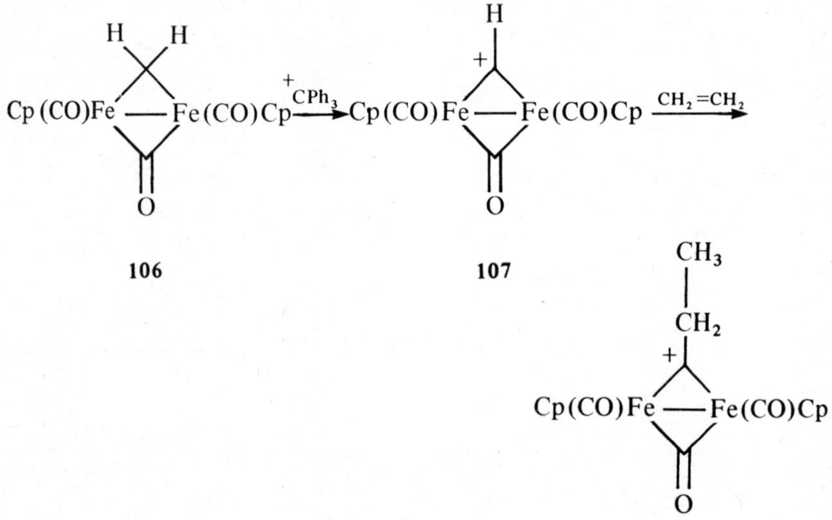

106   107

Bergman and Theopold have presented evidence for a bizarre rearrangement of the bridging 3-pentylidene dicobalt compound **108** to the corresponding 1-pentylidene complex, **109**. The reaction is autocatalytic, indicating that loss of CO is a likely primary reaction step. Isotope labeling studies indicate that both metal/hydrogen migration and π/π-allyl interconversions are occurring in this rearrangement.[103]

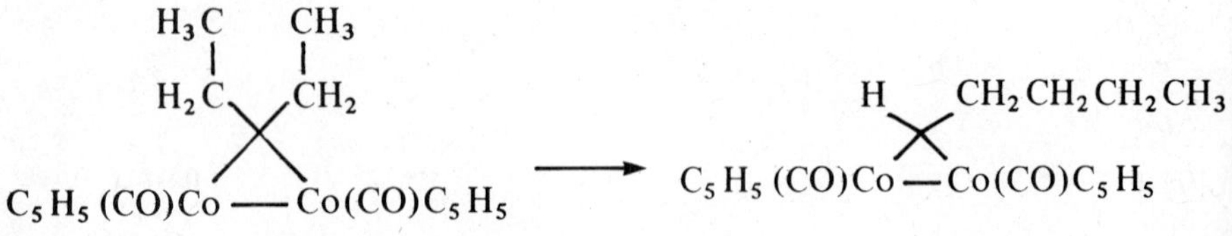

108   109

Katz has shown that $(CO)_5W=C(OCH_3)C_6H_5$ and $(CO)_5W=C(C_6H_5)_2$ can be efficient catalysts for alkyne polymerization.[104] Bridging carbene complexes may also be involved in alkyne polymerization. Rudler has observed 2-butyne insertion into the bridging carbene complex **110**, which gave **111**.[105] Similarly,

Knox has observed insertion of acetylene into the bridging ethylidene complex **112** under photolytic conditions.[106]

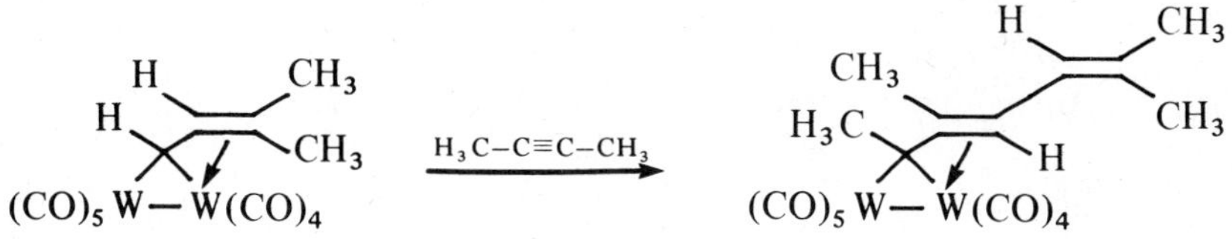

$$ \mathbf{110} \xrightarrow{H_3C-C\equiv C-CH_3} \mathbf{111} $$

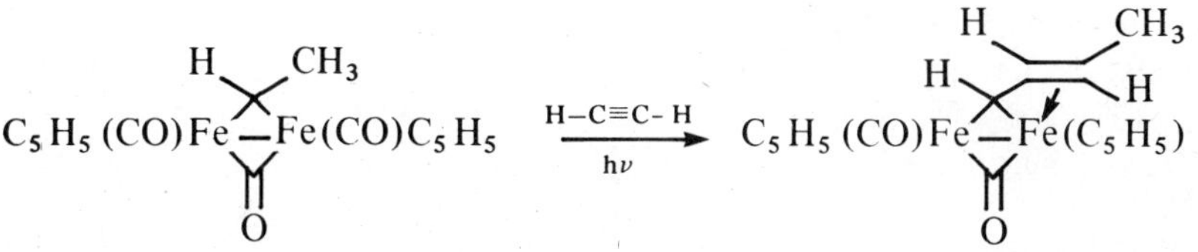

$$ \mathbf{112} \xrightarrow[h\nu]{H-C\equiv C-H} $$

XIV.   NOVEL REACTIONS OF METAL–CARBENE COMPLEXES

Two mechanisms for metal-catalyzed polymerization of ethylene seem able to explain the available data. The Cossee mechanism involves ethylene insertion into a metal–alkyl bond. Green and Rooney proposed an alternate mechanism involving insertion of the metal into an α–CH bond of a metal alkyl, and carbon–carbon bond formation by way of a metallacyclobutane.

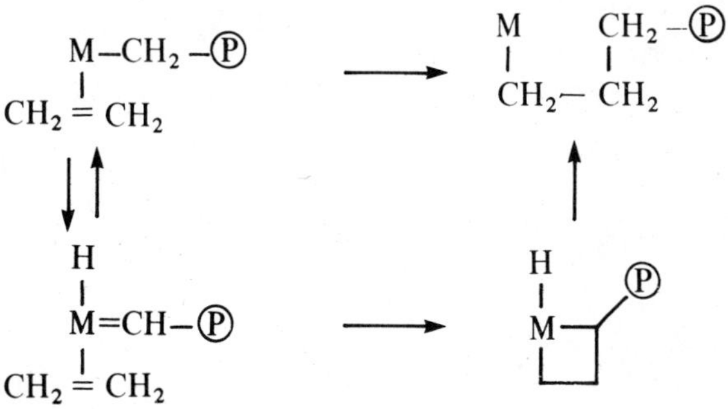

Schrock has now discovered that the tantalum alkylidene hydride, **113**, polymerizes ethylene and that intermediate insertion products are directly observable by NMR.[107] This reaction might proceed either by the Green–Rooney mechanism or by transformation of the alkylidene hydride complex to an alkyl complex prior to carbon–carbon bond formation.

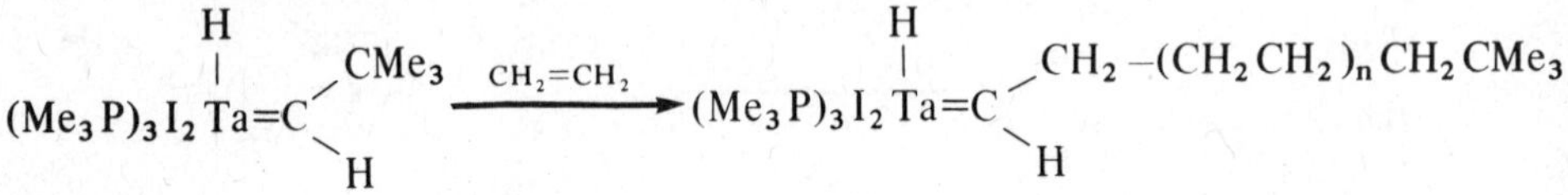

$$
\underset{\textbf{113}}{(Me_3P)_3I_2Ta=C\overset{\overset{\displaystyle H}{|}}{\underset{\displaystyle H}{\diagup}}\diagdown CMe_3}
\quad\xrightarrow{\ CH_2=CH_2\ }\quad
(Me_3P)_3I_2Ta=C\overset{\overset{\displaystyle H}{|}}{\underset{\displaystyle H}{\diagup}}\diagdown CH_2-(CH_2CH_2)_nCH_2CMe_3
$$

In an attempt to distinguish between these two mechanisms for ethylene polymerization, Grubbs has determined the isotope effect on the polymerization of $CH_2=CH_2$ and $CD_2=CD_2$ catalyzed by $(C_6H_5)_2Ti(C_2H_5)Cl$ and $Al(C_2H_5)Cl_2$.[108] The small value of $k_H/k_D = 1.04$ observed is in agreement with the Cossee mechanism; a larger isotope effect would have been expected for the Green–Rooney mechanism which involves insertion into a C–H bond.

Cramer has observed carbon monoxide insertion into the uranium carbene bond of **114** which leads to the η^2-ketene complex **115**.[109] Hegedus has found that photolysis of chromium carbene complexes in the presence of imines leads to the formation of β-lactams such as **116**. This reaction, which may involve carbene–carbon monoxide coupling to give ketenes or ketene complexes, could have applications in the synthesis of biologically active β-lactams.[110]

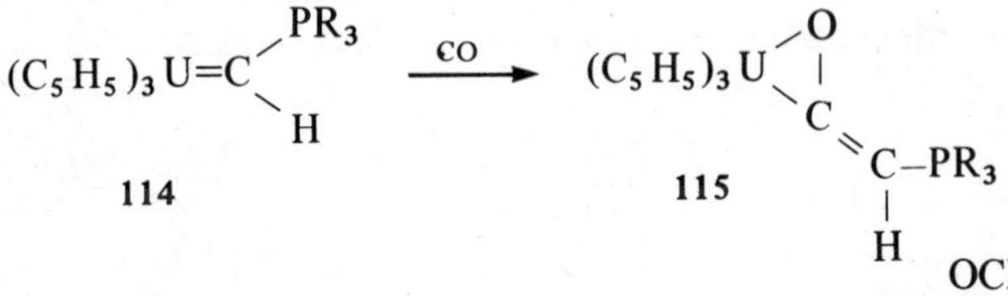

$$
\underset{\textbf{114}}{(C_5H_5)_3U=C\overset{\diagup PR_3}{\diagdown H}}
\quad\xrightarrow{\ CO\ }\quad
\underset{\textbf{115}}{(C_5H_5)_3U}
$$

$$
\underset{}{(CO)_5Cr=C\overset{\diagup OCH_3}{\diagdown CH_3}} \ + \ H-\underset{N}{\overset{S}{\diagup\!\!\diagdown}} \ \longrightarrow \ \underset{\textbf{116}}{}
$$

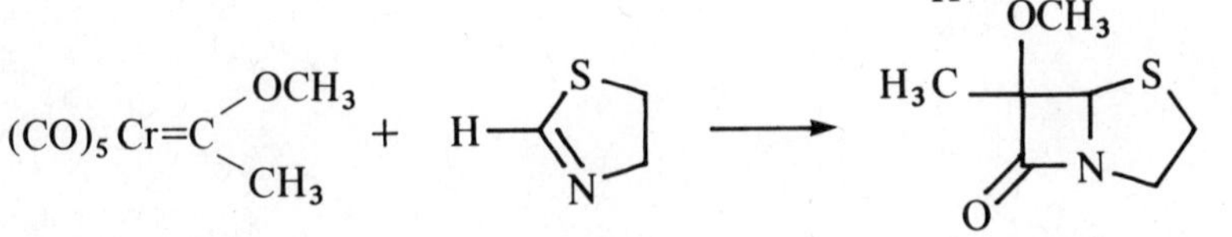

Other interesting new reactions include Watanabe's observation that the vinyl carbene complex **117** reacts with $Fe_2(CO)_9$ to give the ferrole **118**[111] and Roper's synthesis of selenocarbonyl and tellurocarbonyl complexes from the dichlorocarbene complex **119**.[112]

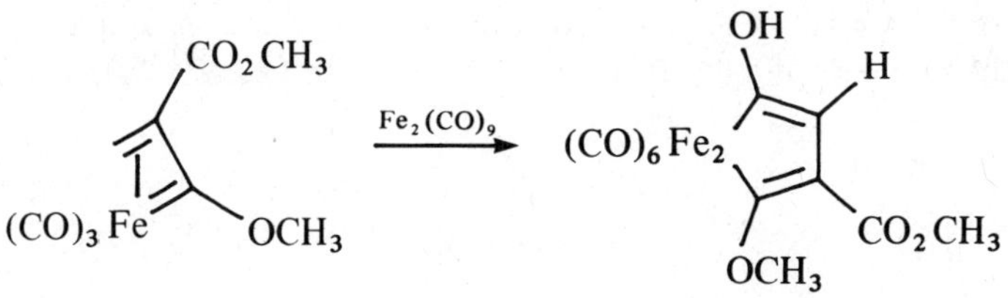

117 **118**

$$(Ph_3P)_2(CO)Cl_2Os{=}CCl_2 \xrightarrow{\text{H-Te}^-} (Ph_3P)_2(CO)Cl_2Os{-}C{\equiv}Te$$

119

XV. NEW SYNTHESES OF METAL–CARBENE COMPLEXES

Fischer has shown that hydride addition to cationic carbyne complexes can lead to carbene complexes such as **120**.[113] Now Schrock[114] and Roper[115] have reported examples of the synthesis of carbene complexes **121** and **122** by protonation of precursor carbyne complexes.

$$C_5H_5(CO)_2\overset{+}{Re}{\equiv}C{-}C_6H_5 \xrightarrow{Et_2AlH} C_5H_5(CO)_2Re{=}C\begin{smallmatrix}H\\ C_6H_5\end{smallmatrix}$$

120

$$Cl_4\overset{-}{W}{\equiv}C{-}CMe_3 \xrightarrow[H_2O]{PEt_3} (Et_3P)_2Cl_2\overset{\overset{O}{\|}}{W}{=}C\begin{smallmatrix}H\\ CMe_3\end{smallmatrix}$$

121

$$(Ph_3P)_2(CO)ClOs{\equiv}C{-}C_6H_4\text{-}p{-}CH_3 \xrightarrow{HCl}$$

$$(Ph_3P)_2(CO)Cl_2Os{=}C\begin{smallmatrix}C_6H_4\text{-}p{-}CH_3\\ H\end{smallmatrix}$$

122

Cooper has found that iminium salts react with metal carbonyl anions and used this method for the synthesis of the very unstable diphenylcarbene molybdenum complex **123**.[116] Schwartz discovered that phosphoranes can be empolyed to transfer methylene groups to early transition metals and used this method in the synthesis of methylene zirconium complex **124**.[117]

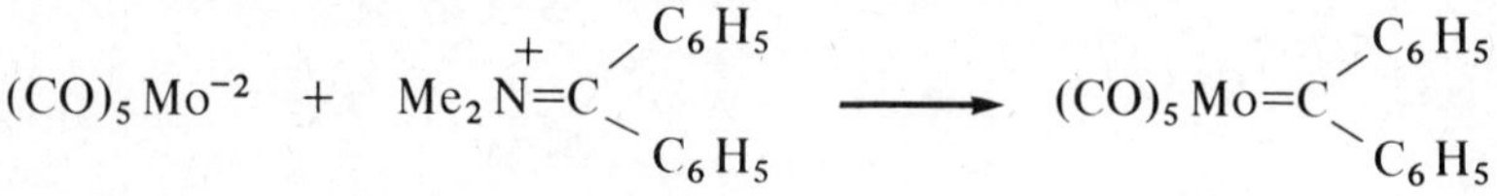

$$(CO)_5 Mo^{-2} \; + \; Me_2 N = \overset{+}{C} \overset{C_6 H_5}{\underset{C_6 H_5}{\diagdown}} \longrightarrow (CO)_5 Mo = C \overset{C_6 H_5}{\underset{C_6 H_5}{\diagdown}}$$

123

$$(C_5 H_5)_2 Zr \overset{P(Ph)_2 CH_3}{\underset{P(Ph)_2 CH_3}{\diagdown}} \; + \; CH_2 = PPh_3 \longrightarrow (C_5 H_5)_2 Zr \overset{CH_2}{\underset{P(Ph)_2 CH_3}{\diagdown}}$$

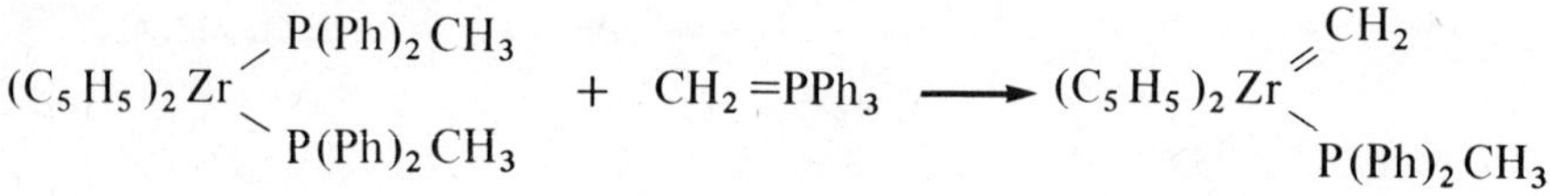

124

Green has observed that HBR_3^- reacts with the cationic molybdenum acetylene complex **125** to produce the metallacyclopropene complex **126**, whose structure was established by X-ray crystallography.[118] The metallacyclopropene is isomeric with a coordinatively unsaturated σ-vinyl complex, and this was suggested as an intermediate in the formation of **126**. However, attack of hydride at the metal or at the acetylene could also lead more directly to **126**. Silicon migration in the related complex, **127**, leads to carbyne complex **128**.[119]

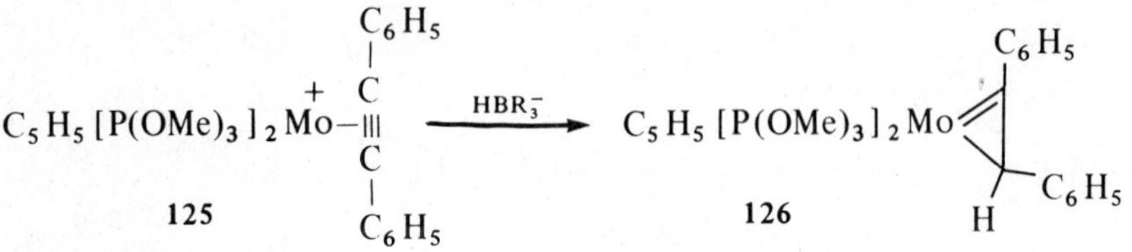

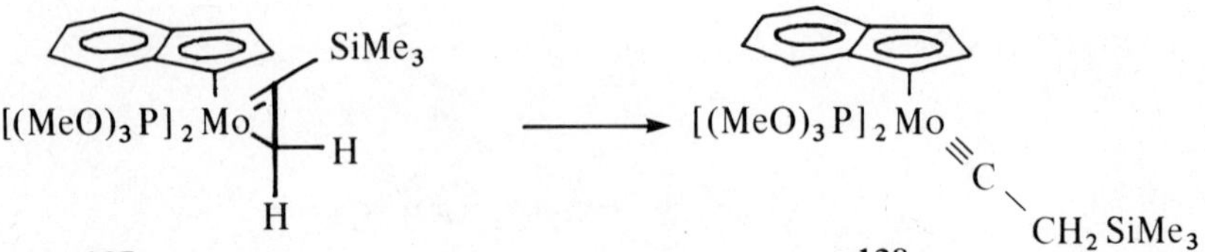

Steinmetz and Geoffroy found that $HB(OR)_3^-$ reduction of $Os_3(CO)_{12}$ gave an unstable metal–formyl complex, which, upon treatment with $H_3 PO_4$ at $0°C$, led to the formation of the bridging methylene complex $Os_3(CO)_{10}(\mu\text{-}CH_2)$ $(\mu\text{-}CO)$ by way of a complex set of disproportionation reactions.[120]

XVI.   NOVEL CARBENE COMPLEXES

Novel carbene complexes continue to be made at a rapid rate. Interesting new complexes include **129**, which contains an electron-donor cyclopropane ring,[121] **130** which has a chelating dicarbene unit,[122] and **131** which is one of the few known M=CHOR complexes.[123] The $(C_5H_5)Mn(CO)_2$ unit is intriguing because it is good at stabilizing carbene complexes with either electron-donating groups or electron-withdrawing groups such as the new complex **132**.[124] Caulton has prepared phenylcarbene complex **133**,[125] which should be interesting in comparison with $(CO)_5W=CHC_6H_5$.

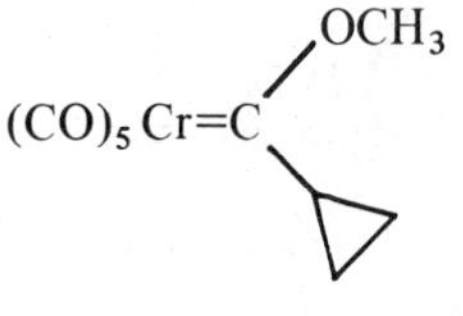

$(CO)_5Cr=C$ — OCH_3

129

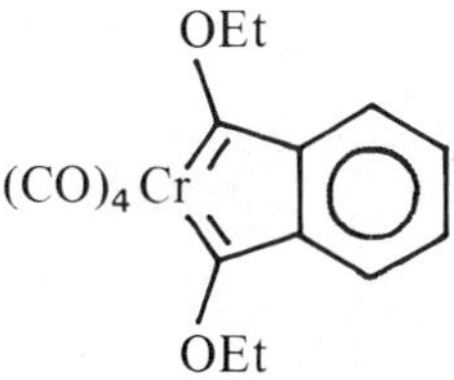

$(CO)_4Cr$ — OEt / OEt

130

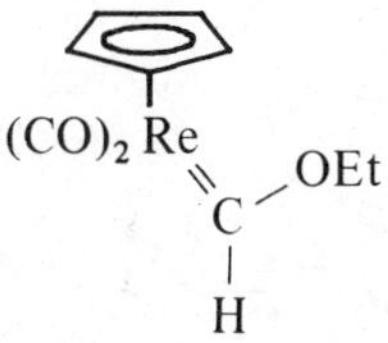

$(CO)_2Re$ — C — OEt / H

131

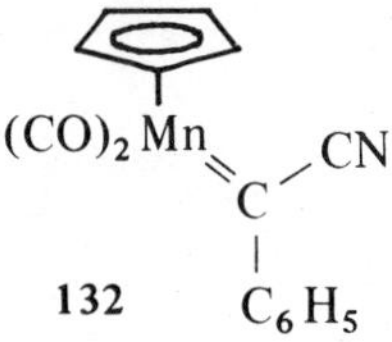

$(CO)_2Mn$ — C — CN / C_6H_5

132

$(C_5H_5)_2W=C$ — H / C_6H_5

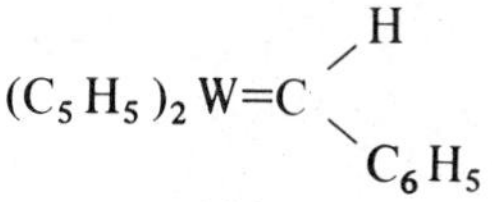

133

Probably the most interesting new carbene complex is the osmabenzene **134**, synthesized by Roper.[126] Reaction of acetylene with thiocarbonyl complex **135** gave adduct **136**, which was alkylated at sulfur by CH_3I.

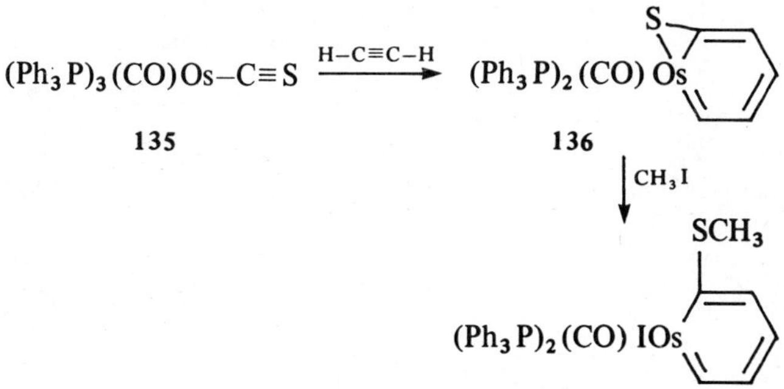

$(Ph_3P)_3(CO)Os-C\equiv S$ $\xrightarrow{H-C\equiv C-H}$ $(Ph_3P)_2(CO)Os$

135   **136**

$\downarrow CH_3I$

$(Ph_3P)_2(CO)IOs$ — SCH_3

134

XVII.   REFERENCES

*1.   W. A. Herrmann, *Adv. Organomet. Chem.,* **20**, 159 (1982).

2.   F. J. Brown, *Prog. Inorg. Chem.,* **27**, 1 (1980).

3.   L. S. Hegedus, *J. Organomet. Chem.,* **237**, 332 (1982).

4.   R. R. Schrock, *Acc. Chem. Res.,* **12**, 98 (1979).

5.   R. R. Schrock, *Science,* **219**, 13 (1983).

6.   D. Mansuy, *Pure Appl. Chem.,* **52**, 681 (1980).

7.   B. A. Dolgoplosk, *J. Polym. Sci., Polym. Symp.,* **67**, 99 (1980).

8.   P. W. Jolly and R. Pettit, *J. Am. Chem. Soc.,* **88**, 5044 (1966).

9.   M. Brookhart, J. R. Tucker, T. C. Flood, and J. Jensen, *J. Am. Chem. Soc.,* **102**, 1203 (1980).

10.   S. Brandt and P. Helquist, *J. Am. Chem. Soc.,* **101**, 6473 (1979).

11.   M. Brookhart and G. O. Nelson, *J. Am. Chem. Soc.,* **99**, 6099 (1977).

12.   M. Brookhart, M. B. Humphrey, H. J. Kratzer, and G. O. Nelson, *J. Am. Chem. Soc.,* **102**, 7802 (1980).

13.   M. Brookhart, J. R. Tucker, and G. R. Husk, *J. Am. Chem. Soc.,* **103**, 979 (1981).

14.   T. Bodnar and A. R. Cutler, *J. Organomet. Chem.,* **213**, C31 (1981).

15.   K. A. M. Kremer, P. Helquist, and R. C. Kerber, *J. Am. Chem. Soc.,* **103**, 1862 (1981).

*16.   C. P. Casey, W. H. Miles, H. Tukada, and J. M. O'Connor, *J. Am. Chem. Soc.* **104**, 3761 (1982).

*17.   K. A. M. Kremer, G-H. Kuo, E. J. O'Connor, P. Helquist, and R. C. Kerber, *J. Am. Chem. Soc.,* **104**, 6119 (1982).

18.   S. E. Kegley, M. Brookhart, and G. R. Husk, *Organometallics,* **1**, 760 (1982).

19.   K. C. Bishop III, *Chem. Rev.,* **76**, 461 (1976).

20.   H. Tanaka, T. Suzuki, Y. Kumagai, M. Hosoya, H. Kawauchi, and R. Noyori, *J. Org. Chem.,* **46** 2854 (1981).

*21.   W. A. Kiel, G-Y. Lin, A. G. Constable, F. B. McCormick, C. E. Strouse, O. Eisenstein and J. A. Gladysz, *J. Am. Chem. Soc.,* **104**, 4865 (1982).

22.   W. Tam, G-Y. Lin, W-K. Wong, W. A. Kiel, V. K. Wong, and J. A. Gladysz, *J. Am. Chem. Soc.,* **104**, 141 (1982).

23.   H. Berke and G. Weiler, *Angew. Chem.,* **94**, 135 (1982).

24.   M. I. Bruce, F. S. Wong, B. W. Skelton, and A. H. White, *J. Chem. Soc., Dalton Trans.,* 2203 (1982).

25.   P. M. Treichel and D. A. Komar, *Inorg. Chim. Acta,* **42**, 277 (1980).

26.   H. Berke, *Z. Naturforsch.,* **35b**, 86 (1980).

27.   B. E. Boland-Lussier, M. R. Churchill, R. P. Hughes and A. L. Rheingold, *Organometallics,* **1**, 628 (1982).

28.   B. E. Boland-Lussier and R. P. Hughes, *Organometallics,* **1**, 635 (1982).

29.   H. Berke, P. Härter, G. Huttner and L. Zsolnai, *Z. Naturforsch.,* **36b**, 929 (1981).

30.   H. Berke, P. Härter, G. Huttner, and J. v. Seyerl, *J. Organomet. Chem.,* **219**, 317 (1981).

31.   H. Berke, P. Härter, G. Huttner, and L. Zsolnai, *Chem. Ber.,* **115**, 695 (1982).

32. J. P. Selegue, *Organometallics*, **1**, 217 (1982).

33. H. Berke and P. Härter, *Angew. Chem., Int. Ed. (Engl.)*, **19**, 225 (1980).

34. N. J. Cooper and M. L. H. Green, *J. Chem. Soc., Chem. Commun.*, 761 (1974).

*35. M. Canestrari and M. L. H. Green, *J. Chem. Soc., Dalton Trans.*, 1789 (1982).

36. Z. Dawoodi, M. L. H. Green, V. S. B. Mtetwa, and K. Prout, *J. Chem. Soc., Chem. Commun.*, 1410 (1982).

37. J. D. Fellmann, H. W. Turner and R. R. Schrock, *J. Am. Chem. Soc.*, **102**, 6608 (1980).

38. A. J. Schultz, J. M. Williams, R. R. Schrock, G. A. Rupprecht, and J. D. Fellmann, *J. Am. Chem. Soc.*, **101**, 1593 (1979).

39. R. J. Goddard, R. Hoffmann, and E. D. Jemmis, *J. Am. Chem. Soc.*, **102**, 7667 (1980).

*40. J. D. Fellmann, R. R. Schrock, and D. D. Traficante, *Organometallics*, **1**, 481 (1982).

*41. R. S. Threlkel and J. E. Bercaw, *J. Am. Chem. Soc.*, **103**, 2650 (1981).

42. M. R. Churchill, H. J. Wasserman, H. W. Turner, and R. R. Schrock, *J. Am. Chem. Soc.*, **104**, 1710 (1982).

*43. S. J. Holmes, D. N. Clark, H. W. Turner, and R. R. Schrock, *J. Am. Chem. Soc.*, **104**, 6322 (1982).

44. P. B. Armentrout and J. L. Beauchamp, *J. Chem. Phys.*, **74**, 2819 (1981).

45. L. F. Halle, P. B. Armentrout, and J. L. Beauchamp, *J. Am. Chem. Soc.*, **103**, 962 (1981).

*46. R. C. Brady and R. Pettit, *J. Am. Chem. Soc.*, **102**, 6181 (1980).

47. R. C. Brady and R. Pettit, *J. Am. Chem. Soc.*, **103**, 1287 (1981).

48. P. R. Sharp and R. R. Schrock, *J. Organomet. Chem.*, **171**, 43 (1979).

49. J. C. Hayes, G. D. N. Pearson, and N. J. Cooper, *J. Am. Chem. Soc.*, **103**, 4648 (1981).

*50. J. C. Hayes, N. J. Cooper, *J. Am. Chem. Soc.*, **104**, 5570 (1982).

51. K. S. Chong and M. L. H. Green, *J. Chem. Soc., Chem. Commun.*, 991 (1982).

52. K. S. Chong and M. L. H. Green, *Organometallics*, **1**, 1586 (1982).

53. W. A. Kiel, G.-Y. Lin and J. A. Gladysz, *J. Am. Chem. Soc.*, **102**, 3299 (1980).

54. D. L. Thorn and T. H. Tulip, *J. Am. Chem. Soc.*, **103**, 5984 (1981).

*55. A. K. Rappé and W. A. Goddard, III, *J. Am. Chem. Soc.*, **104**, 448 (1982).

56. R. R. Schrock, S. Rocklage, J. Wengrovius, G. Rupprecht, and J. Fellmann, *J. Mol. Catal.*, **8**, 73 (1980).

57. S. M. Rocklage, J. D. Fellmann, G. A. Rupprecht, L. W. Messerle, and R. R. Schrock, *J. Am. Chem. Soc.*, **103**, 1440 (1981).

*58. J. H. Wengrovius, R. R. Schrock, M. R. Churchill, J. R. Missert, and W. J. Youngs, *J. Am. Chem. Soc.*, **102**, 4515 (1980).

59. M. R. Churchill and A. L. Rheingold, *Inorg. Chem.*, **21**, 1357 (1982).

60. J. H. Wengrovius and R. R. Schrock, *Organometallics*, **1**, 148 (1982).

61. J. Kress, M. Wesolek, J.-P. LeNy, and J. A. Osborn, *J. Chem. Soc., Chem. Commun.*, 1039 (1981).

*62. J. Kress, M. Wesolek, and J. A. Osborn, *J. Chem. Soc., Chem. Commun.*, 514 (1982).

63.   F. N. Tebbe, G. W. Parshall, and P. W. Ovenall, *J. Am. Chem. Soc.*, **101**, 5074 (1979).

64.   T. R. Howard, J. B. Lee, and R. H. Grubbs, *J. Am. Chem. Soc.*, **102**, 6876 (1980).

65.   J. B. Lee, G. J. Gajda, W. P. Schaefer, T. R. Howard, T. Ikariya, D. A. Straus, and R. H. Grubbs, *J. Am. Chem. Soc.*, **103**, 7358 (1981).

66.   O. Eisenstein, R. Hoffmann, and A. R. Rossi, *J. Am. Chem. Soc.*, **103**, 5582 (1981).

67.   K. C. Ott, J. B. Lee, and R. H. Grubbs, *J. Am. Chem. Soc.*, **104**, 2942 (1982).

*68.   J. B. Lee, K. C. Ott, and R. H. Grubbs, *J. Am. Chem. Soc.*, **104**, 7491 (1982).

69.   K. C. Ott and R. H. Grubbs, *J. Am. Chem. Soc.*, **103**, 5922 (1981).

70.   S. H. Pine, R. Zahler, D. A. Evans, and R. H. Grubbs, *J. Am. Chem. Soc.*, **102**, 3270 (1980).

71.   C. P. Casey and A. J. Shusterman, *J. Mol. Catal.*, **8**, 1 (1980).

*72.   C. P. Casey and A. J. Shusterman, N. W. Vollendorf, and K. J. Haller, *J. Am. Chem. Soc.*, **104**, 2417 (1982).

73.   C. A. Toledano, J. Levisalles, M. Rudler, H. Rudler, J.-C. Daran, and Y. Jeannin, *J. Organomet. Chem.*, **228**, C7 (1982).

74.   C. P. Casey, N. W. Vollendorf, A. J. Shusterman, and K. J. Haller, *Abstracts of Papers*, 184th National Meeting of the American Chemical Society, Kansas City, Mo., Sept. 1982; American Chemical Society, Washington, D.C., 1982; INOR 67.

75.   D. L. Thorn, *Organometallics*, **1**, 879 (1982).

76.   A. J. Schultz, R. K. Brown, J. M. Williams, and R. R. Schrock, *J. Am. Chem. Soc.*, **103**, 169 (1981).

77.   K. Nakatsu, T. Mitsudo, H. Nakanishi, Y. Watanabe, and Y. Takegami, *Chem. Lett.*, 1447 (1977).

78.   K. H. Dötz, *Angew. Chem., Int. Ed. (Engl).*, **14**, 644 (1975).

*79.   H. Fischer, J. Mühlemeier, R. Märkl and K. H. Dötz, *Chem. Ber.*, **115**, 1355 (1982).

80.   K. H. Dötz and B. Fügen-Köster, *Chem. Ber.*, **113**, 1449 (1980).

*81.   W. D. Wulff, P.-C. Tang, and J. S. McCallum, *J. Am. Chem. Soc.*, **103**, 7677 (1981).

82.   K. H. Dötz, I. Pruskil, and J. Mühlemeier, *Chem. Ber.*, **115**, 1278 (1982).

83.   M. F. Semmelhack, J. J. Bozell, T. Sato, W. Wulff, E. Spiess, and A. Zask, *J. Am. Chem. Soc.*, **104**, 5850 (1982).

84.   D. Mansuy, M. Lange, J. C. Chottard, P. Guerin, P. Morliere, D. Brault, and M. Rougee, *J. Chem. Soc., Chem. Commun.*, 648 (1977).

85.   D. Mansuy and J-P. Battioni, *J. Chem. Soc., Chem. Commun.*, 638 (1982).

86.   D. Mansuy, M. Lange, and J. C. Chottard, *J. Am. Chem. Soc.*, **100**, 3213 (1978).

*87.   D. Mansuy, M. Lange, and J. C. Chottard, *J. Am. Chem. Soc.*, **101**, 6437 (1979).

88.   L. Latos-Grazynski, R-J. Cheng, G. N. LaMar, and A. L. Balch, *J. Am. Chem. Soc.*, **103**, 4270 (1981).

89.   H. M. Goff and M. A. Phillippi, *Inorg. Nucl. Chem. Lett.*, **17**, 239 (1981).

90.   W. A. Herrmann, J. M. Huggins, C. Bauer, M. Smischek, H. Pfisterer, and M. L. Ziegler, *J. Organomet. Chem.*, **226**, C59 (1982).

91.   W. A. Herrmann, C. Bauer, and K. K. Mayer, *J. Organomet. Chem.*, **236**, C18 (1982).

92.   C. E. Sumner, Jr., J. A. Collier, and R. Pettit, *Organometallics*, **1**, 1350 (1982).

93. For a recent review, see T. V. Ashworth, M. J. Chetcuti, L. J. Farrugia, J. A. K. Howard, J. C. Jeffery, R. Mills, G. N. Pain, F. G. A. Stone, and P. Woodward, *ACS Symp. Ser.* No. 155, pp. 299–313 (1981).

94. T. V. Ashworth, J. A. K. Howard, M. Laguna, and F. G. A. Stone, *J. Chem. Soc., Dalton Trans.,* 1593 (1980).

95. A. F. Dyke, S. A. R. Knox, K. A. Mead, and P. Woodward, *J. Chem. Soc., Chem. Commun.,* 861 (1981).

*96. C. P. Casey, P. J. Fagan, and W. H. Miles, *J. Am. Chem. Soc.,* **104**, 1134 (1982).

97. E. O. Fischer, T. L. Lindner, H. Fischer, G. Huttner, P. Friedrich, and F. R. Kreissl, *Z. Naturforsch.,* **32B**, 648 (1977).

98. D. C. Calabro, D. L. Lichtenberger, and W. A. Herrmann, *J. Am. Chem. Soc.,* **103**, 6852 (1981).

99. S. F. Xiang, H. W. Chen, C. J. Eyermann, W. L. Jolly, S. P. Smit, K. H. Theopold, R. G. Bergman, W. A. Herrmann, and R. Pettit, *Organometallics,* **1**, 1200 (1982).

100. G. M. Dawkins, M. Green, A. G. Orpen, and F. G. A. Stone, *J. Chem. Soc., Chem. Commun.,* 41 (1982).

101. W. A. Herrmann, J. Plank, D. Riedel, M. L. Ziegler, K. Weidenhammer, E. Guggolz, and B. Balbach, *J. Am. Chem. Soc.,* **103**, 63 (1981).

*102. C. P. Casey and P. J. Fagan, *J. Am. Chem. Soc.,* **104**, 4950 (1982).

*103. K. H. Theopold and R. G. Bergman, *Organometallics,* **1**, 219 (1982).

104. T. J. Katz and S. J. Lee, *J. Am. Chem. Soc.,* **102**, 422 (1980).

105. J. Levisalles, F. Rose-Munch, H. Rudler, J-C. Daran, Y. Dromzee, Y. Jeannin, D. Ades, and M. Fontanille, *J. Chem. Soc., Chem. Commun.,* 1055 (1981).

106. A. F. Dyke, S. A. R. Knox, P. J. Naish, and G. E. Taylor, *J. Chem. Soc., Chem. Commun.,* 803 (1980).

*107. H. W. Turner and R. R. Schrock, *J. Am. Chem. Soc.,* **104**, 2331 (1982).

*108. J. Soto, M. L. Steigerwald, and R. H. Grubbs, *J. Am. Chem. Soc.,* **104**, 4479 (1982).

109. R. E. Cramer, R. B. Maynard, J. C. Paw, and J. W. Gilje, *Organometallics,* *1*, 869 (1982).

110. M. A. McGuire and L. S. Hegedus, *J. Am. Chem. Soc.,* **104**, 5538 (1982).

111. T. Mitsudo, H. Watanabe, K. Watanabe, Y. Watanabe, K. Kafuku, and K. Nakatsu, *Organometallics,* **1**, 612 (1982).

112. G. R. Clark, K. Marsden, W. R. Roper, and L. J. Wright, *J. Am. Chem. Soc.,* **102**, 1206 (1980).

113. E. O. Fischer and A. Frank, *Chem. Ber.,* **111**, 3740 (1978).

114. S. M. Rocklage, R. R. Schrock, M. R. Churchill, and H. J. Wasserman, *Organometallics,* **1**, 1332 (1982).

115. G. R. Clark, K. Marsden, W. R. Roper, and L. J. Wright, *J. Am. Chem. Soc.* **102**, 6570 (1980).

116. R. P. Beatty, J. M. Maher, and N. J. Cooper, *J. Am. Chem. Soc.,* **103**, 238 (1981).

117. J. Schwartz and K. I. Gell, *J. Organomet. Chem.,* **184**, C1 (1980).

118. M. Green, N. C. Norman, and A. G. Orpen, *J. Am. Chem. Soc.,* **103**, 1267 (1981).

119. S. R. Allen, M. Green, A. G. Orpen, and I. D. Williams, *J. Chem. Soc., Chem. Commun.,* 826 (1982).

120.   G. R. Steinmetz and G. L. Geoffroy, *J. Am. Chem. Soc.,* **103**, 1278 (1981).

121.   E. O. Fischer, N. H. Tran-Huy, and D. Neugebauer, *J. Organomet. Chem.,* **229**, 169 (1982).

*122.   E. O. Fischer, W. Röll, N. H. T. Huy, and K. Ackermann, *Chem. Ber.,* **115**, 2951 (1982).

123.   E. O. Fischer, P. Rustemeyer, and K. Ackermann, *Chem. Ber.,* **115**, 3851 (1982).

124.   E. O. Fischer and W. Schambeck, *J. Organomet. Chem.,* **201**, 311 (1980).

125.   J. A. Marsella, K. Folting, J. C. Huffman, and K. G. Caulton, *J. Am. Chem. Soc.,* **103**, 5596 (1981).

*126.   G. P. Elliott, W. R. Roper, and J. M. Waters, *J. Chem. Soc., Chem. Commun.,* 811 (1982).

5

DIRADICALS

WESTON THATCHER BORDEN

Department of Chemistry, University of Washington, Seattle, Washington, 98195

I.   INTRODUCTION

This review, which covers the period 1980–1982, updates the one that appeared in Volume 2 of this series.[1] Like the previous review, this chapter discusses nonconjugated diradicals (e.g., tri- and tetramethylene), antiaromatic annulenes (e.g., cyclobutadiene), and non-Kekulé hydrocarbons (e.g., trimethylenemethane). In all of these areas the coverage given here is selective rather than exhaustive. A collection of reviews of the major topics in diradical chemistry, with coverage through 1981 and some references in 1982, has been published.[2] Hydrocarbon thermal isomerizations, many of which appear to involve diradicals as intermediates or transition states, have been reviewed in a chapter by Berson which appeared in 1980[3] and in a monograph by Gajewski, published in 1981.[4]

II. NONCONJUGATED DIRADICALS[5]

A.   Trimethylene and Related 1,3-Diradicals

As discussed in the previous review of diradicals in this series[1] and in the chapter by Berson,[4] a major unsolved problem in trimethylene chemistry is the discrepancy between thermochemistry and quantum mechanics in the role assigned to this diradical in cyclopropane stereomutations. Benson's thermochemical calculations place trimethylene nearly 10 kcal/mol below the transition state connecting it with cyclopropane.[6] In contrast, a variety of quantum mechanical calculations find little or no barrier to the collapse of trimethylene to cyclopropane.[7] Thus thermochemistry assigns to trimethylene the role of an intermediate, whereas quantum mechanics finds trimethylene to be a transition state for cyclopropane stereomutations.

Doering has recently suggested a resolution of the apparent discrepancy between thermochemistry and quantum mechanics.[8] The thermochemical estimate of the heat of formation of trimethylene is based on the heat of formation of propane and the enthalpy change associated with breaking a C—H bond at each of the terminal carbons. Accepting the premise that the energy required to break the second C—H bond is the same as that for the first, a value for the dissociation energy of a primary C—H bond in propane is required. Doering has argued that the C—H bond dissociation energy used by Benson is too low, based on more recent determinations of the heat of formation of the ethyl radical. Upward revision of the heat of formation of ethyl by 2.6 kcal/mol from the value used by Benson increases the estimated heat of formation of trimethylene from 67.0 to 72.2 kcal/mol.

The heat of formation of the transition state for generation of trimethylene by cyclopropane ring opening is obtained from the heat of formation of cyclopropane and the Arrhenius parameters for its thermolysis. Benson proposed that both geometrical isomerization and rearrangement of cyclopropane to propene involve formation of trimethylene. The activation enthalpy of 63.5 kcal/mol for rearrangement gives a heat of formation of the transition state of 76.3 kcal/mol.

However, subsequent experiments by Rabinovitch indicate that the activation energy for geometrical isomerization of cyclopropane-d_2 is lower than that for rearrangement to propene by 3.7 kcal/mol.[9] A heat of formation of 72.6 kcal/mol for the transition state for geometrical isomerization is then obtained.

The upward revision in the estimated heat of formation of trimethylene and the downward revision of the heat of formation of a transition state that presumably leads to this species leaves only a 0.4 kcal/mol energy difference between them. This difference should certainly not be considered significant in light of the many uncertainties in the two thermochemical estimates. Thus, as noted by Doering these revisions result in the disappearance of the "Benson activation energy" for trimethylene closure to cyclopropane.[8]

Hydrogen shift products, analogous to propene, are formed in the thermal and direct photochemical decomposition of pyrazolines, presumably by way of intermediate singlet 1,3-diradicals. However, sensitized photolyses of the same azo compounds yield cyclopropanes almost exclusively. Doubleday, McIver and Smith have carried out *ab initio* calculations to explain this observation.[10]

Doubleday et al. computed the energies of the lowest singlet and triplet surfaces for trimethylene as a function of the rotation angles for each of the two terminal methylene groups. Restricted Hartree Fock (RHF) calculations were carried out for the triplet, and two-configuration self-consistent field (TCSCF) calculations were performed for the singlet. Doubleday et al. found that the STO-3G minimal basis set failed to describe adequately the long-range interactions between the terminal methylene groups owing to the absence of diffuse functions in this minimal basis set. Therefore a split valence (SV) 3-21G basis set was used.

From the computed surfaces Doubleday et al. obtained the points at which the two surfaces intersect and at which singlet-triplet crossing might, consequently, be anticipated to be most facile. The triplet surface was found to be very flat, and the points of intersection were found to occur at energies only about 0.5 kcal/mol above the shallow minima on this surface. On the singlet surface, the lowest energy intersection was found to be roughly 2 kcal/mol below the energy of the (0.90) geometry, which is shown in Figure 5.1.

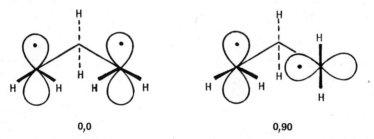

0,0   0,90

FIGURE 5.1.   Trimethylene diradicals.

This geometry serves as the transition state for geometrical isomerization of cyclopropane, and, as noted before, experimental results[9] place the transition state for propene formation 3.7 kcal/mol higher still in energy. Thus the transition state for propene formation was estimated to be more than 5 kcal/mol above those geometries where crossing from the triplet to the singlet surface is expected to be most likely. From these geometries closure to cyclopropane was calculated to proceed without activation on the singlet surface. Therefore, assuming triplet trimethylene to be energetically relaxed before undergoing intersystem crossing to the singlet, cyclopropane formation was predicted on energetic grounds to be much more likely than rearrangement to propene. The authors also argued that closure to cyclopropane should be more favorable entropically than rearrangement to propene.

Doubleday et al. confirmed a number of features of the singlet trimethylene surface that had been found by other calculations.[7] The (0,0) geometry was found to be lower than (0,90) by about 1 kcal/mol.[11] suggesting that racemization of an optically active cyclopropane should be faster than geometrical isomerization. This prediction is in agreement with Berson's experimental results for cyclopropane-d_2^{12}, although substituted cyclopropanes behave in a more stereorandom fashion.[1,3-5]

As in previous studies, the calculations of Doubleday et al. also found that conrotatory closure of (0,0) singlet trimethylene is favored over a disrotatory mode. At the (0,0) geometry through-bond, interaction between the p-pi orbitals on the terminal carbons is stronger than their direct interaction through space.[7] This results in the dominance in the TCSCF wavefunction of the configuration in which the antisymmetric combination of p-pi orbitals is doubly occupied, which, in turn, is responsible for the calculated preference for conrotatory closure.

Doubleday et al. and Goldberg and Dougherty[13] independently arrived at the conclusion that it is the competition between through-bond and through-space effects in (0,0) trimethylene that causes the triplet to be computed to lie below the singlet at this geometry. Both groups found that if the through-space interactions between the terminal methylenes were simulated by pi interactions between two methyl radicals, a singlet ground state was predicted at all distances of separation. However, with increasing C–C–C bond angle in (0,0) trimethylene, the through-space stabilization of the symmetric combination of p-pi orbitals becomes small enough that the through-bond destabilization, which is only weakly dependent on bond angle, begins to dominate. Around the equilibrium bond angle of (0,0) trimethylene these two effects largely cancel. At this geometry, because the singlet is not stabilized strongly by either effect, the smaller electron repulsion energy in the triplet allows it to become the ground state.

Goldberg and Dougherty pursued the question of why the triplet falls below the singlet by recasting the symmetry combinations of p-pi orbitals that emerge from the TCSCF calculations into generalized valence bond (GVB) orbitals. The GVB orbitals, which are the sum and difference of the TCSCF symmetry orbitals, may be physically interpreted as being the orbitals that are each occupied by one electron in the lowest singlet.

Because the through-bond interaction causes the symmetrical combination of p-pi orbitals on the terminal carbons of (0,0) trimethylene to contain a contribution from the central methylene group, the two GVB orbitals each have some density on this group. Unlike the case in the triplet, the electrons in the GVB orbitals of the lowest singlet are not kept by the Pauli principle from simultaneously appearing in this region of space. Thus, even when the exchange integral between the p-pi AOs on the terminal carbons is small, the nonlocalizability of the GVB orbitals into different regions of space confers on the triplet a lower electron repulsion energy than that in the singlet. As Goldberg and Dougherty noted, the criterion of orbital localizability has previously proven extremely useful in predicting the ground states of conjugated diradicals.[7]

Both Doubleday et al. and Goldberg and Dougherty pointed out that the triplet ground state observed for cyclopentane-1,3-diyl(1)[1,3,5,14] is a consequence of the trimethylene moiety that this diradical contains. The calculated singlet-triplet splitting in 1 is essentially the same as that in (0,0) trimethylene when the central C–C–C bond angle in the latter diradical is fixed at the value that it has in 1. The interactions in 1 between the C–H bonds of the ethano group and the p-pi orbitals appear to contribute negligibly to the singlet-triplet splitting.

1   2

Goldberg and Dougherty also carried out calculations on cyclobutane-1,3-diyl (2). The through-space effect in this diradical was computed to be substantial. For two methyl groups interacting in a p-pi fashion at the same distance that separates C–1 and C–3 in 2, the singlet was calculated to lie 15.4 kcal/mol below the triplet. However, through-bond coupling in 2, which involves interaction of the p-pi orbitals with two methylene groups, is sufficiently strong that the triplet was calculated to be the ground state of 2 by 1.7 kcal/mol.

Chang and Dougherty have prepared 2,3-diazabicylo[2.1.1]hex-2-ene (3a) as a possible precursor of 2.[15] They found the activation energy for loss of nitrogen from 3 to be moderately low. It seems likely that destabilizing interactions between the orbitals of the four-membered ring and those of the N–N pi bond[16] are relieved in the transition state for nitrogen loss, thus accounting, at least in part, for the rather low activation energy.

3a , R = R′ = H   4a , R = R′ = H   5a , R = R′ = H
  b, R = D, R′ = H b , R = D, R′ = H   b, R = D, R′ = H
        c , R = H, R′ = D   c, R = H, R′ = D

On pyrolysis of **3a** in solution, only bicyclobutane (**4a**) was formed. However in the gas phase butadiene was also obtained.[17] The ratio of the two products was found to depend on pressure; the amount of butadiene increased with decreasing pressure. As discussed by Chang and Dougherty, this finding suggests that bicyclobutane is formed with excess vibrational energy. Unless a sufficient amount of this excess energy is removed by collisions, the "hot" bicyclobutane undergoes rearrangement to butadiene.

Chang and Dougherty pointed out that this is the second example of a "hot" molecule being formed in the pyrolysis of an azo compound. The first was found by Shen and Bergman, who studied the pyrolyses of **6** and **7**.[18] Although both azo compounds gave rise to spiropentane, only the spiropentane formed from **7** was "hot" enough to undergo further rearrangements. Because Shen and Bergman explained the difference between the behavior of **6** and **7** in terms of the nonequivalence of the C–N bonds in **7**, the formation of vibrationally excited bicyclobutane in the decomposition of **3**, an azo compound with equivalent C–N bonds, is noteworthy.

6   7

Chang and Dougherty examined the stereochemistry of bicyclobutane formation by preparing **3b**. Pyrolysis of **3b** in solution yielded an equal mixture of bicyclobutanes **4b** and **4c**, strongly implicating **2** as an intermediate in this reaction. Interestingly, in the gas phase a 10% excess of the double inversion product (**4c**) was formed. Double inversion has been found to be the predominant pathway in the decomposition of other azo compounds.[5]

Chang and Dougherty also investigated the photolysis of **3a**.[19] In addition to **4a**, a rearrangement product (**5**) and butadiene were also formed. The ratio of products was found to depend on both temperature and whether the photolysis was sensitized. Chang and Dougherty obtained evidence that 5 was formed in the triplet manifold. Since 5 was observed not only in the sensitized photoreaction, where it was the major product, but also in the direct photolysis, 3 apparently undergoes fairly efficient intersystem crossing.

On both direct and sensitized photolysis, 3b gave equal amounts of 5b and 5c, suggesting a freely rotating diradical intermediate rather than a concerted sigmatropic process in the photorearrangement. The bicyclobutane formed in the direct photolysis of 3b showed a 47% excess of 4c, the double inversion product. The interpretation of this result is complicated by the fact that the butadiene and some of the bicyclobutane arose from secondary photolysis of 4-diazo-1-butene, which was apparently also formed as a primary product in the triplet manifold.

Despite the fact that 2 was calculated to have a triplet ground state, photolysis of 3 at 8°K in the cavity of an EPR spectrometer failed to yield a triplet EPR signal. Chang and Dougherty attributed this failure to the apparently effective competition of the rearrangement of triplet 3 with fragmentation of 3 to triplet diradical (2) plus nitrogen. The singlet diradical, if it was formed in the photolysis of 3 at this temperature, apparently closed to 4 faster than it underwent intersystem crossing to the triplet.

McElwee-White and Dougherty have examined the photolysis of another azo compound (8),[20] which is also capable of giving a trimethylene type diradical (9) on deazetation. This diradical might undergo opening of the cyclopropane ring to give a tetraradical (10). The antisymmetric combinations of p-pi orbitals in the two five-membered rings of 10 can mix with each other through spiroconjugation. This mixing splits the resulting molecular orbitals, leaving only the symmetric combinations of p-pi orbitals degenerate. Thus, as the authors pointed out, although 10 is a tetraradical in the structural sense, the energy levels of 10 are those of a diradical.

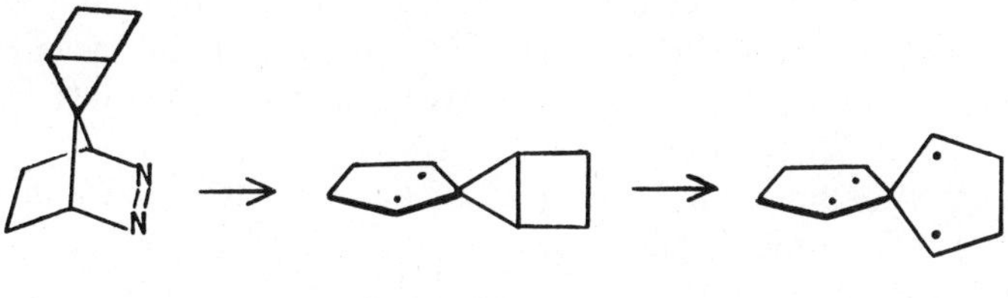

8   9   10

On thermolysis, and on both direct and sensitized photolysis of 8, 11 and 14 were formed. McElwee-White and Dougherty proposed a reasonable mechanism for the formation of 14, which involved rearrangement of 9 to 12. Subsequent stepwise bond cleavage in this highly strained hydrocarbon (by way of 1,4 diradical 13) would afford 14.

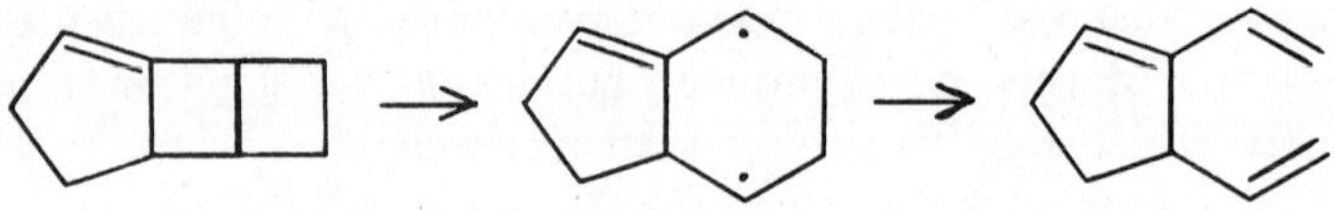

11   **12**   **13**   **14**

The question of whether **10** was involved in the formation of **14** was probed by isotopically labeling **8**. The labeling experiments showed that there was a minor component of the rearrangement in which the two five-membered rings became equivalent, as demanded if **10** lay along the rearrangement pathway. However, the formation of **10** further requires that both faces of the five-membered rings also becomes equivalent. Additional labeling experiments revealed that in the fraction of the product in which the five-membered rings had become equivalent, facial equivalency had only partially been achieved. Although the experimental data permit **10** to play a minor role in the formation of **14**, other possible interpretations of the data were also discussed by the authors.

As noted in a recent review,[5] a unified mechanism capable of predicting the stereochemistry of cyclopropane formation from pyrazoline decomposition is still wanting. A recent example of the extraordinary variability of stereochemical outcome with changes in substituents comes from the work of Schneider and Bippi with optically active *trans*-3,5-diphenylpyrazoline (**15**).[21] On both thermolysis and direct photolysis, optically active *trans*-diphenylcyclopropane (**16**) was formed in 50–73% optical purity with predominant retention of configuration at both chiral centers.

This result contrasts with the finding of Bergman et al. that double inversion was favored over double retention in the decomposition of optically active *trans*-3-ethyl-5-methylpyrazoline.[1,5,22] The major product on thermal decomposition of this latter pyrazoline was the *cis*-cyclopropane, both enantiomers of which were formed in nearly equal amounts. In contrast, **15** yielded only about 10% of the cis isomer of **16**.

Schneider and Bippi also found that on decomposition of **15** with monochromatic light, phenyldiazomethane and styrene were produced by a cycloversion pathway similar to the one that formed 4-diazo-1-butene from **3**. The occurrence of the cycloversion reaction could not be verified in the thermal

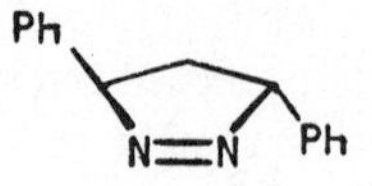

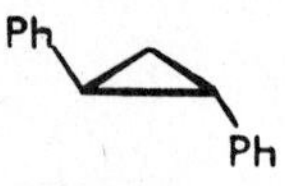

15   **16**

decomposition of **15**. However, if some cyclopropane formation involved a carbene generated by this pathway, it would obviously complicate the stereo-chemical analysis in this and other pyrazoline decompositions.

Substituent effects in cyclopropane stereomutations are somewhat better understood than those in pyrazoline decompositions.[1,3-5] Cleavage of the cyclopropane ring occurs at the bond that yields the most stable trimethylene diradical. Recently published experimental work by Doering and Barsa[23] and by Baldwin and Carter[24] support this generalization.

Doering and Barsa studied optically active 1-cyano-2-methyl-3-(cis-propenyl) cyclopropane (**17**). They found diastereomerization but no racemization when **17** was pyrolyzed at 217.8°C. This result shows that the configuration of one of the asymmetric carbons (that bearing the methyl group) was maintained, suggesting that bond cleavage occurred only between the two carbons bearing the most radical stabilizing substituents. At 258°C slow racemization was observed. The difference in rate constants for diastereomerization and for racemization was found to correlate satisfactorily with the difference between the radical stabilizing abilities of the cyano and methyl groups.

17   18

Baldwin and Carter examined the pyrolysis of **18**, where only the presence of a methyl group favors cleavage of one cyclopropane ring bond over a second. Given the rather modest radical stabilizing effect of a methyl group, it is not surprising that these authors found a nonnegligible contribution to the kinetics from processes involving the rupture of the less substituted of the two bonds. Nevertheless, as expected, diastereomerizations involving cleavage of the most substituted ring bond remained fastest; and the double rotation that necessitates cleavage of the least substituted bond was found to have a rate constant indistinguishable from zero.

As discussed in several reviews,[1,3-5] most substituted cyclopropanes show little evidence of the preferred coupled rotation that is predicted by quantum mechanical calculations[7,10] and found in the unsubstituted molecule.[12] The only substituted cyclopropane that shows a strong preference for coupled methylene group rotations is phenylcyclopropane.[12,25] In contrast, vinylcyclo-propane, which might have been anticipated to behave similarly to the phenyl substituted hydrocarbon, is reported to give stereorandom behavior on pyrolysis.[26] It is to be hoped that this apparent anomaly in trimethylene chemistry will eventually find resolution.

In addition to being postulated as intermediates or transition states in pyrazoline decompositions and in cyclopropane stereomutations, trimethylene diradicals have been invoked as lying along the reaction pathway in the photochemical di-π-methane rearrangement.[27] In a pioneering study, Zimmerman et al. generated diradicals with structure 20 from the corresponding azo compounds (21) as a test of whether this type of diradical is involved in the di-π-methane rearrangements of barrelenes (19) to semibullvalenes (23).[28,29] On pyrolysis, the azo compounds all underwent retrograde homo-Diels-Alder reactions to regenerate the barrelenes. However, on sensitized photolysis, the major product isolated in each case was the semibullvalene.

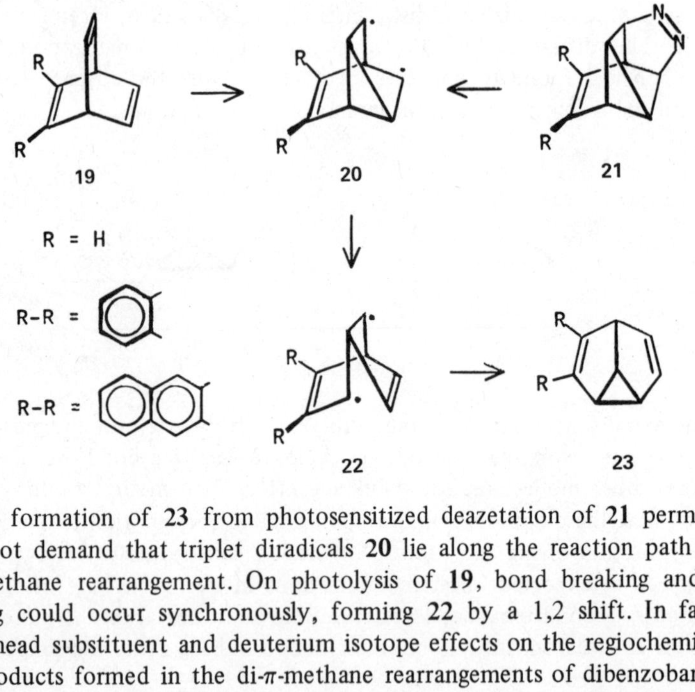

The formation of 23 from photosensitized deazetation of 21 permits but does not demand that triplet diradicals 20 lie along the reaction path in the di-π-methane rearrangement. On photolysis of 19, bond breaking and bond making could occur synchronously, forming 22 by a 1,2 shift. In fact, the bridgehead substituent and deuterium isotope effects on the regiochemistry of the products formed in the di-π-methane rearrangements of dibenzobarrelenes and benzonorbornadienes are most readily interpreted by a mechanism involving a concerted 1,2 shift.[30] If diradicals analogous to 20 are intermediates, the results reported by Paquette and Bay indicate that the diradicals must be generated reversibly, since their formation cannot be the product-determining step.

Evidence for the intermediacy of diradicals similar to 20 has been reported by Schaffner et al.[31] in the di-π-methane rearrangement of 24. On photolysis of 24 at 77 K in a glass, a triplet EPR spectrum was observed which persisted after

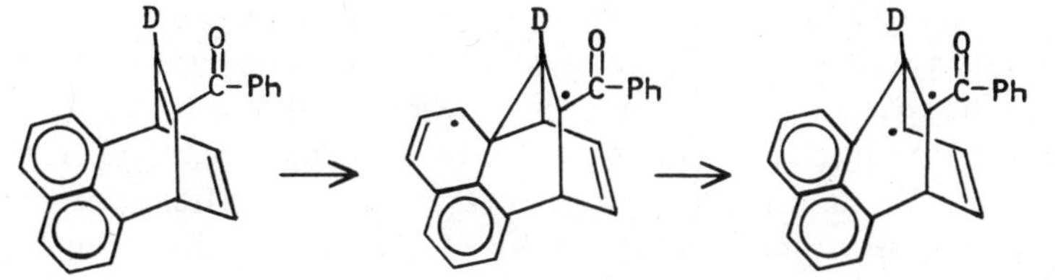

24   25   26

the irradiation was terminated. In addition, new IR bands were observed for the carbonyl group. On warming the sample to 92 K, the EPR signal disappeared and another transient IR band was observed up to 182 K, at which temperature the IR of the carbonyl group of the product began to appear.

Schaffner et al. assigned the EPR and associated low-temperature IR spectra to diradical 25. The transient IR spectrum observed at higher temperatures was attributed to 26, despite the fact that no EPR spectrum was initially detected for this diradical.[31a] Subsequently, evidence was presented for a second EPR active species with an EPR spectrum very much like that of 25, except for the intensities of the peaks.[32b]

The rapid disappearance of this spectrum at 100 K seems inconsistent, however, with the stability of the IR spectrum attributed to 26 at temperatures up to 182 K. It seems at least possible that the IR spectrum that was observed between 92 and 182 K was due to the product, formed with the benzoyl group frozen in a nonequilibrium conformation. In this connection it may be significant that the benzoyl group rotamers in the starting material (24) were reported to be frozen up to temperatures around that at which the second IR transient began to disappear.

The behavior of the diradicals that are postulated to close in the final step of the di-π-methane rearrangement has been investigated by Adam and his co-workers. They generated the diradicals by deazetation of polycyclic azo compounds 27–29, which were each prepared by dipolar cycloaddition of triazolinodiones to the appropriate bicyclic alkene.[32]

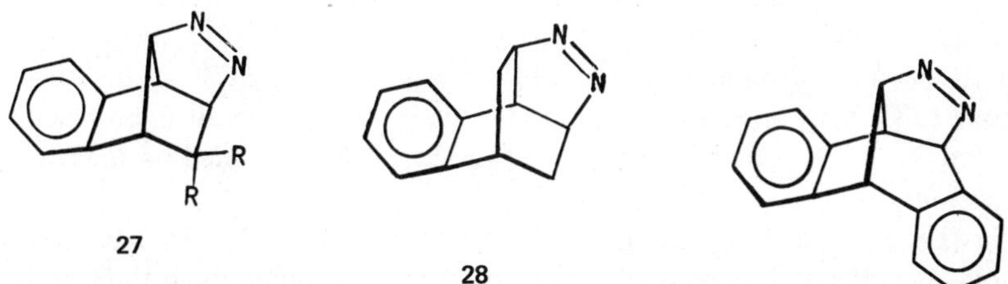

27   28   29

In the case of 27 the course taken by thermolysis and photosensitized photolysis reactions depended on the ability of the substituent group (R_2) to stabilize the adjacent radical center.[33] For example, with R = H rearrangements without nitrogen loss were faster than deazetation. Only on direct photolysis did deazetation occur. By far the major product was that expected from closure of the resulting diradical, although a few percent of benzonorbornadiene, the product of retro-di-π-methane rearrangement, was also formed. In contrast, with R_2 = =CPh_2, only the expected closure product was obtained on thermolysis and on both direct or sensitized photolysis.

Azo compound 28 also underwent deazetation under all three sets of conditions.[34] On thermolysis and direct photolysis, the major product was again that expected from closure of the initially formed diradical, but almost 20% of the retro-di-π-methane product was also obtained. Only the closure product was isolated from sensitized photolysis or from heating in the presence of tetramethyldioxetane. Pyrolysis of the dioxetane yielded triplet acetone, which also served to transfer triplet energy to 28.

Photosensitized deazetation of 29 again led exclusively to the expected closure product, dibenzosemibullvalene.[35] In addition to the semibullvalene, small amounts of retro-di-π-methane products were once more also detected in the pyrolysis and direct photolysis of 29. Sensitized decomposition of an isomer of 29, which could yield the diradical postulated to be formed initially in the di-π-methane rearrangement of dibenzobarrelene to dibenzosemibullvalene, again afforded only this latter compound.

It is of some interest that no retro-di-π-methane products were obtained in the sensitized photolyses of 27–29. If the product triplet diradicals, presumed to be formed in these reactions, undergo intersystem crossing to the singlet diradicals thought to be generated on pyrolysis or direct photolysis, the reactions of the singlet diradicals apparently depend on their origin. The singlets formed by intersystem crossing only close; whereas some of the singlets formed by pyrolysis or photolysis rearrange.

Although closure of the initially formed diradical was the major pathway in the deazetations of 27–29 under all conditions, this was not the case with the diradical (31) formed from azo compound 30.[36] This diradical is a possible intermediate in the oxadi-π-methane rearrangement of norbornenone (32) to 33 and 34. Adam and co-workers found that thermolysis of 30 afforded no 33, and only a few percent of 33 was formed in the direct photolysis of 30. Even triplet-sensitized decomposition of 30 gave only about 30% of the tricyclic ketone (33). Norbornenone (32) was the major product in all these reactions; the amount of the other bicyclic enone product (34) depended on the reaction conditions.

In the sensitized decomposition of 30 the 7:1 ratio of 33–34 was the same within experimental error as the ratio of these two products in the sensitized oxadi-π-methane rearrangement of norbornenone. This result provides permissive evidence for the intermediacy of diradical **31** in this photochemical reaction, and it also suggests that the photochemical 1,3 shift that transforms triplet **32** to **34** actually consists of two sequential 1,2 shifts, with **31** as an intermediate.

30 , R = O
35 , R = H$_2$

31 , R = O
36 , R = H$_2$

32 , R = O
37 , R = H$_2$

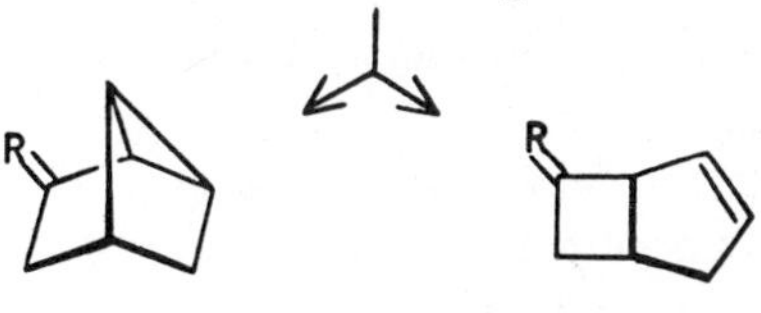

33 , R = O
38 , R = H$_2$

34 , R = O
39 , R = H$_2$

Adam and co-workers have also prepared the azo compound (**35**) which lacks the carbonyl group present in **30**.[37] As was the case with the other hydrocarbon diradicals studied by the Adam group, the major product formed from **36** under all conditions was that (**38**) resulting from cyclopropane ring closure without rearrangement. In the direct photolysis of **35** small amounts of rearrangement products were detected; the only one identified was **39**. Norbornene (**37**) was not detected in any of the deazetation reactions of **35**, despite the fact that norbornenone (**32**) is the major product formed from **30** under all reaction conditions. It appears that the presence of the carbonyl group in **31** not only enhances the rate of rearrangement relative to ring closure but also influences the direction of the 1,2 shift in the rearrangement reaction.

B.   Tetramethylene and Related 1,4-Diradicals

As in the case of trimethylene, Doering has pointed out that upward revision of dissociation energy for a primary C—H bond leads to the near disappearance of the "Benson activation energy" for transformation of tetramethylene into closed-shell products.[8] Quantum mechanical calculations by Segal on the latter diradical had actually predicted the existence of two intermediates on the singlet tetramethylene energy surface, one with a gauche conformation about the central C–C bond and one with a trans conformation.[7,38] Small barriers were calculated to separate the former intermediate from closure to cyclo-butane and the latter from cleavage into two molecules of ethylene.

More sophisticated calculations by Doubleday, McIver, and Page have shown that the barrier to ring closure in gauche tetramethylene is an artifact of the STO-3G basis set used by Segal.[39] Doubleday et al. confirmed that with the STO-3G basis set, TCSCF calculations, which are comparable to the 3×3 CI calculations carried out by Segal, do predict a small barrier to ring closure. However, with the 3-21G basis set, the barrier disappeared. The authors attri-buted this result to the fact that each valence orbital in the latter basis set contains a diffuse function. These diffuse functions are important for repre-senting the bonding interactions between the orbitals on the terminal carbons of tetramethylene at large distances. Their absence in the STO-3G basis set leads to the underestimation of the stabilizing effect of these interactions in the early stages of ring closure.

Borden and Davidson have investigated the ease of terminal methylene group rotation in trans-tetramethylene.[7,40] At this geometry through-bond coupling between the atomic orbitals on the terminal methylene groups should be most important. Indeed, the results of TCSCF calculations showed that through-bond interactions are responsible for favoring a (90,90) conformation, in which these orbitals both eclipse the central C–C bond. However, in the TCSCF wavefunction electron repulsion gives nearly equal weight to the two configurations, only one of which contains stabilizing through-bond effects.[41] Consequently, the effect of through-bond coupling in (90,90) tetramethylene was computed to be only about 1.5 kcal/mol.

The results of Borden and Davidson, combined with those of Segal, suggest that trans-tetramethylene should undergo terminal methylene group rotation almost as fast as it fragments to two ethylenes. However, additional calculations by Doubleday, using an MCSCF wavefunction that correlates two pairs of electrons, have shown that (90,90) trans-tetramethylene can fragment without activation.[42] The crucial difference between this calculation and both Segal's 3×3 CI results and the previous results that Doubleday et al. obtained with a TCSCF wavefunction is the correlation of the pair of electrons in the C–C bond that cleaves. Unlike the case in a closed-shell molecule, fragmentation of tetra-methylene does not demand correlation of the electrons in this bond. However, Doubleday's results indicate that failure to provide such correlation introduces a spurious, albeit small, barrier to this reaction.

The computational results of Doubleday make it appear that there are no diradical intermediates on the singlet potential energy surface for tetramethylene. The surface has large flat regions, but the only stationary points that correspond to diradicals are transition states, one lying on the pathway for fragmentation of cyclobutane and the others being associated with stereoisomerization reactions.

The most recent quantum mechanical results on tetramethylene are thus perfectly analogous to those obtained previously for trimethylene.[7]

The experimental results in 1,4-diradical chemistry have usually been interpreted as supportive of an intermediate or intermediates.[1,3-5] Although Doubleday's recent calculations may have banished 1,4-diradical intermediates from the tetramethylene potential energy surface, such intermediates may nevertheless persist on the free energy surface. Doubleday has pointed out that both cleavage and closure require freezing out several rotational degrees of freedom.[42] Consequently, going to the transition states for these processes results in a decrease in entropy, which can give rise to free energy barriers to product formation.

Closure of trimethylene to cyclopropane also requires freezing out rotational degrees of freedom for the terminal methylene groups. However, trimethylene lacks the central C—C bond of tetramethylene; so that there is one less rotational degree of freedom lost in the transition state for forming closed-shell products from trimethylene. Therefore, a smaller free energy of activation should be required for closure of trimethylene than for cleavage or closure of tetramethylene. One may speculate that this difference accounts for the fact that tetramethylene diradicals behave like true intermediates (i.e. 1,4-diradicals generated by different methods react similarly); whereas trimethylene diradicals do not.

Dervan, whose work with azo compounds has provided the best evidence for 1,4-diradicals as common intermediates[1,3,5] in different reactions, has continued his studies by carrying out the deazetation of **40** and **42**.[43] The 1,1- and 1,2-diazene might be expected to yield the same tetramethylene diradical (**41**). In fact, Schultz and Dervan did find that nearly identical ratios of cleavage (**43**) and closure (**44**) products were formed from both precursors on pyrolysis

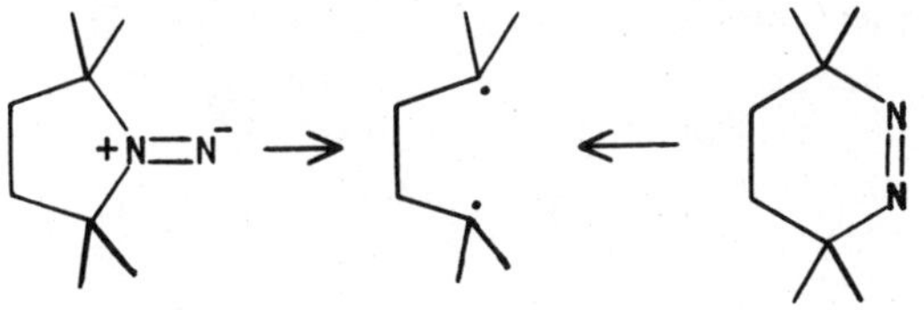

40   **41**   **42**

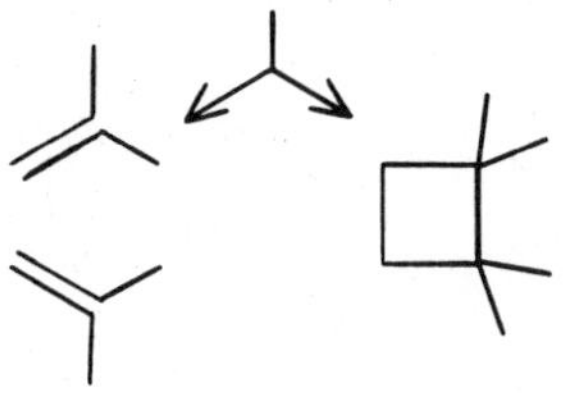

43   **44**

and on direct and photosensitized photolysis. A few percent of the two possible olefins that can result from diradical disproportionation were also isolated.

In both the thermolysis and direct photolysis reactions the ratio of **43–44** was approximately 1:1. Some temperature dependence was reported in the product ratio from the thermolysis of **40**. Nevertheless, the fact that the product ratio from direct photolysis of **40** and **42** at −78°C was almost the same as that in both the thermal decomposition of **40** around 0°C and in the pyrolysis of **42** at 140°C is remarkable. Significantly, the product ratio changed to 3:1 on sensitized decomposition of both azo compounds at −78°C.

The apparent dependence of the product ratio on the spin multiplicity of the putative 1,4-diradical intermediate (**41**) indicates that, as is the case in 1,3-diradicals, intersystem crossing from the triplet produces a singlet diradical that behaves differently from the one generated by direct photolysis or pyrolysis. A possible interpretation of the greater amount of cleavage compared to closure in the sensitized reaction is in terms of population of singlet conformations that favor the former process over the latter. One may speculate that the longer lifetime of triplet **41** allows conformations with trans stereochemistry about the central C−C bond to be populated more than in the singlet, where closure of an initially formed gauche diradical is presumably competitive with rotation about this bond. Alternatively, intersystem crossing might be most favorable at conformations that lead to cleavage rather than closure on the singlet surface.

If intersystem crossing does, in fact, produce a population of conformations different from that generated by direct formation of singlet **41**, it is clear that complete conformational equilibration does not take place before singlet **41** is transformed to **43** and **44**. Thus intersystem crossing, in addition to probably being rate determining for triplet diradicals, may also be product determining. The evidence that singlet diradicals are so short-lived that the conformation in which they are born determines, at least to some extent, the reaction path that they subsequently follow, has been reviewed recently by Sciano.[44] Some of this evidence is discussed later in this section.

Evidence for a longer lifetime for triplet than for singlet 1,4-diradicals comes from the now classic experiment of Bartlett and Porter.[45] They found that the triplet-sensitized deazetation of **45** yielded closure products with significant

45   46   47

loss of stereochemistry. In constrast, the thermolysis and direct photolysis of the 1,2-diazene gave nearly stereospecific ring closure – a result later shown by Dervan to be due to the tertiary radical centers in **46**, which cause rotation about the C–C bonds to be slow relative to closure.[1,3,5]

More recently, Dervan and Schultz found that deazetation of the corresponding 1,1-diazene (**47**) afforded essentially the same stereochemical results.[43] Despite the fact that the reaction temperatures of Dervan and Schultz were all 100–150°C lower than those of Bartlett and Porter, the ratio of cleavage to closure products in each of the three deazetation reactions of **47** was similar, though not identical, to that found in the corresponding reaction of **45**. As was the case with **40** and **42**, both **45** and **47** gave a higher ratio of cleavage to closure products on sensitized photolysis than on direct photolysis or pyrolysis.

Engel and Keys have measured the singlet and triplet lifetimes of a 1,4-diradical by incorporating a "free radical clock" into **49**.[46] On deazetation of **48**, **49** can undergo cyclopropylcarbinyl radical rearrangement to **50**, competitive with closure and cleavage. The known rate of the cyclopropylcarbinyl radical rearrangement provides a clock against which the rates of cleavage and closure can be measured.

48   **49**   **50**

On photosensitized deazetation of **48**, the triplet **49** that was putatively generated underwent substantial rearrangement to **50**. The large separation of the radical sites in **50** and its rather rigid geometry make intramolecular disproportionation difficult. Consequently, **50** can be trapped by CCl_4, and Engel and Keys obtained evidence that at least one of the two disproportionation products arising from **50** was formed intermolecularly.

The fraction of the products derived from **50** increased with increasing temperature, consistent with the fact that the cyclopropylcarbinyl radical rearrangement has an activation energy of 5.9 kcal/mol. Using a value of 2.2×10^7/sec for the rearrangement of **49** to **50** at 25°C, from the ratio of products derived from these two diradicals at this temperature, Engel and Keys obtained a lifetime of about 30 nsec for triplet **49**. Some or all of the cyclopropylcarbinyl rearrangement could occur in a diazenyl diradical precursor of **49**, so that 30 nsec represents an upper limit for the lifetime of **49** at 25°C.

Because singlet **49** need not undergo intersystem crossing to cleave or close, it would be expected to have a shorter lifetime than the triplet. Indeed, on direct photolysis of **48** at 25°C, only about 5% of the product mixture was derived from **50**, giving a lifetime of about 2 nsec for singlet **49** at this temperature. Some or all of the products derived from **50** could have come from triplet **49**, formed by intersystem crossing in excited **48**, or from ring opening in a diazenyl diradical. Consequently, 2 nsec is an upper limit for the lifetime of singlet **49** at 25°C.

The possibility of intersystem crossing in excited **48** is not a problem in the thermolysis of the azo compound. On deazetation of **48** at 230°C, only 2% of product formed from **50** was detected. From this result Engel and Keys estimated a lifetime of 0.9 nsec for singlet **49**. However, this estimate ignores the fact that the cyclopropylcarbinyl rearrangement at 230°C is faster than at 25°C. Using 5.9 kcal/mol as the energy of activation, the rearrangement is calculated to be roughly 60 times faster at 230°C than at 25°C. A lifetime of 15 psec for singlet **49** at 230°C is then obtained. This lifetime is again an upper limit because of the possibility of rearrangement occurring in a diazenyl diradical rather than in **49**.

Wagner and co-workers have employed "radical clocks" to obtain the lifetimes of the triplet diradical intermediates (**52**) formed in the Norrish Type II reactions of aromatic ketones (**51**).[47] With $R = H$ and $R' = C_3H_5$, the radical clock is again the cyclopropylcarbinyl radical rearrangement. With $R = CH_2 CH= CH_2$ and $R' = H$, the radical clock is the rearrangement of 5-hexenyl to cyclopentylmethyl radicals. Assuming that these rearrangements occur in diradical **52** with the same rate constants as in monoradicals, both radical clocks led to lifetimes of about 50 nsec for triplet **52**.

51   52

Sciano has used laser flash photolysis to obtain directly lifetimes of the triplet diradicals formed in Norrish Type II reactions.[48] The lifetimes of the 1,4-diradicals generated from phenylketones in benzene were measured to be around 30 nsec, a value in reasonable accord with that obtained indirectly by Wagner and co-workers using the radical clock technique. This agreement confirms the assumption that diradicals like **52** undergo monoradical reactions (e.g. cyclopropylcarbinyl rearrangement) at the same rate as monoradicals. Encinas, Wagner, and Sciano have also ascertained by direct measurements that such diradicals react intermolecularly with hydrogen atom donors at rates comparable to their monoradical counterparts.[49]

Some of the factors that control the lifetimes of the triplet diradicals formed in Norrish Type II reactions have been discussed by Sciano.[44,48] He has measured the activation parameters for diradical decay and found that the A factors are low and that the activation energies are negligible. From his data Sciano has concluded that intersystem crossing is probably rate determining.

From the data tabulated by Sciano,[48] one might have expected the 1,4-diradical studied by Engel and Keys (49) to have a longer triplet lifetime than the 30 nsec they obtained. For instance, Sciano finds that 1,4-diradicals that lack the phenyl group present in 52 have triplet lifetimes that are at least an order of magnitude longer. Of course, 49 also lacks the hydroxyl group present in the Type II diradicals whose triplet lifetimes have been measured by Sciano. The absence of this structural feature may account for the shorter than expected lifetime of 49.

It should be noted, however, that an oxygen substituent can shorten the lifetime of a triplet 1,4-diradical. Using laser spectroscopy. Freilich and Peters have measured a lifetime of 1.6 nsec for the triplet diradical (53) formed in the Paterno–Buchi reaction of benzophenone with 3,6-dioxacyclohexene.[50] Similar lifetimes have been measured by Caldwell et al. for the diradicals generated by the addition of benzophenone to tetramethylethylene and to ethylvinyl ether.[51] Caldwell and co-workers also generated this type of oxygen-substituted diradical (54) by Type II photoreactions of ethers like 55. Once again, triplet lifetimes an order of magnitude shorter than that of 52 were found. Caldwell speculated that delocalization of one of the unpaired spins into oxygen might be responsible for shortening the triplet lifetime by decreasing the effective distance between the radical centers.

53   54   55

As noted in connection with the dependence on spin state of the ratio of cleavage to closure in 1,4-diradicals generated from azo compounds, Sciano has argued that not only is intersystem crossing rate determining but that it is also product determining.[44,48] Evidence for this postulate comes in part from Sciano's own work on the quenching of Type II triplet diradicals by paramagnetic molecules like triplet O_2 and nitroxides.

For example, in a recently published study, Sciano and co-workers showed that $Cu(acac)_2$ in methanol scavenges the triplet diradical produced in the Type II reaction of γ-methylvalerophenone.[52] In the presence of $Cu(acac)_2$ no new products were formed, but the ratio of cleavage to closure was modified.

In those diradicals that underwent intersystem crossing aided by the paramagnetic $Cu(acac)_2$, the fraction of closure product was doubled.

Sciano has suggested that interaction with the paramagnetic quencher not only facilitates intersystem crossing in the triplet diradicals but also alters the geometries at which intersystem crossing occurs. If the singlet diradicals lived long enough to equilibrate conformationally before cleaving or closing, the product ratio would not depend on whether intersystem crossing was assisted by paramagnetic molecules. Because the product ratio quite clearly does depend on whether or not intersystem crossing is assisted by paramagnetic molecules, Sciano argues that the singlet diradicals must have very short lifetimes to retain some memory of the conformation in which they were born.

An alternative explanation of Sciano's results seems possible, however. Assuming that quenching requires orbital overlap between the quencher and the triplet diradical, it seems likely that at least some singlet diradicals have quencher molecules still in close proximity as they begin to cleave or close. The proximity of the paramagnetic quencher could alter the preferred conformations of the newly born singlet diradicals, thus affecting the product distribution. Sciano's data might therefore alternatively be interpreted to indicate that product formation from the singlet diradicals is faster than diffusion away of the quencher.

Similarly, the dependence of the product ratios on spin state in azo decompositions could be interpreted as indicating that a diazenyl diradical plays a more important role in product formation from one spin state than from another. If nitrogen remained attached to one of the terminal carbons, its presence would almost certainly affect the product ratio. There are several lines of argument that can be pursued to cast doubt on this interpretation, but it cannot be ruled out.

Nevertheless, there is independent evidence that intersystem crossing from triplet diradicals leads to products faster than the singlets undergo conformational equilibration. This evidence comes from the chemically induced dynamic nuclear polarization (CIDNP) which has been observed in Norrish Type I reactions of cycloalkanones by Doubleday,[53-55] and more recently in the Norrish Type II reaction of valerophenone by Kaptein and co-workers.[56] Because the latter reaction involves a 1,4-diradical, it will be discussed first.

On irradiation of valerophenone, the 100 MHz ^1H NMR spectrum of the reaction mixture showed enhanced absorbtions and emissions. The observed "multiplet" effects indicated that hyperfine-assisted intersystem crossing in the initially formed triplet diradical occurred predominantly from T_0. This is the triplet sublevel whose energy is unaffected by the magnetic field.

T_0-S intersystem crossing involves a spin-sorting mechanism in which certain nuclear spin combinations accelerate intersystem crossing. This mechanism is usually seen in radical pairs, where the cage recombination products from a triplet pair have larger than equilibrium populations in those nuclear spin levels that accelerate intersystem crossing. The cage escape products have excess populations in the remaining nuclear spin levels. In the case of a diradical the observation of T_0-S intersystem crossing also requires the existence of a spin-independent escape route that forms the closed-shell products in ratios different from those of the hyperfine-assisted mechanism.

In the Type II reaction of valerophenone, the cleavage and closure products exhibit "recombination" type polarization. "Unreacted" valerophenone, regenerated by hydrogen atom transfer from oxygen back to carbon, exhibits "escape" type polarization. Clearly, the hyperfine-assisted intersystem crossing mechanism favors formation of the products more than does some other intersystem crossing mechanism, for instance, spin-orbit coupling, which apparently favors regeneration of the starting ketone.

These results of Kaptein et al. show that different mechanisms of intersystem crossing favor formation of different closed-shell molecules. If intersystem crossing from the triplet diradical were followed by rapid conformational equilibration in the singlet, the singlet diradicals would presumably partition themselves among the possible products in a way independent of the mechanism by which intersystem crossing occurred. The observation of T_0-S CIDNP shows that this is not the case, indicating that product formation is competitive in rate with conformational equilibration in singlet diradicals.

The first CIDNP evidence that intersystem crossing is to some extent the product-determining step in triplet diradicals came from Doubleday's work on the Type I reaction of phenylcyclohexanone.[53] This reaction involves initial cleavage α to the carbonyl group to form a triplet diradical. Following intersystem crossing, the diradical can collapse back to the reactant or disproportionate to ketene and unsaturated aldehyde.

Doubleday found evidence for T_0-S intersystem crossing in the ^{13}C spectra of the reaction mixtures. The evidence consisted of the magnetic field dependence of the carbonyl signal in all three products, which only in the case of the ketene changed from the emission, expected for T_--S hyperfine-assisted intersystem crossing, to absorption at high magnetic fields. Doubleday showed that the observation of predominant T_0-S CIDNP was consistent with a model in which the mechanism of intersystem crossing is conformationally dependent, provided that collapse to products of the singlet diradicals thus formed is faster than their conformational equilibration.

Subsequently, Doubleday found that the 1,5-diradicals formed from different cyclopentanones showed evidence of predominant crossing from different triplet sublevels, depending on the subtleties of each diradical's structure.[54] In the 1,6-diradicals formed from Type I cleavage of some cyclohexanones, a change from T_+-S to T_--S CIDNP was observed with increasing magnetic field strength.[55] Because the former type of CIDNP is associated with the triplet being below the lowest singlet state and the latter with a singlet below the triplet, it appears that these diradicals have both types of regions on their potential surfaces. This experimental finding is consistent with the results of calculations on 1,3- and 1,4-diradicals,[10,13,40]

Different products displayed different magnetic field dependencies in going from the absorption associated with T_+-S intersystem crossing to the emission associated with T_--S. This observation shows that when magnetic field dependent energy matching between one of the triplet sublevels and the singlet permits intersystem crossing from some triplet conformations, the subsequent partitioning of the resulting singlet diradicals into the closed-shell products occurs faster than complete conformational equilibration. Therefore, these experiments, too, indicate that, to some extent, intersystem crossing is product determining.

These latter studies of Doubleday suggest that there is some conformational dependence in intersystem crossing by he hyperfine-assisted pathway. However, as pointed out by Doubleday,[53] for the T_0-S pathway the conformational dependence can be either in the spin-assisted or spin-independent pathway. It has been shown by de Kanter and Kaptein that a model in which all the conformational dependence is put into a spin-orbit mediated step that leads directly to collapse to closed-shell products is capable of fitting Doubleday's phenylcyclohexanone data, as well as their own data on other ketones.[57]

It seems likely that both hyperfine-mediated and nuclear-spin independent intersystem crossing pathways are dependent on conformation. The CIDNP studies discussed show that each of the two pathways leads selectively to certain closed-shell products. However, it should be noted that CIDNP is a very sensitive technique; it is not yet clear to what extent the mechanism of intersystem crossing has a significant effect on the relative yields of closed-shell products. Attempts to measure a macroscopic effect of magnetic fields on the relative yields of products in diradical reactions have thus far not found any effect outside the limits of experimental error.[44,48] On the other hand, Turro and co-workers have found that magnetic fields do have a measurable effect on the partitioning between recombination and escape products in reactions involving caged radical pairs.[58]

In summary, several lines of evidence, both theoretical and experimental, indicate that the energy barriers to cleavage and closure in singlet 1,4-diradicals are small or nonexistent. As a result, singlet 1,4-diradicals are much shorter lived than their triplet counterparts, which must first undergo intersystem crossing before reacting to give closed-shell products. A number of different types of experiments suggest that singlet diradicals may be so short-lived that intersystem crossing is not only rate but also product determining for triplets.

It may be hoped that in future singlet diradicals will be detected in laser flash photolysis studies on the psec time scale. From the estimate by Engel and Keys of the singlet lifetime of **49**, it appears that this may be possible. Failing the direct detection of a singlet diradical, it would be desirable to use the radical clock approach to determine the lifetime of the 1,4-diradical formed in the Norrish Type II reaction of a ketone that reacts from its n-π* singlet state. A dialkylketone is a likely candidate, since dialkylketones undergo intersystem crossing more slowly than do their aromatic counterparts. In the presence of quenchers to eliminate triplet reactions, dialkylketones have previously been used to study the rates of rotation versus cleavage in singlet 1,4-diradicals.[1,59,60]

III.   ANTIAROMATIC ANNULENES

A.   Cyclobutadiene

A comprehensive review of cyclobutadiene (CBD) chemistry by Balley and Masamune appeared in 1980.[61] Subsequently, the quantum mechanical treatment of CBD has been reviewed, and experimental data that appeared after publication of the Balley–Masamune review discussed.[7] Therefore, in this section attention is focused only on results that are too recent to have been included in either of the foregoing reviews.

In 1980 Whitman and Carpenter published an elegant chemical trapping study which demonstrated that CBD has a nonsquare geometry in solution.[62] CBD–1,4-d$_2$ (57) was generated by thermal deazetation of azo compound 56. Whitman and Carpenter found that with a reactive dienophile, Diels–Alder trapping of 57 was faster than the transformation of 57 to 58 by bond shifting.

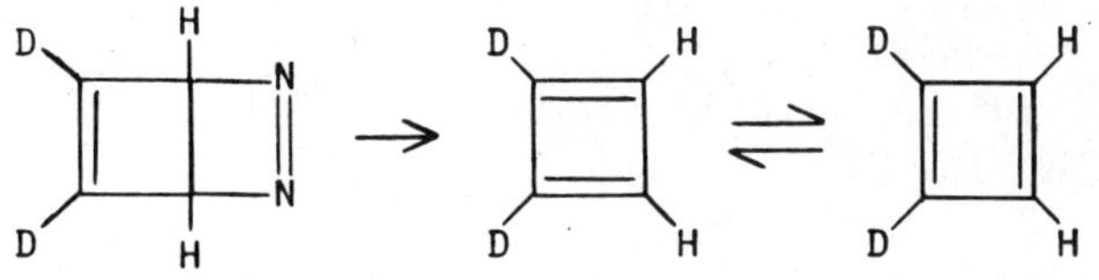

56   57   58

More recently, Whitman and Carpenter have measured the temperature dependence of the ratio of k_2, the rate constant for trapping with methyl 3-cyanoacrylate, to k_1, the rate constant for bond shifting.[63] From their data they obtained the *differences* in the activation parameters for these two processes as $\Delta\Delta H^{\ddagger} = 1.6$ kcal/mol and $\Delta\Delta S^{\ddagger} = 8.3$ cal/mol-°K, with bond shifting having the greater enthalpy and entropy of activation. A lower limit of 1.6 kcal/mol could therefore be placed on $\Delta H^{\ddagger}$ for bond shifting in CBD.

The most interesting aspect of their results, however, is the entropy of activation that is indicated for bond shifting in CBD. Since the Diels–Alder reaction has a very negative entropy associated with it (-25 to -40 cal/mol-

°K), the entropy of activation for bond shifting in CBD is apparently also negative, and substantially so. This is quite a surprising result for a reaction that might have been anticipated to have $\Delta S^{\ddagger} \approx 0$.

Carpenter has suggested that the substantially negative entropy of activation for bond shifting in CBD is due to heavy atom tunneling.[64] Although heavy atom tunneling usually does not contribute to rate constants, the narrowness of the barrier in CBD for going from one set of rectangular bond lengths to another makes possible a large tunneling contribution to bond shifting. In fact, from a model calculation Carpenter estimated that tunneling accounts for >97% of the total reaction below 0°C.

Tunneling not only leads to a negative entropy of activation, but it also allows the enthalpy of activation to be lower than the classical barrier height. Thus with a barrier height of 10.8 kcal/mol in the model calculation, Carpenter computed $\Delta H^{\ddagger} = 4.6$ kcal/mol and $\Delta S^{\ddagger} = -15$ cal/mol-°K. Carpenter pointed out that tunneling may also contribute to the bond-shifting process in other antiaromatic annulenes.

A previous IR study of CBD-d_2 in matrix isolation found that the same spectrum was observed from precursors which should have led initially to **57** and **58** separately.[65] Schaad and Hess have calculated the IR spectrum of each of these molecules. They argue that the observed spectrum is consistent with a 1:1 mixture of **57** and **58** being observed from both precursors.

Although there is now strong experimental evidence that unsubstituted CBD is rectangular, relief of steric interactions between bulky substituents is capable of moving the ring bond lengths along the reaction coordinate for bond shifting toward a square geometry.[7,66] This accounts for the fact that tetra-*tert*-butyl-CBD has almost equal ring bond lengths[67] and that bond shifting in tri-*tert*-butyl-CBD (**59**) cannot be frozen out on the NMR time scale at temperatures as low as -185°C.[68]

$$\text{59a, } R = H \qquad \text{b, } R = H$$
$$\text{60a, } R = D \qquad \text{b, } R = D$$

Of course, when a degenerate rearrangement cannot be frozen out, it is always possible that the putative transition state is lower in energy than either of the classical structures. It may also be the case that the chemical shifts of the two sites being exchanged (in this case the ^{13}C chemical shifts of C-1 and

C–3 of **59**) are accidentally identical. Maier and co-workers have made use of Saunder's elegant isotopic perturbation of equilibrium technique to show that neither of these possibilities is operative in **59**.[69]

Maier et al. prepared **60**, in which one *tert*-butyl groups was replaced by *tert*-butyl-d$_9$. They found that the ^{13}C resonances for C–1 and C–3 were now split. The splitting presumably arises because the smaller effective size of the C$_4$D$_9$ group favors **60b** over **60a**, causing the ring carbon attached to this group to spend more time in a C–1 site than in a C–3 site. The temperature dependence of the splitting in **60** provides evidence that it is indeed due to isotopic perturbation of the 1:1 equilibrium constant which symmetry demands in **59**.

The heavy groups attached to the ring carbons in **59** should diminish the importance of tunneling in this compound. However, as previously noted, the steric bulk of these groups also has the effect of moving the molecule closer to a square geometry than the parent, thus decreasing the width of the barrier and facilitating tunneling. Model calculations by Carpenter suggest that the latter effect may dominate. Consequently, the demonstration of a diminution in the importance of tunneling in CBD by a mass effect must await the preparation of CBD containing "^{56}C" or a similar isotope, or the substitution of the ring with groups that are massive but that have the same effective size as hydrogen atoms.

Somewhat more realistically, confirmation of the importance of tunneling could come from experiments conducted at temperatures elevated sufficiently so that passage over the top of the barrier dominates tunneling through it. The observed activation parameters should be very different under these conditions from the activation parameters in the low-temperature studies of Whitman and Carpenter. Both the entropy and enthalpy of activation should increase.

The destabilizing effect of a CBD ring was apparent in **61**, which was formed by dimerization of 1,5-cyclooctadien-3-yne.[70] Although **61** could be trapped by a Diels–Alder reaction with phenyltriazolinedione, it underwent rapid electrocyclic ring opening to give the annelated octatraene **62**. In contrast, in the unannelated system, the equilibrium lay on the side of the cyclooctatriene instead of the octatetraene.

61   62

Klarner and co-workers have devised a very clever experiment for measuring the destabilization energy due to the antiaromaticity of CBD.[71] They utilized the norcaradiene derivatives **63a** and **63b**. Unlike the compound containing a saturated four-membered ring, which exists exclusively as the cycloheptatriene valence isomer, no trace of **64** was detected. The absence of **64** is presumably due to its containing a CBD ring, which destabilizes this valence isomer.

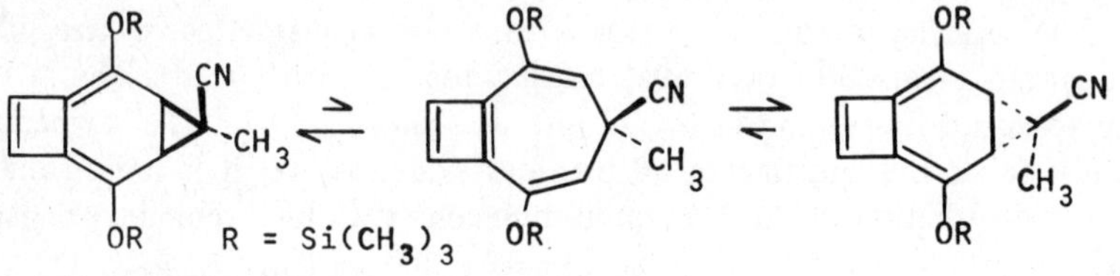

R = Si(CH₃)₃

63a   **64**   **63b**

The activation parameters for the equilibration of **63a** and **63b**, presumably through the intermediacy of **64**, were measured and compared with those of the norcaradiene that has the same substitution but lacks the annelated four-membered ring present in **63**. The data indicated that the unsaturated ring increased the activation energy for cycloheptatriene formation by about 14 kcal/mol. Correcting for the calculated negative resonance energy of the dimethylene-cyclobutene ring present in **63** and the stabilizing effect of the two vinyl groups in **64**, the antiaromaticity of the CBD ring was estimated to destabilize **64** by as much as 21 kcal/mol.

B.   Cyclooctatetraene

Unlike CBD, cyclooctatetraene (COT) is nonplanar. Nevertheless, the energy required for bond shifting in the planar molecule can be obtained from the difference between the energies required for ring inversion with and without bond shifting.[7] Paquette and co-workers have prepared a number of methyl-substituted COT's, and they find that the energy required for bond shifting in the planar molecule decreases with increasing numbers of vicinal methyl groups.[72] In 1,2,3,4-tetramethylcyclooctatetraene the rates of ring inversion with and without bond shifting are essentially the same. As is the case with CBD, relief of steric compression between substituents in COT can apparently drive the molecule along the reaction coordinate for bond shifting.[7]

In the substituted COT's studied by the Paquette group the entropy of activation for bond shifting with ring inversion is negative. However, the entropy of activation for ring inversion is also negative and of about the same magnitude. Thus there is no evidence of substantially negative entropies of activation for the bond-shifting process in these molecules and, hence, no evidence of tunneling.

The entropy of activation for ring inversion with bond shifting in COT itself has been measured recently by Naor and Luz to be -9.7 cal/mol-°K.[73] This value is substantially more negative than those obtained by previous NMR studies of ring inversion without bond shifting in monosubstituted COTs. However, it is also more negative than the entropies of activation that have been previously measured for ring inversion with bond shifting in the same molecules and in the unsubstituted parent.[74] Without further evidence for a negative entropy of activation for the bond-shifting process in the parent COT, there are currently no data to suggest an important role for tunneling in this molecule.

IV.   NON-KEKULÉ HYDROCARBONS

A.   Trimethylenemethane

A study by Claesson et al. of triplet trimethylenemethane (TMM) in a single crystal of methylenecyclopropane has revealed that D, the zero-field parameter that reflects the interaction of the magnetic dipoles of the unpaired electrons in a triplet, is positive.[75] Although the magnitude of D has been known since Dowd first prepared triplet TMM, the sign had not previously been established.[7] Because of the large amount of negative spin density expected at the central carbon of TMM, it was conceivable that D might have been negative. An *ab initio* study of the value of D in TMM showed that the negative spin density in TMM almost halves the value of D that would otherwise be computed.[76]

In their single crystal EPR study of triplet TMM Claesson and co-workers observed temperature-dependent behavior in both D and in the hyperfine coupling parameters.[75] Temperature-dependent, reversible changes in the EPR spectrum of TMM had been previously noted by Dowd and Chow in solid solutions, albeit in a higher temperature range than that explored by Claesson et al.[77] Claesson attributed the changes in D to temperature-dependent oscillation of the TMM molecular plane, and those in the hyperfine parameters to rotation about the threefold symmetry axis in the molecule.

Dowd and Chow have published a full paper detailing their observations of the irreversible disappearance of the EPR triplet signal from TMM at temperatures above $120\,$K.[78] The activation parameters depended on the precursor used, but the largest value measured for the activation energy was less than $8\,$kcal/mol. This is only about half of the singlet-triplet energy separation between $^3A_2$, and $^1B_1$ TMM computed by the best *ab initio* calculations.[1,7]

Because $^1B_1$ TMM is only a local minimum on the singlet potential surface, it is possible that a lower-energy intersection occurs between this surface and the triplet surface. The finding of such a low-energy surface crossing could reconcile the experimental results of Dowd and Chow with computational estimates of the singlet-triplet splitting. To explore this possibility, Feller et al. carried out MCSCF calculations of the potential surface connecting singlet TMM with methylenecyclopropane, the global minimum on the singlet surface.[79]

A two-dimensional projection of the singlet potential surface for the methylenecyclopropane rearrangement was obtained by optimizing the geometry and computing the energy of singlet TMM as a function of two methylene rotational angles. The resulting energies were then Fourier analyzed to yield a closed expression for the energy on the projected singlet surface. A projected triplet surface was obtained by Fourier analysis of the triplet energies obtained at the optimized singlet geometries.

The triplet surface showed only a single minimum, corresponding to a D_{3h} geometry. In contrast, three minima appeared on the singlet surface, corresponding, in order of increasing energy, to methylenecyclopropane (65), $^1B_1$ (66), and $^1A_1$ (67) TMM. The four transition states that connect the minima were also located. As expected from Hammond's postulate, the lowest of these was computed to be the one connecting 65 and 66. Next in energy came the transition state connecting orthogonal (66) and planar (67) TMM. Slightly higher still in energy were the con- and disrotatory transition states connecting 65 and 67.

65   66   67

From the mathematical expressions for the energy on each of the two projected surfaces, the line of surface intersection was located. It was found to occur at energies above that of $^1B_1$ (66) TMM, the singlet used for computing the singlet-triplet energy separation. Therefore there appears to be no low-energy surface intersection and, hence, no obvious way of reconciling Dowd's experimental results with those from *ab initio* calculations.

It should be noted, however, that the calculations by Feller et al. were flawed by their use of the STO-3G basis set to generate the potential surfaces. As observed by Doubleday and co-workers, STO-3G fails to mimic faithfully the long-range interactions between orbitals because of the absence of diffuse functions in this basis set.[10,39] Consequently, the transition states for closure of 66 and 67 to 65, located with this basis set, would be expected to come too late along the reaction pathways.

In fact, the energies calculated with a split-valence (SV) basis set at the

STO-3G transition state geometries showed that these geometries were slightly lower in energy than **66** and **67**. This finding indicates that these geometries lie on the methylenecyclopropane (**65**) side of the transition states that would have been located with the SV basis set. Although the SV energies gave no indication of a low-energy surface intersection, a redetermination of the singlet surface with a basis set of at least SV quality would be desirable.

The singlet potential surface calculated by Feller et al. indicates that methylenecyclopropane (**65**) should form orthogonal $^1B_1$ TMM (**66**) slightly faster than it forms planar $^1A_1$ (**67**). Once formed, **66** can either lead to rearrangement with inversion of configuration in the migrating carbon or to epimerization of the ring carbon that has rotated. One-center epimerization is predicted to be faster than coupled two-center epimerization, which requires formation of $^1A_1$ (**67**). These theoretical results are consistent with previous experimental investigations of the methylenecylcopropane rearrangement.[3,4]

Baldwin and Chang have recently carried out a complete stereochemical analysis of the processes that occur in optically active methylenecyclopropane **68**.[80] They found a good fit to their experimental data by assuming that **68** undergoes methylenecyclopropane rearrangement with complete inversion of configuration, leading to **69** and **70**. This indicates that all of the rearrangement passes through a $^1B_1$ (**66**) intermediate with no contribution from a planar species (e.g. **67**). The ratio of **69**–**70** was 4:1, implying the same preferred direction of ring opening that had been observed in previous studies of other methylenecyclopropanes.

68

69, R = H, R' = D
70, R = D, R' = H

Epimerization of **68** was found to be competitive with rearrangement. The data were best fitted with a nonzero rate constant for coupled rotations at the two methylene ring carbons. This implies that opening of **68** to a planar geometry, presumably that of $^1A_1$ (**67**), did occur, although at a slower rate than opening to **66**. However, formation of a planar TMM intermediate apparently did not lead to rearrangement to **69** or **70**, since a planar TMM would have produced also rearrangement products with configuration maintained at the migrating carbon. Baldwin and Chang also noted that rotation of the unsubstituted ring carbon appears to lead to rearrangement less frequently than to epimerization.

A possible explanation of these results is that a dynamic effect favors keeping rotational energy in a group that is already rotating, rather than transferring it to a nonrotating methylene. This explanation assumes that one-center epimerization and rearrangement occur by way of a common $^1B_1$ (66) intermediate on the potential surface. Alternatively, one could interpret the results of Baldwin and Chang as indicating that there are separate pathways on the potential surface for these two processes, although such an interpretation is not supported by the potential surface calculated by Feller et al.

Gajewski has addressed this question by studying kinetic isotope effects on the stereochemical equilibration of 71 and 72 and on their rearrangement to 73.[81] He and his co-workers found very small isotope effects on the rate of stereochemical equilibration of 71 and 72, but k_H/k_D ratios of about 1.3 were found for the transformation of both stereoisomers to 73. The isotope effect on the latter process arises because the CD_2 group in 71 and 72 must rotate by 90° to form 73.

71 , R = H, R′ = CH₃
72 , R = CH₃, R′ = H

73 , R = H, R′ = CH₃

As Gajewski pointed out, if a common intermediate, for instance, $^1B_1$ (66), were involved in this reaction and in the stereomutation, an inverse isotope effect would have been expected on the rate of equilibration of 71 and 72. An inverse effect is anticipated because, as the rate along the path to 73 is slowed by the presence of deuterium, the putative intermediate should be channeled toward undergoing increased amounts of stereomutation. The failure to observe such an effect can be interpreted as indicating that stereomutation and rearrangement occur by different pathways. The alternative interpretation, that a common intermediate is involved, is not ruled out, but it would require that there be an unexpectedly large normal isotope effect on the formation of such an intermediate.

Crawford and Chang have used isotope effects to show that TMM is not an intermediate in the thermolysis of 4-ethylidenepyrazolines (74).[82] Kinetic isotope effects suggest that the reaction proceeds by cleavage of the C—N bond *anti* to the methyl group to form a diazenyl diradical (75). The preferential cleavage of this C—N bond is presumably due to preferred formation of the least sterically crowded allylic radical.

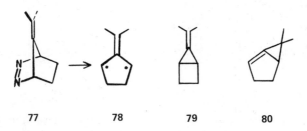

74a , R = H, R′ = D   75   76a , R = H, R′ = D
 b , R = D, R′ = H   b , R = D, R′ = H

The labeling in the major products formed (76a from 74a and 76b from 74b) is difficult to rationalize in terms of subsequent formation of a TMM intermediate. Instead, Crawford and Chang proposed a mechanism in which electrocyclic cyclopropane ring closure is accompanied by nitrogen loss. A similar mechanism was suggested by them for deazetation of 4-methylenepyrazoline, again based on kinetic isotope effects and product labeling.[83]

As Crawford and Chang pointed out, some 4-alkylidenepyrazolines may, in fact, decompose to yield TMM type diradicals. Berson has written an excellent review of the work carried out by his research group on deazetation of bicyclic alkylidenepyrazolines (77) to yield alkylidenecyclopentane-1,3-diyls (78).[84] Reactions of both the singlet and triplet diradicals have been studied; the chemistry of the highly strained, bicyclic, methylenecyclopropane derivatives (79 and 80) that can be formed by ring closure in 78 has also been investigated.

77   78   79   80

Berson's data suggest that 79 and 80 both have negative bond dissociation energies; that is, triplet 78 is lower in energy than either closed-shell hydrocarbon. The singlet-triplet splitting of 13.3 kcal/mol in 78, indicated by Berson's studies, is in reasonable agreement with the results of *ab initio* calculations not only on 78[85,86] but also on the parent TMM.[7] This energy difference is considerably larger than the 8 kcal/mol suggested by Dowd's kinetic results for the parent diradical.

Berson, Rentzepis, and co-workers have carried out picosecond laser flash photolysis studies of the singlet TMM diradical (78) formed from 77.[87] They observed a fluorescence with a lifetime of 280 psec, which they attributed to an excited singlet state of **78**. The authors suggested that the pathway leading to this state involves a two-photon process. Absorption of the first photon leads to deazetation of **77** so rapidly that the **78** formed can absorb a second photon from the same laser pulse. Indeed, a second fluorescence was observed with a lifetime of 38 psec which was attributed to excited **77** undergoing predissociation. The authors suggested that the process that limits the lifetime of excited **78** to 280 psec is intersystem crossing to an excited triplet state.

TMM and derivatives are finding increasing use in organic synthesis. Little has exploited the regioselective trapping of singlet **78** in the synthesis of hirsutene.[88] Palladium complexes of TMM have also been used to construct five-membered rings, and the regiochemistry of additions to a methyl-substituted complex has been explored.[89] The nucleophilicity found for the substituted carbon atom has been rationalized by the results of molecular orbital calculations.[90]

B.   Quinodimethanes

The chemistry of quinodimethanes has been recently reviewed by Platz,[91] so that only studies published after his review are discussed here. A full paper describing Platz's EPR studies of 1,8-naphthoquinodimethane (**81**) and heteroatom analogues has appeared.[92] Berson has also published the details of his group's EPR studies of *m*-quinomethane (**82**) and *m*-naphthoquinomethane (**83**).[93] All three diradicals were found to be ground state triplets, as expected from simple theoretical considerations.[7,91]

81   82   83

Pagni and co-workers have discussed experiments that indicate that isomerization, as well as epimerization, of naphthocyclopropane **84** proceeds by way of the bridged 1,8-naphthoquinodimethane **85**.[94] At temperatures around 150°C, singlet **85** undergoes reclosure to **84** about 36 times faster than it rearranges to phenalene (**86**). This result is quite interesting because Michl and co-workers have previously found that in glassy solutions at temperatures around 125 K, triplet **85** gives **86** rather than **84**.[91,95] This could be another example, and a most dramatic one at that, of intersystem crossing leading to a product distribution different from that formed from a singlet diradical. However, the very different conditions of temperature and phase in the two experiments could also be responsible for the fact that they give different products.

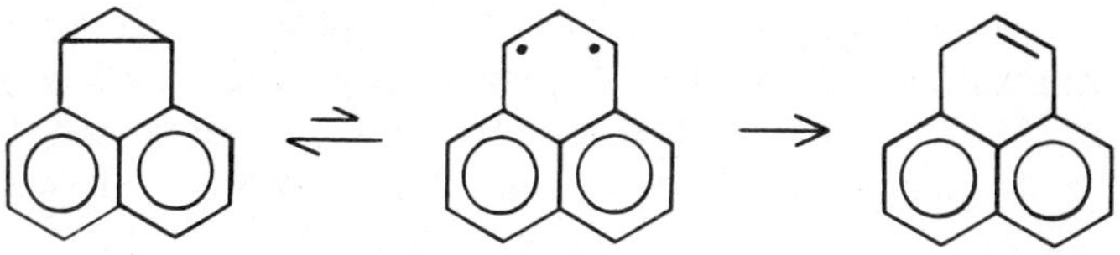

84   85   86

Pagni et al. found that thermally generated singlet **85** did not react with O_2, but the photochemically produced triplet did. Because photolysis of **84** in solution gives amounts of triplet **85** that are too small to be detected spectroscopically, it is not clear whether oxygen simply scavenges a very small amount of triplet that is present or also acts to accelerate intersystem crossing in excited **84**.

Wright and Platz have reported the preparation of *m*-quinodimethane (**87**).[91,96] Their EPR studies indicate a triplet ground state for this diradical, too. A triplet ground state is in accord with naive expectations based on orbital localizability[7,91] and with the results of recent *ab initio* calculations.[97]

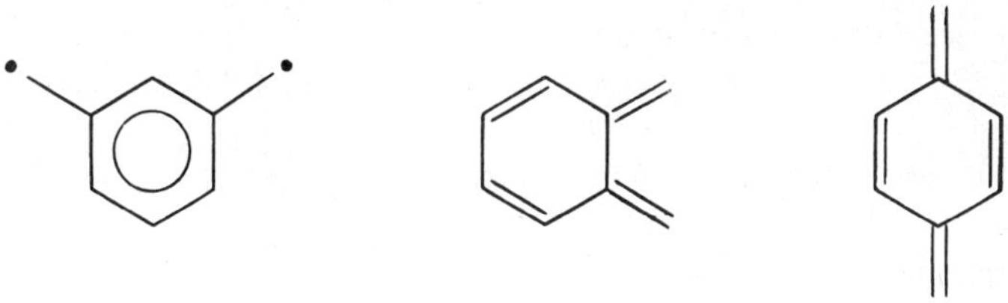

87   88   89

Hehre and co-workers have devised a clever technique for obtaining the energy of a diradical like **87**, relative to that of a closed-shell isomer such as **88** or **89**.[98] The method involves measuring the proton affinities of the isomers and measuring or calculating the relative energies of the cations formed from them. In the case of the isomeric quinodimethanes, the enthalpy of gas-phase deprotonation of the *o*- and *p*-methylbenzyl cations, coupled with *ab initio* calculations of their relative energies, gave the relative energies of **88** and **89**. Unfortunately, none of the bases tried proved strong enough to deprotonate the *m*-methylbenzyl cation in the gas phase. Nevertheless, this failure allowed Hehre et al. to deduce that **87** must be at least 23 kcal/mol less stable than **88** and at least 26 kcal/mol less stable than **89**.

Since gas-phase deprotonation on the time scale of an ICR experiment would be expected to be conservative of the singlet spin state of the carbocation, the lowest singlet state of **87** would presumably be produced initially. This state, $^1A_1$, has been calculated to lie 10 kcal/mol above the triplet ground state.[97] Thus the experimental results of Hehre et al. would place a lower limit of 16 kcal/mol on the energy difference between the triplet ground state of **87** and the singlet ground state of **89**. *Ab initio* calculations yield an energy difference of 24 kcal/mol between the ground states of the two molecules.[97]

V.   SUMMARY AND FUTURE PROSPECTS

The publication in 1982 of a monograph on diradicals[2] and a *Tetrahedron* symposium in print devoted to these reactive intermediates[99] indicates the interest and activity in this area of chemistry. However, the appearance of a monograph in an area of research sometimes signals that activity in the area has peaked and that decline of interest is imminent as researchers begin to move into less mature areas.

Although advances in the field of diradicals have been chronicled here and in other reviews,[1-4] there are as yet too many unanswered questions to warrant calling this field mature. Some of these, discussed in different sections of this review, are (1) the lifetimes of nonconjugated singlet diradicals and whether these species do represent minima on free energy, if not potential energy, surfaces; (2) the extent to which intersystem crossing determines the lifetime of and the products formed from triplet diradicals; (3) the importance of tunneling in the bond-shifting process in antiaromatic annulenes; and (4) whether one-center epimerization in methylenecyclopropane involves the same diradical intermediate implicated in its rearrangement.

Despite the large number and variety of types of diradicals that have been prepared, many synthetic challenges remain. Among these are the generation of singlet diradicals under conditions where they can be detected spectroscopically and their decay channels (i.e., product formation and intersystem crossing) monitored, the preparation of diradicals (other than CBD) in which steric blockading is used to render them persistent at ambient temperatures, and the design of non-Kekulé hydrocarbons with spin multiplicities higher than triplet. Perhaps in future volumes of *Reactive Intermediates* it will prove necessary to retitle this chapter "Polyradicals."

VI.   REFERENCES

1.   W. T. Borden in M. Jones, Jr. and R. A. Moss, Eds., *Reactive Intermediates,* Vol. 2, Wiley-Interscience, New York, 1981, pp. 175–209.

2.   W. T. Borden, Ed., *Diradicals,* Wiley-Interscience, New York, 1982.

3.   J. A. Berson in P. de Mayo, Ed., *Rearrangements in Ground and Excited States,* Vol. 1, Academic, New York, 1980, pp. 311–390.

4. J. J. Gajewski, *Hydrocarbon Thermal Isomerizations,* Academic, New York, 1981.

5. For a recent review see P. B. Dervan and D. A. Dougherty in Reference 2, pp. 107–149.

6. S. W. Benson, *J. Chem. Phys.,* 34, 521 (1961).

7. For a recent review, see W. T. Borden in Reference 2, pp. 1–72.

8. W. von E. Doering, *Proc. Natl. Acad. Sci. U.S.A.,* 78, 5279 (1981).

9. E. V. Waage and B. S. Rabinovitch, *J. Chem. Phys.,* 76, 1695 (1972).

10. C. Doubleday, Jr., J. W. McIver, Jr., and M. Page, *J. Am. Chem. Soc.,* 104, 6533 (1982).

11. The reversal in the relative MCSCF energies of (0,0) and (0,90) that was reported by S. Kato and K. Morokuma, *Chem. Phys. Lett.,* 65, 19 (1979) appears to have been a typographical error. See footnote 16 in Reference 10.

12. J. A. Berson and L. D. Pedersen, *J. Am. Chem. Soc.,* 97, 238 (1975); J. A. Berson, L. D. Pedersen, and B. K. Carpenter, *J. Am. Chem. Soc.,* 98, 122 (1977).

13. A. H. Goldberg and D. A. Dougherty, *J. Am. Chem. Soc.,* 105, 284 (1983).

14. S. L. Buchwalter and G. L. Closs, *J. Am. Chem. Soc.,* 101, 4688 (1979).

15. M. H. Chang and D. A. Dougherty, *J. Org. Chem.,* 46, 4092 (1981).

16. W. L. Jorgensen and W. T. Borden, *J. Am. Chem. Soc.,* 95, 6649 (1973).

17. M. H. Chang and D. A. Dougherty, *J. Am. Chem. Soc.,* 104, 1131 (1982).

18. K. K. Shen and R. G. Bergman, *J. Am. Chem. Soc.,* 99, 1655 (1977).

19. M. H. Chang and D. A. Dougherty, *J. Am. Chem. Soc.,* 104, 2333 (1982).

20. L. McElwee-White and D. A. Dougherty, *J. Am. Chem. Soc.,* 104, 4722 (1982).

21. M. P. Schneider and H. Bippi, *J. Am. Chem. Soc.,* 102, 7363 (1980).

22. T. C. Clarke, L. A. Wendling, and R. G. Bergman, *J. Am. Chem. Soc.,* 99, 2740 (1977).

23. W. von E. Doering and E. Z. Barsa, *Proc. Natl. Acad. Sci. U.S.A.,* 77, 2355 (1980).

24. J. E. Baldwin and C. E. Carter, *J. Am. Chem. Soc.,* 104, 1362 (1982).

25. J. A. Berson, L. D. Pedersen, and B. K. Carpenter, *J. Am. Chem. Soc.,* 97, 240 (1975).

26. M. R. Willcott III and V. H. Cargle, *J. Am. Chem. Soc.,* 91, 4310 (1969).

27. H. E. Zimmerman in P. de Mayo, Ed., *Rearrangements in Ground and Excited States,* Vol. 3, Academic, New York, 1980, pp. 131–166.

28. H. E. Zimmerman, R. J. Boettcher, N. E. Buehler, G. E. Keck, and M. J. Steinmetz, *J. Am. Chem. Soc.,* 98, 7680 (1976).

29. This technique has subsequently been applied to the study of the photochemical rearrangement of norbornadiene to quadricyclane by N. J. Turro, W. R. Cherry, M. F. Mirbach, and M. J. Mirbach, *J. Am. Chem. Soc.,* 99, 7388 (1977).

30. L. A. Paquette and E. Bay, *J. Org. Chem.,* 47, 4597 (1982).

31. (a) M. Demuth, D. Lemmer, and K. Schaffner, *J. Am. Chem. Soc.,* 102, 5409 (1980): (b) M. Demuth, W. Amrein, C. O. Bender, S. E. Braslavsky, U. Burger, M. V. George, D. Lemmer, and K. Schaffner, *Tetrahedron,* 37, 3245 (1981).

32. W. Adam, O. De Lucchi, and I. Erden, *J. Am. Chem. Soc.,* 102, 4806 (1980).

33. W. Adam and O. De Lucchi, *J. Am. Chem. Soc.,* 102, 2109 (1980); *J. Org. Chem.,* 46, 4133 (1981).

34. W. Adam, N. Carballeira, and O. De Lucchi, *J. Am. Chem. Soc.,* 102, 2107 (1980).

35. W. Adam, O. De Lucchi, K. Peters, E.-M. Peters, and H. G. von Schnering, *J. Am. Chem. Soc.,* **104**, 5747 (1982).

36. W. Adam, O. De Lucchi, and K. Hill, *J. Am. Chem. Soc.,* **104**, 2934 (1982).

37. W. Adam, N. Carballeira, and O. De Lucchi, *J. Am. Chem. Soc.,* **103**, 6406 (1981).

38. G. Segal, *J. Am. Chem. Soc.,* **96**, 7892 (1974).

39. C. Doubleday, Jr., J. W. McIver, Jr., and M. Page, *J. Am. Chem. Soc.,* **104**, 3768 (1982).

40. W. T. Borden and E. R. Davidson, *J. Am. Chem. Soc.,* **102**, 5409 (1980).

41. Electron repulsion effects in tetramethylene are discussed within the GVB formalism by Goldberg and Dougherty in Reference 13.

42. C. Doubleday, Jr., R. N. Camp, H. F. King, J. W. McIver Jr., D. Mullally, and M. Page, *J. Am. Chem. Soc.* **106**, 447 (1984). I thank Professor Doubleday for informing me of these results in advance of publication.

43. P. G. Schultz and P. B. Dervan, *J. Am. Chem. Soc.,* **104**, 6660 (1982).

44. J. C. Sciano, *Tetrahedron,* **38**, 819 (1982).

45. P. D. Bartlett and N. A. Porter, *J. Am. Chem. Soc.,* **90**, 5317 (1968).

46. P. S. Engel and D. E. Keys, *J. Am. Chem. Soc.,* **104**, 6860 (1982).

47. P. J. Wagner, K.-C. Liu, and Y. Noguchi, *J. Am. Chem. Soc.,* **103**, 3837 (1981). For a review of diradicals generated by Norrish Type I and II photoreactions of ketones see P. J. Wagner in P. de Mayo, Ed., *Rearrangements in Ground and Excited States,* Vol. 3, Academic, New York, 1980, pp. 381–444.

48. J. C. Sciano, *Acc. Chem. Res.,* **15**, 252 (1982).

49. M. V. Encinas, P. J. Wagner, and J. C. Sciano, *J. Am. Chem. Soc.,* **102**, 1357 (1980).

50. S. C. Freilich and K. S. Peters, *J. Am. Chem. Soc.,* **103**, 6255 (1981).

51. R. A. Caldwell, T. Majima, and C. Pac, *J. Am. Chem. Soc.,* **104**, 629 (1982).

52. J. C. Sciano, C. W. B. Lee, Y. L. Chow, and B. Marciniak, *J. Phys. Chem.,* **86**, 2452 (1982).

53. C. Doubleday, Jr., *Chem. Phys. Lett.,* **64**, 67 (1979).

54. C. Doubleday, Jr., *Chem. Phys. Lett.,* **79**, 375 (1981).

55. C. Doubleday, Jr., *Chem. Phys. Lett.,* **77**, 131 (1981).

56. R. Kaptein, F. J. J. de Kanter, and G. H. Rist, *J. Chem. Soc. Chem. Commun.,* 499 (1981).

57. F. J. J. de Kanter and R. Kaptein, *J. Am. Chem. Soc.,* **104**, 4759 (1982).

58. For a recent review see N. J. Turro and B. Kraeutler in Reference 2, pp. 259–321.

59. L. M. Stephenson, P. R. Cavigli, and J. L. Parlett, *J. Am. Chem. Soc.,* **93**, 1984 (1981).

60. C. P. Casey and R. A. Boggs, *J. Am. Chem. Soc.,* **94**, 6457 (1972).

61. T. Balley and S. Masamune, *Tetrahedron,* **36**, 343 (1980).

62. D. W. Whitman and B. K. Carpenter, *J. Am. Chem. Soc.,* **102**, 4272 (1980).

63. D. W. Whitman and B. K. Carpenter, *J. Am. Chem. Soc.,* **104**, 6473 (1982).

64. B. K. Carpenter, *J. Am. Chem. Soc.,* **105**, 1700 (1983).

65. O. L. Chapman, D. De La Cruz, R. Roth, and J. J. Pacansky, *J. Am. Chem. Soc.,* **95**, 1337 (1973).

66. W. T. Borden and E. R. Davidson, *J. Am. Chem. Soc.*, **102**, 7958 (1980).

67. H. Irngartinger, N. Riegler, K.-D. Malsch, K.-A. Schneider, and G. Maier, *Angew. Chem. Int. Ed. Engl.*, **19**, 211 (1980). A recent redetermination of the structure at low temperature showed, however, that the ring bond lengths are less equal than the initial publication indicated. H. Irngartinger and M. Nixdorf, *Angew. Chem. Int. Ed. Engl.*, **22**, 403 (1983).

68. G. Maier, U. Schafer, W. Sauer, H. Hartan, R. Matusch, and J. F. M. Oth, *Tetrahedron Lett.*, 1837 (1978).

69. G. Maier, H.-O. Kalinowski, and K. Euler, *Angew. Chem. Int. Ed. Engl.*, **21**, 693 (1982).

70. H. Meier, T. Echter, and O. Zimmer, *Angew. Chem. Int. Ed. Engl.*, **20**, 865 (1981).

71. F.-G. Klarner, E. K. G. Schmidt, and M. A. Rahman, *Angew. Chem. Int. Ed. Engl.*, **21**, 138 (1982), **21**, 139 (1982).

72. L. A. Paquette, *Pure Appl. Chem.*, **54**, 987 (1982); L. A. Paquette and J. M. Gardlik, *J. Am. Chem. Soc.*, **102**, 5033 (1980) and references cited therein.

73. R. Naor and Z. Luz, *J. Chem. Phys.*, **76**, 5662 (1982).

74. For a review see J. F. M. Oth, *Pure Appl. Chem.*, **25**, 573 (1971).

75. O. Claesson, A. Lund, T. Gilbro, T. Ichikawa, O. Edlund, and H. Yoshida, *J. Chem. Phys.*, **72**, 1463 (1980).

76. D. Feller, W. T. Borden, and E. R. Davidson, *J. Chem. Phys.*, **74**, 2256 (1981).

77. P. Dowd and M. Chow, *J. Am. Chem. Soc.*, **99**, 9825 (1977).

78. P. Dowd and M. Chow, *Tetrahedron*, **38**, 799 (1982).

79. D. Feller, K. Tanaka, E. R. Davidson, and W. T. Borden, *J. Am. Chem. Soc.*, **104**, 967 (1982).

80. J. E. Baldwin and G. E. U. Chang, *Tetrahedron*, **38**, 825 (1982).

81. J. J. Gajewski, C. W. Brenner, B. N. Stahly, R. F. Hall, and R. I. Sato, *Tetrahedron*, **38**, 853 (1982). Isotope effects on reactions that presumably involve tetramethyleneethane diradicals are also discussed in this paper.

82. R. J. Crawford and M. H. Chang, *Tetrahedron*, **38**, 837 (1982).

83. M. H. Chang and R. J. Crawford, *Can. J. Chem.*, **59**, 2556 (1981).

84. J. A. Berson in Reference 2, pp. 151–194. The thermal and photochemical rearrangements of **78**, which are described as being "unpublished results" in this review, have now appeared in print: S. P. Schmidt, A. R. Pinhas, J. H. Hammons, and J. A. Berson, *J. Am. Chem. Soc.*, **104**, 6822 (1982); M. R. Mazur, S. E. Potter, A. R. Pinhas, and J. A. Berson, *J. Am. Chem. Soc.*, **104**, 6823 (1982).

85. D. A. Dixon, T. H. Dunning, R. A. Eades, and D. A. Kleier, *J. Am. Chem. Soc.*, **103**, 2878 (1981).

86. S. B. Auster, R. M. Pitzer, and M. S. Platz, *J. Am. Chem. Soc.*, **104**, 3812 (1982).

87. D. F. Kelley, P. M. Rentzepis, M. M. Mazur, and J. A. Berson, *J. Am. Chem. Soc.*, **104**, 3764 (1982).

88. R. D. Little and G. W. Muller, *J. Am. Chem. Soc.*, **103**, 5974 (1981).

89. B. M. Trost and D. M. T. Chan, *J. Am. Chem. Soc.*, **103**, 5974 (1981) and references therein.

90. D. J. Gordon, R. F. Fenske, T. N. Nanninga, and B. M. Trost, *J. Am. Chem. Soc.*, **103**, 5974 (1981).

91. M. S. Platz in Reference 2, pp. 195–258.

92. M. S. Platz, G. Carrol, F. Pierrat, J. Zayas, and S. Auster, *Tetrahedron*, **38**, 777 (1982).

93. M. Rule, A. R. Matlin, D. E. Seeger, E. F. Hilsinki, D. A. Dougherty, and J. A. Berson, *Tetrahedron*, **38**, 787 (1982).

94. R. M. Pagni, M. N. Burnett, and H. M. Hassaneen, *Tetrahedron*, **38**, 843, (1982).

95. J.-F. Muller, D. Muller, H. J. Dewey, and J. Michl, *J. Am. Chem. Soc.*, **100**, 1629 (1978).

96. B. W. Wright and M. S. Platz, *J. Am. Chem. Soc.*, **105**, 628 (1983).

97. S. Kato, K. Morokuma, D. Feller, E. R. Davidson, and W. T. Borden, *J. Am. Chem. Soc.*, **105**, 1791 (1983).

98. S. K. Pollack, B. C. Raine, and W. J. Hehre, *J. Am. Chem. Soc.*, **103**, 6308 (1981).

99. J. Michl, Ed., *Tetrahedron*, **38**, 733–867 (1982).

6

CARBOCATIONS

EDWARD M. ARNETT

Department of Chemistry, Duke University, Durham, North Carolina 27706

THOMAS C. HOFELICH

Dow Chemical U.S.A., Midland, Michigan 48640

GEORGE W. SCHRIVER

Department of Chemistry, Tulane University, New Orleans, Louisiana 70118

I. INTRODUCTION

This chapter reports on some areas of progress in carbocation chemistry between 1979 and March 1983. Following the philosophy of this series of volumes, which was well exemplified in the previous chapters on carbocations, we concentrate on a few papers which seem to represent especially clear advances in the field. Not surprisingly, these cover topics that also have been emphasized in the two earlier chapters by Bethell and Whittaker: NMR techniques and thermodynamics of stabilized carbocations in superacid media, thermodynamic comparisons of cation stabilities in the gas phase, theoretical calculations, new evidence regarding the importance of ion-pairing and nucleophilic solvation, and unification of solvolysis mechanisms. There is much less emphasis on new attempts to rationalize complicated kinetic systems, rearrangements, or solvolysis. Likewise, we say nothing about new types of carbocations or synthetic applications, interesting though they might be. For an overall review of carbocation chemistry during the period, we recommend the annual Capon—Rees series on *Organic Reaction Mechanisms* (Interscience Publishers).

Following our predecessors, we have attempted to maintain a high degree of skepticism and criticism throughout. Before proceeding to the discussion of new experimental results, we have been emboldened to make a few generalizations on the rules of inference as they are applied in physical organic chemistry. We then examine the continuing controversy about nonclassical cations from this perspective. In view of the attention that has been given to the topic in previous reviews on carbocations we feel justified in trying a different viewpoint.

The debate about charge delocalization in the 2-norbornyl and cyclopropylcarbinyl cations continues unabated after almost a generation and a half of argument or experiment. It has engaged from its start some of the most powerful minds and personalities in organic chemistry, many of whom are still alive and continue to be the principal debaters of the case, although other figures occasionally come and go from the stage. It is a significant fact that none of the rising younger leaders in physical organic chemistry are jumping into the fray. Their reactions, as expressed to us, range from polite interest to amusement to irritation to boredom. So great is the mass of information to be assimilated, and so subtle are some of the arguments, that it is hard to choose sides even if one wished to. We would like to consider whether the "norbornyl cation problem" has any real chemical significance or if, by Irving Langmuir's criteria, it is an example of "pathological science" similar to the recent "polywater" episode.

Another clue to the marginal relevance of the nonclassical ion controversy is the fact that the chemistry of carbocations has progressed so successfully without its resolution. Those who symbolize the controversial ions with dotted lines seem to proceed as successfuly as those who write them with fixed charges. There is as yet no evidence of opportunity loss because of inappropriate symbolism. This is quite different from the powerful advances that followed the collapse of the phlogiston theory or the resolution of the benzene problem through resonance theory. In our opinion there are some substantive chemical questions to be asked about the 2-norbornyl cation. The failure to get them answered expeditiously reflects only partially the limitations of inference in physical organic chemistry. In the discussion to follow we hope to examine the nonclassical ion controversy from several perspectives, after first presenting some working axioms.

II.   SOME BASIC PRINCIPLES FOR CARBOCATION STUDY

The following precepts have been collected from many discussions with our colleagues. They are part of the working (but usually unstated) policy toward the study of reactive intermediates which forms the ground rules for analyzing many organic chemical problems. Although the rules are not a credo, there is a well-accepted hierarchy of values and an approach toward studying chemistry that underlies many scientific value judgments. We have attempted to articulate here some of those that are most important for the study of processes involving carbocations. The list is not necessarily complete and also could apply to other reactive intermediates.

1.   Organic chemistry is concerned with the relationship between the structures and reactivities of carbon compounds.

2.   Reactivity is expressed in terms of reaction rates which represent the free energy changes required to reach transition states that are, by definition, too unstable for direct observation by spectroscopy, or for isolation.

3.   Thermodynamic comparisons involve the measurable energy changes for conversion of reactants to products. The heat of the reaction[1] can be measured by calorimetry for most processes − the free energy of reaction is derived from

the equilibrium constant in cases where the energy difference between reactants and products is small enough so that they can be observed concurrently at equilibrium. Heats and free energies of reaction can also be calculated exactly from heats and free energies of formation of reactants and products.

4.   Enthalpy and free energy changes for the reaction of similar compounds in similar processes are usually of nearly the same magnitude or are proportional to each other. However, differences between them (differential entropy terms) are common enough so that it is risky to argue fine points about free energy changes from thermochemical data or vice versa. (Plots of ΔG versus ΔH for similar compounds in a given process are usually, but not always linear.)

5.   Since all measures of reactivity by kinetics or thermodynamics involve the conversion of one species into another, the relative energies of the initial states are just as important in any discussion as those of the secondary states (i.e., transition states, excited states, or products). Therefore all comparisons of rates, equilibrium constants, heats of reaction, spectra, and so on, involve at least the comparison of energies of four species — two initial states and two secondary states. The history of carbocation chemistry provides many examples in which comparisons of reaction rates have been discussed entirely in terms of the stabilities of presumed ionic intermediates or transition states.

6.   Since reaction rates are inherently of great interest, an important activity of physical organic chemistry is exploring the relationship between kinetic and thermodynamic measures of reactivity. Hammond's postulate; the Marcus, Brönstead, and related equations; and the reactivity–selectivity principle are a few examples. Useful as these are as guidelines for interpretation and prediction, they are based ultimately on general experience, and their theoretical justification requires many approximations. Carbocation chemistry abounds with examples of species that are kinetically stable (i.e., long-lived) but that are thermochemically quite unstable relative to near homologues.

7.   The methods that are used to interpret the relations between structure and reactivity come down to the judicious use of reasoning by analogy. There are at present no theories of NMR chemical shifts, vibrational or electronic spectra, that can predict exactly what the spectrum of any molecule or ion should be. In the case of proving the structure of an "interesting" new species, its structures and energy will almost always fall outside the usual monotonic range of models based on simple homology. It is exactly the question of their abnormal structures and presumed abnormal energies that makes them "interesting." As a result, it may be difficult to prove the structure of the presumed species (especially if it is quite unstable) because of disagreements on what models should be chosen as reference points for interpreting spectra. Symmetry properties can provide absolute evidence (e.g., isomer number, X-ray crystallographic analysis, NMR splitting patterns), but most spectral and energetic comparisons are based ultimately on analogies.

8.   In view of our dependence on models for structure demonstration, it is usually important to develop an extensive background with a wide range of reasonable models before using a new experimental technique to attack a controversial problem. Unless the technique is more clearly understood than the problem, the results will be ambiguous.

9.   Most carbocation reactions take place in solution where solvation has an enormous influence. The inherent stabilities of the species involved can only be compared in the gas phase, where most methods of structure elucidation cannot be applied. Differential solvation energies can be very large for some kinds of compounds or ions. For carbocations differential solvation energies are mostly surprisingly small (compared to ammonium or oxonium ions, for example), but are by no means always negligible.

10.   Because of the logarithmic relationship between rates and equilibrium constants with free energy terms, an error of only 1 kcal/mol in estimating energies of activation or reaction results in more than a sevenfold error in rate or equilibrium constant. Heats of reaction, which measure *enthalpies* of reaction to about ± 0.4 kcal/mol are inappropriate tools for examining the source of modest rate (*free energy* of activation) differences.

III.   SOME VARIED PERSPECTIVES OF A TEST CASE – THE 2-NORBORNYL CATION PROBLEM

A.   *Much Ado About Nothing?*

Twenty-five years ago the senior author asked one of the leading contributors to a lively symposium on "nonclassical ions" just why there seemed to be such intense emotional involvement in what appeared to be a somewhat arcane and convoluted problem. His answer, "Because we're that kind of guys" was an early clue that something more than detached pure reason was involved. In the ensuing years, as curious spectators and occasional participants in the controversy, we have tried repeatedly to decide what (if any) important scientific questions were at stake. Who, after all, would not prefer to work on an inherently important problem than on an unimportant one? In the following sections we discuss the 2-norbornyl ion in terms of its structure and energy, and a historical perspective on the controversy that has surrounded it.

B.   Concerning the Structure and Energy of the 2-Norbornyl Cation

By the spring of 1983 the principal substantive questions remaining about the 2-norbornyl cation are the following: (1) Is it stabilized to a significant extent by interaction between the electron cloud surrounding C–6 and the open orbital on the 2-carbon? (2) Does this lead to the cation having a symmetrical structure represented by a single energy minimum?

These are important questions because secondary ions in solution are now at the cutting edge of carbocation chemistry. The structure-energy relationships of most tertiary ions are well understood, whereas primary ions are so reactive in solution that they may not even exist *per se* for more than one molecular vibration. Most secondary aliphatic and alicyclic cations also rearrange instantly to tertiary ions. However, a few such as 2-propyl, 2-butyl, cyclopentyl, and 2-norbornyl can be prepared and kept at low temperature in superacid where intermolecular reactions have been minimized. These ions also do not have accessible intramolecular channels for escape to more thermodynamically stable systems, so that they can be examined directly. However, to decide whether or not their structures or energies are abnormal, appropriate models for "normal" secondary ions are required. As stated, at present there are very few such models for comparison in solution under stable ion conditions.

Several new NMR techniques, to be discussed in later sections, have been brought to bear on the structure of the 2-norbornyl cation at low temperature in solution and the (noncrystalline) solid state. Although the methods are still subject to question, they are consistent with a growing body of evidence that if there really is a barrier between rapidly equilibrating cations it must be very low. If the barrier is low enough so that the frequency for equilibration becomes that for a molecular vibration (10^{13}/sec), the structure shall have been settled as that of a single minimum nonclassical ion.

The energetic question, which relates to the structural one, is whether there is evidence for a special degree of stabilization in the 2-norbornyl cation. This immediately takes us into the problem of choosing the right models for comparison. If the free ion is under discussion, then its heat of formation might seem to be the desired property taken relative to the isolated component atoms in whatever standard state is desired. Such heats of formation have been determined in solution and in the gas phase (see below), but there are so few comparable data for model secondary alicyclic ions of unequivocal structure that heat of formation data contribute nothing definitive at present to answering the question.

Historically, the most popular measure of reactivity has been solvolysis rate — a faster S_N1 rate implying a stabler ionic intermediate. In recent years heats of ionization of halide precursors in superacid at low temperature have assumed importance (see Volume 2 of this series). Here the ideal test would be to compare the conversion of 2-*exo*-norbornyl chloride to whatever ion is produced (call it the *exo*-ion) with the corresponding conversion of 2-*endo*-norbornyl chloride to the corresponding *endo* ion. This is impossible at present because of the instantaneous conversion of the putative *endo* ion to the *exo* ion.

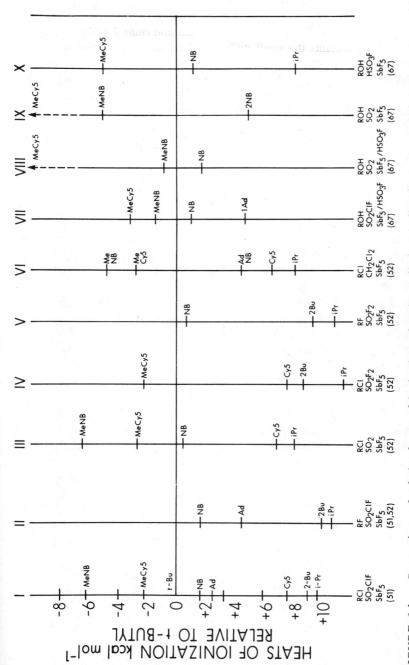

FIGURE 6.1. Comparison of relative heats of ionization of various alkyl and alicyclic compounds in various superacid systems. All values are relative to the *t*-Butyl compound. Symbols: NB = 2-norbornyl, Cy5 = cyclopentyl, MeNB = 2-Methyl-2-Norbornyl, MeCy5 = 2-Methyl-2-Norbornyl, Ad = 1-Adamantyl. Note that in studies VII, VIII, and IX, 2-propanol and cyclopentanol failed to give secondary carbocations under experimental conditions that yielded the 2-norbornyl ion.

195

If one compares the heats of ionization of *exo* and *endo* 2-norbornyl chlorides, one merely obtains the small difference (about 1 kcal/mol) between the heats of formation of the two initial states because they both give the same ion. The problem is reminiscent of, but not identical to, that of making cyclohexatriene to compare its properties to that of benzene.

Figure 6.1 shows the heats of ionization of 10 different systems of varied types of substrates and acids. Although there are clear variations of the relative positions of most of the ions, in the nine systems where both ions may be compared, the 2-norbornyl ion is close to the *t*-butyl ion and, in every case except VI, is formed 4–8 kcal/mol more exothermically than are the secondary ions from their precursors. The variable *relative* energies of ionization for the different precursors should be noted, and although experimental errors of 0.5–1.5 kcal/mol make some large contributions, the principal cause is probably ion pairing, as will be shown later. Although the errors are proportionately large relative to the average secondary-tertiary difference, it is most unlikely that the overall pattern of seven separate data sets would be wrong in placing the 2-norbornyl and *t*-butyl ions so close to each other.

Unfortunately, thermochemical heats of ionization reflect the energetics of the neutral chlorides as much as they do the stabilities of the ions, so it would be better to compare appropriate ions directly. Such a test was performed by N. Pienta[2] using the four-state comparison of the 4-methyl- and 2-methyl-2-norbornyl ions with the 2-butyl and *t*-butyl ions. Using this comparison one may calculate a driving force of 7.5 kcal/mol for the reaction of Figure 6.2. This is a direct test of the difference between the secondary to tertiary isomerization energy of an aliphatic model and a norbornyl ion and strongly implies abnormal stabilization ion the 2-norbornyl ion relative to the 2-butyl cation.

Some limitations of this approach were described carefully in the original publications and further inadequacies in the models based on estimates from molecular mechanics were swiftly pointed out by Fǎrcaşiu,[3] which, if applicable, would reduce the extra stabilization to only a few kilocalories per mol outside experimental error. A critical rebuttal of these calculations by Schleyer[4] was published back-to-back in the same journal issue, bringing the disputed stabilization energy up to 6 ± 1 kcal/mol. Without question, one may suggest

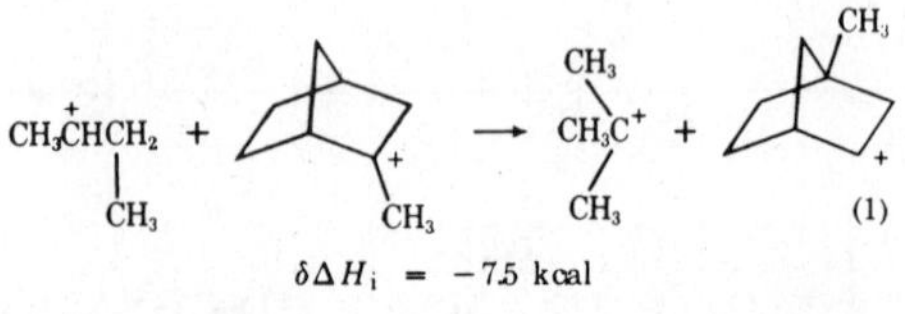

$$\delta\Delta H_i = -7.5 \text{ kcal}$$

FIGURE 6.2.   Heat of reaction of 2-butyl cation with 2-methylnorbornyl cation in SbF_5/SO_2ClF at low temperature as an experimental comparison of the "extra stability" of norbornyl and aliphatic cations.

better aliphatic ions than the 2-butyl and better norbornyl ones than the 4-methyl to eliminate some of the shortcomings in the foregoing test for "extra stability." Unfortunately, none of the proposed secondary ions are long-lived enough to test these proposals. The reader is left again with the choice between an actual experimental test and arguments *pro* or *con* based on the estimated behavior of systems that are too unstable for direct observation.

Because all ions are stabilized by solvation, it would be nice to make some of the foregoing comparisons in the gas phase. Again, there are few relevant data, but those shown in Figure 6.3 imply a much greater degree of charge stabilization for the 2-norbornyl system than is found in solution. Note that these are heats of formation so that no initial states are involved. The value for the 2-methyl-2-norbornyl cation appears to be anomalous when compared to the others. Such data are derived from appearance potentials and ionization potentials of the hydrocarbon radicals and are not above question,[5] especially because the structures of the ions in the gas phase are hard to confirm. Taken at face value, these data suggest a large degree of extra stabilization in the norbornyl system. However, this need not be due to non-classical delocalization. The large degree of internal polarization of hydrocarbon frameworks in the gas phase gives an inherent stabilization to cations which is related closely to their carbon number and degree of branching close to the charge (as is shown by the aliphatic ions in Figure 6.3). However, this factor does not seem to be

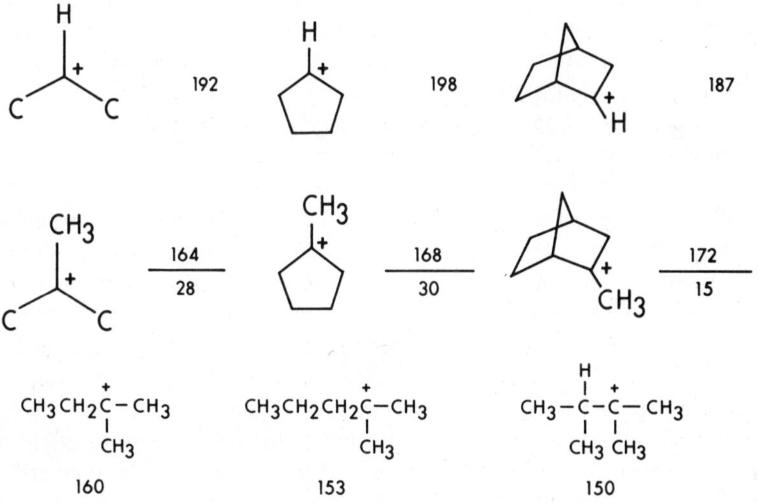

FIGURE 6.3.   Comparison of gas-phase heats of formation of aliphatic and norbornyl cations. Data from D. H. Aue and M. T. Bowers, *Gas Phase Ion Chemistry*, Volume 2, Academic Press, 1979, Chapter 9.

worth anything like 15 kcal/mol and, if the gas phase data stand the test of time, they are a strong argument for stabilization through σ-bridging under conditions where there should be maximum demand for it.

A gas-phase hydride ion equilibration of cations and hydrocarbons by Solomon and Field[6] showed that the t-butyl cation readily abstracted a hydride ion from norbornane but totally failed to react with propane or cyclopentane. Although some of the gas-phase hydride transfers are slow, these authors concluded that the norbornyl cation had 10 kcal/mol more stability than would be expected for a secondary species.

In our opinion, the accumulated evidence from kinetics and thermodynamics indicates a modest but persistent ability of the 2-norbornyl system to stabilize charge relative to the few available models. This "extra stability" is consistent with overwhelming spectroscopic evidence that bonding in the ion is unusual. However, at this time, there is no proven relationship between the energetic and spectroscopic evidence that would require a symmetrical charge delocalized ion. Nor is there any evidence for a "resonance energy" comparable in magnitude to that of benzene.

Any system that has as small an energy difference between a double minimum and a single minimum as does the norbornyl system may be expected to shift one way or the other depending on the balance between available internal and external means for charge delocalization. Thus a continuum between single-minimum and double-minimum hydrogen bonding is well recognized. In some cases such systems can be shifted from one to the other merely by changing conditions.[7] Again, there is strong evidence (see below) for a continuum of solvolysis transition states between S_N1 and S_N2. By analogy, one might expect the demand for stabilization by the σ-electrons of the 1,6 bond in the 2-norbornyl ion to vary in competition with external stabilization from the solvent or counterion leading to different degrees of σ-participation.[8]

A somewhat different perspective on the 2-norbornyl problem has been articulated over the years by Nobel Laureate H. C. Brown, who has remained steadfastly sceptical of the evidence advanced to support the nonclassical formulation of carbocations. Without doubt, his criticism has stimulated the development of new techniques and has forced a much closer examination of the evidence than might otherwise have been required. But for his persistence and prestige, the case for the nonclassical structure of the 2-norbornyl ion would have been accepted as settled beyond reasonable doubt long ago.

In many carefully written reviews, including his well-known debate in print with Professor Paul Schleyer,[9] Professor Brown has described kinetic and stereochemical evidence that the rate difference between solvolysis of *exo* and *endo* norbornyl substrates might be assigned to steric effects without having to invoke a special electronic stabilization factor in the ion. Using the familiar Goering–Schewene analysis, which removes differences in free energies between *exo* and *endo* initial states, Brown estimates from the relative heights of the *exo* and *endo* transition states that the putative *endo* cation should be somewhat more than 6 kcal/mol less stable than the *exo* if the latter were stabilized by σ-bridging.

The structure of the 2-nobornyl system makes it almost uniquely qualified for detecting σ-bridging through solvolytic ionization. The 1,6 bond is ideally positioned for interaction with developing charge on the 2-carbon of 2-*exo* substrates while a totally different electronic environment lies behind the bond from the 2-carbon to the departing *endo* group. However, as Brown has emphasized, the large rate difference between 2-*exo* versus *endo* norbornyl solvolyses might as well be due to the *endo* compounds being abnormally slow as to the *exo* compounds being abnormally fast. His extensive researches on steric effects on the *exo/endo* rate ratio have been directed toward demonstrating that the observed factor in secondary 2-norbornyl substrates has a steric rather than an electronic origin. As the energy gap in the solvolysis transition states is a thoroughly substantiated phenomenon, the question then remains "How large a difference should exist between the *exo* and *endo* ions, and what is its source?" There could be large differential steric factors in the two transition states, but because these should be lacking in the free ions, any electronic factor such as σ-bridging must be tested by comparison of the free stable ions.

To approach this question directly, thermochemical measurements have been made in our laboratory under stable ion conditions and have been discussed. However, because (as stated) all attempts to isolate the *endo* ion are probably doomed to failure, we are left with various indirect comparisons of substituted ions as models to infer the energetics of the *endo-exo* energy gap. As we have shown (Figure 6.1), direct comparison of the 2-norbornyl system with other secondary and tertiary systems places it much closer to the latter. However, by comparing the effect of methyl and phenyl substitution on these same systems, Brown has argued that the 2-norbornyl compounds behave in a normal manner. Because his arguments are less direct than those we have given, they are less convincing to us.

We are also skeptical of the basic assumption that the *exo* and *endo* transition states can be readily modeled by stable ionic species. As shown in a later section, many subtle gradations of timing are possible for reactions that take place through very unstable intermediates, such as those for secondary norbornyl substrates. To assign the *exo/endo* solvolysis rate ratio to any single cause such as σ-bridging or steric effects, the two transition states (or related intermediates) would have to be alike in all other ways, such as their interaction with solvent or counterion. This seems most unlikely to us, considering their drastically different structural features.

Let us return to our original proposition — that in the last analysis structural factors cannot be proved or disproved by energy measurements such as solvolysis rates or thermochemistry, although large energetic differences between closely similar structures are of great chemical interest and require a structural answer. Regardless of its historical antecedents, the principal question behind the nonclassical ion controversy is the structure of the 2-norbornyl ion, not the intricacies of the solvolysis data.

C.   A Historical Perspective

By March of 1983, more than thirty years after the beginning of the nonclassical ion controversy, the remaining champions on both sides in the lists have announced that the matter is essentially settled, However, each side insists that the bottom line comes out in their favor. Like Mark Twain's obituary, the announcement of the death of the argument is probably premature. Because there even remains disagreement on how to pose the question there is little chance of getting a concensus on the answer.

We expect that future historians of organic chemistry will be puzzled why this, of all problems in organic chemistry, should have absorbed such a quantity of effort toward its questionable solution. It has been an important working rule of science to challenge new *ad hoc* explanations to reduce the proliferation of unnecessary "effects." This useful policy does not necessarily imply that "Nature is simple" but does recognize the requirements of intellectual discipline if we are to think about it productively. Now, if, in fact, the proposal of nonclassical delocalization in the 2-norbornyl ion is just another such *ad hoc* explanation, one wonders why with all the other *ad hoc* explanations that have been proposed by physical organic chemists, this one has received so much attention.

The original formulations of σ-bridged bicyclic cations by Winstein, Roberts, Bartlett, and Ingold[10] were extraordinary in affecting what many other chemists did and thought. Surely this is one sociological test of the importance of a problem. In retrospect, with most of the mass of significant data at hand, one must be impressed with the sagacity of these chemists in choosing interesting systems to study, and that their original guesses about the existence of abnormal bonding in the norbornyl cation were very probably correct. However, because their inference was drawn from solvolysis and stereochemical evidence which clearly could have other explanations, they were at best right for insufficient reasons — a situation with plenty of historical precedent.

The flood of papers which subsequently has explored every conceivable subtlety of the norbornyl question seems to be more of a tribute to the stature of the proponents of the nonclassical hypothesis, and their adversaries, than to the inherent merits of the question. The flames of controversy were fanned repeatedly during the 1960s with acrimonious confrontations at major symposia. Gradually, a sense of puzzlement and ennui developed on the part of listeners, referees and, at last, reviewing panels of granting agencies. By 1970 the Golden Age of solvolysis was over and with it a good deal of support for more important research on carbocation chemistry.

In 1953 Irving Langmuir, in a totally different context, defined "pathological science" as the "science of things that aren't so." Professor Felix Franks has examined the recent case of "polywater" in the light of Langmuir's criteria.[11] The nonclassical ion controversy shows some of the same characteristics of great emotional intensity over an elusive phenomenon, so that it is interesting to compare the two cases. As quoted by Franks, the six criteria of pathological science are as follows:

1. The maximum effect that is observed is produced by a causative agent of barely detectable intensity, and the magnitude of the effect is substantially independent of the intensity of the cause.
2. The effect is of a magnitude that remains close to the limit of detectability, or many measurements are necessary because of the very low statistical significance of the results.
3. Great accuracy is claimed.
4. A fantastic theory contrary to experience is put forth.
5. Criticisms are met by *ad hoc* excuses.
6. The ratio of supporters to critics rises to somewhere near 50% and then falls gradually to oblivion.

At this point the nonclassical ion controversy really does not fit any of these criteria with the possible exceptions of 1 and 2.

Franks also refers to the sociological evaluation of the polywater controversy in the M.A. thesis of Simon Schaffer.[12] Schaffer's criteria are the following:

1. Hostile attitudes by the scientific establishment, which in turn affect financing of research.
2. The quality and quantity of press exposure.
3. The lack of communication and secretiveness adopted by the researchers.
4. The amount of hasty and shoddy research.

These are more damning of the polywater affair and happily bear little correspondence to the performance of the physical organic chemistry community throughout the nonclassical ion argument. The establishment (with the notable exception of Professor Brown) was, if anything, unduly enthusiastic. What little press exposure the controversy has aroused has been excellent. Communication has been wide open, and most of the research on the problem has been of the highest quality.

In conclusion, the question should be asked: "How much harm, if any, has the nonclassical ion controversy done, and what lessons can be learned from it?" In our opinion the ebb and flow of the problem has mostly reflected the roller-coaster boom-or-bust economics of research funding in the United States. The problem took off on the spring tide of academic funding during the early 1950s. By the 1960s it had reached a degree of elaboration that was out of proportion either to its inherent merits or the ability of available methods to solve it. As the larger community of chemistry and science began to lose patience, its priority fell, and when funds became short support was mostly terminated.

It has been charged, and may be true, that the controversy did irreparable harm to the image of physical organic chemistry in general and carbocation chemistry in particular. As things stand now, we doubt this assertion. We believe that the greater community of organic chemistry responded realistically within reasonable time to the limitations of solvolysis research, and the remaining vestiges of the original problem are now focused much more sharply and consume few funds and little effort, except those of the principal adversaries. No young newcomers are building their careers in this area. We see no major opportunities that are hung up until there is unanimous agreement on the structure of the 2-norbornyl ion. The whole field of reactive intermediates has probably benefitted by the development of new techniques and the example of critical thinking applied to this testing ground.

Whatever damage the controversy has done lies mostly in the past and may carry a warning to other fields. For a while the problem attracted more researchers and more support than it probably should have. It became over-personalized by its leaders, who attracted followings that were too large and uncritical. The questions of relevance or broader significance were rarely addressed. Those most actively engaged failed to consider adequately the overall value or impact of the field from a detached external viewpoint. Eventually, the problem took on a life of its own which could only be curbed by outside forces. The most important warning to other areas of chemistry is the price of getting too overgrown and ingrown.

IV.   NUCLEAR MAGNETIC RESONANCE AND STRUCTURE IN SOLUTION

Brown, Periasamy, Kelly, and their colleagues[13-21] have proposed a new substituent parameter σ^{C+} to correlate the cationic ^{13}C shifts for a wide range of meta and para substituted tertiary carbocations under stable ion conditions. They found that the shifts of a series of substituted cumyl cations correlate with the established parameter σ_{m^+} ($\rho^+ = -18.18$, $r = 0.990$, n = 9), and also with σ_{p^+} ($\rho^+ = -34.9$, $r = 0.970$, n = 7). However, when both meta and para substituents were placed on the same correlation, a systematic deviation for the para derivatives was observed, becoming much larger as the electron-withdrawing ability of the substituent increased. Therefore, as in the development of the σ^+ constants,[14] the correlation line for the meta substituent was used to calculate new values for the para substituents according to the equation

$$\Delta\delta C^+ = \rho^{C+}\sigma^{C+}$$

where $\Delta\delta C^+$ represents the NMR shift difference (in parts per million) between the unsubstituted and substitued cumyl cations. Table 6.1 gives a list of the σ^{C+} constants and contrasts them with the σ^+ constants.

In subsequent studies,[15–21] the σ^{C+} correlation was extended to a wide variety of other tertiary carbocations. The 2-aryl-2-norbornyl cations present "certain discrepancies."[22,23] The shifts of strongly electron-withdrawing substituents, [e.g. p-CF_3 and 3,5-$(CF_3)_2$], deviate from the σ^{C+} correlation. Subsequent studies[21] of 2-aryl-exo-5,6-trimethylene-2-norbornyl, 2-aryl-endo 5,6-trimethylene-2-norbornyl, 3-aryl-3-nortricyclyl, and 1-aryl-1-cyclopropyl-ethyl, as well as 1-aryl-1-phenylethyl cations show similar deviations. These deviations have been taken by others to indicate the onset of σ-bridging.[22,23] However, other systems, for which σ-bridged structures have also been suggested, failed to show such responses to strongly electron-withdrawing substituents. These systems included the cyclobutyl, 2-bicyclo[2.1.1]hexyl, 2-bicyclo[2.2.2] octyl, and 6-bicyclo[3.2.1]octyl cations. Anomalous shifts were rationalized in terms of steric inhibition of resonance and the greater intrinsic stability of ions which fail to exhibit the behavior of "normal" systems. No satisfying explanation for the discrepancies has yet been found.

TABLE 6.1.   Carbocation substituent constants[a]

Substituent	σ^+_m	σ^+_p	σ^{C+}_m	σ^{C+}_p
$-OCH_3$	0.047	−0.778		−2.02
$-CH_3$	−0.066	−0.311	−0.13	−0.67
$-CH(CH_3)_2$	−0.060	−0.280	−0.14	
$-H$	0.00	0.00	0.00	0.00
$-F$	0.352	−0.073	0.35	−0.40
$-Cl$	0.399	0.114	0.36	−0.24
$-Br$	0.405	0.150	0.33	−0.19
$-CF_3$	0.52	0.612	0.56	0.79
3,5-$(CF_3)_2$	1.04		1.03	
3,5-Cl_2	0.798		0.66	
$m-CH_2CH_2-O-p^b$	−0.984		−2.4	

[a] Data taken from Reference 13.
[b] Coumaranyl, Reference 19.

The observed correlations are remarkably good, considering the wide variety of structures studied, but a thorough understanding of ^{13}C chemical shifts has yet to be achieved.[24,25] The failure of σ^{C+} to correlate with enthalpies of ionization for a series of substituted cumyl cations[26] should be kept in mind whenever ^{13}C chemical shifts are used as a criterion of carbocation stability. There is currently no satisfactory basic theory that can be applied generally to the detailed interpretation of NMR chemical shifts, let alone for ^{13}C shifts of carbocations.

The reports of Saunders and co-workers[27-30] that the degeneracy of a rapidly rearranging carbocation under stable ion conditions can be lifted by deuterium substitution has received further attention in the last three years. The technique is based on the observation of the shift of the deuterated cationic carbon. Large, temperature-dependent shifts are indications of equilibrium processes, whereas smaller, temperature-independent shifts indicate an isotopic perturbation of resonance. Figure 6.4 shows some representative structures and their changes in chemical shift upon deuterium substitution. Again, the interpretive problem here lies in the suitable choice of models for estimating the shifts that might be expected for nondegenerate, nonequilibrating ions.

A number of other "more interesting" structures have now been studied, in particular, the 2-norbornyl and the cyclopropylcarbinyl cations which lie at the heart of the nonclassical ion problem.

The 69.7 MHz ^{13}C NMR spectrum of the 2-D-2-norbornyl cation[30] under stable ion conditions shows only broadening of the C_6, C_1, and C_2 resonances. The $C_1 - C_2$ peak was observed to have a width at half-height of 2.3 ppm, making

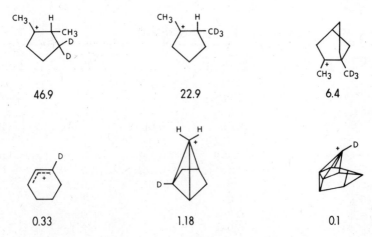

FIGURE 6.4.   Changes in ^{13}C NMR shifts upon deuteration (parts per million per D).

this an upper limit on the isotopic splitting. The implication is that 2-norbornyl is a static, nonequilibrating ion under these conditions.

A high-field (395 MHz ^{1}H and 50 MHz ^{13}C NMR) study, over the range from -80 to $-160°C$, gives forceful evidence to the conclusion that the ion is either bridged (nonclassical) or that there is such a low barrier between equilibrated classical ions to render their distinction from the nonclassical formulation operationally meaningless.[31]

The spectrum of the monodeuterated cyclopropylcarbinyl cation is much more surprising.[32] The observed shifts resulting from the substitution of an α–D atom have, until this point, been all upfield. The splitting of the ^{13}C peak for the α–D cyclopropylcarbinyl cation shows both an upfield and a downfield splitting. These were observed to be temperature dependent, suggesting the existence of an equilibrium between two species with very different bonding characteristics. The observation of an equilibrium process probably eliminates the tricyclobutonium ion as the main species present. The authors favor a rapid, degenerate equilibrium between a bicyclobutonium ion (because it contains a pentacoordinated carbon, thus offering a rationale for the downfield splitting) and either the cyclopropylcarbinyl or bent cyclobutyl cation.

The technique has been applied to other systems[33−35] including a fivefold degenerate dication.[33] Sorensen, in particular, has used these criteria to examine a series of δ-substituted cyclooctyl and ϵ-substituted cyclodecyl cations to demonstrate a pleasing progression of σ-bridging.[35,36] Saunders' technique is appealing and undoubtedly will develop increasing value as its limits for discriminating between equilibrating and delocalized ions are tested more extensively.

An important paper by Servis and Shue[37] presents an extensive list of deuterium-isotope effects on the ^{13}C NMR shifts of a variety of carbocations under stable ion conditions. Representative classical static, classically delocalized, and putatively bridged ions are included, as well as their precursors (mostly alcohols).

The precursors all show small upfield shifts upon α and β deuteration as a result of an increase in electron density at the attached and adjacent carbons. By way of explanation: because the C–D bond has a lower zero-point vibrational energy, its length will be, on the average, shorter, leading to an increase in electron density at the attached carbon. All the ions that are α-deuterated show a slight upfield shift. However, the β-deuterium isotope effects differ and are related to the type of carbocation.

Classical aliphatic carbocations all show negative β-deuterium isotope effects (i.e., a downfield shift is observed) which may be explained by means of a hyperconjugative mechanism if the C–D bond adjacent to the empty p-orbital is acting to decrease electron delocalization. These effects are similar to the observed β-deuterium isotope effects in solvolysis studies.[38]

In contrast, the β-deuterium effects on carbocations that are stabilized

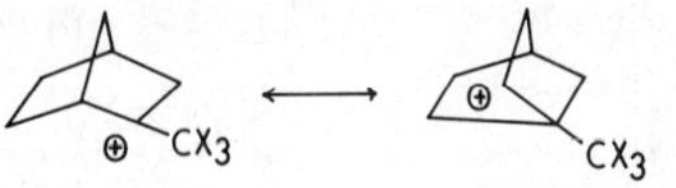

FIGURE 6.5.   Canonical structures of a 2-substituted 2-norbornyl cation.

by an aryl, cyclopropyl, or allylic double bond all show negligible shifts, demonstrating the lessened demand upon hyperconjugation for stabilization.

Those ions to which "nonclassical" σ-bridging, or π-bridging have been attributed, all show shifts similar in magnitude and direction to their precursors, that is, "normal" downfield shifts. These observed shifts may be understood by looking at the contributing canonical structures, such as those for 2-methyl-2-norbornyl cation (see Figure 6.5.) If the canonical form on the right becomes more important upon deuterium substitution, a decrease in electron density at C_1 and an increase in electron density at C_2 and C_3 would be predicted in accordance with the observed spectrum: C_2 shifts -2.3 ppm, C_1 shifts 0.4 ppm, and C_3 shifts -0.3 ppm. Alternatively, a different weighting of rapidly equilibrating classical structures may be assigned to the ions.

Following earlier work with stable carbocations,[39] the solid-state ^{13}C NMR spectra of the sec-butyl[40] and 2-norbornyl[41] cations have recently been determined. The ions were prepared from isotopically enriched halide precursors and SbF_5 using Saunders' codeposition technique.[42] Well-resolved spectra required the use of magic-angle spinning and ^{1}H–^{13}C cross-polarization.

Although the ions thus produced were amorphous solids, precluding correlation of the spectra with structural information derivable from such tools as X-ray diffraction, the method has promising features. The absence of solvent allows the use of low temperatures. In the present case, spectra were obtained at temperatures as low as 5 K.[43] However, the dynamics of the rearrangement processes might change considerably on going from the solution to the solid state. For example, in an earlier study,[44] the degenerate Cope rearrangement of semibullvalene was found to occur 500 times more slowly in the solid state than in CF_2Cl_2 solution, at 163 K.

The sec-butyl cation suffered complete scrambling of its ^{13}C label on formation, but showed no coalescence of peaks due to the corresponding skeletal rearrangement, even at the highest temperature used, 213 K. This means that such a rearrangement is considerably slower than the corresponding solution-phase process.[45] By contrast, at the lowest temperature, 83 K, there was no evidence for a freezing out of the 2,3-hydride shift. If a comparable inhibition of this rearrangement is operating, the upper limit for its activation barrier in solution may be considerably lower than the 2.4 kcal/mol value recently determined.[46]

The spectrum of the 2-norbornyl cation was studied in the temperature

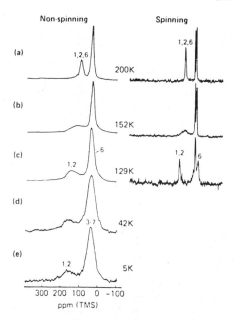

FIGURE 6.6.   Observed ^{13}C CP-NMR (with and without MAS) spectra of the 2-norbornyl cation at low temperatures (From Reference 43. Copyright 1982 American Chemical Society).

range 5–200 K (Figure 6.6). At the highest temperatures, its spectrum resembles that obtained in solution. Full lineshape analysis gave an activation energy for the 6,1,2-hydride shift of 6.1 ± 0.5 kcal/mol, again comparable to that observed in solution. Below 130 K, all resonances appear to broaden at the same rate. The lack of greater broadening of the peak due to carbons 1 and 2 is interpreted to mean that no classical-type conformation of the norbornyl cation is being frozen out, even at these low temperatures. A simulated spectrum for such exchange is shown in Figure 6.7. If classical exchange theory applies to these unprecedented conditions, the activation barrier for interconversion would only be 0.2 kcal/mol and operationally almost indistinguishable from a σ-bridged ion.

This experiment is probably the most compelling single piece of structural evidence yet presented for the nonclassical formulation of the 2-norbornyl ion. A vitally needed development in this field is the performance of similar experiments on solids of known intermolecular environment, such as crystalline compounds, so that correlation can be made between intermolecular forces and changes in the rates of rearrangement processes. However, its agreement with the low-temperature solution phase results referred to before[27,30] provides

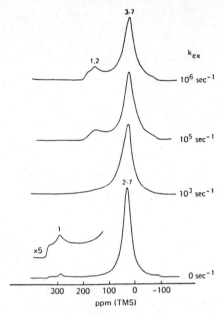

FIGURE 6.7.   Simulation of two-site chemical exchange of a classical 2-norbornyl cation (From Reference 43. Copyright 1982 American Chemical Society).

strong evidence that, under these conditions, at least, the controversy over the double-minimum versus single-minimum structure of the ion is either settled or virtually meaningless.

V.   REARRANGEMENTS

Koptyug and Shubin have summarized the available information on the rates of degenerate rearrangements of carbocations.[47] These comprise 35 instances of proton migration, 18 methyl migrations, 19 aryl migrations, and 17 different migrating groups in substituted hexamethylbenzenonium ions.

Although the activation free energies for the rearrangements of similar ions can be correlated with the charge deficiency of the cationic center, as measured by its ^{13}C NMR chemical shift, comparison of different types of cations is less successful. It is suggested that better results could be obtained if changes in the extent of delocalization of the π systems were considered. To model this, the migration process is considered as the sum of three separate steps: (1) The charge in the ion is localized at the carbon that is the target of the migration.

(2) The migration occurs to this carbon, leaving behind a localized charge. (3) That charge is then delocalized throughout the π system. The delocalization energies can be computed by molecular orbital methods, and the barrier to rearrangement of the localized ion is taken to be constant. Clearly, the effects of steric hindrance and molecular strain are completely neglected by this approach, a justifiable simplification in view of the present paucity of experimental data bearing on their importance.

Where the rearrangement is not degenerate, the authors introduce a simplified Marcus type equation to describe the rearrangement process,

$$\Delta G_{ij}^* = \frac{(\Delta G_{ii}^* + \Delta G_{jj}^* + \Delta G_{ij}^\circ)}{2}$$

where ΔG_{ij}^* corresponds to the activation free energy for the nondegenerate process, ΔG_{ii}^* and ΔG_{jj}^* correspond to degenerate model reactions, and ΔG_{ij}° is the thermodynamic driving force for the reaction. Using their charge-deficiency formalism to estimate ΔG_{ii}^*, they obtain reasonable agreement for the rates for acid-catalyzed isomerization for 14 aromatic systems, spanning four orders of magnitude in reactivity.

Extensive work has been done on the solvolyses of neopentyl systems. By studying model systems in which the migrating carbon can be distinguished from the others at the quaternary center, Ando and co-workers[48] and Shiner and Tai[49] were able to conclude that the small γ-d_9 isotope effect in these compounds was the superposition of a normal isotope effect for the migrating group and an inverse isotope effect for the nonmigrating methyls.

Shiner studied the cyclic compounds I, II, and III (Figure 6.8) where deuterium was incorporated either α to the sulfonate ester, as a CD_3 group, or in the two methylenes flanking the quaternary carbon. The rearranged products of I showed both methyl migration and ring enlargement; II and III gave almost exclusive and exclusive ring enlargement, respectively. By analysis of the various

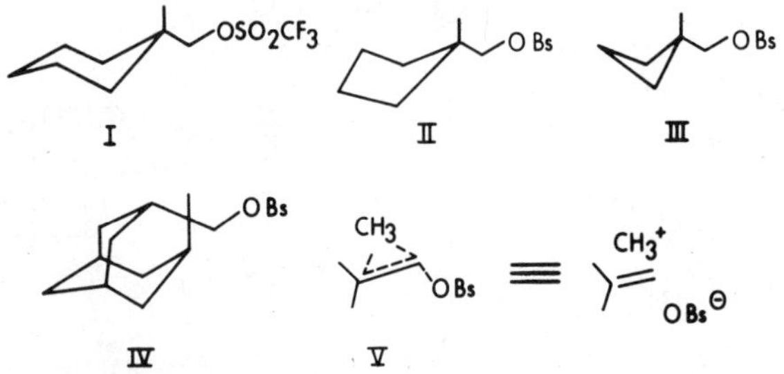

FIGURE 6.8.   Deuterium-substituted sulfonate esters studied by Shiner and Tai (From Reference 49).

deuterium isotope effects and extrapolation to neopentyl systems, the literature value for the CH_3, CD_3 relative migratory aptitude in partially deuterated neopentyl solvolyses was rationalized.

The magnitudes of the inverse isotope effects suggest that the methyl group bears less charge than a migrating hydrogen or aryl group would. This conclusion seems to be opposite to that of Ando et al.,[48] who interpret similar isotope effects in the acetolyses of deuterated derivatives of IV (Figure 6.8) as consistent with a transition state involving weak bridging of the migrating group, coupled with pronounced weakening of the bond to the migrating group and enhanced double-bond character, that is, structure V in Figure 6.8.

VI.   THE ENERGETICS OF CARBOCATION FORMATION

Petro and Arnett[50,51] used the calorimetrically determined heats of ion-ization for a wide variety of alkyl and aralkyl halides in superacidic media as a primary measure of cation stabilities relative to their neutral precursors. These

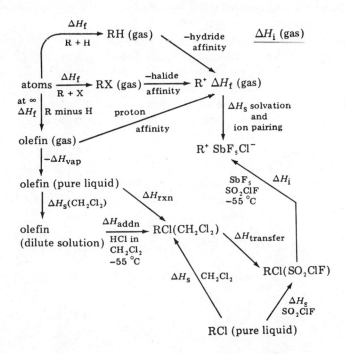

FIGURE 6.9.   Scheme for relating enthalpy terms for reaction and formation of cations in the gas phase and solution (From Reference 52. Copyright 1980 American Chemical Society).

measurements have now been extended[1,52] and a more rigorous comparison of
ion stabilities through the actual heats of formation of the carbocations themselves has been made.[52] To calculate the heat of formation, many enthalpy
terms are needed to relate the carbocation in solution in superacid at low temperature back to the standard state of gaseous atoms. The scheme of Figure 6.9
was used to accomplish this.

The key to realizing this scheme, experimentally, involved measuring the
heats of hydrochlorination of the appropriate olefins in CH_2Cl_2 solution to
yield the respective tertiary alkyl chlorides. Since the heats of formation, and
heats of vaporization and combustion are known for a great many olefins, all
of the thermodynamic steps to relate the component atoms in the gas phase to
the ions in solution are complete (i.e., all initial-state differences due to structural features in the precursor chlorides have been removed). The resulting heats
of formation of the carbocations in solution are now directly comparable to
those measured in the gas phase,[53] devoid of solvent effects. A good linear
relationship was observed between the heats of formation of the ions in the
gas phase and in superacid as is shown in Figure 6.10. These heats of formation
are also directly comparable to the results of theoretical calculations. Using
previous correlations,[54] it is therefore finally possible, in principle, to estimate
the rate constant for solvolysis of some alkyl chlorides from *ab initio* quantum
mechanical calculations. Consideration of Figure 6.9 however, should be a
reminder of the long chain of experiments and the accumulation of errors that
may be involved.

Another important result is that the differential heats of solution of the

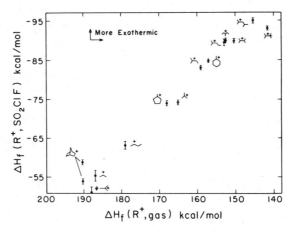

FIGURE 6.10.   Comparison of the heats of formation of cations in solution
and in the gas phase (From Reference 52. Copyright 1980 American Chemical
Society).

carbocations are negligible in SO_2ClF, SO_2F_2, SO_2, and CH_2Cl_2. One can infer from this observation that Born-type charging of the solvent is the most important energetic factor in carbocation solvation (in these relatively nonbasic solvents). Two widely discordant values for the heat of combustion of norbornene are responsible for the two points for the 2-norbornyl cation in Figure 6.7. Happily, this problem should be solved more easily and unequivocally than are most of the other arguments about the stability of this ion.

Most of the substrates studied by Petro and Pienta in the senior author's laboratories produced relatively unstable secondary and tertiary aliphatic or alicyclic cations. It would obviously be desirable to extend the series to include the entire range of kinetically stable carbocations such as those thoroughly studied by means of acidity function methods by Deno and others.[55] Unfortunately, the halide precursors to the more stable carbocations are difficult to prepare and maintain in the highly pure form needed for good calorimetry (the more stable the ion, the less stable the precursor chloride). After an extensive search for the optimum combination of precursor and conditions, the heats of ionization of a series of alcohols in SO_2ClF with $1-1$ (molar) $SbF_5:HFSO_3$ (to avoid ion-pairing effects, see below) were shown to be proportional to the previously measured heats of ionization of the alkyl and aralkyl halides in SO_2ClF and SbF_5.[26] The heat of ionization scale was then extended to include a wide variety of carbinols whose pK_R's in aqueous sulfuric acid were known.[56] A fairly good correlation with Deno's pK_R scale (slope $= -1.65$; $r = 0.95$; 12 points) was observed. In addition, these heats of ionization were shown to correlate better with σ^+ ($r = 0.96$) than with σ^{C+} ($r = 0.91$).

Unfortunately, the set of ions does not include for comparison any allylic systems for which there are many data available. Many of the alcohols that we attempted to study were too insoluble, dissolved too slowly, or gave side reactions that prevented all attempts to measure accurate heats of ionization.

Kramer[57] has studied the free energy changes for the transfer of a hydride ion between t-butyl cation and a number of other carbocations. The t-butyl cations were generated in a variety of superacid systems: $AlBr_3/CH_2Cl_2$, $AlBr_3/CH_2Br_2$, and $GaCl_3/CH_2Cl_2$. The equilibrium constants were determined by means of NMR measurements and are shown in Table 6.2. The trends in these free energy data are in the same direction as those observed in the previous enthalpic measurements (Figure 6.1) and are, by the nature of equilibration studies, much more precise. Again the 2-norbornyl ion is classed with the tertiary ions in its stability. One striking observation is that the adamantyl ion is ergonically favored over the t-butyl ion in the $GaCl_3$ or $AlBr_3$ superacid systems, whereas the opposite order of stability is found from enthalpic measurements in SbF_5/SO_2ClF (chloride precursor)[50] or $1-1$ $SbF_5:HFSO_3$ in SO_2ClF (alcohol precursor).[1] Kramer attributes this reversal to steric hindrance to solvation, but, if everything else were equal, this requires that the solvent

TABLE 6.2.   Hydride-transfer equilibria in solution and vapor phases[a]

$$t-C_4H_9^+ + RH \rightleftharpoons i-C_4H_9H + R^+$$

| RH | $\Delta G°$ (kcal/mol) | | | | |
	$AlBr_3/$ CH_2Cl_2 (223 K)	$AlBr_3/$ CH_2Br_2 (223 K)	$GaCl_3/$ CH_2Cl_2 (223 K)	$AlBr_3/$ SO_2FCl (223 K)	Gas (300 K)
i-Pentane	−1.0		−0.4	−0.8	−2.5
2,3-Dimethylbutane	−0.8		−1.1		−5.4
Adamantane	−3.3	0.0	−0.1	−0.6	
Norbornane	+0.7			+1.2	−2.3
Methylcyclopentane	−1.2		−1.8		−5.5

[a] Data from Reference 57

effect manifest itself primarily through entropic factors. The reversal is most apparent in the $AlBr_3/CH_2Cl_2$ system. In view of the likely intrusion of ion-pairing factors in some carbocation equilibria in superacids (see Section VII), it is important to note the differences in counterion (from the Lewis acid) and solvent when comparing the two sets of data. We suspect that a different degree of ion pairing in the two systems is a more likely cause of the reversal in stability order than differential entropic factors due to steric hindrance to solvation. A simple thermochemical experiment with Kramer's system should settle the matter.

VII.   ION PAIRING

Ion pairing was of great interest to the early pioneers of carbocation research[58,59] to establish the reality of conducting species with charge on carbon. Later, Winstein[60] and his co-workers demonstrated that a number of different ion pairs might play a significant role in the mechanism of uni-molecular solvolyses:

$$R-X \rightleftharpoons \{R^+ X^-\} \rightleftharpoons R^+ \parallel X^- \rightleftharpoons R^+ + X^-$$

Each type of ion pair shown, as well as the free ions, could display different reactivities to solvent or added nucleophile.

The question, whose answer has long eluded workers in the field, is, "Just how different are ion pairs from free ions in their thermodynamic stabilities and/or their kinetic reactivities?" Further, "How do the different types of ion pairs differ among themselves?"

The best way to attack such a problem is to study the thermodynamic and kinetic behavior of bona fide ion pairs and free ions separately. Unfortunately, experimental conditions where such studies could be carried out are not achieved easily.

In systems of stable ions, conductance measurements are the ultimate experimental diagnostic for the presence of free ions. Ion-pairing constants are derived through complicated mathematical treatments of the concentration dependence of conductance data. Again, in the study of solvolyses that generate free carbocationic intermediates, complicated mass law effects on kinetics are used as evidence for ionic intermediates.

Although the special salt effect, as well as many stereochemical studies and isotopic scrambling experiments, has been rationalized in terms of various types of ion pairs, differentiation among the relevant types of ion pairs by well-established structural techniques has not fared well. Spectral differences (ultraviolet, visible) are minimal[61,62] in systems of very stable carbocations.

A novel approach to studying the kinetic behavior of an ion pair has been made by Ritchie and Hofelich.[63] A zwitterionic model for an intramolecular ion pair was synthesized (Figure 6.11), and rate constants for its reaction with a set of nucleophiles in water at 23°C were measured and compared to rate constants determined for reactions of a number of free carbocations with the same set of nucleophiles. These nucleophiles had been used to determine the N_+ scale of nucleophilicities. No reversals of any kind were observed between the reactions of the model ion pair and those of the free ions. Charged as well as uncharged nucleophiles behaved as one would predict based on free ion reactivity patterns. This study would, then, lead to the belief that ion pairs and free ions follow kinetically similar patterns in discriminating among nucleophiles. However, there exists a marked difference in the reactivity patterns of the carbocations studied by Ritchie and the relative reactivities of free ions obtained by the trapping of solvolysis intermediates.[64,65] The reason for these reversals in reactivities remains to be found.

Kessler and Feigel[66] have recently reviewed the use of dynamic NMR techniques to obtain free energies of activation for the collapse of carbocation ion pairs to covalent substrates. The measured barriers are remarkably high, as is shown in Table 6.3.

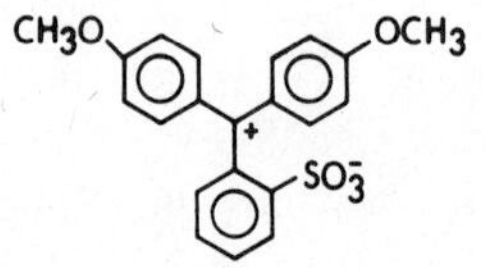

FIGURE 6.11.   Zwitterionic model of an intramolecular carbocation: sulfonate ion-pair (From Reference 63).

TABLE 6.3.   Free energies of activation for ion pair recombination reactions in $CD_2Cl_2{}^a$

Ion Pair	$\Delta G^{\ddagger}$ (kcal/mol)
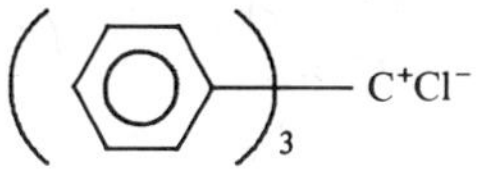 $\left(\ \bigcirc\!\!-\!\!\right)_3 - C^+Cl^-$	9.9
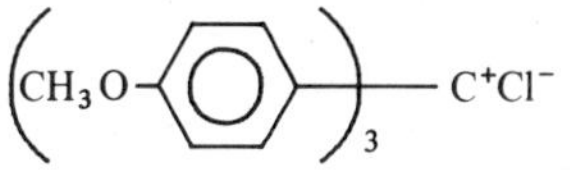 $\left(CH_3O-\bigcirc\!\!-\!\!\right)_3 - C^+Cl^-$	7.7
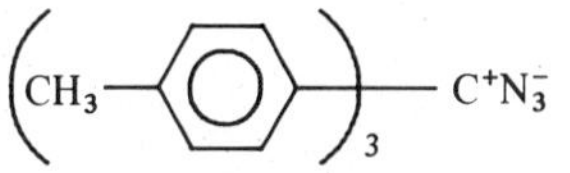 $\left(CH_3-\bigcirc\!\!-\!\!\right)_3 - C^+N_3^-$	11.5

^{a}Data from Reference 66.

Unfortunately, the restrictions on the experimental conditions for the observation of the ion pair–covalent equilibria (i.e., $\Delta G^{\circ} \approx 0$) severely limits the range of any systematic study of the collapse of carbocation ion pairs with a standard set of counterions in a single solvent.

Although the study of the thermodynamics of ion pairing of very stable carbocations in liquid SO_2 received a great deal of attention in the early days of carbocation research, there is little information about the less stable ions in superacid media.

The possibility that ion pairs may play an important role in superacid solutions of alkyl carbocations has been suggested by Kramer in the previously discussed studies of hydride transfer equilibria.[57] The chemical shifts of the t-butyl cation's protons in CH_2Cl_2 were found to change from 83 Hz, with $AlBr_3$ as the Lewis acid, to 78.8 Hz using $GaCl_3$, as though the cations were shielded by different counterions in close association.

Another striking observation[67] is the change with structure of the heats of ionization of a series of carbinols when ionized with SbF_5 in SO_2, compared to the corresponding alkyl chloride precursors ionized with SbF_5 in SO_2ClF; see Figure 6.12. The alcohols in SO_2 show a much greater sensitivity to structural change than do the chlorides in SO_2ClF. In addition, those ions that enjoy stabilization by means of charge delocalization fall off of the correlation line for the alkyl ions. The effect is reproduced using $HSO_3F{:}SbF_5$ as the ionizing acid in SO_2. Alkyl halides ionized in SO_2 with SbF_5 behave identically to those ionized with SbF_5/SO_2ClF. When the solvent is changed to SO_2ClF, however, the alcohols behave like the halides, showing a correlation with unit slope with no deviations. This effect is most easily rationalized in terms of ion pairing; the increase in slope implies that the cation is still influenced thermochemically by

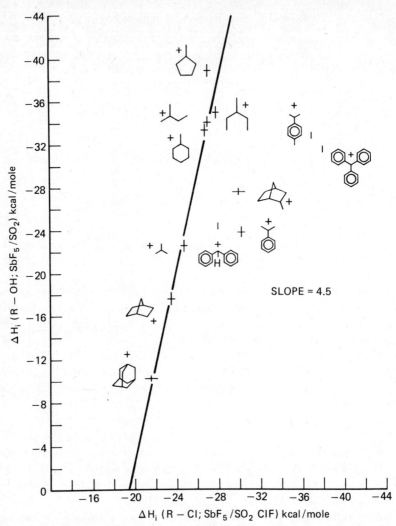

FIGURE 6.12.   Comparison of the heats of ionization of alcohols with SbF_5 in SO_2 with the ionization of the corresponding chlorides with SbF_5 in SO_2ClF.

the counterion produced from the Lewis acid and the leaving group as though ionization had occurred, but there was little or no dissociation. Further studies of the conductance behavior, as well as high-field NMR studies should be done to test this hypothesis. A molecular beam study of the gas-phase reaction of alkyl halides with SbF_5 indicates that a wide range of complex ions can be formed, depending on the stoichiometry and nature of the halide substrate.[68]

The possibility of ion pairing in these superacid media should have a pronounced influence on the future study of the structure and equilibria of carbocations. The fact that species other than free ions, solvent, and Lewis acid may be present in superacids presents a new complication in what had previously seemed to be a reasonably simple picture. Inevitably these complications will also provide an escape route from some of the impasses produced by the collision of current speculations that assume the formation of free ions in superacid.

VIII.   THE S_N1–S_N2 SPECTRUM OF MECHANISMS AND THE LIFETIME OF CHEMICAL INTERMEDIATES

The Hughes–Ingold classification of mechanisms placed those nucleophilic substitution reactions at aliphatic carbon that go through a solvent-equilibrated, free carbocationic intermediate, and that are first order in substrate and zeroth order in nucleophile (solvent or other), into the S_N1 pigeonhole. Those reactions that exhibit bimolecular kinetics and are first order in both substrate and nucleophile go into the other, S_N2, pigeonhole.

A perennial question is, "How well separated are these compartments?" The questions can be asked many ways. As an example, the solvolysis of methyl tosylate may be considered a benchmark S_N2 reaction, and the solvolysis of trityl chloride an S_N1 reaction, but what about all the other, less stable tertiary and secondary systems in between? The question of intermediates in the solvolyses of secondary systems has long been a controversial issue.[69-73]

A correct solution to such a problem often depends on asking the right question, and Jencks[74,75] has recently posed it in terms of the lifetimes of the alleged intermediates. An intermediate may be said to exist when its lifetime is longer than a molecular vibration, about 10^{-13} sec. The question then becomes, what happens as an intermediate's stability decreases toward the point where it is too unstable to even exist for 10^{-13} sec? As Jencks puts it, "When is an intermediate not an intermediate?"[74]

One experimental approach to the answer has been taken by Bentley, Schleyer, and their co-workers.[76,77] This involves an attempt at correlation of the rates of solvolysis of a wide variety of secondary and tertiary alkyl tosylates, in a variety of solvents, through the linear free energy relationship:

$$\log (k/k_0)_{\text{ROTs}} = (1 - Q') \log (k/k_0)_{2\text{-PrOTs}} + Q' \log (k/k_0)_{2\text{-AdOTs}}$$

Here, k refers to the solvolysis rate constant in any solvent, k_0 is the rate constant for solvolysis in 80% ethanol–water (v/v), and Q' is an adjustable "blending" parameter. It is postulated that all of the solvolysis reactions treated in the study (and perhaps all others) have mechanisms that lie along a continuum between the two extreme models. Then Q' represents the fraction of S_N1–like character in the transition state and *ipso facto* $(1 - Q')$ is the fraction of S_N2 character. The solvolysis of 2-adamantyl tosylate is taken (and convincingly shown) to be the model for limiting S_N1 behavior, and 2-propyl tosylate is taken as the model S_N2 substrate. The equation does a remarkably good job of correlating rate constants (with an average standard deviation of 0.16 in log k) ranging over eight orders of magnitude for 16 substrates in 11 solvents. The gradual change in Q', furthermore, models quite reasonably the gradual change in mechanism from classical S_N2 behavior (nucleophilic solvent assistance) to limiting S_N1 behavior as a function of change in substrate. The two mechanisms may therefore be not only necessary, but also even sufficient to correlate all solvolyses. It remains to be seen whether other parameters are eventually required.

Jencks, in a natural extension of his work on enforced catalyses,[78] has postulated and shown strong evidence in support of a substitution mechanism in which a preassociation of nucleophile and substrate is required as the stability of the incipient carbocationic intermediate is decreased. This approach has been reviewed recently;[74,75] however, a brief description is given here.

Figure 6.13 shows the solvolysis of an alkyl chloride. As the typical Winstein solvolysis proceeds from A to B to C, the carbocations involved may either react with added nucleophile or solvent to form stable products. As the stability of the intermediate carbocation decreases, a point is eventually reached where the ion pair B collapses to starting material faster than the nucleophile (solvent) can attack. That is, $k_{-1} > k_d$ (NUC). In the transition state for nucleophilic attack, then, the nucleophile must be already present before the heterolysis of A can give ion-pair intermediate F. Thus the preassociation of solvent and ion pair is required by the short lifetime of the intermediate and the principle of microscopic reversibility. As the lifetime of the intermediate approaches 10^{-13} sec, a change in mechanism should occur from an S_N1 to an S_N2 type of displacement.

This mechanistic postulate was put to the test recently in a study by Jencks and Richard,[79,80] in which the trapping ratios (for added azide ion and solvent) of the incipient intermediates were obtained in the solvolysis of a series of

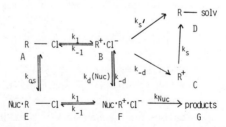

FIGURE 6.13.   Ion pair solvolysis scheme.

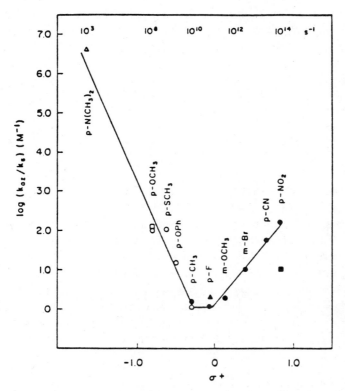

FIGURE 6.14.   Trapping ratios for added azide ion and solvent versus σ^+ for a series of 1-phenylethyl chlorides (From Reference 79. Copyright 1982 American Chemical Society).

substitued 1-phenylethyl chlorides. Figure 6.14 shows a plot of the selectivities (trapping ratios) versus the substituent parameter σ^+. In addition, the following observations were made: those substrates whose substituents placed them in the $\sigma^+ \leqslant -0.32$ range exhibited kinetics that were zeroth order in added azide ion; those with substituents placing them in the $\sigma + \leqslant -0.08$ range showed first-order dependence on azide ion up to 0.6 M. Virtually no dependence on leaving group was observed for those compounds to the left side of the break in the structure-activity plot, whereas a strong dependence was observed for those to the right.

These results are consistent with a gradual change from an S_N1 to an S_N2 or preassociation mechanism as the lifetime of the intermediate decreases. The estimated lifetimes of the intermediate carbocations are plotted at the top of the figure and were calculated using a diffusion-controlled rate constant of 5×10^9/mol (sec) for the reactions of azide ion with those substrates to the left of the break,[81] and a simple extrapolation for those very unstable cations on the right of the break.

Three distinct areas of reactivity-selectivity are observed:

1. The area of $\sigma^+ \leqslant -0.32$ is that where the rate constant for attack by azide ion is just at the diffusion-controlled limit, $k_{Azide} = 5 \times 10^9/\text{sec-mol}$, and this is much larger than k_{Sol}, the rate constant for reaction with solvent. Here, a solvent-equilibrated ion (or ion pair) is formed and reacts with the added nucleophile at a diffusion-controlled rate and with the solvent at an activation-controlled rate. This is the region responsible for the well-known correlation between reactivity and selectivity noted by Sneen[82] and Schleyer.[83]

2. In this section, $-0.32 \leqslant \sigma^+ \leqslant -0.08$, the rate for attack by solvent competes with the rate for attack by azide ion $[N_3] k_{Azide}; k_{Sol} = 5 \times 10^9/\text{sec-mol}$ at the diffusion limit. Here the carbocation's stability is barely enough to allow it to form before it is snatched up by solvent or added nucleophile.

3. In the region of $\sigma^+ \geqslant -0.08$, an upward break in the selectivity is observed. This implies a change in mechanism and is consistent with the onset of either an S_N2 or a preassociation mechanism (both exhibit second-order kinetics). A very thorough discussion by Richard and Jencks[80] provides strong evidence that this second-order reaction with azide ion is concerted (S_N2), but the arguments are beyond the scope of this review.

It is well worth noting that these three regions of reactivity-selectivity behavior, as well as one additional region, where the lifetime of the carbocation is greater than 10^{-3} sec [i.e., $k_{Azide} \ll 10^9/\text{sec-mol}$], reconcile the constant selectivities observed for carbocations by Ritchie,[84] and the apparent reactivity-selectivity relationships observed by others for less stable carbocations.[82,83,85]

IX.   THEORETICAL STUDIES

Molecular quantum mechanics takes as its start the Schrödinger equation, which is exact, and proceeds to introduce approximations until computationally tractable methods are reached. For systems of different size, that point is reached at different stages. A question of great practical importance revolves around the degree of sophistication required to calculate molecular properties to specified accuracy.

Pople and co-workers have provided a valuable parameterization along these lines.[85] They have calculated the structures and energies of a number of cations containing 1, 2, and 3 carbon atoms. Beginning with Hartree–Fock calculations using a large, polarized basis set, designated (6–31G*), they have explored the effects of added polarization functions, and treatment of electron correlation using perturbation theory at different levels.

Typically, the minimal-energy geometries were found using the 6–31G* basis, which includes two independent sets of basis functions for each valence orbital and a set of d-symmetry functions on carbon. These latter functions can interact with p-symmetry functions to produce a polarization effect, the movement of electron density from an atomic center. At those optimum geometries, energies were recalculated with larger basis sets and treatment of electron correlation at some level. It frequently happens that some minima (such as that labeled * in Figure 6.15, trace A) disappear at higher levels of theory (trace B). It is equally true that the remaining minima will shift in position by some (it is hoped) small amount. Unfortunately, the great amount of computation required to optimize geometries at the highest levels of theory precludes their use for all but the smallest systems.

In some cases the geometry optimization was performed with treatment of electron correlation, and the expected results were obtained. Three-center two-electron bonds, with their diffuse electron density, were found to be shorter. Thus in protonated methane the three-center C–H length shrank from 1.228 Å (optimized at the 6–31G* basis) to 1.18 Å (at the same basis, with second-order perturbation corrections for the effects of electron correlation). In the ethyl cation, the nonbridged structure only appears to be a local minimum when electron correlation is neglected. In a more dramatic example, optimization using correlated wavefunctions lowers the length of the corresponding C–H bond in C–H-protonated ethane from 2.7 to 1.3 Å.

Overall, the best treatments employed in this study reproduced experimental values for the relative energies of the carbocations within about 5 kcal/mol. Further, the ions treated here display some of the structural features (such as bridging hydrogens and corner-protonated cyclopropanes) likely to be important in larger cations and suggest the magnitudes for corrections due to the effect of electron correlation on lower-level calculations for such species. Caution is still in order since the importance of bridging in the ethyl cation is likely to be much greater than it is in the 2-butyl cation, to which it bears a structural resemblance.

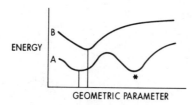

FIGURE 6.15.   Change in energy curves resulting from increasing the level of a theoretical calculation from A to B (i.e., increasing the size of the basis set).

TABLE 6.4.   Relative 2-norbornyl cation energies in kilocalories per mole[a]

Calculation	Classical	Nonclassical	Protonated Nortricyclane
4-21G	0.0	0.2	17.3
Large basis	0.0	1.0	17.8
4-21G + polarization	0.0	− 0.2	12.4

[a] Data taken from reference 86.

Goddard, Osamura, and Schaefer[86] have reported an ambitious effort to determine the optimum structure for the 2-norbornyl cation. The doubling in size from $C_3H_7^+$, the largest ion considered in the foregoing work, to $C_7H_{11}^+$ forces a retreat to simpler computational methods. A classical and nonclassical structure, as well as a protonated nortricyclane, were studied using the 4–21G basis set (somewhat smaller than 6–31G*, the smallest basis in the previous study,[85] and lacking its polarization functions). Using the geometries obtained at that level, energies were redetermined using a large basis set and also the 4–21G basis plus polarization functions. All three levels of theory predicted the classical and nonclassical structures to lie close in energy with the protonated nortricyclane significantly higher (Table 6.4).

As noted, the effects of electron correlation would be expected to provide greater stabilization to the nonclassical species in view of its more delocalized bonding. This extra stabilization can often amount to several kilocalories per mol, but in the present case it is likely to be less. Inspection of the calculated geometries of the classical and nonclassical ions (Figure 6.16) shows that bridging is far advanced in the classical structure. Although not as far along as the non-classical ion, the C_1-C_6 bond is extremely long, and the C_1-C_2 bond is closer to the length of C–C double bond. This "classical" structure must certainly be regarded as asymmetrically bridged at the very least, although its relevance to the true structure of the norbornyl cation has yet to be decided.

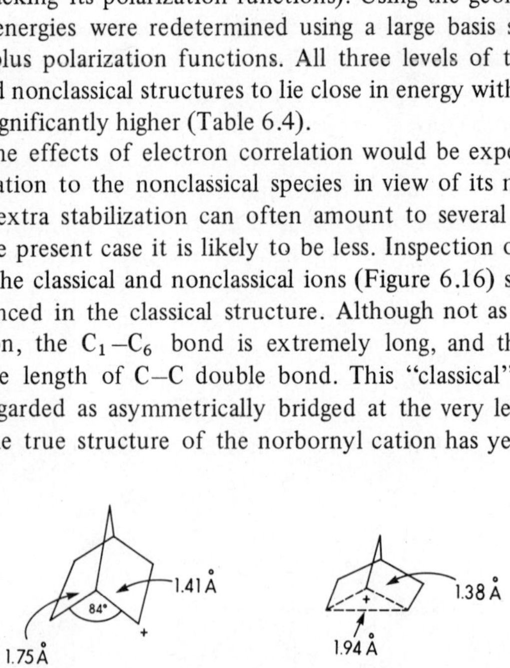

FIGURE 6.16.   Calculated geometries of classical and nonclassical 2-norbornyl cations (Data from Reference 86).

It is interesting that the energetic results of this calculation are in accord with the recent low-temperature study[43] of this ion, which indicates a very low barrier (if any) between the two proposed structures.

The great amount of effort required to obtain credible results in even medium-sized molecules has led to a variety of semiempirical methods for the determination of energies and geometries of molecular species. Dewar's MINDO/3 method[87] has proven as successful as any, and is especially parameterized to reproduced heats of formation.

The first systematic study of the applicability of this method to carbocation stabilities has been done by Harris and co-workers.[88] They have calculated the heats of formation of 42 carbocations for which experimental values have been determined. At first glance, the results are discouraging, with several values in error by amounts in excess of 20 kcal/mol. The average error is about 13 kcal/mol. However, some half-dozen ions account for the bulk of the discrepancy. These include vinyl, propargyl, and norbornyl cations. Were these types of cations to be excluded, the correlation would improve markedly. Yet, there would be no surety that other types of carbocations would not give unexpected discordant results. A more useful approach is the realization that several structural features in the ions introduce constant errors in their calculated heats of formation. For instance, heats of formation of tertiary acyclic ions are consistently calculated to be too high. This same trend is noted in the related hydrocarbons, and such systematic errors are well-known in semiempirical methods. In the present case, empirical corrections can be applied based on these recognizable structural features, or better, the heat of formation of the cation can be compared with the calculated heat of formation of the corresponding hydrocarbon to give a heat of hydride abstraction. This can be combined to yield a heat of formation for the ion which incorporates a correction for the particular structural features involved. Applying this procedure reduces the errors to about one-half the preceding values.

A further step can be taken to account for a possible intrinsic error in the calculations due to the presence of the cationic center itself. Consideration of the following hypothetical hydride transfer reaction places all of the prospective sources of systematic error in position to cancel. This technique, the method of

$$R_1^+ + R_2H \rightleftharpoons R_1H + R_2^+$$

isodesmic reactions (widely used in computational theoretical chemistry to minimize the effects of the neglect of electron correlation), reduces the mean absolute error in the foregoing set of carbocation heats of formation to about 2 kcal/mol, which is about as good as the experimental values to which they are being compared.

X.   ACKNOWLEDGMENTS

We are glad to express our appreciation to the Petroleum Research Fund, administered by the American Chemical Society, and to the National Science Foundation for support of our own research carbocations. Professors O. T. Benfey, Cheves Walling, Ned Porter, Pelham Wilder, and Jerome Berson were kind enough to read and criticize the first part of this chapter. Their viewpoints on the 2-norbornyl ion do not necessarily agree with ours.

XI.   REFERENCES

1.*   E. M. Arnett and T. C. Hofelich, *J. Am. Chem. Soc.,* **105**, 2889 (1983).

2.   E. M. Arnett, N. Pienta, and C. Petro, *J. Am. Chem. Soc.,* **102**, 398 (1980).

3.*   D. Fǎrcasiu, *J. Org. Chem.,* **46**, 223 (1981).

4.*   P. von R. Schleyer and J. Chandrasekhar, *J. Org. Chem.,* **46**, 227 (1981).

5.   These data are all taken from the tabulation by Aue and Bowers, *Gas Phase Ion Chemistry*, Volume 2, Academic, New York, 1979, Chapter 9. Experts in the field express skepticism about many of these values. Rather than trying to arbitrarily cite the most recent values as best, we chose to simply use a (common) critical tabulation for them all in the hope that errors will be small and compensatory.

6.   J. J. Soloman and F. H. Field, *J. Am. Chem. Soc.,* **98**, 1567 (1976).

7.   D. F. DeTar and R. W. Novak, *J. Am. Chem. Soc.,* **92**, 1361 (1970).

8.*   C. Y. Grob, *Angew. Chem. Int. Ed. Engl.,* **21**, 87 (1982).

9.   H. C. Brown, "The Nonclassical Ion Problem", Plenum Press, New York, 1977. (See also, H. C. Brown, *Acc. Chem. Res.,* **16**, 432 (1983).)

10.   P. D. Bartlett, *Nonclassical Ions,* W. A. Benjamin, New York, 1965. The preface to this volume provides a clear statement of why the problem was important from the perspective of a leading figure in American physical organic chemistry at the time. Essentially all of the evidence available in 1965 was kinetic or stereochemical. A brief chapter by H. C. Brown emphasized the role of four-state comparisons and steric effects. Brown anticipated herein his subsequent research on *endo-exo* comparisons and clearly identified the norbornyl systems as the principal future bone of contention.

11.   F. Franks, *Polywater,* The MIT Press, Cambridge, Mass., 1981.

12.   S. Schaffer, M. A. Thesis, Harvard University (cited in Reference 11).

13.* H. C. Brown, D. P. Kelly, and M. Periasamy, *Proc Natl. Acad. Sci. USA,* **77**, 6956 (1980).

14.   H. C. Brown and Y. Okamato, *J. Am. Chem. Soc.,* **80**, 4979 (1958).

15.   H. C. Brown, M. Periasamy, and K.-T. Liu, *J. Org. Chem.,* **46**, 1646 (1981).

16.   D. P. Kelly, M. J. Jenkins, and R. A. Mantello, *J. Org. Chem.,* **46**, 1650 (1981).

17.   H. C. Brown and M. Periasamy, *J. Org. Chem.,* **46**, 3161 (1981).

18.   H. C. Brown, D. P. Kelly, and M. Periasamy, *J. Org. Chem.,* **46**, 3170 (1981).

19.   H. C. Brown and M. Periasamy, *J. Org. Chem.,* **47**, 5 (1982).

20.* H. C. Brown, M. Periasamy, D. P. Kelly, and J. J. Giansiracusa, *J. Org. Chem.,* **47**, 2089 (1982).

21.   J. J. Giansiracusa, M. L. Jenkins, and D. P. Kelly, *Aust. J. Chem.*, **35**, 443 (1982).

22.   D. G. Farnum, R. E. Botto, W. T. Chambers, and B. Lam, *J. Am. Chem. Soc.*, **100**, 3847 (1978).

23.* G. A. Olah, A. L. Berrier, and G. K. S. Prakash, *Proc. Natl. Acad. Sci. USA*, **78**, 1998 (1981).

24.   G. A. Olah, G. K. S. Prakash, and G. Liang, *J. Am. Chem. Soc.*, **99**, 5683 (1977).

25.   See also Chapter 4 of the first volume (1978) in this series for a discussion of chemical shifts and charge distribution.

26.   E. M. Arnett and T. C. Hofelich, *J. Am. Chem. Soc.*, **105**, 2889, (1983).

27.   M. Saunders, L. Telkowski, and M. R. Kates, *J. Am. Chem. Soc.*, **99**, 8070 (1977).

28.   M. Saunders and M. R. Kates, *J. Am. Chem. Soc.*, **99**, 8071 (1977).

29.   M. Saunders, M. R. Kates, K. B. Wiberg, and W. Pratt, *J. Am. Chem. Soc.* **99**, 8072 (1977).

30.* M. Saunders and M. R. Kates, *J. Am. Chem. Soc.*, **102**, 6867 (1980).

31.* G. A. Olah, G. K. S. Prakash, M. Arvanaghi, and F. A. L. Anet, *J. Am. Chem. Soc.* **104**, 7105 (1982).

32.* M. Saunders and H.-U. Siel, *J. Am. Chem. Soc.*, **102**, 6869 (1980).

33.   H. Hogeveen and E. M. G. A. van Kruchten, *J. Org. Chem.*, **46**, 1350 (1981).

34.   L. R. Schmitz and T. S. Sorensen, *J. Am. Chem. Soc.*, **102**, 1645 (1980).

35.   R. P. Kirchen, K. Ranganayakula, A. Rauk, B. P. Singh, and T. S. Sorensen, *J. Am. Chem. Soc.*, **103**, 588 (1981).

36.   R. P. Kirchen, N. Okazawa, K. Ranganayakula, A. Rauk, and T. S. Sorensen, *J. Am. Chem. Soc.*, **103**, 597 (1981).

37.* K. L. Servis and F. F. Shue, *J. Am. Chem. Soc.*, **102**, 7233 (1980).

38.   C. J. Collins and N. S. Bowman, Eds., *Isotope Effects in Chemical Reactions*, A.C.S. Monograph No. 167, 1970.

39.   J. R. Lyerla, C. S. Yannoni, D. Bruck, and C. A. Fyfe, *J. Am. Chem. Soc.* **101**, 4770 (1979).

40.   P. C. Myhre and C. S. Yannoni, *J. Am. Chem. Soc.*, **103**, 230 (1981).

41.* C. S. Yannoni, V. Macho, and P. C. Myhre, *J. Am. Chem. Soc.*, **104**, 907 (1982).

42.   M. Saunders, D. Cox, and J. R. Lloyd, *J. Am. Chem. Soc.*, **101**, 6656 (1979).

43.* C. S. Yannoni, V. Macho, and P. C. Myhre, *J. Am. Chem. Soc.*, **104**, 7380 (1982).

44.   R. D. Miller and C. S. Yannoni, *J. Am. Chem. Soc.*, **102**, 7396 (1980).

45.   M. Saunders, P. Vogel, E. L. Hagen, and J. Rosenfeld, *Acc. Chem. Res.*, **6.**, 53 (1973).

46.   M. Saunders and M. R. Kates, *J. Am. Chem. Soc.*, **100**, 7082 (1978).

47.* V. A. Koptyug and V. G. Shubin, *Russ. J. Org. Chem.*, **16**, 1685 (1981); *Zh. Org. Khim.*, **16**, 1977 (1980).

48.   T. Ando, H. Yamataka, H. Morisaki, J. Yamawaki, J. Kuramochi, and Y. Yukawa, *J. Am. Chem. Soc.*, **103**, 430 (1981).

49.   V. J. Shiner, Jr. and J. J. Tai, *J. Am. Chem. Soc.*, **103**, 436 (1981).

50.   E. M. Arnett and C. Petro, *J. Am. Chem. Soc.*, **100**, 5402 (1978).

51.   E. M. Arnett and C. Petro, *J. Am. Chem. Soc.*, **100**, 5408 (1978).

52.* E. M. Arnett and N. J. Pienta, *J. Am. Chem. Soc.*, **102**, 3329 (1980).

53.   D. H. Aue and M. T. Bowers, in M. T. Bowers, Ed., *Gas Phase Ion Chemistry*, Academic, New York, 1979.

54.   E. M. Arnett, C. Petro, and P. von R. Schleyer, *J. Am. Chem. Soc.*, **101**, 522 (1979).

55.   N. C. Deno, J. J. Jaruzelski, and A. Shriesheim, *J. Am. Chem. Soc.*, 77, 3044 (1955).

56.   V. Gold and D. Bethell, *Carbonium Ions, an Introduction*, Academic, London, 1967.

57.*   D. Mirda, D. Rapp, and G. M. Kramer, *J. Org. Chem.*, 44, 2619 (1979).

58.   P. Walden, *Chem. Ber.*, 35, 2018 (1902).

59.   M. Gomberg, *Chem. Ber.*, 35, 2405 (1902).

60.   S. Winstein, P. E. Klinedinst, Jr., and G. C. Robinson, *J. Am. Chem. Soc.* 83, 885 (1961).

61.   A. G. Evans, I. H. McEwan, A. Price, and J. H. Thomas, *J. Chem. Soc.*, 3098 (1955).

62.   J. W. Bayles, J. L. Cotter, and A. G. Evans, *J. Chem. Soc.*, 3104 (1955).

63.*   C. D. Ritchie and T. C. Hofelich, *J. Am. Chem. Soc.*, 102, 7039 (1980).

64.   Z. Rappoport and J. Greenblatt, *J. Am. Chem. Soc.*, 101, 1343 (1979).

65.   R. E. Royer, G. H. Daub, and D. L. Vander Jagt, *J. Org. Chem.*, 44, 3196 (1979).

66.*   H. Kessler and M. Feigel, *Acc. Chem. Res.*, 15, 2 (1982).

67.*   E. M. Arnett and T. C. Hofelich, *J. Am. Chem. Soc.*, 104, 3522 (1982).

68.*   L. Lee, J. A. Russell, R. T. M. Su, R. J. Cross, and M. Saunders, *J. Am. Chem. Soc.* 103, 5031 (1981).

69.   R. A. Sneen, *Acc. Chem. Res.*, 6, 46 (1973).

70.   A good discussion of a particular case can be found in C. D. Ritchie, *Physical Organic Chemistry: The Fundamental Concepts*, Marcel Dekker, Inc. New York, 1975, pp. 76–80.

71.   D. J. Raber, J. M. Harris, R. E. Hall, and P. von R. Schleyer, *J. Am. Chem. Soc.*, 93, 4821 (1971).

72.   T. W. Bentley and P. v. R. Schleyer, *Adv. Phys. Org. Chem.*, 14, 1 (1977).

73.   R. A. Sneen and H. M. Robbins, *J. Am. Chem. Soc.*, 94, 7868 (1972).

74.*   W. P. Jencks, *Acc. Chem. Res.*, 13, 161 (1980).

75.*   W. P. Jencks, *Chem. Soc. Rev.*, 10, 345 (1981).

76.*   T. W. Bentley, C. T. Bowen, D. H. Morten, and P. von R. Schleyer, *J. Am. Chem. Soc.*, 103, 5466 (1981).

77.*   T. W. Bentley and G. E. Carter, *J. Am. Chem. Soc.*, 104, 5741 (1982).

78.*   W. P. Jencks, *Acc. Chem. Res.*, 9, 425 (1976).

79.*   J. P. Richard and W. P. Jencks, *J. Am. Chem. Soc.*, 104, 4689 (1982).

80.*   J. P. Richard and W. P. Jencks, *J. Am. Chem. Soc.*, 104, 4691 (1982).

81.   See the text in Reference 52, as well as P. R. Young and W. P. Jencks, *J. Am. Chem. Soc.*, 99, 8238 (1977).

82.   R. A. Sneen, J. V. Carter, and P. S. Kay, *J. Am. Chem. Soc.*, 88, 2594 (1966).

83.   D. J. Raber, J. M. Harris, R. E. Hall, and P. von R. Schleyer, *J. Am. Chem. Soc.*, 93, 4821 (1971).

84.   C. D. Ritchie, *Acc. Chem. Res.*, 5, 348 (1972).

85.*   K. Raghavachari, R. A. Whiteside, J. A. Pople, and P. von R. Schleyer, *J. Am. Chem. Soc.*, 103, 5469 (1981).

86.*   J. D. Goddard, Y. Osamura, and H. F. Schaefer, III, *J. Am. Chem. Soc.*, 104, 3258 (1982).

87.   R. Bingham, M. J. S. Dewar, and D. H. Lo, *J. Am. Chem. Soc.*, 97, 1285 (1975).

88.*   J. M. Harris, S. G. Schafer, and S. D. Worley, *J. Comput. Chem.*, 3, 208 (1982).

7

FREE RADICALS

LEONARD KAPLAN

Department of Chemistry, Cleveland State University, Cleveland, Ohio 44115

Professor Kaplan's present address is: Department of Chemistry, Case Western Reserve University, Cleveland, Ohio 44106.

Free radical chemistry is changing. New kinds of events are being observed. New techniques permit the direct observation of radicals and the determination of their properties and behavior.

I.   STRUCTURE OF PROTOTYPAL RADICALS

A.   Methyl

"In spite of . . . numerous investigations . . . there are still large uncertainties as to the molecular structure . . . of the methyl radical."[1a] This is a refreshingly realistic statement in view of the prevalent climate of currently fashionable belief.

Two similar sets of vibration-rotation frequencies of the methyl radical have been analyzed by use of somewhat different potentials and Hamiltonians and have been fitted satisfactorily to potential energy curves corresponding to a planar structure.[1] Although aspects of the results of the approximate analyses[1a,b] were taken to be incompatible with a double-minimum potential function, that is, a nonplanar structure, no indication was given of the sensitivity of the goodness of fit to departures from planarity. One would like to know the degree to which the observed frequencies could be fitted satisfactorily to equations optimized for a structure that is significantly nonplanar, say, 5° (one-quarter of the way to "tetrahedral").

For very recent observations of frequencies of methyl, see Reference 2.

B.   General Comments on the Structure of Alkyl Radicals

We have been interested for some time in questions related to the structure of alkyl radicals.[3,4a,5a,6a] The questions asked about structure have proven difficult to answer, even with the advent of new techniques. Indeed, a hard look at what has actually been established "beyond a reasonable doubt" leads us to the belief that we do not know now much more than we thought we knew twenty years ago. Furthermore, over the last several years, we have been drawn increasingly strongly to the belief that it is not just that some important questions are difficult to answer, but rather that there are no answers to them because the questions are not meaningful: speaking of the "inversion (or pyramidal bending)" or the "internal rotation (or torsion)" of an alkyl radical as if it were an isolable, distinguishable, solitary event is simplistic, if not wrong. Methyl may not be a prototypal alkyl radical.

We have called attention[6b] to the difficulties introduced into the interpretation of results related to structure as a consequence of the simultaneity of occurrence of bending and internal rotational motions in an alkyl radical. Such indications have become common:[7]

The frequency for the umbrella inversional motion of *tert*-butyl is undoubtedly quite low and it may well be similar to the methyl torsion frequency, in which case these two motions will be coupled.[9a]

. . . as the [ethyl] radical center undergoes a pyramidal type vibration the characteristically large amplitude of this mode easily places the radical center in a region on the energy surface where free rotation . . . will readily occur.[9b]

The calculations indicate that the planarity or non-planarity of a radical center is controlled by rotations of the methyl groups which in turn are separated by very small energy differences . . . In addition, it is found that the rotation of the methyl groups not only profoundly changes the geometry of the radical center but also creates dramatic changes in the potential energy for the pyramidal distortion of the radical.[9c]

. . . the rate of interconversion between the two conformations of the isobutyl radical is competitive with the time required for a vibration. . . . the pyramidal bending motion [in isobutyl and neopentyl radicals] should be coupled with the rotation about the α—CC bond to produce a complicated motion. . . .[9d]

. . . the torsional motion of the methyl groups in the *tertiary*-butyl radical are low and coupled with the pyramidal bending motion of the radical center. These motions most likely drive each other, and it is a . . . mechanism by which inversion of the radical center occurs . . . The most likely reason for the low barrier [for internal rotation] is the relative ease with which the radical center distorts pyramidally . . . as rotation occurs in the ethyl radical away from the staggered geometry, the repulsive interaction between the CH bonds is alleviated by the ease with which the radical center undergoes a pyramidal distortion. . . . the rotation-pyramidal bending motion is coupled and dramatically decreases the energy differences between the minima and maxima on the rotational potential function.[9e]

. . . in spite of the very low . . . barrier for internal rotation . . . [in Et·], rather dramatic geometric changes take place about the radical center.[9f]

. . . the dramatic geometric changes that occur about the radical center as the almost free internal rotation proceeds. . . . The rather low barrier for internal rotation in the neopentyl radical could be a result of a complicated, coupled pyramidal bending-internal rotation motion that is facilitated by the relative ease with which the radical center undergoes the pyramidal distortion. . . . Both of the [equilibrium] structures [of the isobutyl radical] have nonplanar geometries about the radical center that changes dramatically when internal rotation takes place.
. . . the rather complex coupling that takes place in alkyl radicals between the internal rotation motion and other vibrational modes . . .[9g]

Compounding all of the foregoing is the further complication that, even when a combined pyramidal bending/internal rotation process is considered, the motions and energetics determined in a particular investigation may not be transferable to other environments. For example, see References 6c, 10a, and 11.

The coupling of bending and internal rotational motions does not bear only on questions of structure and internal dynamics. Since determinations of the heats of formation of alkyl radicals can depend upon the results of modeling of the internal motions of the radical,[6d] uncertainties and problems concerning structure can lead to uncertainties and problems concerning energetics.[8a,12] Thus the great sagas of structure and energetics[10b] of alkyl radicals are made even more complex by their interdependence.

C.   Ethyl

A report of more complete and detailed work on the infrared spectrum and the structure of the ethyl radical in a solid argon matrix has appeared.[9f,13] The "ethyl radical"[14] and its deuterated analogues, generated by photolysis of the corresponding propionyl peroxides at low temperature, were observed and their infrared frequencies reported and assigned. These results were then discussed in a narrow context — solely in terms of a structure of the symmetry of that which emerged from an "ab initio calculation," that is, a structure in which "the radical center is nonplanar [and] the angle between CC axis and the plane formed by the CH_2 group is 6.19°."[15] The ethyl radical is considered to have "a methylene group that easily undergoes a pyramidal (umbrella) type motion."

D.   t-Butyl

The infrared spectrum of products of the pyrolysis of $Bu^t N=NBu^t$ at 700 K in the gas phase was obtained in an argon matrix at an unspecified temperature, presumably 8 K, and was assigned to $Bu^t \cdot$ on the basis of undefinitive, but moderately strongly indicative, indirect evidence.[13d,16a] From a rather overextended analysis of the spectrum, it was concluded that C_{3h}, C_3, and C_s geometries for the t-butyl radical are excluded. A C_{3v} structure was not excluded.

Photolysis of $Bu^t NO$ in the gas phase led to a transient, considered arbitrarily to be $Bu^t \cdot$, whose infrared spectrum in the gas phase was broad and poorly defined and had its "main peak" at approximately $2833\ cm^{-1}$, with "a hint of a much weaker sideband at $\sim 2750\ cm^{-1}$" and apparently "a weak but distinguishable high frequency sideband at $\sim 2950\ cm^{-1}$."[17]

II.   CHARACTERISTICS OF PROTOTYPAL RADICALS

A.   Methyl and Allyl

The characteristics of methyl radicals are covered in recent literature by the following: "New Electronic States in CH_3, Observed Using Multiphoton Ionization,"[18a] and "Multiphoton Ionization of CH_3 Radicals in the Gas Phase."[18b]

Vacuum pyrolysis of $CH_2=CHCH_2 X$ (X=Cl, Br, I, $SiMe_3$, $CH_2CH=CH_2$) at 800–1000°C produced the allyl radical, which was condensed in an Ar matrix at 12 K. Its infrared spectrum in the matrix had bands at 3107(m), 3019(m), 1602(m), 1477(m), 1463(m), 1389(s), 983(s), 972(m), 801(vs), and 510(s) cm^{-1}.[18c]

B.   The Cation Radical of Carbon Monoxide: Magnetic Properties

The esr spectra, obtained at 4 K "in the $g = 2$ region where numerous impurity signals are often observed," of products of photoionization of ^{12}CO and ^{13}CO in neon in the gas phase under conditions ("such a potentially 'dirty' source as an open tube neon discharge") under which "background impurity gases such as $O_2(g)$, $CO_2(g)$, or $OH(g)$ . . . are likely to be present" and which led also to "weak . . . signals attributable to such neutral radicals as HCO, CH_3, H, and N atoms" were assigned to $^{12}CO^{+\cdot}$ and $^{13}CO^{+\cdot}$, the corresponding cation radicals of carbon monoxide.[19a] Values of $g = 2.000-2.001$, $a_C = 561\,G$, and $a_{dipolar} = 16\,G$ were determined. Assignment of the spectrum to a CO-derived product is secure, and the results are consistent with that product being $CO^{+\cdot}$.

A discharge current was passed through a gaseous mixture of ^{13}CO and He to generate $^{13}CO^{+\cdot}$.[19b] Analysis of the microwave spectrum yielded $a_C = 532-543\,G$ and $a_{dipolar} = 17\,G$. Values of $a_C = 535-545\,G$ and $a_{dipolar} = 5-16\,G$ had earlier been extracted from the structure of the electronic absorption spectrum of $^{13}CO^{+\cdot}$ which had been generated in the gas phase by electron impact on CO.[20] The agreement of these results with those obtained recently by use of esr[19a] strongly supports the assignment to $CO^{+\cdot}$ in the latter work and weakens an earlier[21a,b] assignment.

Earlier[5c] we discussed esr results obtained for the isoelectronic series $H_2CO^{+\cdot}$, $H_2CN\cdot$, and $H_2BO\cdot$. Results are available also for molecules isoelectronic with $CO^{+\cdot}$: $CN\cdot$ has been extensively studied[22] and $a_C = 210\,G$[22a] has been reported. There are also reports of the esr spectra of $BO\cdot$,[23] $AlH^{+\cdot}$,[24] $BeF\cdot$,[25] and $MgH\cdot$.[26]

C. Benzoyl: Optical Spectrum

A $0.01\,M$ solution of $PhCOBu^t$ in 3-methyl-3-pentanol was photolyzed at $13°C$.[27] A spectrum with bands at 368 and approximately $460\,m\mu$ was observed and was assigned to $Ph\dot{C}O$ on the basis that its carrier was a transient, but was not $Bu^t\cdot$. The assignment is weak and is not strengthened by the report that photolysis of PhCOPh in an ether–isopentane–ethanol glass at $77\,K$ led to a species that had a band at approximately $350\,m\mu$, but that was "fairly stable" in solution at room temperature and disappeared only "gradually" on exposure to air.[28a]

It had been observed earlier that photolysis of PhCOBr followed by condensation of products at $70\,K$ led to an orange-red material whose color and esr signal disappeared simultaneously on warming.[28b]

D. Allene and Cyclopropane Radical Cations

The esr spectrum (q, $a = 13\,G$) of a product of the γ-irradiation of allene in $CFCl_3$ at $77\,K$ was not inconsistent with that anticipated for the cation radical of allene.[29,84a]

The esr spectra of the mixtures resulting from irradiation at $4\,K$ of cyclopropane in SF_6 (recorded at 4 and $77\,K$),[196a] $CFCl_2CF_2Cl$ (recorded at 4 and $77\,K$),[196a] and in $CFCl_3$ (recorded at 4, 77, and $140\,K$)[196a,b] were considered to be of the cation radical of cyclopropane.

E. $Li_3C\cdot$

The product of the reaction of Li vapor with graphite was bombarded with high-energy electrons in a mass spectrometer to produce Li_3C^+.[30a] From a determination of the ion current of Li_3C^+ as a function of electron energy, a number (4.6 eV) was obtained that was considered to be the ionization potential of $Li_3C\cdot$. It has been stated that "on the basis of ionization potentials, CLi_3^+ is the most stable substituted methyl cation known to date."[30b]

F.   Acyloxy Radicals, $RCO_2\cdot$: Magnetic Properties

It has been suggested that esr spectra of $-CH=\overset{|}{C}-CO_2\cdot$ radicals ($g = 2.0146$) have been observed, the first such observation of a $RCO_2\cdot$ in solution.[31a] Don't bet too heavily on it.

By use of an indirect method that relies strongly on fitting and simulation based on approximate models, it has been estimated that $g(CH_3CO_2\cdot) \approx 2.009$.[31b]

G.   Radical Cations of Ethers, Amides, and Sulfoxides

γ-Irradiation of a solid solution of Me_2O in $FCCl_3$ at 77 K gave a seven-line spectrum [$g = 2.0085, a = 43.0\,G$ (recorded at 97 K);[32a] $a = 42.3\,G$ (recorded at 120 K)[32b]] consistent with expectation for $Me_2O^{+\cdot}$. Similar treatment of tri-methyleneoxide gave a spectrum ($a = 10.0, 65.0\,G$; recorded at 133 K), which was not inconsistent with expectation for the corresponding radical cation.[32f] Similar treatment of tetrahydrofuran (t of t with $a = 40\,G, 89\,G$ at 77 K; warming to 155 K produced a continuous and reversible change to a five-line spectrum of $a = 65\,G$), 2-methyltetrahydrofuran ("the spectrum at 77 K can be interpreted as a double triplet with $a = 83\,G$ and $42\,G$ which changes reversibly into another double triplet of $a = 78\,G$ and $49\,G$ upon warming, e.g., to 146 K"), cis-2,5-dimethyltetrahydrofuran (spectrum at 77 K can be viewed as a doublet of doublets plus a triplet, $a = 35\,G, 95\,G$), and trans-2,5-dimethyltetrahydrofuran (t, $a = 97\,G$, 77 K) gave spectra that were not inconsistent with expectation for the corresponding radical cations.[32b] See also Reference 33.

Esr spectra have been assigned to $Me_2S^{+\cdot}$[34a] and $Bu_2^tS^{+\cdot}$.[34b,c] Optical spectra have been assigned to $Me_2S^{+\cdot}$,[34d] $Bu_2^tS^{+\cdot}$,[34e-g] and (tetrahydrothiophene)$^{+\cdot}$.[34g]

γ-Irradiation of $HCONMe_2$ and of $MeCONMe_2$ in $FCCl_3$ at 77 K gave products whose esr spectra were reasonably assigned to the corresponding action radicals. CH_3SOCH_3 was treated similarly, and it was stated that the observed "very small proton coupling constant . . . is expected [for $CH_3SOCH_3^{+\cdot}$] . . ."[32c,d]

H. t-Butoxy and Methoxy

"In the photodecomposition of $Me_3COOCMe_3$ . . . weak signals peaking at 320 nm are observed;" an assignment to t-butoxy was implied.[35]

$CH_3O\cdot$ and $CD_3O\cdot$ were produced by reaction of CH_3OH and CD_3OD, respectively, with the 2450 MHz microwave discharge products of CF_4. A C–O bond length of 1.36 Å in $CH_3O\cdot$ was calculated, as compared to the corresponding bond length of 1.42 Å in CH_3OH.[197]

I.   Radical Anions Comprised of Tetracoordinated Carbon Only

Work on the title compounds, for example, the radical anions of haloalkanes $(RX^{-\cdot})$, and the related radical . . anion complexes $(R\cdot/X^-)$ has continued. Since this is an ongoing story of some complexity, the reader should first read our earlier discussion.[6e]

a.   *Haloalkane Radical Anions*

The esr spectrum (recorded at 87 K) of a solution of $(CF_3)_3CI$ of unstated purity in a 2-methyltetrahydrofuran glass irradiated at 77 K was assigned to $(CF_3)_3CI^{-\cdot}$ on the reasonable, but undefinitive, basis of consistency in detail of the spectrum with expectation.[36] On the basis of this observation, the assumption that $(CF_3)_3C\cdot$ is "nearly planar" (?), and of an irrelevant microscopic reversibility[37] argument the "explanation of the factors governing the stability of RX^- radical anions" that has been advanced by Symons[6f] was disputed.

Symons has qualified significantly and expanded his position[38a,n] on the question of which molecules form σ^*-radicals more easily than others and "why" they do so.[38b,c] However, he continues to put forward an unchanged view[6h] of the existence of $RX^{-\cdot}$ and $R\cdot/X^-$.[39a,b]

It has been briefly reported that "preliminary results [of experiments with 1-bromoadamantane] give well-defined features that seem to be characteristic of a σ^* bromine radical"[40a] and that "it seems probable" that "1-bromo adamantane . . . [forms] a stable σ^* anion rather than an $R\cdot/Br^-$ adduct."[40b]

b.   *Radical . . Anion $(R\cdot/X^-)$ Complexes*

There have been further reports of esr spectra of alkyl $R\cdot/X^-$.[38k,41]

c.   *Ethylene Oxide*

Additional reports of esr spectra attributed to $\cdot CH_2CH_2O^{-}$[42a] and tentatively attributed to the radical anion of ethylene oxide,[42b] formed by reaction of "$O^{-\cdot}$" and ethylene, have appeared.

J.   Comparative Chemistry of Radicals in Different Electronic States

We have discussed reports of observations that were interpreted in terms of chemical reactions of the "same" radical in more than one electronic state.[6j] These reports concerned amidyl and imidyl radicals; additional reports have appeared.[43,44]

Similar reports have appeared concerning aroyloxy and acyloxy radicals.

The comparative complex multiproduct, undefinitively defined chemistry of the $ArCO_2^-/CH_3NO_2/OH^-/TiCl_3/S_2O_8^=/H_2O/EDTA$ and $ArCO_3H/CH_3NO_2/OH^-/H_2O/TiCl_3/EDTA$ systems has been rationalized, based on narrow thinking,[46] in terms of the involvement of two $ArCO_2 \cdot$ radicals:[47] "a σ-radical . . . , with the unpaired electron in an in-plane σ-orbital" and a "π-type zwitterion."[47c] See also Reference 48.

The photoinitiated bromination of butane by several acyl hypobromites (RCO_2Br) in $CFCl_3$ in the presence of added Br_2 and of added $Cl_2C{=}CH_2$ has been described.[49,50a] Based on the observations that the ratio of 2-bromobutane/1-bromobutane is invariant with concentration of butane and with temperature and that the ratio is different (greater) in the presence of Br_2 than in the presence of $Cl_2C{=}CH_2$, it is stated that "two different carboxylate radicals are involved: those generated . . . [in the presence of Br_2] . . . and those generated . . . [in the presence of $Cl_2C{=}CH_2$] . . ." It is later stated that "these [antecedent?] results require two hydrogen abstracting carboxylate radicals." This conclusion does not follow from the results that are reported; additional information, not reported in the Communication, is necessary. It must be demonstrated that the ratio of bromobutanes in the product is equal to the ratio of rates of formation of the corresponding n- and sec-butyl radicals[51] and that H is abstracted from butane only $RCO_2 \cdot$.[52a] These points are not even addressed,[52b] let alone demonstrated.

The communication addresses also the question of the relative rates of decarboxylation of $RCO_2 \cdot$.[49] That aspect of the work is discussed in Section XI.C.

III.   ENERGETICS OF PROTOTYPAL RADICALS

A.   Thermochemistry of t-butyl $[\Delta H_f(t\text{-}Bu\cdot), D(Me_3C{-}H)]$

a.   Thermolysis of Bu^tR

Tsang[53a] has reiterated his earlier report that the results of shock-tube studies of the decomposition of Bu^tR lead to $\Delta H_{f,300}(Bu^t\cdot) \cong 12.5\,kcal/mol$[6k] and has now presented a range of 11–13 kcal/mol as accommodating in a mutually consistent way the variety of results discussed earlier.[53b]

The confirmation of some aspects of the experiments and their analysis which led to the emergence of $\Delta H_{f,298}(Bu^t\cdot) \cong 10$ kcal/mol from studies of the oxidation of Bu^tBu^t [54a] has been reported.[54b]

The oxidation of Bu^tPr^i has been studied[55] in a manner analogous to that of Bu^tBu^t.[61] However, the chemistry of this system is much dirtier, and about all that can validly be stated is that the results are not inconsistent with the value of the heat of formation obtained from the study of Bu^tBu^t. A value of $\Delta H_{f,298}(Bu^t\cdot) = 9 \pm 0.5$ kcal/mol was recommended.

The "azomethane sensitized" pyrolysis of $(CH_3)_3CH$ was studied in the narrow range 311–331°C and measurements were made of the net rates of appearance of H_2 and $(CH_3)_3CC(CH_3)_3$ from which $k(Me_3C\cdot \rightarrow Me_2C{=}CH_2 + H\cdot)/k^{1/2}(2Me_3C\cdot \rightarrow Me_3CCMe_3)$ was calculated;[56] only weak experimental justification of the validity of doing so was provided. Combination of that result with the rate constant for dimerization of $Bu^t\cdot$[57a] gave a value of $k(Me_3C\cdot \rightarrow H\cdot)$. The reaction of $Me_2C{=}CH_2$ with $H\cdot$ at 25–290°C was also studied, and its overall rate constant was estimated from a loose and assumption-laden analysis of the variation of $[H\cdot]$ with time and was then combined with an earlier "determination" of $k(Me_2C{=}CH_2 + H\cdot \rightarrow Me_2CHCH_2\cdot)/k(Me_2C{=}CH_2 + H\cdot \rightarrow Me_3C\cdot)$ to yield $k(Me_2C{=}CH_2 + H\cdot \rightarrow Me_3C\cdot)$. Combination of the component rate constants of $Me_3C\cdot \rightleftarrows Me_2C{=}CH_2 + H\cdot$ yielded an equilibrium constant which was transformed into a $\Delta H_{f,300}(Bu^t\cdot)$ of 10.6 kcal/mol. Similar transformations of a variety of earlier results and of quantities estimated on the basis of earlier results[57b–57h] led to $\Delta H_{f,300} = 9.5, 10.8, 10.1, 8.1, 11.3$ kcal/mol. A value of 10.5 ± 1 kcal/mol, which is approximately 2 kcal/mol higher than the "traditional" value,[6d] was put forth. This value is strongly dependent on the rate constant for dimerization of $Bu^t\cdot$[10c] and on the quantitative modeling of $Bu^t\cdot$[10d] for the purpose of estimating its thermochemical characteristics.[57i]

From a combination of the treatment of the results of a study of the very-low-pressure pyrolysis of $PhCH_2Bu^t$ from 645–791°C with an estimated value of $k(PhCH_2\cdot + Bu^t\cdot \rightarrow PhCH_2Bu^t)$ and with thermochemical characteristics of $Bu^t\cdot$ obtained from modeling its internal dynamics, a value of $\Delta H_{f,298}(Bu^t\cdot) = 10$ kcal/mol was obtained.[58] Although this work is no less reliable than that of others in the area, the authors had the good sense to assign an "error limit" (± 2 kcal/mol) which approaches reality and to lump together other recent determinations into the range 8–12 kcal/mol for the purpose of comparison.

The arguments presented in the foregoing papers for the validity of particular values of $\Delta H_f(Bu^t\cdot)$ often contain, as corroborative support, statements that analysis of other results yield similar values. The following results and comparison should assist in putting such reasoning into perspective: A study[59] of the extent of occurrence of the dissociative electron capture process, $Bu^tCl +$

$e^- \rightarrow Cl^- + Bu^t \cdot$, as a function of energy of the electron yielded $\Delta H_f(Bu^t \cdot) = 3.0$ kcal/mol, which was compared to values of 4.1 and 6.7 kcal/mol calculated from earlier reports of $\Delta H_f(Bu^{t+})$ and of the ionization potential of $Bu^t \cdot$.

b.   *Equilibria Involving* $Bu^t \cdot$

Values of K for the reactions[10i] $Me \cdot + RI \rightarrow R \cdot + MeI$ were used to calculate $\Delta H_{f,300}(R \cdot)$, that of $Me \cdot$ being used as a standard.[60] It was necessary to know $\Delta H_f(RI)$, and also, as in the previous section, the thermochemical characteristics of $R \cdot$, such as its S^o, to calculate its ΔH_f. So, here too, it was necessary to use the results of the judgmental modeling of $Bu^t \cdot$ to obtain the value of $\Delta H_f(Bu^t \cdot)$, 9.4 kcal/mol. This value corresponds to $D(Me_3C-H) \approx 94$ kcal/mol.

c.   *Problems, and a Status Report*

In a discussion of one of the more recent studies[57g] which gave a value of $\Delta H_f(Bu^t \cdot) = 8.4$ kcal/mol, that is, a "traditional" value, we indicated the sensitivity of the result to the way in which $Bu^t \cdot$ was modeled: "use of a different reasonable value for the out-of-plane bending frequency . . . alone would have increased ΔH_f^o by 1.1 kcal/mol."[6m] Indeed, the results of a recent infrared study[10e,13d] have been applied to the original results, and $\Delta H_{f,300}^o = 10.5$ kcal/ mol has been calculated.[13d]

Given the types of measurements that are extant,[6d,10b] even if they were perfectly accurate $\Delta H_f(Bu^t \cdot)$ would not be known with confidence until $S^o(Bu^t \cdot)$ is. Unfortunately, the calculation of S^o requires a degree of knowledge of $Bu^t \cdot$'s internal motions and their dynamics that is not currently in hand. And although various techniques, such as those that involve photoelectron, ion cyclotron resonance, and photoionization spectroscopy, are not in this category, they have their own problems. Griller, et al., recommend use of their value of $\Delta H_f = 9.4$ kcal/mol (Section III.A.b) "until more accurate methods of measurement become available."[60b]

  . . . The strength of the method is that a series of related experiments yield a complete set of values for $\Delta H_{f,300}(R \cdot)$. Its weakness is that it relies upon thermodynamic data drawn from the literature. Despite the weakness in the technique, it is clear that our measurements of K_{300} adequately characterize the properties of the . . . systems and can always be used to recompute values of $\Delta H_f(R \cdot)$ as thermodynamic data for properties such as $S^o(R \cdot)$ are revised.

The approach used in this work, in common with all other methods currently available, yields measurements of $\Delta H_f(R \cdot)$ and BDE (R−H), which are subject to fairly high limits of error. At present, it seems that the only way in which such values can be refined is by combination and comparison of results from a variety of experiments that effectively define acceptable ranges for these quantities.[60b]

Although we might opt for a slightly higher value of $\Delta H_f(Bu^t\cdot)$, their attitude is quite sensible.

d. *Impact*

We have concentrated in this section on $Bu^t\cdot$. However, we caution the reader that similar, but attenuated, "old-new" dichotomies exist regarding the energetics of other alkyl radicals, for example, $Et\cdot$ and $Pr^i\cdot$.

One of the reasons we devote the degree of attention that we do to the structure, energetics, and characteristics of prototypal radicals is the ripple effect which anything concerning prototypal systems brings about throughout free radical chemistry. For example:

Prior to 1974, heats of formation of the centrally important primary, secondary, and tertiary radicals were based largely on kinetic studies by Benson and his collaborators of iodination and bromination of hydrocarbons and on other bond-fission reactions. . . .

Mechanistic organic chemists have been understandably reluctant to question the accuracy and precision of these values.[61] Experimental techniques for determination of bond dissociation energies are difficult and have fallen by tradition in the province of physical chemistry. If questioning the well-known heats of formation has seemed vain, . . . the obligation nonetheless has persisted to note the vulnerability of the values to possible change in the future as well as the consequences . . .

Impetus for upward revision in the heats of formation of the ethyl (primary), isopropyl (secondary), and *t*-butyl (tertiary) free radicals has gained sufficient momentum in the last 6 yr to prompt an illustrative reassessment of the far-ranging implications for interpretation of thermal reorganizations. Previous obstacles to consideration of diradicals as transition states in rearrangements of cyclopropanes and cyclobutanes are removed. The energetic relation of cyclohexa-1,4-diyl as an intermediate diradical in the Cope rearrangement to the transition state of a concerted mechanism is clarified.[62]

Interested readers can find many other likely subjects for possible reassessment in the recent review of Berson [in P. de Mayo, Ed., *Rearrangements in Ground and Excited States*, Academic, Vol. 1, 1980, pp. 324–334] and the extensive thermochemical-kinetic analyses developed by O'Neal and Benson [H. E. O'Neal and S. W. Benson, *J. Phys. Chem.*, **72**, 1866 (1968); *Int. J. Chem. Kinet.*, **2**, 423 (1970)] . . .[62]

Note that the higher, 1981 values of the crucial heats of formation of ethyl, isopropyl, and *t*-butyl radicals are not established. In the end, internal consistency among many bits of information may be the decisive factor. In this sense, a thorough exploration of the consequences of higher values is justified.[62]

Also:

. . . the new bond energies provide a sound basis for understanding cis-trans isomerization and small ring decyclization processes.

. . . it is now clear that the biradical mechanism does provide an accurate quantitative basis for predicting the activation energy for the isomerization of small ring compounds.[63]

B. $D(HC\equiv C-H)$

On the basis of estimates, some of which make use of voodoo thermochemical kinetics, it was concluded that the more recent reports of $D(HC\equiv C-H)$ are wrong and that earlier determinations are more closely correct.[64a] Note that Benson recently cited the earlier value.[64b]

The significance of this paper is not the details of its contents but that it serves to flag a problem: $D(HC\equiv C-H)$, too, may be up for grabs.

IV.   EXPERIMENTAL TECHNIQUES

A.   Magnetic Resonance of Free Radicals as Manifested in other Phenomena or Species

a.   Optical Detection of ESR of Free Radicals in Solution

I.   *Basis of the technique*

The method is based on detection of changes in the luminescence intensity of radical pair reaction products caused by absorption of resonance microwave radiation by the radical pairs in an external magnetic field.[65a]

. . . an electron spin transition corresponding to one of the EPR lines of a radical ion is excited selectively by microwave irradiation in . . . [an] EPR spectrometer. Sequential excitation of the EPR lines is achieved, as usual, by sweeping the field of the EPR magnet. Such selective irradiation of one partner of the radical pair destroys the coherence between the two unpaired electrons, thereby increasing the rate of singlet-triplet mixing. For pairs formed and reacting via the singlet state an increase in singlet-triplet mixing rate produces a decrease in the reaction probability.

. . . ion pairs eventually react by reverse electron transfer to form fluorescent excited singlet states. Those pairs whose singlet states have been perturbed by microwave irradiation will therefore produce fewer excited products and exhibit decreased fluorescence. Furthermore, the magnitude of the fluorescence decrease will depend on the extent to which the e^- spins have been perturbed by the microwave field, i.e., on the EPR intensities. If the microwave frequency is constant, therefore, a plot of fluorescence decrease vs. magnetic field will correspond to the EPR spectrum of the radical ion intermediate.[65b]

II.   *Scope of the technique.*   "The ESR spectra of anion and cation radicals can be detected . . . provided the radicals are spin-correlated partners of an ion-radical pair"[65c] and ". . . the recombination product or the product formed when the radical leaves the cage"[65d] "yields a detectable excited state."[65e]

The method involves "a combination of the selectivity of a magnetic resonance measurement with the sensitivity of fluorescence detection."[65f] It "allows one to detect ion-radical pairs at their mean concentration of a few tens per sample . . . [and] allows detection of the spectra of short-lived radicals with lifetimes as small as several nanoseconds. So short a lifetime presents an insuperable difficulty in taking resolved spectra by means of conventional ESR spectroscopy."[65c] ". . . this technique can [therefore] open up quite new possibilities in studying the nature, reactions, and molecular dynamics of short-lived radical pair intermediates in various chemical processes."[65a] ". . . one can observe radical cations and anions in a wide range of chemical compounds, which earlier could not be observed at all on account of their instability. Similar radicals were stabilized only under special conditions, for instance, at cryogenic temperatures."[65g]

III.   *Observations.*   A plot of the dependence on magnetic field strength of the intensity of fluorescence of products of the positron irradiation of $0.01\ M$ naphthalene in squalane yielded a single-line ($w_{1/2} = 14\,\text{G}$) spectrum that was attributed to (naphthalene)$^+$/(naphthalene)$^-$.[65a,d,h,i;66a] The average concentration of radical pairs was estimated to be about $10^1 - 10^2$ per sample as compared to the sensitivity of an esr spectrometer of about $10^{11} - 10^{12}$ spins per sample.

The spectrum ($a = 2.7\,\text{G}$) of products of the electron irradiation of $0.005\ M$ diphenyl in squalane was attributed to and is consistent with that expected for (diphenyl)$^+$/(diphenyl)$^-$.[65a,k] The average concentration of radical pairs was estimated to be approximately 10^2 per sample. A plot of the dependence on magnetic field strength of the intensity of products $0.5\ \mu\text{sec}$ after the electron irradiation of $10^{-3}\,M$ biphenyl in cyclohexane yielded a spectrum ($a = 2.5\,\text{G}$) that was attributed implicitly to a radical ion involving biphenyl.[65e,65l,67]

A spectrum, consistent with that of $C_6F_6^-$ including the hyperfine structure, was observed upon electron irradiation of $0.002M\ C_6F_6$ and $0.005\,M$ anthracene in squalane.[65j,o;68] A plot of the dependence on magnetic field strength of the intensity of fluorescence $0.1\ \mu\text{sec}$ after the electron irradiation of $0.002M$ C_6F_6 and $0.001\,M$ anthracene in cyclohexene also yielded a spectrum consistent with that of $C_6F_6^-$, including the hyperfine structure.[65p]

Electron irradiation[65q] and X-irradiation[65o,r] of $0.02M$ durene alone[65q] in squalane (6 and $-50°C$) and in the presence of $0.5M$ C_6F_6[65q] in squalane ($-1°C$) or of $0.001M$ perdeuteroanthracene in squalane ($-4°C$)[65o,q,r] or in decalin (-123 to $-38°C$)[65q] led to a spectrum ($a = 10.8\,G$) consistent with that expected based on experiment for durene[+·], including the hyperfine structure.[69a,b]

Electron irradiation of $0.012M$ p-xylene in squalane in the presence of $0.001M$ deuteroanthracene ($8°C$) or $0.5M$ C_6F_6 ($-1°C$) led to a spectrum (five broad lines, presumed to be the central part of a binomial septet, $a = 16.7\,G$) considered to be reasonable for p-xylene[+·], whose spectrum is described as being recorded for the first time.[65q;69a,b]

Electron irradiation of $0.02M$ hexamethylbenzene in decalin in the presence of $0.012M$ perdeuteroanthracene at $-68°C$ led to a spectrum ($a = 6.5\,G$) consistent with that expected based on experiment for hexamethylbenzene[+·].[65q,69b]

A plot of the dependence on magnetic field strength of the intensity of fluorescence approximately $0.1\,\mu sec$ after the electron irradiation of $10^{-6}-10^{-1}\,M$ pyrene in decalin was a single broad line attributed to a radical ion of pyrene.[65f]

Electron irradiation ($-4°C$)[65q] and X-irradiation[65g,r] of $0.01M$ Et_3N in squalane in the presence of $0.001M$ perdeuteroanthracene led to a spectrum (seven lines, $a = 21.6\,G$) consistent with that expected for Et_3N[+·], whose spectrum is described as being recorded for the first time.[69b]

See also Reference 65m.

IV. *Further potential of the technique*

. . . the EPR spectra can reveal the presence of radical exchange processes, suggesting that this technique can prove useful in the elucidation of charge transport mechanisms . . .[65m]

. . . potentially useful for the study of aromatic ion recombination in complex systems such as surfactant micelles or model biological membranes.[65f]

. . . [a] possibility for the use of the method is the study of ultrafast reactions of paramagnetic ion radicals, e.g., electron or proton transfer reactions.[65g]

. . . time-resolved fluorescence detection of magnetic resonance, with its submicrosecond time resolution, its sensitivity to radical ions without interference from other paramagnetic species, and its wealth of information concerning the structure and kinetics of these radical ions, constitutes a unique tool for elucidating the behavior of radical ion species in pulse radiolysis.[65m]

. . . the method . . . opens unique possibilities for studying short-lived radical pairs. Here, the procedure itself used for the generation of radical pairs is not of principal importance.[65d] Such reactions are characteristic not only of radiation chemical processes, but they extend to photochemistry, in particular to processes of photosynthesis [Reference 70], and even to a series of thermal reactions, e.g. oxidation reactions.[65g]

b.   Optical Detection of NMR of Free Radicals in Solution

The possibility of using the title technique has been pointed out:

The optical detection of the EPR spectra is based on a change in the quantum yield of luminescence under the influence of an alternating magnetic field in the UHF range on resonance frequencies of the electron spins of the system . . . The luminescence intensity can also be influenced by an alternating magnetic field in the radiofrequency range, if the frequency of this field coincides with the resonance frequencies of spins of the magnetic nuclei in the system . . . . Therefore, it is reasonable to expect that radiofrequency pumping will also change the luminescence intensity of recombination products of radical pairs in the way of resonance . . . .[71]

c.   NMR-Detected Nuclear Resonance of Transient Radicals

I.   Basis of the technique
A necessary condition for the realization of this method is the observation (in the course of . . . [a] radical process) of the effects of chemical polarization of . . . nuclei (CPN) in . . . reaction products outside . . . [a] cage. This last feature means that the intermediate free radicals R, polarized as a result of singlet-triplet transitions in radical pairs, have a sufficiently short lifetime τ for the nuclear polarization to be preserved at the moment of formation of the final reaction products . . . . If in time τ the saturation [or selective perturbation[65b]] of the resonance transitions in the NMR spectrum of the intermediate radical R is carried out, the intensity of the corresponding polarized signals in the NMR spectrum of the final reaction products changes. Thus the dependence of the intensity of the CPN on the frequency of the saturating radiofrequency field will represent the NMR spectrum of the intermediate radical.[72a]

II.   Observations.   This technique has been applied successfully to several radicals, demonstrating that the product molecules can be used quantitatively in this way as probes for the observation and characterization of their free radical precursors.

A plot of the intensity of the polarized NMR signal assigned to $PhCH_2CH_2Ph$, observed during the photolysis of $PhCH_2COCH_2Ph$, versus the frequency of a saturating rf field passed through a maximum. The spectrum yielded a hyperfine coupling constant in agreement with that of $PhCH_2 \cdot$.[72a]

The pulse radiolysis of CH_3OD in D_2O was studied. A plot of the intensity of the polarized NMR signal assigned to $DOCH_2CH_2OD$ versus the frequency of the rf field showed extrema, corresponding to transitions between nuclear energy levels of $\cdot CH_2OD$, at frequencies in agreement with expectation for $\cdot CH_2OD$.[72b-e] Similar treatment of CH_3OD/CH_3I in D_2O showed that application of an rf field appropriate for $\cdot CH_2OD$ affected the NMR signals assigned to $DOCH_2CH_3$, $DOCH_2CH_2OD$, and $DOCH_2D$ and that application of an rf field appropriate for $CH_3\cdot$ affected those assigned to CH_3H, CH_3I, CH_3CH_2OD, and CH_3CH_3.[72b,e]

Similarly, the intensities of the NMR signals assigned to XCH_2CH_2X and $(^-O_2C)_2CH{\rightarrow}_2$ were monitored as a function of rf frequency during the pulse radiolysis of CH_3X ($X = CO_2^-$, $CONH_2$)[72c] and $^-O_2CCH_2CO_2^-$ (in the presence and absence of N_2O),[65b,72c] respectively, and showed extrema at frequencies in agreement with expectation for $XCH_2\cdot$ and $(^-O_2C)_2CH\cdot$, respectively.

III.   Potential of the technique

The . . . method has a number of advantages . . . compared with the usual ESR method. Firstly, the sensitivity of this method is greater by several orders of magnitude than that of the ESR method. This advantage follows from the familiar fact that the CPN effects can be observed in those cases where low steady-state concentrations of the radicals make them impossible to detect by the ESR method. Secondly, compared with ESR, the . . . method has a higher resolving power . . . Finally, . . . the method . . . is distinguished by the simplicity . . . of the assignment of the lines: The observed lines in the NMR spectrum of the final reaction product correspond to a definite group of equivalent nuclei in the precursor radical.[72a]

. . . this technique provides easy access to hyperfine coupling constants. . . . In addition, details of radical reaction mechanisms are obtainable. . . . Furthermore, radical kinetics with $1-2\,\mu s$ time resolution can easily be carried out. . . . The accuracy of rate constants . . . should compare quite favorably with those obtained by usual optical or fast EPR methods.[72d]

. . . it seems clear from even preliminary experiments that . . . [the technique] offers a unique opportunity for establishing cause and effect relationships between reactive free radicals and the products derived from them.[65b]

d.   NMR-Detected Electron Resonance of Transient Radicals

In this method, dynamic polarization is induced in nuclei belonging to free radical intermediates . . . by microwave pumping. This polarization is then "transferred" to the . . . products of the reaction and can thus be registered in the NMR spectra of these products. Thus, the dependence of the intensity of the signals in the NMR spectra of the end products on the frequency of the microwave field (or on the intensity of the constant external magnetic field) must correspond to the ESR spectrum of the radical intermediates.[73]

This technique was applied to the intermediates in the photolysis of PhCHO in CCl_4. The NMR spectrum of PhCHO was observed while the solution was pumped with microwaves of constant frequency. The intensities of all the NMR signals passed through a minimum, that is, "resonated," as the strength of the magnetic field varied. The resulting plot of NMR signal intensity versus magnetic field strength "is in good agreement (with respect to g-factor and half-width) with the unresolved ESR spectrum of the PhCHOH radical" and is thus consistent with the formation of PhCHOH and its reversion to PhCHO.[73]

It was estimated that this technique is a few orders of magnitude more sensitive than the direct esr method.[73]

e.   *Acoustical Detection of ESR*

"The ESR signal of diphenylpicrylhydrazyl was recorded by detecting the . . . absorbed [modulated] microwave power with a gas-coupled microphone," which provides an acoustic signal due to the periodic heating of the gas caused by the periodic heating of the sample with which it is in contact as a consequence of the absorption of the modulated microwaves. The acoustically detected signal was very similar to the corresponding conventionally detected one and was much less dependent on the tuning of the spectrometer. "By no means is acoustical detection a substitute for conventional ESR, for its sensitivity is . . . [much] lower."[74]

B.   Modulation Optical Spectroscopy and Kinetics in Solution

We have mentioned the great impact on free radical chemistry of the ability to follow transient free radicals directly.[10f] A variety of techniques, often involving the use of home-built equipment unique to a particular laboratory, has been introduced recently. The totality of this work has changed free radical spectroscopy and kinetics. Indicative of particular utility is the construction and use in one laboratory[75a] of the equipment developed in another:[27, 75b]

The principles rest on the pioneering work of Hunziker who has applied the technique to study free radicals in the gas phase, and on the extensions of Günthard and Paul to liquid phase ESR investigations.

Free radicals are generated by UV.-photolysis of suitable substrates with light intensities varying sinusoidally in time. Simultaneously, the absorbance of the solution is observed and analyzed by phase sensitive detection on the light modulation frequency. Substrates, transient intermediates and products contribute to the modulated absorbance with amplitudes and phases depending differently on the modulation frequency. Analysis of these dependencies yield the kinetic information. Spectra are obtained from the wavelength dependence of the modulated absorbance at constant frequency.[75b]

The most recent advance in the field of modulation spectroscopy has been made by Huggenberger *et al.* who have constructed a spectrometer for the study of free radicals *in solution*. With this instrument, radicals are generated using sinusoidally modulated light and are detected optically. In this work, we have used such a spectrometer in a kinetic study of the self-reactions of some α-aminoalkyl radicals.[75a]

For a recent application of the technique developed earlier for use in the gas phase see Reference 76.

C.   Equilibria Involving Radicals

Additional reports of the direct measurement of the equilibrium constant of a reaction involving a very reactive organic free radical have appeared.[6gg]

a. $R' \cdot + RI \rightleftarrows R'I + R \cdot$

Equilibrium constants K for the title series of reactions have been determined by direct monitoring of $R' \cdot$ and $R \cdot$ in $Bu^t ON{=}NOBu^t/(Ph_3 As$ or $Ph_3 B)/RI/R'I/$ isooctane and $Me_3 SnSnMe_3/RI/R'I/isooctane/h\nu$ reaction mixtures by use of esr spectroscopy.[60] "Extrapolation of these data to the start of the experiment gave the relative radical concentrations at a point where the concentrations of the iodides were known precisely. . . . Two conditions must be fulfilled in order to calculate values of K from these results. First, the reaction of phenyl radicals with alkyl iodides . . . [Reference 77a] must be essentially irreversible. Second, the reactions of alkyl radicals with alkyl iodides must be faster than the reactions that remove radicals from the system [the removal for $R \cdot$ and $R' \cdot$, not the net removal of radicals] . . . ." We believe that the first condition is probably met, although it is unnecessary, and that there is evidence that the modified second condition is met. Values of K were calculated for $R' \cdot / R \cdot$ pairs involving $CH_3 \cdot$, $CH_3 CH_2 \cdot$, $CH_3 CH_2 CH_2 \cdot$, $(CH_3)_2 CH \cdot$, $CH_3 \dot{C} HCH_2 CH_3$, $(CH_3)_3 C \cdot$, and $c{-}C_5 H_9 \cdot$ and "were independent of both relative and absolute concentrations of the two iodides when these were varied by a factor of 10."[60]

b. $R \cdot + O_2 \rightleftarrows RO_2 \cdot$

The flash photolysis of $CH_2{=}CHCH_2 \text{)}_2$ in the presence of various concentrations of O_2 was studied at several concentrations in the gas phase at $109{-}180°C$ and the intensity of the UV absorption at $223 \, m\mu$, which was assumed to have arisen solely from $CH_2{=}CHCH_2 \cdot$ and $CH_2{=}CHCH_2 O_2 \cdot$, was followed as a function of time. The data were force-fitted to equations corresponding to

$$(CH_2{=}CHCH_2)_2 \rightleftarrows CH_2{=}CHCH_2 \cdot \underset{k_{-1}}{\overset{O_2, \, k_1}{\rightleftarrows}} CH_2{=}CHCH_2 O_2 \cdot$$

as the sole chemistry occurring. Out of this process emerged numbers, considered to be k_1 and k_{-1}, from which an equilibrium constant and thermodynamic parameters were calculated.[78a] Such information was obtained also by use of another technique whereby species of m/e 41 were followed by use of a photoionization mass spectrometer.[78a,b] The equilibrium constants obtained agreed with expectation based on extrapolation of results of the UV work, a cross-check which adds support to the validity of both methods.

The reaction of CH_4 with O_2 in the gas phase was studied, and the equilibrium constant of $CH_3 \cdot + O_2 \rightleftarrows CH_3O_2 \cdot$ was reported, based on an avowedly superficial and approximate treatment of the data. It was assumed that (1) $[CH_3 \cdot]$ could be determined simply from the net rate of accumulation of ethane and from literature values of the rate constants of decomposition of ethane to methyl and recombination of methyl to ethane; (2) $[CH_3OO \cdot]$ could be determined simply from the intensity of the esr signal of radicals frozen out of a "slip stream" taken from the reactor; and (3) the equilibrium constant was equal to $[CH_3O_2 \cdot]/[CH_3 \cdot][O_2]$. "$[CH_3O_2 \cdot]/[CH_3 \cdot][O_2]$" varied from 1.1×10^3 to 1.5×10^2/atm between 433–513°C.[78c]

$$\text{c.} \quad Cl \cdot + CH_4 \rightleftarrows HCl + CH_3 \cdot$$

Work that elaborates upon and confirms earlier work on the title reactions has been reported.[79]

D. Miscellany

1.   "Sample Magnetization Using Immobilized Free Radicals for Use in Flow NMR Systems."[80]

2.   (a)   "CIDNP Assisted Assignment of Carbon-13 NMR Lines in α-Tetralone and α-Indanone."[81a]

   (b)   "Assignment of Carbon-13 N.M.R. Lines using CIDNP."[81b]

V.   MECHANISTIC TECHNIQUES

A.   Generation and Trapping of Radicals

The transformation,

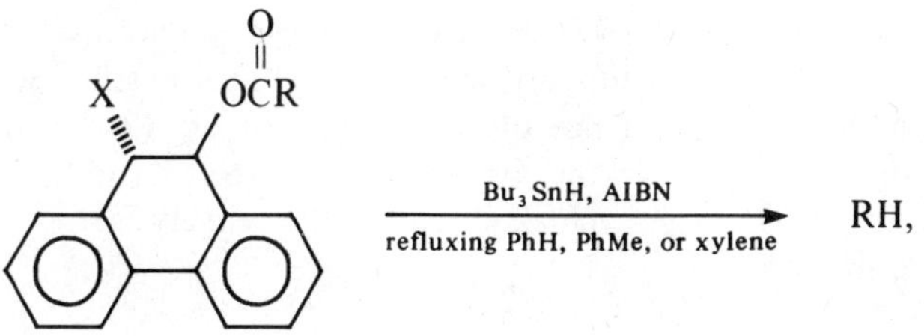

was described for $X = Cl$, PhS and for $R–H = –CH_2–H$, $CH_3\overset{|}{C}H–H$, $-\overset{|}{\underset{|}{C}}-H$,
$Ar–H$, $-\underset{\underset{O}{\|}}{C}H–H$.

A free radical mechanism,

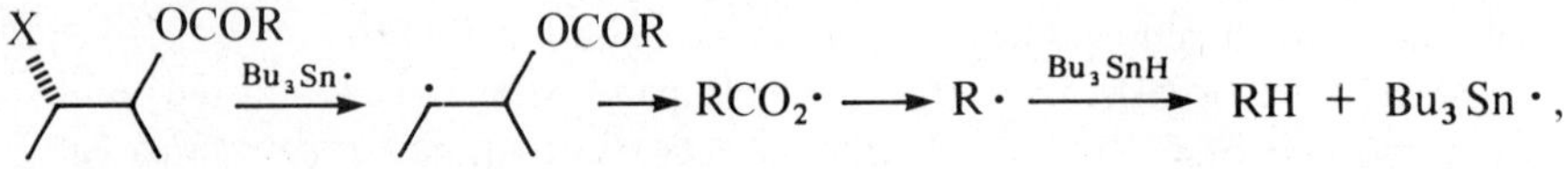

$$\underset{}{\overset{X\quad OCOR}{\bowtie}} \xrightarrow{Bu_3Sn\cdot} \overset{OCOR}{\bowtie} \longrightarrow RCO_2\cdot \longrightarrow R\cdot \xrightarrow{Bu_3SnH} RH + Bu_3Sn\cdot ,$$

was presented, but its basis was not discussed.[82]

See also Reference 10g.

It has long been believed that $>$PH are highly efficient donors of hydrogen to free radicals.[83a] Di(cyclohexyl)phosphine, $(c–C_6H_{11})_2$PH, has recently been presented as a free radical trapping agent.[83e—i,k, 83l]

The reaction of radicals with organotin hydrides[5f,6n] and the cyclization of the 5-hexenyl radical to the cyclopentylcarbinyl radical[6n] play central roles in the trapping of radicals and in the study of free radical kinetics and mechanisms. The results of a recent study have led to

$$k(R\cdot + n\text{-}Bu_3SnH \rightarrow RH + n\text{-}Bu_3Sn\cdot) = 2 \times 10^6/\text{mol (sec)} \tag{7.1}$$
$$(\text{room temperature}, R = 1°, 2°, 3° \text{ alkyl and cycloalkyl}),$$

$$\log k(1° \text{ alkyl}) = 9.1 - 3.7/2.3RT(/\text{mol (sec), kcal/mol}), \tag{7.2}$$

and

$$\log k(5\text{-hexenyl}\cdot \rightarrow \text{cyclopentylcarbinyl}\cdot) = 10.4 - 6.9/2.3RT(/\text{sec}), \tag{7.3}$$

Eq. (7.3) being based on Eq. (7.2).[85] We agree[85] that these rate constants, rather than those reported earlier, should be used in quantitative studies in the future. Most probably they are "correct." However, we do have a slight lingering concern based upon (1) the fact that the accuracy, as distinct from the often-illusory formally derived precision, of these numbers rests on a rather complex system of chemical support, with the attendant possibility of systematic error; and (2) the wide range of applicability of Eq. (7.1) and the indistinguishability (in our view) of the Arrhenius parameters (A and E) for R = Me, 1° alkyl, 2° alkyl, 3° alkyl, cycloalkyl and of the primary isotope effects for R = Me, Et, n-Bu.[86]

B.   CIDNP in the Gas Phase[87]

Cycloheptanone $(1-4\,\mu l)$ was placed in a quartz tube whose volume was 0.4 ml before it was sealed. The tube was placed into the probe of an NMR spectrometer and was heated to $120-220°C$. The NMR spectrum was recorded with and without simultaneous irradiation with a mercury lamp "under conditions of both complete sample evaporation and vapor-liquid equilibrium. There was no essential difference in the ^1H NMR spectra in both cases."

"Studies of the temperature dependence of the NMR signals provided additional evidence for the ^1H NMR spectra belonging to gas-phase molecules. When excess ketones were used, a rise in temperature resulted in a monotonous increase in the NMR signal intensity which corresponding to the equilibrium between the saturated vapors and the liquid. Under complete evaporation of ketone, as expected, the temperature dependence of the NMR signal intensity reached a plateau at $T > T_{evap}$."[87b]

Irradiation led to a decrease in intensity of the signals of approximately 50%, a result attributable to a CIDNP effect. Since this is only a quantitative change, it is not evidence as strong as the observation of a qualitative change would have been. In support of the attribution to a CIDNP effect, rather than to an effect of unknown physical or physicochemical origin, the results that the effect (1) is not observed with methylethylketone and (2) is qualitatively similar to what is observed in $CHCl_3$ solution can be cited.[88a]

VI.   NEW RADICAL PHENOMENA

A.   Radioemission from Chemical Reaction Systems

a.   *Introduction*

Compounds having abnormal populations of the Zeeman levels in the sense that there is an excess population of the upper levels may be generated in free radical reactions and exhibit certain CIDNP and CIDEP effects. The excess energy of such systems is usually dissipated as heat. However, conditions under which the energy is emitted also as radio-frequency radiation can exist or be created, that is, chemical polarization of nuclei or electrons can lead to the emission of radio-frequency energy.[89] "On the basis of the calculations and estimates carried out in this investigation there are excellent grounds for considering that the creation of radio-frequency chemical and photochemical masers is technically possible and of very considerable theoretical and practical interest."[89a] The chemically induced emission of radio waves has since been detected experimentally.

b.   *Observations*

A solution of $1-5$ mM porphyrin and p-benzoquinone in $CHCl_3$ containing CCl_3COOH was photoirradiated at 546 mμ in a 23500 G magnetic field. The radiophysical effect was manifested as a high-frequency current in a coil surrounding the sample.[89b,c] The energy difference between the nuclear Zeeman levels corresponds to radio-frequency radiation having a wavelength of approximately 3 meters. "The . . . [electron transfer from porphyrin to quinone] is practically reversible, and so the substance is not consumed and radio-frequency generation may continue for a rather long time."[89c]

The thermal decomposition of $0.5 M$ benzoyl peroxide in cyclohexanone was studied at $21°C$ in the 14600 G magnetic field of the probe of an NMR spectrometer. Radio-frequency fields were produced and monitored. The amplitude of the generated emf was approximately $2 \mu V$, corresponding to an amplitude of the magnetic component of the rf field of approximately $0.2 \mu G$.[89d]

c.   *Potential*

Thus the products of radical-type chemical reactions emit coherent radio signals . . . , i.e., they behave like quantum generators in the radio-frequency range, or rasers. This makes it possible to develop both new methods for studying the chemical reactions themselves, and new radio-engineering equipment in which the polarized products of these reactions serve as the working medium. For example, a prototype of a magneto-meter for measuring the earth's magnetic field has been developed and tested at the . . . Siberian Branch of the USSR Academy of Sciences, under the direction of Yu. N. Molin . . . The working medium is the polarized products of a reversible photochemical reaction, which gives this magnetometer substantial advantages over those presently known. This is one of the first examples of the technical applications of the chemical polarization of nuclei.

If the products of a chemical reaction can be a source of radio-frequency radiation, then the following questions arise. Can this radiation not in turn affect the course of the reaction? Is an electromagnetic field in the radio-frequency region capable of affecting the spin systems of the reacting particles in such a way as to alter the rate of the chemical reaction or the ratio of products? What are the conditions for such "chemical radio reception"? . . . there opens up the tempting possibility of transmitting information on the molecular level about chemical and biochemical reactions by means of radio-frequency radiation — for example, when one reaction emits radio waves and another receives them and "remembers" the information obtained. If necessary, "chemical radio communication" could be supplemented by the usual radio-type amplifiers . . . .

"Chemical radio reception" is a very important scientific and technical problem, the solution of which may have far-reaching consequences.[89c,90]

VII.   FREE RADICAL CHEMISTRY OF MUONIUM AND CHARACTERISTICS OF MUONIC RADICALS

A.   Muonium (Mu, μ^+e^-)[91,92a]

High-energy (typically $\sim 10^{10}$ kcal/mol) protons (p) enter into nuclear reactions such as $^9Be + p \rightarrow {}^{10}Be + \pi^+$, in which pions ($\pi^+$) are produced. Positive pions rapidly ($t_{1/2} \sim 10^{-8}$ sec) give a muon (μ^+, spin 1/2) and a neutrino. The muons are produced with high polarization (relatedness of the direction of its spin and its momentum). A beam of muons of high spin polarization can be obtained from pions at rest, which produce muons of a single energy, or by selecting muons according to their momentum as they are formed from nonmonochromatic pions. The muons are themselves unstable, giving ($t_{1/2} \sim 10^{-6}$ sec) positrons (e^+), whose angular distribution is anisotropic with respect to the direction of the spin of the muons, along with neutrinos and antineutrinos. When the μ^+ encounters matter, it acquires an electron, thus producing the muonium atom (Mu, μ^+e^-). Muonium is a "light isotope" of H; its Bohr radius, ionization potential, and mass are 1.0043, 0.9957, and 0.11315, respectively, that of H. Of the series Mu, H, D, T (relative mass 0.11, 1, 2, 3), two (Mu, T) are radioactive. Since μ^+ and e^- each have spin 1/2, Mu has "singlet" and "triplet" states.

B.   Muon Spin Research[91,94a]

In a weak magnetic field perpendicular to the direction of spin of the μ^+ (a "transverse" field), Mu, and hence the anisotropic $\mu^+ \rightarrow e^+$ decay pattern, precesses with a characteristic frequency (the "Larmor frequency," 1.39 MHz/G for Mu; that of μ^+ is 3.18 times that of H$^+$). Information about the Mu can therefore be obtained from the positrons, which can be monitored by use of appropriate detectors; information about the magnitude, precession, and any decay of the polarization of the muons can be obtained from the rate at which positrons strike a detector at a location of known relationship to the direction of polarization of the muons and from the dependence on time of that rate.

C.   Application of Muon Spin Research to Chemical Reactions and to Compounds Containing Muonium[96]

Numbers which are considered to be the rate constants k for reaction of Mu with various substrates S are obtained by fitting the observed dependence on time of the decay of the polarization to equations corresponding to approximate models which are considered to describe those mechanisms of depolarization which are selectively considered to be operative. The relaxation rate is sub-divided into several terms independent of the concentration of substrate, which are irrelevant to our present purposes, and a term containing $k[S]$, which is assumed to be the way in which $[S]$ enters into the relaxation rate constant and in which it is usually further assumed that S increases the rate of relaxation solely by reacting chemically with Mu. Note that in such experiments the identity of the product, although commonly specified, actually is unknown and its buildup is not followed. These are not limitations inherent in the nature of muon spin research, but rather are simply characteristic of recent work, whose scope and sophistication has been rapidly increasing, that is, they are a manifestation of the kinds of "breadth versus depth" choices and trade-offs typically made in the early stages of a field, particularly in one subject to severe limitations on the existence of and access to experimental facilities. For example: "reduction or abstraction which converts Mu to diamagnetic states [e.g., μ^+, MuH] can enhance [the polarization at the (very different) frequency of precession of such states], whereas addition reactions . . . result in the muon remaining in a paramagnetic species [which has a frequency of precession which is very different yet — see below]."[93e]

Muon spin research does indeed permit the "direct" observation of organic free radicals containing Mu, that is, the observation of the characteristics of the decay of a muon contained in an organic free radical. These experiments are usually conducted at field strengths greater than those normally used to observe Mu. Just as, for example, the proton hyperfine coupling constants of, for example, $CH_3CH_2 \cdot$ are much less than that of $H \cdot$, the hyperfine coupling constant of muonium in a muonic organic free radical is much less than that of free muonium. In muonic radicals that contain magnetic nuclei in addition to the muon, the spin angular momenta of the electron, the muon, and the other nuclei are all coupled. Because of that and because of the Zeeman interactions of the electron, the muon, and the other nuclei, the polarization of the muon is distributed over many frequencies. Consequently, at low fields (where Mu is usually studied) the detection of muonic radicals can be difficult. However, in high fields, where the Zeeman interaction of the electron is much greater than the hyperfine interaction of the muon (which is typically greater than that of the other nuclei), the spin of the muon is effectively decoupled from those of the other nuclei. The result is that, in high fields, muonic radicals can be treated as a spin system analogous to Mu, that is, a system that exhibits two precession frequencies, from which the absolute values of the muon-electron hyperfine coupling constants can be obtained.

Specific chemical reactions and compounds containing muonium are discussed in the sections that follow. All of the foregoing problems aside, we believe that the workers in this area have described the significance of their work in a far too limited way. It has usually been justified in terms such as (1) providing indirect information about the chemistry of $H \cdot$, (2) providing information about diffusion in liquids, and (3) testing theories of chemical reaction (for example, explaining the "origin" of isotope effects). We believe the work to be of fundamental importance. Its justification is not the fallout from it. It justifies itself.

D.   The Organic Free Radical Chemistry of Muonium

Studies[98] have been reported of the kinetics of the reaction of muonium with substrates such as acetone,[97b,99a] acrylamide,[97d] acrylic acid,[97d] acrylonitrile,[97d] benzene,[99b] bromoacetic acid,[97b] 2-bromopropionic acid,[97b] 3-bromopropionic acid,[97b] 2-butanol,[93c,97b,99a] ethanol,[93i] ethylene,[93a,99c] formate,[97b,f;99a] fumaric acid,[97b,99a] maleic acid,[97b,f;99a,d] methanol,[97e,o] methyl methacrylate,[97d] naphthalene,[97c] p-nitrophenol,[97m] phenol,[97c,e,m,m] 2-propanol,[93c,97b,99a] styrene,[97d] and tetramethylethylene.[99b] See also Reference 100.

E.   Characteristics of Muonic Radicals

Muon hyperfine coupling constants have been reported for the products of reaction of Mu with styrene,[93f;97d,h,k;101] olefins,[93d;95b;97a,i;97k] arenes,[93f;95b;97a,h—j,l;102] dienes,[97a,i—k] ketones,[93f,97a,i] unsaturated ethers,[97k] unsaturated alcohols,[97k] unsaturated esters,[93f,97k,101] unsaturated nitriles,[97k] heteroaromatics,[93f,97a] and alkylbenzenes.[93f,97a] A "spectrum" is shown in Figure 7.1.[97a] In some cases the dependence of the coupling constant on temperature was determined.[95b,97a,h,k]

In this work, the structure of the muonic radical that is being observed is typically *assumed*, based on an assumed preferred mode of reaction of Mu with

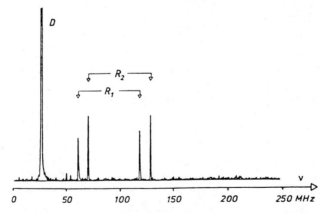

FIGURE 7.1.   Muon precession frequencies in isoprene at 2 kG and ambient temperature, D: muons in diamagnetic environments; R_1 and R_2: muonic radicals.

the substrate. Since conditions (high field) that lead to a very simple pattern of hyperfine interactions are deliberately created, the qualitative aspects of the pattern cannot be used as a basis of assignment of structure. Comparison with known proton couplings is of limited value because the ratio of muon and proton couplings is not generally equal to the ratio of their magnetic moments. The assignments are supported by the internal consistency of an increasingly sizable body of assignments. In very limited instances, assignments were supported further by experiments at lower field strength, which revealed additional couplings in the molecules.[93d,102a]

F.   Reactions of Muonic Radicals

"[A] promising field of investigation with muonic radicals . . . is radical kinetics. . . . Muonium [could] . . . merely act as a tracer atom [a "spin probe"], and could find application in radical reactions difficult . . . to study by other means."[93b] The chemistry of a muonic radical can be studied by performing experiments on the product of reaction of Mu with a substrate (i.e., on the reactant muonic radical) which are analogous to those performed on Mu (Section VII.C) or by determining the line width of the "peaks" in the "spectrum" of the muonic radical as a function of the concentration of its reaction partner.

Reports have appeared of the rate constant for (1) reaction with styrene of the radical produced by reaction of Mu with styrene,[97h] and (2) reaction with p-benzoquinone of the radical produced by reaction of Mu with benzene.[93g,97j]

VIII.   ISOTOPE ENRICHMENT via MAGNETIC FIELD EFFECTS ("MAGNETIC ISOTOPE EFFECTS")

For introduction and background and a discussion of the "origin" of the effect, see our description of work in this area in Volume 2 of this series.[60] We suggest that this be read first.

A.   Reports of the Effect

a.   Dibenzyl Ketone

The work that was discussed earlier[6p] on the enrichment of ^{13}C in the $(PhCH_2)_2CO$ remaining after photolysis has been reported also elsewhere.[103] In addition to results very similar to those presented earlier,[6q] it is reported that the evolved CO was depleted 25% in ^{13}C.

Work on dibenzyl ketone has continued. We summarize those findings that we find to be of significance and of interest within the scope of the present discussion.

A more detailed presentation, substantiation, and refinement of and elaboration upon results discussed earlier[60] has appeared.[104a] See also References 104b–104g.

One of the products of the photolysis of $(PhCH_2)_2CO$ in micellar solution is $PhCH_2CO(p-C_6H_4CH_3)$, presumably formed by way of coupling of $PhCH_2\overset{.}{C}O$ to $PhCH_2\cdot$ at its p-position. Photolysis to 80% conversion of $5\,mM\,(PhCH_2)_2CO$ containing ^{13}C in natural abundance in $0.05\,M$ aq. hexadecyltrimethylammonium chloride at $25°$ led to $PhCH_2CO(p-C_6H_4CH_3)$ in an amount equal to 5–10% of the $PhCH_2CH_2Ph$ produced. The contents of ^{13}C at the carbonyl and methylene carbons of the product were 23 and 17%, respectively, in excess of the contents of ^{13}C at natural abundance; any enrichment at other positions was much less. These results are consistent with expectations.[6r] A similar experiment in which the starting ketone contained 47.6% ^{13}C at the carbonyl carbon gave product containing approximately 14–23% excess ^{13}C, predominantly at the carbonyl carbon, at 29–91% converstion the chemical yield of $PhCH_2CO$ $(p-C_6H_4CH_3)$ decreased strongly and its percent excess ^{13}C increased strongly as the temperature was increased to $70°C$.[104b–d,105]

b.   Other Ketones

Photolysis to 90% conversion of $5\,mM\,PhCH_2COPh$ in $0.05\,M$ aq. hexadecyltrimethylammonium chloride led to an enrichment of residual ketone of 140%, based on enrichment of a single carbon.[105a]

Similar treatment of $PhCO(1\text{-adamantyl})$ led to enrichment of the PhCO and adamantyl fragments of residual ketone of approximately 160 and approximately 180%, respectively, based on enrichment of a single carbon.[105a] Since the benzoyl and 1-adamantyl radicals have high ^{13}C hyperfine coupling constants, the substantial enrichment of both fragments is consistent with expectation.[6r]

c.   Benzoyl Peroxide

The work on the triplet-sensitized photolysis of benzoyl peroxide discussed earlier[6s] has been described elsewhere[106a,b] and in greater detail.[106c] It has also been reported that its thermolysis gave benzene with 3% enrichment of ^{13}C.[103]

d.   Autoxidation of Hydrocarbons

Earlier[6t] we described work on the autoxidation of ethylbenzene in which the residual O_2 was selectively enriched in ^{17}O (versus ^{16}O and ^{18}O), that is, in the isotope *intermediate* in mass. We presented this as an elegantly simple experiment the *qualitative* nature of whose results required that is was not a "classical" kinetic isotope effect that had been observed. More recently, work along similar lines has been done on the polypropylene hydroperoxide-, benzoyl peroxide-, and cumyl peroxide-initiated autoxidation of solid polypropylene. The experiments with Bz_2O_2 (0.1 mol/kg) at $130°C$ and an initial pressure of O_2 of

600 mm gave, after $< 15\%$ conversion of the polymer, enrichments of the residual O_2 in ^{17}O and ^{18}O of 120 and 10%, respectively, after 80% consumption of the O_2.[107]

It has recently become apparent to us that our earlier statement was not correct, that is, in these cases, as in all others, the operation of a "classical" isotope effect cannot be dismissed by use of a qualitative argument alone. Here, too, the case for a "magnetic" origin must be argued in terms of the magnitude of the effect because it now appears that there can be circumstances under which a "classical" isotope effect can indeed lead to selective enrichment in the isotope intermediate in mass.[108]

e. Aromatic Endoperoxides

The work discussed earlier[6u] on the enrichment of ^{17}O in the $^3O_2$ produced upon decomposition of 9,10-diphenylanthracene endoperoxide has been reported in greater detail.[109]

f. Trimethyltin Hydride

The work discussed earlier[6v] on the AIBN-induced decomposition of Me_3SnH also has been presented elsewhere.[106b]

B. Other Environments

Photolysis at $0°C$ of $(PhCH_2)_2CO$ in a plastic crystal of cyclohexanol led to $(^{13}C/^{12}C$ in residual ketone$)/(^{13}C/^{12}C$ in initial ketone$) = 1.8$ at 99.1% conversion, as compared to 1.6 at the same conversion in liquid cyclohexanol at $25°C$.[110a]

"Micelles do not seem to be unique in furnishing suitable cages for a high ^{13}C enrichment in dibenzyl ketone: for example, preliminary experiments indicate photolyses of solutions of dibenzyl ketone impregnated in porous glass . . . and in polymethacrylate polymer films . . . give substantial ^{13}C enrichments . . ."[104a] which drop "significantly . . . when the . . . photolyses are conducted in a strong magnetic field."[110b]

Photolysis of $(PhCH_2)_2CO$ adsorbed onto a surface of a monolayer of dodecyl groups bonded to a support of silica gel led to $(^{13}C/^{12}C$ in residual ketone$)/(^{13}C/^{12}C$ in initial ketone$) = 1.9$ at 86% conversion.[110c]

C. Effect of Solvent

For studies of the photolysis of dibenzyl ketone in several solvents see Table 7.1. During the photolysis of PhCO(1-adamantyl) also, the percent enrichment was observed to increase slightly on going from cyclohexane (viscosity $= 0.9$ cp) to n-dodecane to cyclohexanol as solvent.[105a]

TABLE 7.1.   Enrichment of dibenzyl ketone in various solvents

Solvent	Viscosity (cp)	Conversion (%)	$^{13}C/^{12}C$ in residual ketone / $^{13}C/^{12}C$ in initial ketone	Ref.
		25°C		
Benzene	0.6	96	1.06	104a
Toluene	0.6	96[a]	1.1	110a
		99.1[a]	1.2	110a
n-Dodecane	1.4	92	1.06	104a
3-Pentanol	4.0	99.1[a]	1.3	110a
Cyclohexanol	60	86	1.10	104a
		86	1.2	110a
		99.1	1.6	110a
		1°C		
Bu^{t}OH/glycerol,[b]				
70/30(v/v?)	1700	56	1.02	113a,
		77	1.02	113c
		96	1.05	
50/50	2400	35	1.01	
		67	1.02	
		82	1.06	
40/60	2900	37	1.01	
		56	1.02	
		77	1.03	
20/80	5100	43	1.01	
		64	1.01	
		66	1.02	

[a]Interpolated.

[b]Use of a Bu^{t}OH/glycerol mixture (viscosity = 540 cp, measured at an unreported temperature) of unreported composition was reported to lead to an enrichment of ^{13}C in the residual ketone at 90% conversion almost as great as has been observed in micellar solutions.[113b] This extraordinary report is not cited in Reference 113a or 113d.

The results with $(PhCH_2)_2CO$ and with $PhCO(1\text{-adamantyl})$ were attributed to changes in viscosity. Experiments in a series of alkanes, for example, would be more informative as would clearer demonstrations of the effect.

D.   Effect of Field Strength

The extent of enrichment of ^{13}C in residual $(PhCH_2)_2CO$ during photolysis in aq. soap solution is not the decreasing monotonic function of magnetic field strength that may be surmised from the results reported earlier[6w] at three field strengths. The extent of enrichment actually is at a maximum at a field strength of a few hundred gauss.[104e,111a,b] The explanation offered[104e] for this observation is predicated upon and begins with an "origin" of the magnetic isotope effect such as was discussed earlier[6o] and, indeed, the observation supports such an "origin." The explanation involves an idealized picture of one particular mechanism for triplet-singlet transitions:[111c]

The triplet-singlet conversion of radical pairs that revert to $(PhCH_2)_2CO$, the key step in determining the magnitude of the magnetic isotope effect (i.e., the extent of enrichment),[6o] is effected by means of a hyperfine interaction mechanism. "When the hyperfine coupling constant is much greater than the Zeeman splitting [which is proportional to magnetic field strength], [i.e., at "low" field strength], all three triplet [sub]levels [T_+, T_-, T_0 in the usual notation] can undergo hyperfine induced intersystem crossing to the . . . singlet."[104e] This applies to both ^1H and ^{13}C interactions in the radical pair; it does not apply to ^{12}C, which has no magnetic moment. The behavior of ^{13}C-relative to ^{12}C-containing radical pairs provides the sorting that leads to enrichment.[6o] Triplet $\rightarrow$ singlet intersystem crossing brought about by ^1H hyperfine interaction, viewed as a process independent of that involving ^{13}C, causes the radical pair to revert to $(PhCH_2)_2CO$ indiscriminately, that is, does not cause the ^{13}C-containing pairs to revert selectively, as does the ^{13}C hyperfine interaction. Thus, to the extent that ^1H hyperfine interaction is effective at inducing intersystem crossing, it serves to decrease the efficiency of enrichment of ^{13}C in residual $(PhCH_2)_2CO$. That is, by providing a mechanism of reversion of the radical pair independent of ^{13}C-content, it decreases the fraction of radical pairs that revert selectively (^{13}C versus ^{12}C) to residual $(PhCH_2)_2CO$. "When the hyperfine coupling constant is much less than the Zeeman splitting, T_+ and T_-, but not T_0, are inhibited from undergoing hyperfine induced intersystem crossing."[104e] (The energies of the T_+ and T_- levels diverge from that of the singlet as the magnitude of the Zeeman splitting, i.e., the field strength, increases; that of the T_0 level does not) Thus (?) the efficiency of hyperfine induced intersystem crossing decreases with increasing field strength. As the field strength increases from "zero," ^1H hyperfine interaction would become a less effective mechanism of intersystem crossing and, consequently, would become less effective at decreasing the efficiency of ^{13}C. Since the ^1H hyperfine interaction constants are relatively low, this decrease in effectiveness would occur at relatively low field strengths, that is, at relatively low field strengths the magnitude of an effect that decreases the efficiency of enrichment of ^{13}C is substantially decreased. Since the hyperfine interaction constants of ^{13}C are relatively high, the effect of increasing field strength in decreasing the efficiency of ^{13}C-hyperfine induced intersystem crossing, that is, in decreasing the effectiveness of the mechanism that leads to enrichment of ^{13}C, is felt to a substantial degree at relatively high field strengths. Hence the extent of enrichment of ^{13}C increases with increasing field strength: "Since the hyperfine interaction constant for ^{1}H hyperfine-induced intersystem crossing is smaller than the hyperfine interaction constant for ^{13}C hyperfine-induced intersystem crossing, smaller values of the applied field (~ 10–$100\,G$) suffice to 'quench' ^{1}H hyperfine-induced intersystem crossing from the T_+ and T_- levels of the radical pair than are needed (200–$300\,G$) to 'quench' ^{13}C hyperfine-induced intersystem crossing . . . At still higher fields ($> 500\,G$), the T_+ and T_- levels are substantially inhibited from intersystem crossing, and the . . . [extent of enrichment of ^{13}C] decreases to a value even lower than that found in the earth's field."[104e]

This is a reasonable argument. However, as has been pointed out, ". . . H possesses a magnetic moment that is several times larger than D. As a result, for chemically equivalent positioning in a radical pair, H will induce intersystem crossing faster than D."[114] In view of this, and in the absence of an invocation of the superimposed consequences of a "conventional" isotope effect on the relative competitive kinetics, we would have difficulty reconciling the argument with the observation[110b,112a,b] that the extent of enrichment of ^{13}C in residual $(PhCD_2)_2\,CO$ is a decreasing monotonic function of field strength and, at "zero" field, is the same as that of $(PhCH_2)_2\,CO$.

The situation is complicated further by the fact that approximate models and calculations[115] lead to the expectation that there can be circumstances under which, as a natural consequence of the hyperfine interaction mechanism of intersystem crossing, the efficiency of triplet-singlet transitions passes through a maximum as a function of field strength, that is, there may be no need to invoke any effect of the proton hyperfine interactions.

It is possible that the present level of understanding of these systems is not adequate to permit explanations of the dependence of extent of enrichment on field strength.

E.   Miscellany

a.   *Observations*

The emulsion polymerization of styrene in aq. sodium dodecyl sulfate was studied. "the solution was stirred vigorously with a magnetic stirring bar *that was driven by a magnetic stirrer.* The magnetic field generated . . . was of the order of $150\,G$." "under comparable conditions $(PhCH_2)_2\,CO$ as [photo]initiator produces a 52% yield of polymer (ave. mol. wt. $= 4.0 \times 10^6$) in 5 hr, whereas $(Ph^{13}CH_2)_2\,CO$ . . . produces a 13% yield . . . (ave. mol. wt. $= 1.4 \times 10^6$) . . . We interpret this to mean that recombination of $C_6H_5^{13}CH_2 \cdot$ $\cdot^{13}CH_2C_6H_5$ radical pairs in micelles is more efficient than that of $C_6H_5^{12}CH_2 \cdot$ $\cdot^{12}CH_2C_6H_5$ radical pairs."[116]

b. *Comments*

It has been pointed out that

> Magnetic field and magnetic isotope effects provide a complement to CIDNP for probing mechanisms of radical reactions. In contrast to CIDNP which required NMR measurement during reaction, magnetic field and magnetic isotope effects reveal themselves in terms of chemical products which may be analyzed at the chemist's convenience by conventional methods.[110b,114,117]

and that

> It should also be noted that unusual isotope effects, that have been interpreted in terms of conventional mass effects, should be reconsidered as possible magnetic isotope effects and subjected to the magnetic field criterion.[110b,114,117]

Earlier we indicated by implication that the possibility of an explanation other than a "magnetic isotope effect" should be addressed (i.e., at least considered) for some of the results of photochemical reactions attributed to a magnetic isotope effect, namely, that an explanation involving an isotope effect on intersystem crossing (by means of an effect on vibronic interactions) versus deactivation versus chemical reaction should not be dismissed out of hand, that is, without argument.[6x] H- versus D-isotope effects on the photochemistry at low temperature of several compounds have indeed been interpreted in this way, and, even more interesting, it has been indicated that work directed toward the observation of such isotope effects with heavier elements is underway.[106a]

c. *Potential*

> Even higher [than in micelles] enrichment levels can be expected for one-dimensional diffusion . . . This situation can be realized in practice by carrying out a chemical reaction within channels 20–30 Å in diameter (zeolites, porous solids, etc.).[118]

> The magnetic isotope effect and the isotope enrichment can be enhanced by the interaction of an external radio-frequency field with the electronic or the nuclear spins of the radicals in the radical pairs. Radio-frequency radiation accelerates the relaxation of the electrons and increases the rate of triplet-singlet evolution of the pairs. In principle it should be possible to irradiate selectively the magnetic pairs, but this approach suffers from technical difficulties . . . Nevertheless the method became a practical possibility following the discovery of a new method of detecting short-lived radicals in radio-frequency spectroscopy. [Ref. 10h] The r.f. stimu-

lation of isotope enrichment has great practical interest, but it is not yet fully understood and it needs further theoretical analysis and experimental study.[118]

The expectation that "appreciable . . . magnetic isotope effects [can be observed] in gas-phase radical reactions" has been expressed.[88b]

The enrichment of molecular oxygen in [17]O isotope in chain oxidation may have important consequences. . . . the geochemical and cosmological consequences are the most important. The global process of oxidation of organic matter in nature must enrich the molecular oxygen of the atmosphere in [17]O isotope. The fact that the isotope composition of atmospheric oxygen is stationary means that there must be certain processes to balance the isotope enrichment. Previously, before the magnetic isotope effect was discovered, it was considered that the distribution of the oxygen isotopes ^{16}O, [17]O, and [18]O in any mass-dependent fractionation process must follow the pleiad rule. Deviations from this rule were considered to be the result of nuclear processes, i.e., the result of isotope conversions rather than isotope separation. Now, after the discovery of the magnetic isotope effect, this tradition must be reexamined. Obviously, the evolutional geochemistry of the [17]O isotope has become an independent problem. Here we should keep in view that the magnetic isotope effect and the separation of isotopes may take place not only in the interaction of peroxyl radicals (this is only a special case), but also in recombination (or disproportionation) of other oxygen-containing active species (OH radicals, O_2^-, oxygen atoms and ions, etc.). Moreover, the mechanisms we have considered for the separation of isotopes may also be manifested in the oxidation of inorganic substances. This means that investigation of [17]O distribution can in principle give important information for evolutionary geochemistry.

An analogous situation occurs in cosmology. For example, it has been established that in certain meteorites . . . , in contrast to . . . terrestrial rock, the ratio of the isotopes ^{16}O, [17]O, and [18]O does not follow the pleiad rule. This anomaly in the isotope composition has been explained by mixing of two components of cosmic substance in various proportions, these components differing in nucleogenetic origin. When the magnetic isotope effect is taken into consideration, an alternative explanation of this anomaly can be offered. It is clear that this same situation applies to other possible anomalies in the isotopy of oxygen and other elements.[119]

. . . we wish to point to the unique possibility presented by the discovery of the magnetic isotope effect for the separation of nuclear isomers. As we know, nuclear isomers may differ with respect to both spin and hyperfine interaction energies. This difference may be very substantial. Thus, for example, spins of the tin isotope Sn^{119} and its isomer Sn^{119m} (used as an absorber and emitter in the Mössbauer spectroscopy) are 1/2 and 11/2, respectively. Admittedly, according to the shell model, the magnetic moments of Sn^{119m} and Sn^{119} are expected to be quite close to each other — as well as in the cases of two remaining known Mössbauer isomer sources (Kr^{83m} — Kr^{83}, Pt^{195m} — Pt^{195}). Nevertheless, in many cases not only the spins but also the magnetic moments of ground and excited isomer nuclear states can strongly differ. Thus, . . . one would expect a selection of nuclear isomers with respect to different products in the radical reactions. On the other hand, experimental studies of the effectiveness of nuclear isomer separation in these reactions may, evidently, provide additional means for determining the magnetic moments of nuclei in excited states.[120]

F.   See Also

Reference 121.

G.   A Caution

Soon after the discovery of a phenomenon, attempts are usually made to describe what is going on by use of physical, mechanistic, or theoretical models to provide (1) coherence to the observations; (2) the sense of well-being that accompanies "understanding" something; (3) a basis for prediction and for the conception of future experiments. These models are usually necessarily approximate, as are their mathematical formulations. The models are used to provide a basis for describing and "explaining" what is going on. Example: "Since a consequence of an approximate mathematical description of a simple, approximate, possibly incomplete/incorrect model [reference provided] is that, if A is occurring, X [a measure of the magnitude of the phenomenon] increases as Y [some variable] decreases, our observation that X does indeed increase as Y decreases means that it may be useful to think in terms of the possibility that A is occurring." This is a hypothetical example; you won't find one like it. What you will find is more like, "Since X increases as Y decreases, A is occurring." Then, sometimes years later and after several questions of mechanism have been "answered" by use of the models (often by superficial and overextended use of the models), it is "discovered" that the actual situation need not be so simple. (Of course, this is really an illusory "discovery," since there never was a proper basis for believing that the actual situation *was* so simple).

Work on the magnetic isotope effect is now in the early stages of evolving along these lines. So, do your homework — and be careful out there.

IX.   NEW RADICAL PROCESSES

A. S_H Reaction at Saturated Carbon

We have reported some reactions which may be visualized as being of the homo-S_H' type $(n = 1)$,[77b]

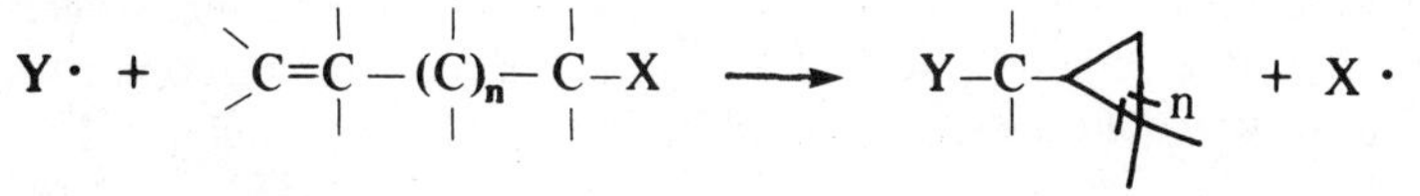

$$Y\cdot + \quad \overset{|}{\underset{|}{C}}{=}\overset{|}{\underset{|}{C}} - (\overset{|}{\underset{|}{C}})_n - \overset{|}{\underset{|}{C}}{-}X \quad \longrightarrow \quad Y{-}\overset{|}{\underset{|}{C}}{\underset{\raisebox{-3pt}{n}}{\bigtriangleup}} \quad + \; X\cdot \qquad (7.4)$$

Our purpose was to incorporate closure of a γ-iodopropyl radical to a cyclopropane into a sequence of free radical processes, namely, to cause such a step (an S_H reaction, a homolytic displacement at saturated carbon[77c,122]) to occur, sequentially or concertedly, along with radical addition to a double bond. As part of an extensive series of studies of homolytic displacement at carbon[123] Dr. Michael Johnson and his co-workers have investigated reactions which they describe[124] as proceeding as in Eqn. 7.4 with $Y = CCl_3$,[125a–c] $NCCCl_2$,[125a] $p{-}MeC_6H_4SO_2$,[125a,c] Me_2NSO_2,[125d] $NCCHBr$,[125b] CBr_3[125c] and $X = Co(dimethylglyoxime)_2 py$,[125a,b,c] $Co(dimethylglyoxime)_2(imidazole)$,[125a] $Fe(CO)_2$ (cyclopentadienyl),[125c] Br[125b] for $n = 1$ and $Y = CCl_3$[125e] and $X = Co(dimethylglyoxime)_2 py$[125e] for $n = 3$.[126]

Although intramolecular displacement at saturated carbon by a carbon-centered radical has long been known, observation of the corresponding intermolecular reaction would be novel.

A mixture of CF_3COCF_3 and $C(CH_3)_4$ was photolyzed at $300°C$.[127] CF_3CH_3, CF_3CF_3, and CH_3CH_3 were produced, among other products. Most of the 1.5% yield[128a] of CF_3CH_3 was considered to have been produced by means of the reaction $CF_3\cdot + C(CH_3)_4 \rightarrow CF_3CH_3 + (CH_3)_3C\cdot$, that is, by means of an intermolecular S_H reaction at saturated carbon, on the basis of the following evidence:

1.   "Isobutane [presumably formed from $\cdot C(CH_3)_3$] *appears to be* [emphasis added] a *primary* product . . . as is shown by a plot of isobutane formed against time, which indicates a finite rate of production . . . at zero time . . . ."[127a,128b] In our view, the qualitative shape of this plot is critically dependent upon the accuracy of the least accurate analysis of isobutane.[128c]

2.   "In four experiments . . . , values of 31–61 [Reference 128e] for this ratio $\{[CF_3CH_3]^2/[C_2H_6][C_2F_6]\}$ were obtained . . . . This indicates that . . . the S_H2 reaction is producing 1,1,1-trifluoroethane." Although the chemistry of $CH_3\cdot/CF_3\cdot$ systems[129] is not as simple as is presumed,[127] we do agree that, if the reactions $2 CH_3\cdot \rightarrow$, $2 CF_3\cdot \rightarrow$, and $CH_3\cdot + CF_3\cdot \rightarrow$ are the only reactions that involve C_2H_6, C_2F_6, and CH_3CF_3, ratios in the observed range would not

be expected. Is the ratio high because C_2H_6 is less stable than CF_3CH_3 under the reaction conditions? Because $CH_3 \cdot + CF_3 \cdot \rightarrow$ is not the only route to CH_3CF_3? Since reports in the literature[130] indicate that the rate constant for reaction of $CF_3 \cdot$ with C_2H_6 at 300°C is approximately 30 times that of $CF_3 \cdot$ with CF_3CH_3, the first possibility cannot be dismissed out of hand. Consideration of possible routes to CF_3CH_3 other than the cross-coupling of $CF_3 \cdot$ and $CH_3 \cdot$ is difficult, because very low yields of CF_3CH_3 are obtained; therefore, a consideration of its origin need not be confined to preferred reaction pathways. One possibility would be an S_H2 reaction.[131a] Others would include, for example,[132] routes wherein $CF_3CH_2 \cdot$, perhaps formed from $-\overset{|}{\underset{|}{C}}-\overset{|}{\underset{|}{C}}-CH_2CF_3$, is its precursor.

Although no supporting evidence was reported, it was concluded that the reaction

$$ROR' + HSiCl_3 \xrightarrow{\;(Bu^tO)_2\;} RH$$

proceeds by way of $\cdot SiCl_3 + ROR' \rightarrow R \cdot + R'OSiCl_3$, that is, an S_H2 reaction at an ether oxygen atom.[133]

X.   ENERGETICS OF PROTOTYPAL REACTIONS[134]

The activation energy for thermal decomposition of C_6H_5At was reported to be approximately 45 kcal/mol and was taken to be an estimate of the corresponding astatine–carbon bond dissociation energy.[135]

XI.   RATES OF PROTOTYPAL REACTIONS

Techniques for the generation, observation, and determination of transient radicals have very rapidly become more numerous and more sensitive. This has made it possible to observe and measure various characteristics of the radicals in fluid media. It has also transformed the field of free radical kinetics. It has become commonplace for studies of the kinetics and mechanism of free radical reactions to include determinations of the dependence on time of the concentrations of the transient free radicals that are present. It is difficult to exaggerate the importance of such determinations and the magnitude of the role they play in the proper "determination" of the mechanism of a free radical reaction and in the proper determination of the rates of elementary free radical processes. The field has come a long way from the (recent) days when the time-dependence of the concentrations of only "stable" species (e.g., reactants, products, an occasional intermediate) was studied.

A.   Hydrogen Abstraction Reactions

a. $R \cdot + R'H \rightarrow RH + R' \cdot$

Work has been reported[136] that was motivated by the perceptions that "there is no quantitative data in the literature of the kinetics of reactions of secondary and tertiary alkyl radicals with alkanes and their derivatives in the liquid phase"[136a] and "completely absent from the literature are any quantitative data on the reactivity of secondary and tertiary alkyl radicals in reactions of hydrogen abstraction from hydrocarbons and hydrocarbon derivatives"[136b] and that was directed toward a determination of the rate constant for abstraction of hydrogen from neat cyclohexane by 3-methylcyclohexyl radical. The radicals were generated by reaction of the corresponding iodide with an acyl or alkyl peroxide.[77b-g] The experimental counterpart of a back-of-an-envelope calculation yielded a rate constant of $10^{11.6} \exp(-9.5/RT)/M(h)$.

b. $HC \equiv C \cdot + RH \rightarrow HC \equiv CH + R \cdot$

$CF_3C \equiv CH$ was photolyzed at $180 \, m\mu$ in the presence of CH_4 and of C_2H_6 at $24°C$. The absorption at $152 \, m\mu$ was monitored with time and was attributed solely to product $HC \equiv CH$, postulated to have been formed solely by abstraction of hydrogen from CH_4 and from C_2H_6 by thermal $HC \equiv C \cdot$ and to be stable under the reaction conditions. $k(HC \equiv C \cdot + CH_4 \rightarrow HC \equiv CH + CH_3 \cdot) = 7 \times 10^8/M$ (sec) and $k(HC \equiv C \cdot + C_2H_6 \rightarrow HC \equiv CH + C_2H_5 \cdot) = 4 \times 10^9/M(sec)$ were obtained.[137a]

The effect of CH_4 on the dependence on time of the concentration of chemiluminescent products of the photolysis at $193 \, m\mu$ of mixtures of O_2 with $HC \equiv CX$ (X = H) (in the absence of CH_4, the dependence on time was the same for X = H, Br, and CHO) at $27°C$ was analyzed in terms of a simple, postulated reaction scheme. $k(HC \equiv C \cdot + CH_4 \rightarrow HC \equiv CH + CH_3 \cdot) = 3 \times 10^9/M$ (sec) was obtained.[137b]

c. $R \cdot + R'CO_2H \rightarrow RH + R'CO_2 \cdot$

"The extensive literature describing kinetic investigations of hydrogen transfer reactions of small alkyl radicals contains only one reference to the study of attack on an organic acid: the work of Ausloos and Steacie on the photolysis of CH_3COOD. They determined Arrhenius parameters for CH_3 attack on the acetyl group but not on the deuterium atom, and thus there are no results extant for abstraction of a carboxy hydrogen atom."[138a,b]

B.   Reaction of Radicals with Oxygen

a. $(CH_3)_3C \cdot + O_2$

The rate of decrease of the concentration of a product, reasonably considered to be $(CH_3)_3C \cdot$, of the photolysis of a mixture of $(CH_3)_3CI$ and O_2 in the gas phase was followed by use of photoionization mass spectrometry for several concentrations of O_2. The data were analyzed in terms of a simplified reaction model; several approximations and assumptions were made. $k(Bu^t \cdot + O_2 \to) = 1.4 \times 10^{10}/M(sec)$ was obtained.[139,140a]

b. $PhCH_2 \cdot + O_2$

The fluorescence resulting from optical excitation of a product, reasonably considered to be $PhCH_2 \cdot$, of the photolysis of a mixture of $PhCH_2Cl$ and O_2 in the gas phase was monitored as a function of time at several temperatures and pressures. The rate constant $k(PhCH_2 \cdot + O_2) = 6 \times 10^8/M(sec)$ was obtained at 22–99°C.[141a]

The same rate constant was obtained by following the decay of the optical spectrum of a product, reasonably considered to be $PhCH_2 \cdot$, of the photolysis of a mixture of $PhCH_2Br$ and O_2 in the gas phase at room temperature.[141b,c]

c. $R\dot{C}O + O_2$

$PhCOCH_3$ was flash-photolyzed in the presence of O_2 in the gas phase. The rate of decay of a species having the mass of PhCO was followed in a photo-ionization mass spectrometer as a function of the pressure of O_2. $k(Ph\dot{C}O + O_2) = 3 \times 10^9/M$ (sec), impliedly at 27°C, was reported.[142] A similarly determined value of $k(CH_3CO + O_2) = 1 \times 10^9/M$ (sec), impliedly at 27°C, also was reported.[142] Its consistency with expectation based on presumably less accurate earlier reports[143] of $k(CH_3\dot{C}O + O_2)/k(CH_3\dot{C}O \to CH_3 \cdot + CO)$ and recent values[76b,144] of $k(CH_3\dot{C}O \to CH_3 \cdot + CO)$ is not very weak.

C.   Decarboxylation of Acyloxy Radicals, $RCO_2 \cdot \to R \cdot + CO_2$

This discussion continues in Section II.J, which should be read first.

A "kinetic analysis" based on an idealized scheme involving the reactions

$$RCO_2 \cdot + C_4H_{10} \xrightarrow{k_a} RCO_2H + C_4H_9 \cdot$$

and

$$RCO_2 \cdot \xrightarrow{k_d} R \cdot + CO_2$$

gave the relationship

$$\frac{(RCO_2Br)_{initial}}{(C_4H_9Br)_{final}} - 1 = \frac{k_d}{k_a[C_4H_{10}]}$$

It is stated that "This relationship was valid in all cases $[R = Et, Pr^i, Bu^t]$, over a 10-fold variation of butane concentration, as illustrated . . . for pivaloxy."[49] The data are not strongly consistent with the relationship[145] and, moreover, questions would still remain as to the degree to which strong consistency would validate the relationship and the degree to which a strongly valid relationship would in turn validate the reaction pathways from which the relationship was derived. Thus we have serious concerns about whether the data illustrated could validly be used as an element in a quantitative determination of relative values of k_d, even if all of the problems discussed in Section II.J were to be ignored. However, *if* the *postulate* that the chain carrier is $RCO_2 \cdot$, as shown in the preceding reactions, is accepted, the possibility that the analysis does indicate the qualitative result that k_d/k_a decreases on going from $R = Et$ to Pr^i to Bu^t ought to be considered seriously. Since it is reasonable to anticipate that k_d/k_a would vary primarily because of changes in k_d, this result would mean that the more branched acyloxy radical, that is, the one whose enthalpy of decarboxylation presumably is lower, decarboxylates more slowly. This is an intriguing idea, as is the explanation offered in terms of steric retardation of the opening of the OCO angle in the transition state en route to linear CO_2. Should the idea reach a stage when it is more strongly supported by experiment, it ought to be reconciled with results on the decomposition of diacyl peroxides which were interpreted in terms of the decarboxylation of $EtCO_2 \cdot$ being faster than that of $MeCO_2 \cdot$[146] and of that of $Pr^iCO_2 \cdot$ being faster than that of $EtCO_2 \cdot$.[146d] A discussion of the idea with reference to the report that $EtO\dot{C}O$ decarboxylates much more slowly than $Bu^tO\dot{C}O$[147] would be of interest.

D.   Radical Addition to Carbonyl Groups

Although intermolecular attack at the carbonyl carbon of formaldehyde,[148a—n] $RCHO$,[148h,p—s] and R_2CO[148t] by alkyl,[148h,q—t] cycloalkyl,[148c] $\cdot CH_2OH$,[148b,i—m] $RCHOH$,[148b,d—g] $R_2\dot{C}OH$,[148b,d,p] and $R\dot{C}HOR$[148a,e] radicals has often been reported, measurement of a rate has not been reported until recently.

a. $CH_3CH_2CH_2 \cdot + CH_2O \rightarrow CH_3CH_2CH_2CH_2O \cdot$

Photolysis of 97/3 $CH_3CH_2CH_2N{=}N/Pr^iN{=}NPr^n$ (0.3 mM) + CH_2O (0.6–3.4 mM) in the gas phase at 60 and 90°C led to a variety of products, including $CH_3CH_2CH_2CH_2OH$ (typically 10%), $C_4H_9OC_3H_7$ (typically 0.5%) and $n{-}C_3H_7CHO$ (typically 2%), the amounts of which were fitted to the equations derived from a postulated reaction scheme. After some "known" rate constants were introduced, $k(CH_3CH_2CH_2 \cdot + CH_2O \rightarrow CH_3CH_2CH_2CH_2O \cdot) = (1{-}2) \times 10^3/M$ (sec) emerged.[149a]

b. $HOCH_2 \cdot + CH_2 O \rightarrow HOCH_2 CH_2 O \cdot$

γ-Irradiation of solutions of formaldehyde in methanol at 130–200° led to conversion of the formaldehyde to ethylene glycol almost exclusively.[148o] The dependence of the dose-rate-normalized yield[150a] of ethylene glycol on the estimated[150b] concentration of "free" CH_2O was analyzed in terms of the following reaction scheme:

$$\cdot CH_2 OH + CH_2 O \longrightarrow HOCH_2 CH_2 O \cdot \xrightarrow{CH_3 OH} HOCH_2 CH_2 OH + \cdot CH_2 OH;$$

$$2 \cdot CH_2 OH \longrightarrow HOCH_2 CH_2 OH$$

From this analysis and a literature value of $k(2 \cdot CH_2 OH \rightarrow HOCH_2 CH_2 OH)$, an implausible Arrhenius expression emerged.

c. $CH_3 \cdot + CD_3 COCD_3 \rightarrow CH_3 (CD_3)_2 CO \cdot$

The reactions in the gas phase of $CD_3 COCD_3$ with $CH_3 N{=}NCH_3$ and $(Bu^t O)_2$ at 240–290 and 140°C, respectively, have been studied. By use of a procedure for which "we fully recognize the possibility of systematic errors caused by the assumptions involved and the low concentrations of products following reaction $[CH_3(CD_3)_2 CO \cdot \rightarrow CH_3 COCD_3 + CD_3 \cdot]$" compared with other products," the relative amounts of N_2, CH_4, $CH_3 D$, $CD_3 H$, and CD_4 from the first reaction and methylethyl ketone and the various methanes and ethanes from the second were used to calculate $k(CH_3 \cdot + CD_3 COCD_3 \rightarrow CH_3(CD_3)_2 CO \cdot)/k(CH_3 \cdot + CD_3 COCD_3 \rightarrow CH_3 D + CD_3 COCD_2 \cdot)$, which was combined with an estimated $k(CH_3 \cdot + CD_3 COCD_3 \rightarrow CH_3 D + CD_3 COCD_2 \cdot)$ to obtain $\log \ k(CH_3 \cdot + CD_3 COCD_3 \rightarrow CH_3(CD_3)_2 CO \cdot)(/M \sec) = 10.5 - 11.7/4.6 \, T$.[149b]

d. $CH_3 \cdot + CH_3 CH_2 COCH_3 \rightarrow CH_3 CH_2 (CH_3)_2 CO \cdot$

Reaction of $CH_3 N{=}NCH_3$ with $CH_3 CH_2 COCH_3$ at 290–340°C in the gas phase led in part to CH_4 and $CH_3 COCH_3$, whose relative amounts were introduced into an approximate treatment of equations resulting from the combination of some "facts" and a postulated reaction scheme. Out of this came a number, $2 \times 10^5/M(\sec)$, which was considered to be the rate constant of the title reaction at 290°C.[149c]

E.   Rearrangement of Allylcarbinyl to Cyclopropylcarbinyl Radical

The parameters $\log A(\sec^{-1}) = 10.4$ and $E = 9.1$ kcal/mol ($k = 5 \times 10^3/\sec$ at 25°C) have been reported for the title reaction, studied by use of esr spectroscopy and competitive trapping by Bu_3SnH. When combined with those determined earlier for the reverse reaction,[5g] these results yield $\Delta H°$ (allylcarbinyl $\rightarrow$ cyclopropylcarbinyl) $= -3$ kcal/mol   and $\Delta S° = 8$ e.u. $(K = 2.6 \times 10^4$ at $25°C)$.[151,152]

XII.   SYNTHETIC APPLICATIONS

A.   Asymmetric Induction

Work on asymmetric induction in organic radical addition reactions[6z] has continued.[153]

a.   Addition of an Achiral Compound to a Prochiral Substrate in the Presence of a Chiral Material

Another report of the work described earlier[6z] has appeared[154] as has a report of further work[155a] on the addition of thiolacetic acid to *cis*-2-octene. For example,

$$CH_3COSH + CH_3CH=CHC_5H_{11} \xrightarrow[\text{hexane, 45°}]{\textit{l}\text{-menthol, AIBN}} CH_3COSCHC_6H_{13}$$

$$\underset{CH_3}{\overset{|}{}}$$

45%

$$+ CH_3COSCHC_5H_{11}$$

$$\underset{C_2H_5}{\overset{|}{}}$$

40%

$[\alpha]_0^{25}$ (benzene) is now reported as $-1.82°$ ($\leqslant \sim 28\%$ optical yield)[156a] and $-0.84°$, respectively.[156b]

Other reactions include, for example, the following:

$$CH_3COSH + \textit{trans}\text{-}CH_3CH=CHCO_2CH_3 \xrightarrow[\text{2) LiAlH}_4]{1)\ \textit{l}\text{-menthyl acetate, AIBN, 78°C}}$$

53% $HSCH(CH_3)CH_2CH_2OH$, $[\alpha]_D^{25} -1.41°(CHCl_3)$, optical yield 9.6% (Ref. 155b)

$$HSCH_2CO_2H + \textit{trans}\text{-}CH_3CH=CHCO_2CH_3 \xrightarrow[\text{benzene, 80°C}]{d\text{-camphor, AIBN}}$$

67% $HO_2CCH_2SCH(CH_3)CH_2CO_2CH_3$, $[\alpha]_{400}^{25}$
$-0.11°(MeOH)$, optical yield 0.7% (Ref. 155b)

$$CH_3(CH_2)_{11}SH + CH_2=\overset{\overset{\displaystyle CH_3}{\displaystyle |}}{C}COCH_3 \xrightarrow[\text{AIBN, toluene, } 60°C]{\text{poly}(d\text{-bornyl methacrylate})}$$

79% $CH_3(CH_2)_{11}SCH_2CH(CH_3)COCH_3$, $[\alpha]_D^{20}$

$+ 2.21°$ (MeOH), optical yield 14%[156c] (Ref. 155c)

$$BrCCl_3 + PhCH=CH_2 \xrightarrow[\text{EtOH/benzene, } 80°C]{[(-)\text{-diop}]RhCl^{156d}} 26\% \text{ PhCHBrCH}_2\text{CCl}_3,$$

$[\alpha]_D - 22.5°$ (benzene), optical yield $\geqslant 32\%$ (Ref. 155d)

Another report is entitled, "Apparent Asymmetric Induction by the Radical Copolymerization of Isobutyl Vinyl Ether with Maleic Anhydride in *l*-Menthol."[157]

b. *Addition of a Chiral Compound to a Prochiral Substrate*

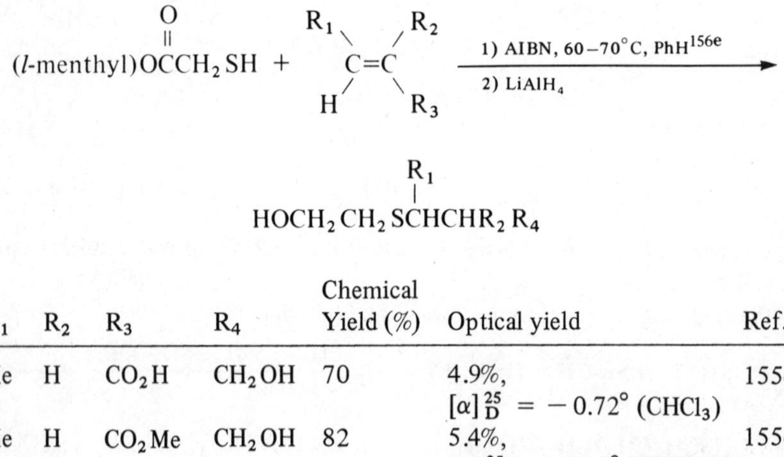

$$(l\text{-menthyl})\overset{\overset{\displaystyle O}{\displaystyle ||}}{O}CCH_2SH + \underset{H}{\overset{R_1}{\diagdown}}C=C\underset{R_3}{\overset{R_2}{\diagup}} \xrightarrow[\text{2) LiAlH}_4]{\text{1) AIBN, } 60-70°C, \text{PhH}^{156e}}$$

$$HOCH_2CH_2SCHCHR_2R_4 \overset{\overset{\displaystyle R_1}{\displaystyle |}}{}$$

R_1	R_2	R_3	R_4	Chemical Yield (%)	Optical yield	Ref.
Me	H	CO_2H	CH_2OH	70	4.9%, $[\alpha]_D^{25} = -0.72°$ (CHCl$_3$)	155e
Me	H	CO_2Me	CH_2OH	82	5.4%, $[\alpha]_D^{25} = -0.78°$ (CHCl$_3$)	155e
H	Me	Ph	Ph	60[156f]	?, $[\alpha]_{400}^{25} = +0.41°$ (CHCl$_3$)	155f
Ph	H	Ph	Ph	63	?, $[\alpha]_D^{25} = +0.64°$ (CHCl$_3$)	155g
Et	Me	Me	Me	87	?, $[\alpha]_D^{25} = +0.20°$ (MeOH)	155g

c. *Addition of an Achiral Compound to a Chiral Substrate*

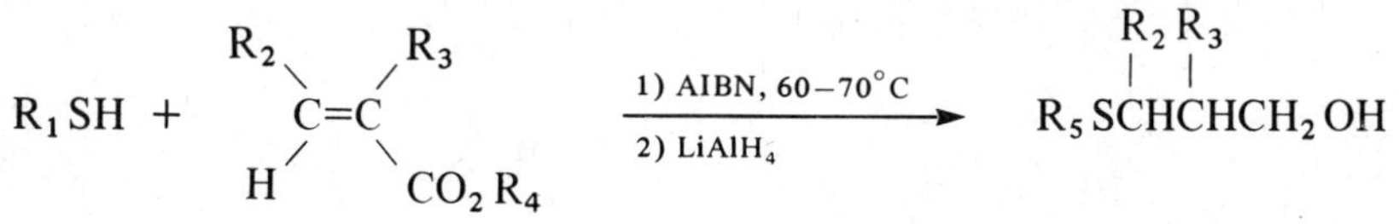

$$R_1SH + \underset{H}{\overset{R_2}{\diagup}}C=C\underset{CO_2R_4}{\overset{R_3}{\diagdown}} \quad \xrightarrow[\text{2) LiAlH}_4]{\text{1) AIBN, 60--70°C}} \quad R_5SCHCHCH_2OH$$

R_1	R_2	R_3	R_4	R_5	Chemical Yield (%)	Optical Yield, $[\alpha]_D^{25}(CHCl_3)$	Ref.
HO_2CCH_2	Me	H	*l*-menthyl	$HOCH_2CH_2$	78[156f]	15%, −3.09°	155h
MeCO	Me	H	*l*-menthyl	H	70[156f]	18%,[156g] + 3.49°	155h
MeCO	H	Me	*l*-menthyl	H	55[156f]	?, + 1.7°	155f
MeCO	Me	H	*iso*bornyl	H	40	19%, −1.35°	155e
MeO_2CCH_2	Me	H	*l*-menthyl	$HOCH_2CH_2$	81[156f]	13%, −2.76°	155h
$n-C_4H_9$	H	Me	*l*-menthyl	$n-C_4H_9$	70[156f,h]	?, −0.16°	155f

Similar reactions involving the addition of main group IV hydrides have long been known. For recent examples, see Reference 158.

The following reactions have been reported:[159]

$$\overset{O}{\overset{\|}{PhCCO_2}}(l\text{-menthyl}) \xrightarrow[\substack{(a)\ h\nu,\ 60°C \\ (b)\ 20°C \\ (c)\ MeOH,\ 65°C}]{Bu_3SnH} ?\% \ \overset{OSnBu_3}{\overset{|}{PhCHCO_2}}(l\text{-menthyl}) + ? \xrightarrow{LiAlH_4}$$

optical yield (a) 5%
(b) ?%
(c) 11%

(a) 68%
(b) 65%
(c) 45%

$$\overset{OH}{\overset{|}{PhCHCH_2OH}}, \text{ optical yield (S isomer)}[160a]$$

(a) 5%
(b) 17%
(c) 10%

B. Interconversions of Functional Groups

Earlier[5h,6aa] we discussed the replacement of thiyl and hydroxyl by another group according to the transformations

$$\overset{OO}{\overset{\|\ \|}{ROCCOOBu^t}} \xrightarrow{CCl_4 \text{ or } BrCCl_3} RCl\ (Br) \qquad (1)$$

$$RSH \xrightarrow{CO} RH \qquad (2)$$

$$\underset{\text{ROCCl}}{\overset{\displaystyle O}{\overset{\|}{}}} \xrightarrow{R'_3SnH} RH \qquad\qquad (3)$$

$$\underset{\substack{\text{ROCR}' \\ (4)}}{\overset{\displaystyle O}{\overset{\|}{}}} \text{ or } \underset{(5)}{\overset{\displaystyle O}{\overset{\|}{\text{ROCCl}}}} \xrightarrow{R''_3SiH} RH \qquad (4, 5)$$

and we referred to reports of the transformations,

$$\underset{\text{ROCR}'}{\overset{\displaystyle O}{\overset{\|}{}}} \xrightarrow{R''_3SnH} RH \qquad (6), \qquad RSR' \xrightarrow{R''_3SnH} RH$$
$$(7)^{161a}$$

$$\underset{\text{ROCR}'}{\overset{\displaystyle S}{\overset{\|}{}}} \xrightarrow{R''_3SnH} RH \qquad (8),^{162a} \quad \underset{\text{ROCSR}'}{\overset{\displaystyle S}{\overset{\|}{}}} \xrightarrow{R''_3SnH} RH$$
$$(9)^{163a}$$

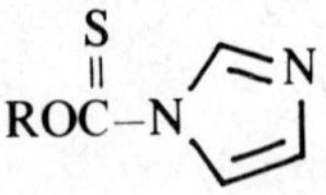

$$\underset{\text{ROC--N}}{\overset{\displaystyle S}{\overset{\|}{}}}\!\!\!\!\!\overset{\displaystyle \diagup{=}N}{\diagdown\diagup} \xrightarrow{R'_3SnH} RH \qquad (10),^{164a} \quad \underset{\text{ROCH}}{\overset{\displaystyle S}{\overset{\|}{}}} \xrightarrow{R'_3SnH} RH$$
$$(11)^{165a}$$

Work along these lines has continued. The following transformations have been reported,

$$\underset{\text{ROCSR}'}{\overset{\displaystyle O}{\overset{\|}{}}} \xrightarrow{R''_3SnH} RH \quad \text{(Ref. 166a)} \qquad ROCH_2SR' \xrightarrow{R''_3GeH} ROCH_3$$
$$\text{(Ref. 166a)}$$

$$\underset{\underset{|}{\text{RN--CSR}'}}{\overset{|\ \ |}{}} \xrightarrow{R''_3SnH} \underset{|}{\overset{|\ \ |}{\text{RN--CH}}} \quad \text{(Ref. 166b)} \qquad \underset{\text{ROCOR}'}{\overset{\displaystyle S}{\overset{\|}{}}} \xrightarrow{R''_3SnH} RH$$
$$\text{(Refs. 166c, d)}$$

$$\underset{\text{ROCH}}{\overset{\displaystyle O}{\overset{\|}{}}} \xrightarrow{(Bu^tO)_2} RH \qquad \text{(Refs. 166e, 167a)}$$

as have additional examples categorizable as (5),[168a] (6),[168b] (7),[168c,d] (8),[166d;168e,f] (9),[168e,g-n,166d] (10),[166d,168e,o-q] and (11).[166a,168r]

Barton and McCombie's reactions (8)–(10) have become established in the methodology of organic synthesis.

In an earlier discussion of the generation and trapping of radicals, we described reactions wherein "an allyl group in $R_3SnCH_2CH=CH_2$ may serve the same function as does H in R_3SnH."[51] Along those lines, a reaction analogous to reaction (10) has been reported:[169]

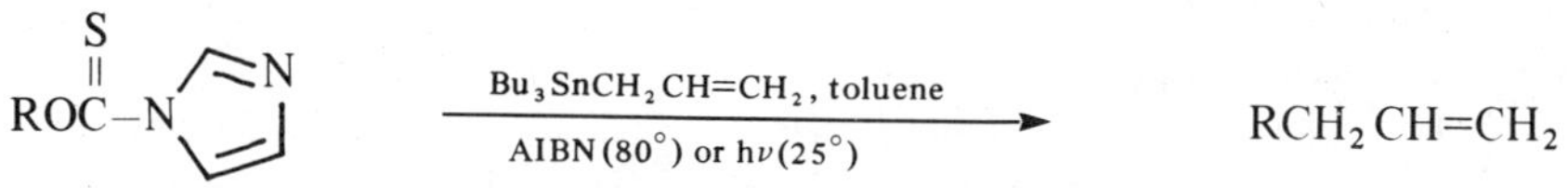

$$ROC-N \xrightarrow[\text{AIBN}(80°)\text{ or }h\nu(25°)]{Bu_3SnCH_2CH=CH_2,\text{ toluene}} RCH_2CH=CH_2$$

Finally, we caution readers tempted to speculate about the mechanisms of the reactions in this section to beware of overconfidence regarding knowledge of the mechanisms of reactions that begin as free radical but that have as a product of the radical reaction a compound that can function as an electrophilic catalyst.[170]

XIII. ADDENDA

A good example of "problem solving" is found in "Autoxidation of Biological Molecules. 1. The Antioxidant Activity of Vitamin E and Related Chain-Breaking Phenolic Antioxidants in Vitro."[171]

Times change: For many years, the observation of the esr spectrum of an alkoxy radical was an object of research.[6bb] Now, the problem can be its suppression.[172]

A.   Reviews

"Structure and Properties of Free Radicals,"[173]
Tabulations of optical absorption spectra of radicals.[174]

B.   Awards

For contributions to the structural theory of organic chemistry:
  H. Knof and D. Krafft, *Adv. Mass Spectrom.*, **7A**, 699 (1978); *Pract. Spectrosc.*, **3** (*Mass Spectrom., Part B*), 338 (1980).
For candor and for eschewing the macho approach to chemistry:
  M. C. R. Symons, "Radical Products from the Radiolysis of 2,2,3,3-Tetramethylbutane," *Chem. Phys. Lett.*, **69**, 198 (1980).

For candor and telling it like it is:

> *a.* P. Gray, "Fundamental Aspects of Free-Radical Reactivity and Chain Reactions," in C. Capellos and R. F. Walker, Eds., *Fast Reactions in Energetic Systems,* Reidel, 1981, p. 153.
>
> *b.* R. Hiatt and V. G. K. Nair, *Can. J. Chem.,* **58**, 450 (1980).

For a demonstration that an ounce of prevention is worth a pound of cure:

> H.-G. Korth, H. Trill, and R. Sustmann," [1-^{2}H] Allyl Radical: Barrier to Rotation and Allyl Delocalization Energy," *J. Am. Chem. Soc.,* **103**, 4483 (1981).

For heroic experimentation:

> D. Mihelcic, D. H. Ehhalt, J. Klomfass, G. F. Kulessa, U. Schmidt, and M. Trainer, "Measurements of Free Radicals in the Atmosphere by Matrix Isolation and Electron Paramagnetic Resonance," *Ber. Bunsenges. Phys. Chem.,* **82**, 16 (1978).

Rip Van Winkle Awards:

> *a.* H.-D. Beckhaus, M. A. Flamm, and C. Rüchardt, *Tetrahedron Lett.,* **23**, 1805 (1982).
>
> *b.* J. M. Tedder, *Angew. Chem. Int. Ed.,* **21**, 401 (1982).

Pollyanna Awards:

> *a.* Yu. N. Kozlov, A. P. Moravskii, A. P. Purmal', and V. F. Shuvalov, *Zh. Fiz. Khim.,* **55**, 764 (1981).
>
> *b.* Y. Takemura and T. Shida, *J. Chem. Phys.,* **73**, 4133 (1980).
>
> *c.* H. Bock and W. Kaim, *Acc. Chem. Res.,* **15**, 9 (1982).

The Lamont Cranston/Scarlet Pimpernel Award:

> *Winner*: W. J. Bouma, J. K. MacLeod, and L. Radom, *J. Am. Chem. Soc.,* **104**, 2930 (1982).
>
> *Runner-up*: J. L. Holmes, F. P. Lossing, J. K. Terlouw, and P. C. Burgers, *J. Am. Chem. Soc.,* **104**, 2931 (1982).

XIV.   POSTSCRIPTS TO VOLUME 1

The numbering and labeling of the following sections correspond to that of the respective sections in Volume 1 of this series.

II.   Characteristics of Prototypal Radicals

A.   *Ultraviolet Spectra of Isopropyl and t-Butyl*

Since there is a low barrier to bending/internal rotation[10d] in radicals such as isopropyl and *t*-butyl, their uv spectra could show significant matrix effects.

Photolysis of 0.03 *M* diisopropyl ketone in hexadecane at 38°C produced a transient molecule whose uv spectrum, assigned to $(CH_3)_2CH\cdot$ on the basis of the behavior of its carrier as a transient, was a "structureless band with onset at 290 nm and maximum at 220 nm."[75b]

Photolysis of $(CH_3)_2CHN=NCH(CH_3)_2$ in the gas phase at room temperature led to the observation of a broad continuous transient absorption in the region $225-260\,m\mu$, consisting of a more intense absorption at approximately $235-240\,m\mu$ and a weaker one at approximately $245-250\,m\mu$, which was assigned to $(CH_3)_2CH\cdot$ on the basis of suggestive arguments. It was pointed out that "isopropyl peroxy radicals . . . have a maximum absorbance at 240 nm with an extinction coefficient almost identical to that of the isopropyl radical."[175a]

"isopropyl radicals were generated by . . . photolysis of azoisopropyl and . . . [have a] transition at 235 nm."[175b]

Photolysis of $0.02M$ $(CH_3)_3CCOC(CH_3)_3$ in 3-methyl-3-pentanol at $10°C$ produced a transient molecule whose uv spectrum, assigned to $(CH_3)_3C\cdot$ largely on the basis of the behavior of its carrier as a transient, was a broad continuous absorption with an onset at approximately $350\,m\mu$ and consisted of a more intense absorption at approximately $220\,m\mu$ and a weaker one at approximately $307\,m\mu$. The spectrum had the same "'shape'" in cyclohexane and methanol as well.[75b]

E.   The Isoelectronic Series $H_2CN\cdot$, $H_2CO^{+\cdot}$, $H_2BO\cdot$

"Cations of Aldehydes and Ketones: An ESR Study."[176] (Acetone and Cyclohexanone)

III.   Energetics of Prototypal Radicals

C.   The 3-Cyclopropenyl Radical

A value for $D[(3\text{-cyclopropenyl})\text{-H}] \approx 91$ kcal/mol [the same as is reported for $D(CH_2=CHCH_2-H)$] was reported[177] which, if correct, would have serious implications with respect to the quoted statement[5j] to the effect that "no large amount of stabilization is detectable in the cyclopropenyl radical from these data." However, it is acknowledged by implication that the calculation of that value is based upon the unsubstantiated assumption that the cation radical produced by electron impact upon cyclopropene was that of cyclopropene.

V.   Mechanistic Techniques

C.   Recent Examples of the Generation and Trapping of Radicals

Additional examples of the reaction $RX + Bu_3SnCH_2CH=CH_2 \rightarrow RCH_2CH=CH_2$, a system that both generates and traps radicals by use of the same reagent, have been reported.[169,181]

VII.   New Radical Processes

A.   *1,2-Shift of Hydrogen from Carbon to Carbon*

A recurrence: "Thermal Decomposition of 2,2-Propane-d$_2$. CH$_3$CDCH$_3$ Radical Isomerization."[178]

B.   *Disproportionation of Radicals via α-Abstraction*

"UV Absorption Spectrum and Photochemistry of CF$_2$ClBr."[179a]

"Disproportionation Reactions between CF$_2$H and C$_2$H$_5$ Radicals in the Gas Phase."[179b]

IX.   Rates of Prototypal Reactions

A.   *Dimerization of Radicals*

For additional work on dimerization of radicals see the following:

Reference 6cc.

Ethyl.[180]

Isopropyl.[75b,175a]

t-Butyl.[17a,b,75b]

XV.   POSTSCRIPTS TO VOLUME 2

The numbering and labeling of the following sections correspond to that of the respective sections in Volume 2 of this series.

I.   Structure of Prototypal Radicals

A.   *Isopropyl*

Work very similar to that discussed in Volume 2 has been reported.[182] A composite ir spectrum attributed to the isopropyl radical is presented. The reaction conditions apparently differed slightly and, although the frequencies assigned to the absorptions are somewhat different from those reported earlier, the earlier conclusions are not negated; neither are they supported further.

B.   *t-Butyl: ESR Spectroscopy*

"Lack of Orbital Following: Its Significance in the Interpretation of EPR Data."[183a]

"Me$_3$C· is *probably* [emphasis added] . . . *close to* [emphasis added] planarity."[183b] Do we detect a softening? Perhaps. But perhaps not: "My own case has always been staunchly in favor of effective [?] planarity . . ." and "I still consider that the esr data favor planarity . . ."[183c]

II.   Characteristics of Prototypal Radicals

H.   *Olefin Radical Cations*

Additional reports of optical[184a,h,i] and esr[184b-g,j,k] spectra of olefin radical cations have appeared.

There have also been reports of esr spectra that can be reasonably and consistently assigned to radical anions of tetraalkyldiboranes, $R_2 BBR_2^-$.[184e,185] This completes a series of isoelectronic compounds: $>C=C<^{+\cdot}$, $>B-\dot{C}<$ (Reference 6dd), $>B-B<^{-\cdot}$.

III.   Energetics of Prototypal Radicals

B.   *D(Cyclopropyl-H)*

C.   *The Phenoxy Radical: D(PhO−H), ΔH_f(PhO·)*

D(PhO−H) = 87 kcal/mol and D(cyclopropyl-H) = 106 kcal/mol have been reported,[177] in good agreement with the values of Volume 2.

D.   *The Anilino Radical: D(PhNH−H), ΔH_f(PhNH·)*

D(PhNH−N) was reported to be 85 kcal/mol,[186] in good agreement with the value of Volume 2.

H.   *Electron Affinity of Alkoxyl Radicals*

The electron affinities of ethyoxyl, isopropoxyl, and *t*-butoxyl were determined, by way of that of oxygen atom as a reference, to be 39.8 ± 0.8, 42.4 ± 0.7, and 44.1 ± 1.2 kcal/mol, respectively, from the photoelectron spectra of the corresponding alkoxide anions, which had been produced by subjecting the corresponding alcohols to an electrical discharge.[194] The value for $Bu^tO\cdot$ agrees well with that determined earlier.[101]

IV.   Experimental Techniques

A.   *Equilibria Involving Radicals*

Work concerned with "a direct observation of the equilibrium $C_2 H_6 \rightleftarrows 2CH_3$" had been described in Reference 187. We apologize for our oversight.

B.   *Use of Radicals/ESR to Study Quite Another Phenomenon*

In volume 2 we discussed an example of the use of esr spectroscopy to study complexation of cations by "spin-labeled" crown ethers. Other examples have since been reported:

"ESR Studies of the Alkali Metal Complexes of Galvinoxyl-Labeled Benzo-15-Crown-5."[188a]

"ESR Studies of the Alkali Metal Complexes of Verdazyl-Labeled Benzo-15-Crown-5."[188b]

"ESR Investigation of Alkali Metal Complexes of Galvinoxyl-labeled Benzo-15-Crown-5 in Frozen Solution."[188c]

"Synthesis of Two Stereochemically Distinct Chiral Di-nitroxide Crown Ethers."[188d]

"Three Novel Spin-Labeled Crown Ethers."[188e]

For another type of use see Reference 188f.

V.   Mechanistic Techniques

B.   *Trapping of Radicals*

The technique and the terminology have been refined. For a review see Reference 189.

C.   *Spin Trapping of Radicals*

The following additional work has been reported on spin trapping of radicals:

"Substituent Effects in the Kinetic Analysis of Free Radical Reactions with Nitrone Spin Traps."[190a]

"Spin Traps in Radiation Chemistry (Review)."[190b,190c]

"Kinetic Studies of Spin Trapping Reactions. III. Rate Constants for Spin Trapping of the Cyclohexyl Radical."[190e]

VI.   Persistent Radicals

A.   *The Other Side of Persistence*

Unusually long C–C bonds: Reference 191.

VIII.   Rates of Prototypal Reactions

A.   *Rate Constants for the Scavenging of Radicals by Iodine*

"Optoacoustic Method of Measuring Reaction Rates of the Radicals CH_3, CD_3, C_2H_5 and CH_2I with I and I_2."[195]

B.   *Rate Constants for Reaction of $Bu^tO\cdot$*

The laser flash photolysis/optical absorption spectroscopy technique for the determination of rate constants for reaction of $Bu^tO\cdot$[6ee] has been applied in the following:

"A Kinetic Study of the Reactions of *tert*-Butoxyl with Alkenes: Hydrogen Abstraction vs. Addition."[192a]

"Absolute Rate Constants for the Reactions of *tert*-Butoxyl Radicals and Some Ketone Triplets with Silanes."[192b]

"Laser Photolysis Study of the Reactions of Alkoxyl Radicals Generated in the Photosensitized Decomposition of Organic Hyponitrites."[192c]

"Absolute Rate Constants for the Reactions of *tert*-Butoxyl with Ethers: Importance of the Stereoelectronic Effect."[192d]

"Reaction of Benzophenone Triplets with Allylic Hydrogens. A Laser Flash Photolysis Study."[192e]

"A Laser Flash Photolysis Study of t-Butoxyphosphoranyl Radicals. Optical Spectra and Kinetics of Formation, Fragmentation, and Rearrangement."[192f]

"Reaction of *tert*-Butoxy Radicals with Phenols. Comparison with the Reactions of Carbonyl Triplets."[192g]

"Studies on the Spiro[2.5]octadienyl Radical and the 2-Phenylethyl Rearrangement."[192h]

"Absolute Rate Constants for the Reactions of *tert*-Butoxyl, *tert*-Butylperoxyl, and Benzophenone Triplet with Amines: The Importance of a Stereoelectronic Effect."[192i]

"Detection of Trialkylstannyl Radicals Using Laser Flash Photolysis."[192j]

"Rate Constants and Arrhenius Parameters for the Reactions of Primary, Secondary, and Tertiary Alkyl Radicals with Tri-*n*-butyltin Hydride."[85]

The flash photolysis/esr technique[6ff] has been applied to "An Indirect Measurement of the Absolute Rate Constant of the Self-Reaction of *tert*-Butoxy Radicals."[192k]

Based upon a conjectural analysis of the results of "A Thermal Lensing Study of a Photolysis of Di-*t*-Butyl Peroxide," rate constants for abstraction of hydrogen by $Bu^tO\cdot$ from toluene and cyclohexane have been reported.[192l]

See also "Arrhenius Parameters for the *tert*-Butoxy Radical Reactions with Trimethylsilane in the Gas Phase."[192m]

IX.   Synthetic Applications

C.   *Host-Guest Interactions: Macrocyclically Complexes Radicals and Radical Transfer Agents*

"Submicrosecond Pulse Radiolysis Conductivity Measurements in Aqueous Solutions — II. Fast Processes in the Oxidation of Some Organic Sulfides."[34g]

"Electrochemical Study of Iodine and Bromine in the Presence of Cryptands."[193a]

A paper on the "Effect of *Dibenz* [emphasis added]-18-crown-6 on the Reactivity of Undecyl Radicals in Reactions Involving Removal of Hydrogen and Chlorine Atoms" has appeared.[193b] This work is probably not an example of what we were discussing.

XVI.   REFERENCES AND NOTES

1.[*]   (a)   C. Yamada, E. Hirota, and K. Kawaguchi, *J. Chem. Phys.*, **75**, 5256 (1981).
        (b)   V. Špirko and P. R. Bunker, *J. Mol. Spectrosc.*, **95**, 381 (1982).
        (c)   E. Hirota and C. Yamada, *J. Mol. Spectrosc.*, **96**, 175 (1982).

2.   (a)   M. E. Jacox, *Chem. Phys.*, **69**, 407 (1982).
        (b)   T. Amano, P. F. Bernath, C. Yamada, Y. Endo, and E. Hirota, *J. Chem. Phys.*, **77**, 5284 (1982).
        (c)   T. Amano, P. F. Bernath, and C. Yamada, *Nature*, **296**, 372 (1982).

3.   D. E. Applequist and L. Kaplan, *J. Am. Chem. Soc.*, **87**, 2194 (1965).

4.   L. Kaplan, in J. K. Kochi, Ed., *Free Radicals*, Vol. 2, Wiley-Interscience, New York, 1973, (a) p. 361; (b) p. 369, footnote †.

5.   L. Kaplan, in M. Jones, Jr. and R. A. Moss, Eds., *Reactive Intermediates. A Serial Publication*, Vol. 1, Wiley-Interscience, New York, 1978, (a) p. 164; (b) p. 165; (c) p. 166; (d) p. 183; (e) p. 175; (f) V.B; (g) IX.B; (h) p. 184; (i) p. 174; (j) p. 168.

6.[*]   L. Kaplan, in M. Jones, Jr. and R. A. Moss, Eds., *Reactive Intermediates. A Serial Publication*, Vol. 2, Wiley-Interscience, New York, 1981, (a) p. 253; (b) p. 254, comments from refs. 5b, c; (c) p. 254–255; (d) III.A; (e) II.E–G; (f) p. 262–263; (g) p. 263; (h) p. 262; (i) II.G; (j) II.I; (k) p. 266; (l) p. 267; (m) p. 268; (n) IV.B; (o) IX.D; (p) ref. 117h; (q) Tables 7.4 and 7.5; (r) p. 292; (s) ref. 117k; (t) p. 293; (u) ref. 117m; (v) ref. 117r; (w) p. 295; (x) p. 294; (y) p. 298; (z) p. 288; (aa) p. 289; (bb) p. 256; (cc) VIII.C; (dd) p. 264; (ee) p. 285–286, refs. 99a and 99b; (ff) p. 287, ref. 99c; (gg) IV.A.

7.   (a)   This should not have come as a "surprise." As was pointed out,[8a] it has long been known that the internal rotational and inversional motions in CH_3NH_2 are strongly coupled.[8b-d]
        (b)   We are quoting extensively because the structure of a species is its most basic attribute and because of the less than universal awareness of these indications.
        (c)   See also Ref. 9h

8.   (a)   G. L. Pratt and D. Rogers, *J. Chem. Soc. Faraday I*, 2751 (1981).
        (b)   D. R. Lide, Jr., *J. Chem. Phys.*, **22**, 1613 (1954).
        (c)   D. Kivelson and D. R. Lide, Jr., *J. Chem. Phys.*, **27**, 353 (1957).
        (d)   T. Nishikawa, T. Itoh, and K. Shimoda, *J. Chem. Phys.*, **23**, 1735 (1955).

9.[*]   (a)   D. Griller, K. U. Ingold, P. J. Krusic, and H. Fischer, *J. Am. Chem. Soc.*, **100**, 6750 (1978).
        (b)   J. Pacansky and H. Coufal, *J. Chem. Phys.*, **72**, 5285 (1980).
        (c)   J. Pacansky and M. Dupuis, *J. Chem. Phys.*, **73**, 1867 (1980).
        (d)   J. Pacansky, D. W. Brown, and J. S. Chang, *J. Phys. Chem.*, **85**, 2562 (1981).
        (e)   M. Yoshimine and J. Pacansky, *J. Chem. Phys.*, **74**, 5168 (1981).
        (f)   J. Pacansky and M. Dupuis, *J. Am. Chem. Soc.*, **104**, 415 (1982).
        (g)   J. Pacansky and W. Schubert, *J. Chem. Phys.*, **76**, 1459 (1982).
        (h)   R. E. Overill and M. F. Guest, *Mol. Phys.*, **41**, 119 (1980).

10. Present chapter, (a) Ref. 14 and Section XIV.II.a; (b) III.A; (c) XIV.IX.a; (d) I.B; (e) I.D; (f) XI; (g) XII.B; (h) IV.A.c; (i) IV.C; (j) V.A; (k) IV.A.b; (l) III.H.b.

11. K. Toriyama, M. Iwasaki, K. Nunome, and H. Muto, "An ESR and ENDOR study of tunneling rotation of a hindered CH_3 group in C_2H_5 radicals trapped in xenon matrices at 4.2 K: Environment effect on internal rotation," *J. Chem. Phys., 75*, 1633 (1981).

12. G. L. Pratt and D. Rogers, *J. Chem. Soc. Faraday I*, 1694 (1980).

13. (a) See also refs. 9b, 9d, 13b–d.
    (b) J. Pacansky and H. Coufal, *J. Mol. Struct., 60*, 255 (1980).
    (c) J. Pacansky and H. Coufal, *J. Chem. Phys., 71*, 2811 (1979).
    (d) J. Pacansky and J. S. Chang, *J. Chem. Phys., 74*, 5539 (1981).
    (e) An earlier communication,[13f] in which it was concluded on the basis of a "preliminary"[9f] analysis that "the electronic structure of the CH_2 group in the ethyl radical probably exists in an sp^2 configuration," was discussed earlier: "A comparison of [the] . . . absorption bands" of "a transient molecule, reasonably considered to be $CH_3CH_2\cdot$" "with those of $CH_2 = CD_2 \ldots$ combined with the reasoning that similarity of force constants indicates similarity of electronic structure, which indicates similarity of bond angles and distances, leads to a consistent assignment to planar $MeCH_2\cdot$ as the carrier."[5b]
    (f) J. Pacansky, G.P. Gardini, and J. Bargon, *J. Am. Chem. Soc., 98*, 2665 (1976).

14. In the text and tables, "ethyl radical" is represented as CH_3CH_2. However, on the composite spectra it is represented as $CH_3CH_2 \cdot \overset{O\ O}{\underset{O\ O}{C\ C}} \cdot CH_2CH_3$. This uncertainty, in representation and in reality, is not unimportant: Pyramidal bending/internal rotation barriers (see above) are low, and hence the structure would be anticipated to be very susceptible to "matrix effects" and other aspects of the environment of the radical (see also ref. 6c).

15. (a) Lest this be thought to be nearly planar, recall that it is about one-third of the way to being "tetrahedral."
    (b) The structure in terms of which the experimental results were discussed was changed,[13e] apparently in response to the outcome of the calculation. This is the converse of the situation commented upon by Symons.[4b]

16.* (a) See also ref. 16b.
    (b) J. Pacansky, J. S. Chang, and D. W. Brown, *Tetrahedron, 38*, 257 (1982).

17.* (a) D. S. Bethune, J. R. Lankard, P. P. Sorokin, A. J. Schell-Sorokin, R. M. Plecenik, and Ph. Avouris, *J. Chem. Phys., 75*, 2231 (1981).
    (b) Ph. Avouris, D. S. Bethune, J. R. Lankard, P. P. Sorokin, and A. J. Schell-Sorokin, *J. Photochem., 17*, 227 (1981).
    (c) Compare with the frequencies of 1825 and 2931 cm^{-1} in an argon matrix at low temperature.[13d, 16a] The similarities are mutually supportive with regard to assignment of the carrier.

18.* (a) T. G. DiGiuseppe, J. W. Hudgens, and M. C. Lin, *J. Chem. Phys., 76*, 3337 (1982).
    (b) T. G. DiGiuseppe, J. W. Hudgens, and M. C. Lin, *J. Phys. Chem., 86*, 36 (1982).
    (c) A. K. Mal'tsev, V. A. Korolov, and O. M. Nefedov, *Izv. Akad. Nauk SSSR*, 2415 (1982).

19.* (a) L. B. Knight, Jr. and J. Steadman, *J. Chem. Phys., 77*, 1750 (1982).

(b)   N. D. Piltch, P. G. Szanto, T. G. Anderson, C. S. Gudeman, T. A. Dixon, and R. C. Woods, *J. Chem. Phys.*, **76**, 3385 (1982).

20.   A. Carrington, D. R. J. Milverton, and P. J. Sarre, *Mol. Phys.*, **35**, 1505 (1978).

21.   (a)   J. C. Vedrine and C. Naccache, *Chem. Phys. Lett.*, **18**, 190 (1973).
      (b)   See also refs. 21c, 21d.
      (c)   N. D. Chuvylkin, G. M. Zhidomirov, and V. B. Kazansky, *Chem. Phys. Lett.*, **26**, 180 (1974).
      (d)   D. P. Vercauteren, J.-M. André, and E. G. Derouane, *Chem. Phys. Lett.*, **36**, 322 (1975). With regard to this reference, see ref. 4b.

22.   (a)   W. C. Easley and W. Weltner, Jr., *J. Chem. Phys.*, **52**, 197 (1970).
      (b)   E. L. Cochran, F. J. Adrian, and V. A. Bowers, *J. Chem. Phys.*, **36**, 1938 (1962).
      (c)   F. J. Adrian and V. A. Bowers, *Chem. Phys. Lett.*, **41**, 517 (1976).
      (d)   F. Owens, R. A. Breslow, and O. R. Gilliam, *J. Chem. Phys.*, **54**, 833 (1971).

23.   (a)   L. B. Knight, Jr., W. C. Easley, and W. Weltner, Jr., *J. Chem. Phys.*, **54**, 1610 (1971).
      (b)   L. B. Knight, Jr., M. B. Wise, E. R. Davidson, and L. E. Murchie, *J. Chem. Phys.*, **76**, 126 (1982).

24.   L. B. Knight, Jr., R. L. Martin, and E. R. Davidson, *J. Chem. Phys.*, **71**, 3991 (1979).

25.   L. B. Knight, Jr., M. B. Wise, A. G. Childers, E. R. Davidson, and W. R. Daasch, *J. Chem. Phys.*, **73**, 4198 (1980).

26.   L. B. Knight, Jr. and W. Weltner, Jr., *J. Chem. Phys.*, **54**, 3875 (1971).

27.* C. Huggenberger, J. Lipscher, and H. Fischer, *J. Phys. Chem.*, **84**, 3467 (1980).

28.   (a)   H. Murai and K. Obi, *J. Phys. Chem.*, **79**, 2446 (1975).
      (b)   U. Schmidt, K. H. Kabitzke, and K. Markau, *Angew. Chem. Int. Ed. Eng.*, **4**, 355 (1965).

29.* Y. Takemura and T. Shida, *J. Chem. Phys.*, **73**, 4133 (1980).

30.* (a)   C. H. Wu and H. R. Ihle, *Chem. Phys. Lett.*, **61**, 54 (1979).
      (b)   J. Chandrasekhar, J. A. Pople, R. Seeger, U. Seeger, and P. v. R. Schleyer, *J. Am. Chem. Soc.*, **104**, 3651 (1982).

31.   (a)   Yu. N. Kozlov, A. P. Moravskii, A. P. Purmal', and V. F. Shuvalov, *Zh. Fiz. Khim.*, **55**, 764 (1981).
      (b)   S. V. Rykov, E. D. Skakovskii, and E. A. Turetskaya, *Zh. Prikl. Spektrosk.*, **31**, 1079 (1979).

32.* (a)   J. T. Wang and F. Williams, *J. Am. Chem. Soc.*, **103**, 6994 (1981).
      (b)   H. Kubodera, T. Shida, and K. Shimokoshi, *J. Phys. Chem.*, **85**, 2583 (1981).
      (c)   D. N. R. Rao and M. C. R. Symons, *Chem. Phys. Lett.*, **93**, 495 (1982).
      (d)   However, in an earlier publication an esr spectrum is assigned to $CH_3SOCH_3^{+\cdot}$, and it is stated that the observed "proton coupling of *ca.* 12 G is reasonable [for $CH_3SOCH_3^{+\cdot}$]."[32e]
      (e)   M. C. R. Symons, *J. Chem. Soc. Perkin II*, 908 (1976).
      (f)   T. Shida, H. Nakatsuji; K. Ota, and K. Ushida, *'81-nen Bunshi Kozo Touronkai Yoshishu ('81 Molecular Structure Colloq.)*, Abstr. 1D16, p. 138.

33.   (a)   M. C. R. Symons and B. W. Wren, *J. Chem. Soc., Chem. Commun.*, 817 (1982).
      (b)   L. D. Snow, J. T. Wang, and F. Williams, *J. Am. Chem. Soc.*, **104**, 2062 (1982).

(c) K. Ushida and T. Shida, *J. Am. Chem. Soc.*, **104**, 7332 (1982).

(d) P.-O. Samskog, L. D. Kispert, and A. Lund, *J. Chem. Phys.*, **77**, 2330 (1982).

(e) D. N. R. Rao, M. C. R. Symons, and B. W. Wren, *Tetrahedron Lett.*, **23**, 4739 (1982).

(f) T. Shida, K. Ushida, and H. Kubodera, *20th ESR Symp.*, Osaka, Oct. 6–8, 1981, Abstr. 8A06, p. 109.

34.   (a) J. T. Wang and F. Williams, *J. Chem. Soc., Chem. Commun.*, 1184 (1981). Corrigendum: 980 (1983).

(b) R. L. Petersen, D. J. Nelson, and M. C. R. Symons, *J. Chem. Soc. Perkin II*, 225 (1978).

(c) W. B. Gara, J. R. M. Giles, and B. P. Roberts, *J. Chem. Soc. Perkin II*, 1444 (1979).

(d) Y. L. Chow and K. Iwai, *J. Chem. Soc. Perkin II*, 931 (1980).

(e) K.-D. Asmus, in G. E. Adams, E. M. Fielden, and B. D. Michael, Eds., *Fast Processes in Radiation Chemistry and Biology*, Wiley, New York, 1975, p. 40.

(f) K.-D. Asmus, D. Bahnemann, M. Bonafačić, and H. A. Gillis, *Faraday Disc. Chem. Soc.*, **63**, 213 (1977).

(g) E. Janata, D. Veltwisch, and K.-D. Asmus, *Radiat. Phys. Chem.*, **16**, 43 (1980).

35.   C. Chatgilialoglu, K. U. Ingold, J. C. Scaiano, and H. Woynar, *J. Am. Chem. Soc.*, **103**, 3231 (1981).

36.[*]   J. T. Wang and F. Williams, *J. Am. Chem. Soc.*, **102**, 2860 (1980).

37.   (a) The same argument was presented, without being labeled as a microscopic reversibility argument, and elaborated upon in ref. 37b.

(b) J. T. Wang and F. Williams, *Chem. Phys. Lett.*, **72**, 557 (1980).

(c) The principle of microscopic reversibility is being applied to observations to which it is not applicable. What Symons is trying to explain has apparently been misconceived.[37d] A thesis is set up, and then, on the basis of the principle of microscopic reversibility, it is promptly destroyed. This thesis, Symons' explanation as applied[36,37b] to the "positive affinity or directional tendency of a chemical reaction," that is, to the position of an equilibrium, is unfortunately attributed to Symons. That thesis is indeed "palpably false." In common jargon, a transition state effect alone indeed cannot be invoked to explain a change in the position of an equilibrium because, according to the principle of microscopic reversibility, it would affect equally the rates of forward and reverse reactions. The transition state, and anything about it, is simply irrelevant. Symons really applied his explanation to the variation with R of the rate of decomposition of RX^{-}.[37d]

(d) This statement is neutral as to the merits of Symons' arguments as an explanation of what he *is* trying to explain; we have discussed that elsewhere. It is neutral also as to the degree to which that Symons is trying to explain corresponds to what has been observed.

38.   (a) His original position, against which we argued too delicately,[6g] has since been characterized as "intuitively reasonable."[37b]

(b) See our discussion in ref. 6f.

(c) ". . . in most well defined σ^* radicals in which the excess electron is largely confined to carbon–halogen σ^* orbitals, the radical component, R·, has a structure such that there is little change in shape and in carbon s-p orbital hybrizidation on dissociation. . . . Clearly, when there is a change from s-p hybridization towards pure p-orbitals, the tendency to dissociate must be enhanced. This is not, of course [38d], the only factor involved, as is illustrated by the e.s.r. data for $[(CF_3)_3C \cdot I]^-$ anions, which clearly have a σ^* structure [36], despite the fact that $\cdot C(CF_3)_3$ radicals are planar [38e] at the central carbon atom. . . . Another major factor contributing to the extreme instability of the alkyl halide anions is the large difference in electronegativity between carbon and the halogens."[38h] "Obviously [38d], there are several other factors [in addition to the degree of change in shape and in orbital hybridization] that need to be taken into account, a major one being the way in which the three electrons are disposed within the carbon—halogen bond for any potential σ^* anion. . . . Clearly [38d] my argument, based on shape change and orbital re-hybridization becomes less and less significant as the electron affinity of the R group . . . increases."[38i]

"I suggest that there are two major factors, working together. One is . . . the effective electronegativity difference of the bonding atoms . . . and the other is . . . [the shape/hybridization argument]."[38j]

"One reason . . . [is the shape/hybridization argument]. . . . Another contributory factor is the . . . electron affinity of . . . [R]."[38k,l]

". . . there are three factors involved: (1) [the shape argument]. (2) . . . as a radical becomes planar, its tendency to form covalent bonds falls . . . (3) . . . the greater the electronegativity of the groups R-, the greater the stability of the σ^* anion, and the less significant will be the potential changes in the configuration of the group R-."[38l,m]

(d) Of course!? Obviously!? Clearly!? Until now, in four publications submitted over a period of at least 15 months, although it was implied that other factors were operative, the shape/hybridization factor was the only one invoked.

(e) Symons is the last player in a game of "telephone." It was concluded that $\cdot C(CF_3)_3$ has "a more planar geometry" than $\cdot C(CH_3)_3$, to which a "nearly tetrahedral equilibrium geometry" was attributed.[38f] It was next stated that "the $C(CF_3)_3$ radical is judged to be nearly planar on the basis of . . ." and ref. 38f is referenced.[36,38g] Symons next goes on to state "the fact that $\cdot C(CF_3)_3$ radicals are planar" in the same sentence as he refers to ref. 36.

(f) P. J. Krusic and R. C. Bingham, *J. Am. Chem. Soc.*, **98**, 230 (1976).

(g) It is unclear what is being referenced, the judgment or the basis of the judgment.

(h) M. C. R. Symons, *J. Chem. Res. (S)*, 160 (1981).

(i) M. C. R. Symons, *Pure Appl. Chem.*, **53**, 223 (1981).

(j) M. C. R. Symons, *Proc. 6th Intl. Congr. Radiat. Res.*, Japanese Assoc. of Radiation Res. (Toppan Printed, Tokyo), 1979, p. 238.

(k) M. C. R. Symons, *Radiat. Phys. Chem.*, **15**, 453 (1980).

(l) See also reference 37d of ref. 6 g.

(m) M. C. R. Symons, *Chem. Phys. Lett.*, **72**, 559 (1980).

(n) In the context of a discussion of electron trapping and hole trapping an apparently contrasting view was presented [M. C. R. Symons, *Ultramicroscopy*, **10**, 15(1982)]: "Chemical trapping may give an electron excess center, in which case trapping is only effective if the acceptor relaxes extensively on electron addition . . . RCO_2^- . . . is . . . locally planar, but electron capture gives $R\dot{C}O_2^{2-}$, which is pyramidal, so the electron remains trapped. . . . Very similar considerations apply [to hole trapping]. To fix AB^+ a shape change [from AB] is required."

39. (a) See Ref. 6i for notation.

(b) For example: "In my view, the fact that these adducts [between radicals and halide ions, i.e., $R \cdot /X^-$] are stable at low temperatures rules out the possibility of stable intermediate σ^* radicals $[RX^{-\cdot}]$ – since the reactants are in contact and there are no obvious barriers to reaction, why doesn't it occur?"[38i] He is saying that there cannot be an intermediate in a dissociative process in any phase, that is, that there cannot be more than one energy minimum, exclusive of that of the freely dissociated products, on the corresponding free energy surface. For example, if $RX^{-\cdot}$ and $(R \cdot + X^-$, statistically free of each other) exist, then $R \cdot /X^-$ cannot; if $R \cdot /X^-$ and $(R \cdot + X^-)$ exist, then $RX^{-\cdot}$ cannot. However, "the existence of a double-minimum potential [against which he is arguing] is not prohibited a priori]"[6h] If Professor Symons maintains that it is prohibited, that it can be ruled out at the start, then he has the responsibility to demonstrate the validity of his contention in a way more substantial than simply by stating that it is not obvious to him how he could be mistaken.

40. I. G. Smith and M. C. R. Symons, unpublished results, cited in (a) ref. 38j; (b) ref. 38k.

41. (a) M. C. R. Symons and I. G. Smith, J. Chem. Soc. Faraday I, 2701 (1981).

(b) M. C. R. Symons and I. G. Smith, J. Chem. Soc. Perkin II, 1180 (1981).

42. (a) J. F. Hemidy and A. J. Tench, J. Catal., 68, 17 (1981).

(b) M. Anpo and Y. Kubokawa, J. Catal., 75, 204 (1982).

43. (a) R. W. Yip, Y. L. Chow, and C. Beddard, "Flash Photolysis of N-Bromo-3,3-dimethylglutarimide: Detection of the Imidyl Radical," J. Chem. Soc., Chem. Commun., 955 (1981).

(b) R. L. Tlumak, J. C. Day, J. P. Slanga, and P. S. Skell, "Excited-State σ Succinimidyl and Glutarimidyl Radicals: Reversible Ring Opening," J. Am. Chem. Soc., 104, 7257 (1982).

(c) R. L. Tlumak and P. S. Skell, "Reactions of a Graded Set of Radicals with N-Bromosuccinimide; Two Transition States," J. Am. Chem. Soc., 104, 7267 (1982).

44. Since we no longer view the title observations as being novel, [45a] we will not discuss work merely because it is in that category but will apply the additional criterion that the radicals and chemistry involved be important and significant in free radical chemistry. We believe that $RCO_2 \cdot$ (see below), but not amidyl or imidyl, meet this criterion.

45. (a) As different electronic states in their respective energy minima can have the same structure[45b] only coincidentally and as different structures, each at an energy minimum, can generally have the same energy only coincidentally, terms such as "different electronic states" and "different structural isomers" can be equivalent operationally. Which language is used seems to be determined largely by custom, by the genesis and chemistry of the material, and by the magnitude of the difference in structure. In principle, however, cis and trans isomers, for example, could be described as different electronic states (which, of course, have structures that are not identical, i.e., are different) and the "ground state" and "first excited singlet state" of a compound, for example, could be described as different structural isomers (which, of course, have different energies) of that compound. Energy and structure are inseparable, each being a valid, and incomplete, descriptor of a molecule. Once this is

recognized, it must also be recognized (which we didn't at the time) that there need not be anything new in principle about the type of observations discussed earlier,[6j] that is, about two structural isomers (which, of course, have different energies) being intermediates in similar reactions.

(b) In the term "structure," we are including bond lengths and angles, that is, we are referring to a quantitative description, not merely the formal qualitative relative dispositions of the component atoms of a molecule.

46. For example, that a particular proposed intermediate would not behave identically in different systems (chemical environments, i.e, the nature and concentrations of the various stable and transient compounds) is ignored.

47.* (a) B. Ashworth, B. C. Gilbert, R. G. G. Holmes, and R. O. C. Norman, *J. Chem. Soc. Perkin II*, 951 (1978).
(b) R. O. C. Norman, *Chem. Soc. Revs.*, **8**, 1 (1979).
(c) B. C. Gilbert and R. O. C. Norman, *Can. J. Chem.*, **60**, 1379 (1982).

48. S. A. Glover, S. L. Golding, A. Goosen, and C. W. McCleland, "Intramolecular Cyclizations of Biphenyl-2-carboxyl Radicals: Evidence for a Π-State Aroyloxyl Radical," *J. Chem. Soc. Perkin I*, 842 (1981).

49.* D. D. May and P. S. Skell, *J. Am. Chem. Soc.*, **104**, 4500 (1982).

50. (a) In a seemingly incidental footnote[49], an error in an earlier publication[50b] is pointed out. Nothing is said about the consequences of this error which, in actuality, devastates the principal thrust of the earlier report. Therefore, we reluctantly refrain from discussing it.
(b) P. S. Skell and D. D. May, "π- and σ-Acetoxy Radicals," *J. Am. Chem. Soc.*, **103**, 967 (1981).

51. It must first be demonstrated that the ratio of bromobutanes in the product is equal to the selectivity for formation of 1- and 2-bromobutanes (contrary to their statement, it is the former, not the latter, that is reported). This requires a demonstration that the bromobutanes are stable to the reaction conditions. Then it must be demonstrated that the selectivity for formation of $CH_3CH_2CH_2CH_2Br$ and $CH_3CH_2CHBrCH_3$ is equal to the selectivity for formation of $CH_3CH_2CH_2CH_2 \cdot$ and $CH_3CH_2\dot{C}HCH_3$, that is, either that the butyl radicals "quantitatively" go on to bromobutane or that the selectivities with which they go to bromobutane and enter into side reactions are the same for both butyl radicals. At a minimum, good material balances are needed. How good? In general, the material not accounted for should be much less than the smallest amount of product whose quantity enters into an argument in an important way. In the present case, that amount of product is the quantity of 1-bromobutane, which varies from approximately 1 to 4% of the amount of RCO_2Br as R is varied. Since the ratio of 2- to 1-bromobutane in the presence of Br_2 is about twice that in the presence of $Cl_2C = CH_2$, doubling (or halving) the amount of 1-bromobutane would equalize the ratios, that is, a change in the amount of one product by an amount equivalent to a few percent of the amount of RCO_2Br (a few hundredths or tenths of a percent of the amount of butane, which was present in an amount $10^1 - 10^2$ times that of RCO_2Br) would negate the conclusions drawn. So, what material balances were obtained? Nothing is said about the butane and 5–10% of the RCO_2Br is not accounted for.

52. (a) What about $R \cdot$, $Br \cdot$, and $\cdot CCl_2F$?
(b) H-abstraction by $R \cdot$ is disputed reasonably by D. D. May, Ph.D. thesis, Pennsylvania State University, 1982, p. 12, and that by $Br \cdot$, free or complexed, is disputed weakly.

53.   (a) W. Tsang, in A. Lifshitz, Ed., *Shock Waves in Chemistry*, Dekker, New York, 1981, p. 90.
       (b) Reference 55a in ref. 6k.

54.[*] (a) References 55c, in ref. 61.
       (b) R. R. Baldwin, M. W. M. Hisham, A. Keen, and R. W. Walker, *J. Chem. Soc. Faraday I*, 1165 (1982).

55.[*] R. R. Baldwin, R. W. Walker, and R. W. Walker, *J. Chem. Soc. Faraday I*, 825 (1980).

56.[*] C. E. Canosa and R. M. Marshall, *Int. J. Chem. Kinet.*, **13**, 303 (1981).

57.   (a) Reference 7b in ref. 5d.
       (b) Reference 55c in ref. 61.
       (c) R. Hiatt and S. W. Benson, *Int. J. Chem. Kinet.*, **5**, 385 (1973).
       (d) R. Hiatt and S. W. Benson, *J. Am. Chem. Soc.*, **94**, 25 (1972).
       (e) R. Hiatt and S. W. Benson, *J. Am. Chem. Soc.*, **94**, 6886 (1972).
       (f) D. F. McMillen, D. M. Golden, and S. W. Benson, *J. Am. Chem. Soc.*, **94**, 4403 (1972).
       (g) Reference 55f in ref. 6m.
       (h) R. M. Marshall, J. H. Purnell, and P. D. Storey, *J. Chem. Soc. Faraday I*, 85 (1976).
       (i) It is argued, unconvincingly, that neither of these dependencies can introduce a significant error.

58.[*] D. A. Robaugh, B. D. Barton, and S. E. Stein, *J. Phys. Chem.*, **85**, 2378 (1981).

59.   J.-S. Wang and J. L. Franklin, *Int. J. Mass Spectrom. Ion Phys.*, **36**, 249 (1980).

60.[*] (a) A. L. Castelhano, P. R. Marriott, and D. Griller, *J. Am. Chem. Soc.*, **103**, 4262 (1981).
       (b) A. L. Castelhano and D. Griller, *J. Am. Chem. Soc.*, **104**, 3655 (1982).

61.   Not Professor Doering. In 1964 (!), when presented by us with the possibility that the entire network of heats of formation of free radicals might be wrong, his response was, "That wouldn't surprise me one bit."

62.[*] W. von E. Doering, *Proc. Natl. Acad. Sci. USA*, **78**, 5279 (1981).

63.   W. Tsang, National Bureau of Standards Special Publication **572**, 43 (1980).

64.   (a) Yu. P. Yampol'skii and V. V. Zelentsov, *React. Kinet. Catal Lett.*, **17**, 347 (1981).
       (b) S. W. Benson, *Thermochemical Kinetics*, 2nd ed., Wiley, New York, 1976, p. 299.

65.[*] (a) Yu. N. Molin, O. A. Anisimov, V. M. Grigoryants, V. K. Molchanov, and K. M. Salikhov, *J. Phys. Chem.*, **84**, 1853 (1980).
       (b) A. D. Trifunac and R. G. Lawler, *Mag. Res. Rev.*, **7**, 147 (1982).
       (c) A. B. Doktorov, O. A. Anisimov, A. I. Burshtein, and Yu. N. Molin, *Chem. Phys.*, **71**, 1 (1982).
       (d) O. A. Anisimov, V. M. Grigoryants, V. K. Molchanov, and Yu. N. Molin, *Dokl. Akad. Nauk SSSR*, **248**, 380 (1979).
       (e) J. P. Smith and A. D. Trifunac, *J. Phys. Chem.*, **85**, 1645 (1981).
       (f) A. D. Trifunac and J. P. Smith, *Chem. Phys. Lett.*, **73**, 94 (1980).
       (g) Yu. N. Molin, *Priroda*, No. 11, 24 (1981).
       (h) O. A. Anisimov, V. M. Grigoryants, V. K. Molchanov, and Yu. N. Molin, *Chem. Phys. Lett.*, **66**, 265 (1979).

288　　　Leonard Kaplan

(i)　Yu. N. Molin, *Abstr. Lect. Posters 4th Spec. Colloq. Ampere Dyn. Processes Mol. Syst. Stud. rf Spectrosc.*, 64 (1979).

(j)　O. A. Anisimov, V. M. Grigoryants, and Yu. N. Molin, *Chem. Phys. Lett.*, **74**, 15 (1980).

(k)　O. A. Anisimov, V. M. Grigoryants, and Yu. N. Molin, *Pis'ma Zh. Eksp. Teor. Fiz.*, **30**, 589 (1979).

(l)　A. D. Trifunac and J. P. Smith, in J. H. Baxendale and F. Busi, Eds., *The Study of Fast Processes and Transient Species by Electron Pulse Radiolysis* (*NATO Adv. Study Ser.*, **C86**). D. Reidel Publishing Co., 1982, p. 179.

(m)　J. P. Smith, S. Lefkowitz, and A. D. Trifunac, "Observation of Short-Lived Radical Ion Pairs by Pulse Radiolysis of Alkane Solutions by Time-Resolved Fluorescence-Detected Magnetic Resonance. Electron Paramagnetic Resonance Spectra of Alkane Radical Cations," *J. Phys. Chem.*, **86**, 4347 (1982).

(n)　J. P. Smith and A. D. Trifunac, *Chem. Phys. Lett.*, **83**, 195 (1981).

(o)　Yu. N. Molin, *Vestn. Akad. Med. Nauk SSSR*, 12 (1981).

(p)　J. P. Smith and A. D. Trifunac, unpublished work cited in ref. 65b.

(q)　V. M. Grigoryants, O. A. Anisimov, and Yu. N. Molin, *Zh. Strukt. Khim.*, **23**, 4 (1982).

(r)　O. A. Anisimov, V. M. Grigoryants, V. I. Korsunskii, V. I. Melekhov, and Yu. N. Molin, *Dokl. Akad. Nauk SSSR*, **260**, 1151 (1981).

(s)　M. K. Bowman, D. E. Budil, G. L. Closs, A. G. Kostka, C. A. Wraight and J. R. Norris, *Proc. Natl. Acad. Sci. USA*, **78**, 3305 (1981).

66.　(a)　With regard to the assignment it should be noted that, during the electron-irradiation of C_6F_6 in squalane, a "$\approx$ 18 G wide line . . . supposedly associated with the recombination of the ion-radical pairs resulting from the capture of electrons and holes by the aromatic impurities in squalane" was observed.[65j] "Another factor responsible for this line is perhaps the signal from a cation (squalane⁺) affording an exchange-narrowed line due to a fast hole transfer between the solvent molecules."[65j,b]

　　　(b)　See also ref. 65m.

67.　See also ref. 65n.

68.　"Since the quantum yield of C_6F_6 fluorescence is low (4%), it was difficult to detect the ESR spectra . . . by optical means. To enhance the luminescence signal, we added some anthracene whose luminescence quantum yield is rather high (36%)."[65j]

69.　Also observed in the corresponding experiments were (a) a broad line attributed to $C_6F_6^{-\cdot}$, and (b) a single line ($w_{1/2} \sim 5$ G; $w_{1/2} \sim 14$ G when perprotioanthracene was used) attributed to (perdeuteroanthracene)⁺·/(perdeuteroanthracene)⁻·

70.　See also ref. 65s.

71.*　S. V. Vosel', P. A. Purtov, and K. M. Salikhov, *Teor. Eksp. Khim.*, **17**, 406 (1981).

72.*　(a)　R. Z. Sagdeev, Yu. A. Grishin, A. Z. Gogolev, A. V. Dushkin, A. G. Semenov, and Yu. N. Molin, *Zh. Strukt. Khim.*, **20**, 1132 (1979).

　　　(b)　R. H. D. Nuttall and A. D. Trifunac, *Chem. Phys. Lett.*, **81**, 151 (1981).

　　　(c)　R. H. D. Nuttall and A. D. Trifunac, *J. Phys. Chem.*, **86**, 3963 (1982).

　　　(d)　A. D. Trifunac and W. T. Evanochko, *J. Am. Chem. Soc.*, **102**, 4598 (1980).

　　　(e)　A. D. Trifunac, in J. H. Baxendale and F. Busi, Eds., *The Study of Fast Processes and Transient Species by Electron Pulse Radiolysis* (*NATO Adv. Study Inst. Ser.*, **C86**), D. Reidel Publishing Co., 1982, p. 347.

73.* Yu. A. Grishin, A. Z. Gogolev, E. G. Bagryanskaya, A. V. Dushkin, R. Z. Sagdeev, and Yu. N. Molin, *Dokl. Akad. Nauk SSSR*, **255**, 1160 (1980).

74.* H. Coufal, *Appl. Phys. Lett.*, **39**, 215 (1981).

75.* (a) P. R. Marriott, A. L. Castelhano, and D. Griller, *Can. J. Chem.*, **60**, 274 (1982).
   (b) C. Huggenberger and H. Fischer, *Helv. Chim. Acta*, **64**, 338 (1981).

76. (a) D. A. Parkes, "The Ultraviolet Absorption Specta of the Acetyl Radical and the Kinetics of the CH_3 + CO Reaction at Room Temperature," *Chem. Phys. Lett.*, **77**, 527 (1981).
   (b) C. Anastasi and P. R. Maw, "Reaction Kinetics in Acetyl Chemistry over a Wide Range of Temperature and Pressure," *J. Chem. Soc. Faraday I*, 2423 (1982).

77. (a) For the generation of R• by means of the reaction Ph• + RI, see refs. 77b–g.
   (b) L. Kaplan, *J. Chem Soc., Chem. Commun.*, 754 (1968).
   (c) R. F. Drury and L. Kaplan, *J. Am. Chem. Soc.*, **95**, 2217 (1973).
   (d) L. Kaplan, *J. Am. Chem. Soc.*, **89**, 1753 (1967).
   (e) L. Kaplan, *J. Am. Chem. Soc.*, **89**, 4566 (1967).
   (f) L. Kaplan, *J. Org. Chem.*, **32**, 4059 (1967).
   (g) L. Kaplan, *J. Org. Chem.*, **33**, 2531 (1968).

78.* (a) C. A. Morgan, M. J. Pilling, J. M. Tulloch, R. P. Ruiz, and K. D. Bayes, *J. Chem. Soc. Faraday II*, 1323 (1982).
   (b) R. P. Ruiz, K. D. Bayes, M. T. Macpherson, and M. J. Pilling, *J. Phys. Chem.*, **85**, 1622 (1981).
   (c) L. A. Khachatryan, O. M. Niazyan, A. A. Mantashyan, V. I. Vedeneev, and M. A. Teitel'boim, *Int. J. Chem. Kinet.*, **14**, 1231 (1982).

79. S. P. Heneghan, P. A. Knoot, and S. W. Benson, *Int. J. Chem. Kinet.*, **13**, 677 (1981).

80. D. Bruck, R. Dudley, C. A. Fyfe, and J. Van Delden, *J. Magn. Res.*, **42**, 51 (1981).

81. (a) H. M. Maurer and J. Bargon, *Org. Magn. Res.*, **13**, 430 (1980).
   (b) H. M. Maurer, G. P. Gardini, and J. Bargon, *J. Chem. Soc., Chem. Commun.*, 272 (1979).

82.* D. H. R. Barton, H. A. Dowlatshahi, W. B. Motherwell, and D. Villemin, *J. Chem. Soc., Chem. Commun.*, 732 (1980).

83. (a) Examples: $\supset PCH_2\dot{C}HR$ + $\supset PH$ → $\supset PCH_2CH_2R$ + $\supset P$• "occurs with considerable ease"[83b]; ". . . the P-H hydrogen of . . . alkyl phosphines participate in chain transfer reactions with great ease . . ."[83c]; "the phosphines . . . [are] active transfer agents . . . ."[83d]
   (b) M. M. Rauhut, H. A. Currier, A. M. Semsel, and V. P. Wystrach, *J. Org. Chem.*, **26**, 5138 (1961).
   (c) C. M. Starks, *Free Radical Telomerization*, Academic, 1974, p. 198.
   (d) J. Pellon, *J. Polymer Sci.*, **43**, 537 (1960).
   (e) H. G. Kuivila and G. F. Smith, *J. Org. Chem.*, **45**, 2918 (1980).
   (f) M. S. Alnajjar and H. G. Kuivila, *J. Org. Chem.*, **46**, 1053 (1981).
   (g) G. F. Smith, H. G. Kuivila, R. Simon, and L. Sultan, *J. Am. Chem. Soc.*, **103**, 833 (1981).
   (h) M. Newcomb and M. G. Smith, Office of Naval Research Technical Report, Contract N00014-79-C-0584, Task No. NR 053-714, report no. 2 (1981), NTIS order no. AD-A105433.

(i) See also ref. 83j.

(j) E. C. Ashby, J. N. Argyropoulos, G. R. Meyer, and A. B. Goel, *J. Am. Chem. Soc.*, **104**, 6788 (1982).

(k) E. C. Ashby and R. DePriest, *J. Am. Chem. Soc.*, **104**, 6144 (1982).

(l) H. G. Kuivila and M. S. Alnajjar, *J. Am. Chem. Soc.*, **104**, 6146 (1982).

84. (a) See also ref. 84b.

(b) R. Bolze, H. Eierdanz, K. Schlüter, W. Massa, W. Grahn, and A. Berndt, "Tetra-*tert*-butylallene and Its Radical Cation," *Angew. Chem. Int. Ed. Eng.*, **21**, 924 (1982).

85.* C. Chatgilialoglu, K. U. Ingold, and J. C. Scaiano, *J. Am. Chem. Soc.*, **103**, 7739 (1981).

86. This concern is perhaps more personal than scientific, as we have come to be wary, rather than enamored, of apparent good fortune.

87.* (a) A. V. Dushkin, A. V. Yurkovskaya, and R. Z. Sagdeev, *Zh. Fiz. Khim.*, **53**, 2643 (1979).

(b) A. V. Dushkin, A. V. Yurkovskaya, and R. Z. Sagdeev, *Chem. Phys. Lett.*, **67**, 524 (1979).

88. (a) A common misconception is conveyed in the reports of this work: ". . . CIDNP arises in liquid-phase radical processes, appreciable cage effects being a necessary condition to observe it."[88b] Similar statements are made in ref. 87. The condition for the observation of CIDNP, as distinct from the conditions under which it has thus far been observed, is not that recombination compete with diffusion; that is just a way in which the condition can be met. At one extreme, the condition is that recombination compete with a process that does not depend on spin. More generally, the condition is that recombination compete with a process that depends on spin in a way different from the way in which recombination depends on spin.

(b) R. Z. Sagdeev, Yu. N. Molin, K. M. Salikhov, Yu. A. Grishin, A. V. Dushkin, and A. Z. Gogolev, *Bull. Magn. Reson.*, **2**, 66 (1981).

89.* (a) V. L. Berdinskii, A. L. Buchachenko, and A. D. Pershin, *Teor. Eksp. Khim.*, **12**, 666 (1976).

(b) A. G. Zhuravelev, V. L. Berdinskii, and A. L. Buchachenko, *Pis'ma Zh. Eksp. Teor. Fiz.*, **28**, 150 (1978).

(c) A. L. Buchachenko and V. L. Berdinskii, *Vestn. Akad. Nauk SSSR*, No. 1, 91 (1981).

(d) V. L. Berdinskii and A. L. Buchachenko, *Zh. Fiz. Khim.*, **55**, 1827 (1981).

(e) V. L. Berdinskii and A. L. Buchachenko, *Zh. Fiz. Khim.*, **54**, 676 (1980).

90. (a) Mental telepathy? Thought control? Psychokinesis?

(b) See also refs. 10h and 10k.

91. CAUTION: In this section, I am uncritically and superficially summarizing what I have read.

92. (a) For review and general discussion oriented to chemistry, see refs. 92b–d and 93.

(b) J. H. Brewer, D. G. Fleming, and P. W. Percival, in A. G. Marshall, Ed., *Fourier Hadamard, and Hilbert Transforms in Chemistry*, Plenum, 1982, p. 345.

(c) D. Richter, in Kaufman and Shenoy, Eds., *Nuclear and Electron Resonance Spectroscopies Applied to Materials Science*, Elsevier North Holland, 1981, p. 233.

(d) D. W. Cooke, *A. I. P. Conf. Proc. Polariz. Phenom. Nucl. Phys.-1980*, Pt. 2, **69**, 1057 (1981).

93.[*] (a) D. G. Fleming, in S. Datz, Ed., *Physics of Electronic and Atomic Collisions*, Elsevier North-Holland, 1982, p. 297.

(b) P. W. Percival, *Radiochim. Acta*, **26**, 1 (1979).

(c) D. G. Fleming, D. M. Garner, L. C. Vaz, D. C. Walker, J. H. Brewer, and K. M. Crowe, in H. J. Ache, Ed., *Positronium and Muonium Chemistry*, A. C. S. Adv. Chem. Ser. **175**, 1979, p. 279.

(d) E. Roduner and H. Fischer, *Chem. Phys.*, **54**, 261 (1981).

(e) D. C. Walker, *J. Phys. Chem.*, **85**, 3960 (1981).

(f) A. Hill, G. Allen, G. Stirling, and M. C. R. Symons, *J. Chem. Soc. Faraday I*, 2959 (1982).

(g) H. Fischer, *Chimia*, **35**, 85 (1981).

(h) B. Lévay, *Atomic Energy Review*, **17**, 413 (1979).

(i) P. W. Percival, H. Fischer, E. Roduner, M. Camani, F. N. Gygax, A. Schenck, and H. Graf, *Proc. 4th Int. Conf. Positron Annihilation*, Helsingar (Denmark), Aug. 1976.

94. (a) For general discussion and review oriented to chemistry, see refs. 92b–d, 93, 94b, and 95a–d.

(b) P. W. Percival, E. Roduner, and H. Fischer, *Chem. Phys.*, **32**, 353 (1978).

95.[*] (a) P. W. Percival and H. Fischer, *Chem. Phys.*, **16**, 89 (1976).

(b) E. Roduner, *J. Mol. Struct.*, **60**, 19 (1980).

(c) J. H. Brewer, K. M. Crowe, F. N. Gygax, R. F. Johnson, D. G. Fleming, and A. Schenck, *Phys. Rev. (A)*, **9**, 495 (1974).

(d) H. Kaji and Y. Ito, *Kagaku*, **36**, 732 (1981).

96. For general discussion and review, see refs. 93a–e, 93h, 95a–d, and 97.

97.[*] (a) E. Roduner, in K. Crowe, J. Duclos, G. Fiorentini, and G. Torelli, Eds., *Exotic Atoms '79*, Plenum, 1980, p. 379.

(b) P. W. Percival, E. Roduner, and H. Fischer, in H. J. Ache, Ed., *Positronium and Muonium Chemistry*, A. C. S. Adv. Chem. Ser. **175**, 1979, p. 335.

(c) Y. C. Jean, B. W. Ng, J. M. Stadlbauer, and D. C. Walker, *J. Chem. Phys.*, **75**, 2879 (1981).

(d) J. M. Stadlbauer, B. W. Ng, D. C. Walker, Y. C. Jean, and Y. Ito, *Can. J. Chem.*, **59**, 3261 (1981).

(e) Y. Ito, B. W. Ng, Y.-C. Jean, and D. C. Walker, *Can. J. Chem.*, **58**, 2395 (1980).

(f) B. W. Ng, Y. C. Jean, Y. Ito, T. Suzuki, J. H. Brewer, D. G. Fleming, and D. C. Walker, *J. Phys. Chem.*, **85**, 454 (1981).

(g) H. Fischer, *Hyperfine Interactions*, **6**, 397 (1979).

(h) S. F. J. Cox, A. Hill, and R. DeRenzi, *J. Chem. Soc. Faraday I*, 2975 (1982).

(i) E. Roduner, P. W. Percival, D. G. Fleming, J. Hochmann, and H. Fischer, *Chem. Phys. Lett.*, **57**, 37 (1978).

(j) E. Roduner, *Hyperfine Interactions*, **8**, 561 (1981).

(k) E. Roduner, W. Strub, P. Burkhard, J. Hochmann, P. W. Percival, H. Fischer, M. Ramos, and B. C. Webster, *Chem. Phys.*, **67**, 275 (1982).

(l) E. Roduner, *Chem. Phys. Lett.*, **81**, 191 (1981).

(m) Y. C. Jean, J. Brewer, D. G. Fleming, D. M. Garner, R. J. Mikula, L. C. Vaz, and D. C. Walker, *Chem. Phys. Lett.*, **57**, 293 (1978).

(n) Y. C. Jean, J. H. Brewer, D. G. Fleming, D. M. Garner, and D. C. Walker, *Hyperfine Interactions*, **6**, 409 (1979).

(o) E. Roduner and H. Fischer, *Hyperfine Interactions*, **6**, 413 (1979).

98. We are not citing some earlier work which, although seminal with respect to discoveries and techniques, is not as quantitatively meaningful as is more recent work.

99.* (a) E. Roduner, P. W. Percival, H. Fischer, M. Camani, F. N. Gygax, and A. Schenck, *Mezony Veshchestve Tr. Mezhdunar. Simp. Probl. Mezzonnoi Khim. Mezomol. Protsessov Veshchestve [Mesons in Matter, Proc. Intl. Symp. on Meson Chemistry and Mesomolecular Processes in Matter*, Dubna, July 1977], V. N. Pokrovskii, Ed., JINR, Moscow, 1977, p. 326.

(b) P. W. Percival and J. Hochmann, *Hyperfine Interactions*, **6**, 421 (1979).

(c) D. G. Fleming, D. M. Garner, and R. J. Mikula, *Hyperfine Interactions*, **8**, 337 (1981).

(d) P. W. Percival, E. Roduner, H. Fischer, M. Camani, F. N. Gygax, and A. Schenck, *Chem. Phys. Lett.*, **47**, 11 (1977).

100.* (a) C. Bucci, G. Guidi, G. M. De'Munari, M. Manfredi, P. Podini, R. Tedeschi, P. R. Crippa, and A. Vecli, "Direct Evidence for Muonium Radicals in Water Solutions," *Chem. Phys. Lett.*, **57**, 41 (1978).

(b) C. Bucci, P. R. Crippa, G. M. de'Munari, G. Guidi, M. Manfredi, R. Tedeschi, A. Vecli, and P. Podini, "Interaction of Muonium with Molecules of Biological Interest in Water Solution," *Hyperfine Interactions*, **6**, 425 (1979).

(c) Y. C. Jean, B. W. Ng, and D. C. Walker, "Chemical Reactions between Muonium and Porphyrins," *Chem. Phys. Lett.*, **75**, 561 (1980).

(d) M. P. Balandin, A. L. Buchachenko, V.S. Evseev, T. N. Mamedov, V. S. Roganov, O. P. Tkacheva, and M. V. Frontas'eva, "Depolarization of Negative and Positive Muons in Nitroxyl Stable Radicals, Triacetonamine, and Methylcyclohexanol," *Khim. Vys. Energ.*, **15**, 552 (1981).

(e) C. Bucci and A. Vecli, "Muonium in Condensed Matter: Applications to Problems of Biological Interest," *Nucl. Sci. Applicns.*, **1**, 81 (1980).

101.* A. Hill, unpublished work, reported in ref. 97k.

102.* (a) E. Roduner, G. A. Brinkman, and P. W. F. Louwrier, *Chem. Phys.*, **73**, 117 (1982).

(b) G. A. Brinkman and P. W. F. Louwrier, *Chem. Mag.*, The Hague, June, 363 (1980).

103.* A. D. Pershin and A. L. Buchachenko, in *Magnetic Resonance and Related Phenomena [Proc. 20th (1978) AMPERE Congr.*, Tallinn (Estonia)], E. Kundla, E. Lippmaa, and T. Saluvere, Eds., Springer-Verlag, 1980, p. 157.

104.* (a) N. J. Turro, M.-F. Chow, C.-J. Chung, and B. Kraeutler, "Magnetic and Micellar Effects on Photoreactions. 1. ^{13}C Isotopic Enrichment of Dibenzyl Ketone via Photolysis in Aqueous Detergent Solutions," *J. Am. Chem. Soc.*, **103**, 3886 (1981).

(b) B. Kraeutler and N. J. Turro, "Photolysis of Dibenzyl Ketone in Micellar Solution. Correlation of Isotopic Enrichment Factors with Photochemical Efficiency Parameters," *Chem. Phys. Lett.*, **70**, 266 (1980).

(c) B. Kraeutler and N. J. Turro, "Probes for the Micellar Cage Effect. The Magnetic ^{13}C-Isotope Effect and a New Cage Product in the Photolysis of Dibenzyl Ketone," *Chem. Phys. Lett.*, **70**, 270 (1980).

(d) N. J. Turro, M.-F. Chow, and B. Kraeutler, "The Dynamics of the Photodecarbonylation of Dibenzylketone in Micellar Detergent Solution: Effect of Temperature on the Absolute Quantum Yields and on ^{13}C Enrichment," *Chem. Phys. Lett.*, **73**, 545 (1980).

(e) N. J. Turro, D. R. Anderson, M.-F. Chow, C.-J. Chung, and B. Kraeutler, "Magnetic and Micellar Effects on Photoreactions. 2. Magnetic Isotope Effects on Quantum Yields and Magnetic Field Effects on Separation Efficiency.

Correlation of ^{13}C-Enrichment Parameters with Quantum Yield Measurements," *J. Am. Chem. Soc.,* **103**, 3892 (1981).

(f) J. C. Scaiano, E. B. Abuin, and L. C. Stewart, "Photochemistry of Benzophenone in Micelles. Formation and Decay of Radical Pairs," *J. Am. Chem. Soc.,* **104**, 5673 (1982).

(g) N. J. Turro, "Methods Employing Magnetic Isotope Effects," U.S. Patent 4351707 (1982).

105.* (a) N. J. Turro, D. R. Anderson, and B. Kraeutler, *Tetrahedron Lett.,* **21**, 3 (1980).

(b) N. J. Turro, C.-J. Chung, R. G. Lawler, and W. J. Smith, III, *Tetrahedron Lett.,* **23**, 3223 (1982).

106. (a) Yu. N. Molin and R. Z. Sagdeev, in *Problemy Khimicheskoi Kinetiki, M.,* "Nauka," Moscow, 1979, p. 185.

(b) Yu. N. Molin, R. Z. Sagdeev, T. V. Leshina, A. V. Podoplelov, A. V. Dushkin, Yu. A. Grishin, and L. M. Weiner, *Proc. 20th Congr. AMPERE*, Tallinn (Estonia), 1978, p. 49.

(c) A. L. Buchachenko, R. Z. Sagdeev, and K. M. Salikhov, in *Magnitnye i Spinovye Effekty v Khimicheskikh Reaktsiyakh,* "Nauka," Sib. Otd.: Novosibirsk, 1978, p. 177.

107.* (a) A. L. Buchachenko, L. L. Yasina, S. F. Makhov, V. I. Mal'tsev, A. V. Fedorov, and É. M. Galimov, *Dokl. Akad. Nauk SSSR,* **260**, 1143 (1980).

(b) See also ref. 107c.

(c) V. A. Belyakov, V. I. Mal'tsev, and A. L. Buchachenko, "Magnetic Isotope Effect and Enrichment of Oxygen in ^{17}O Isotope in Chain Oxidation Reaction. 2. Kinetics," *Izv. Akad. Nauk SSSR*, 1022 (1982).

108. S. Skaron and M. Wolfsberg, *J. Chem. Phys.,* **72**, 6810 (1980).

109.* (a) N. J. Turro and M.-F. Chow, *J. Am. Chem. Soc.,* **102**, 1190 (1980).

(b) N. J. Turro, M.-F. Chow, and J. Rigaudy, *J. Am. Chem. Soc.,* **103**, 7218 (1981).

110.* (a) L. Sterna, D. Ronis, S. Wolfe, and A. Pines, *J. Chem. Phys.,* **73**, 5493 (1980).

(b) N. J. Turro and B. Kracutler, *Acc. Chem. Res.,* **13**, 369 (1980).

(c) G. A. Epling and E. Florio, *J. Am. Chem. Soc.,* **103**, 1237 (1981).

111. (a) This statement rests on one point only; if the one point at "zero" field were absent, the plot could not be said to show a maximum.

(b) The results that show a maximum appear also in refs. 110b and 112.

(c) This argument is being presented in detail because it serves as an illustration of the type of thinking typically involved in discussions of magnetic isotope effects.

112.* (a) N. J. Turro, M.-F. Chow, C.-J. Chung, G. C. Weed, and B. Kraeutler, *J. Am. Chem. Soc.,* **102**, 4843 (1980).

(b) N. J. Turro, J. Mattay, and G. F. Lehr, in S. L. Holt, Ed., *Inorganic Reactions in Organized Media, A. C. S. Symp Ser.* **177**, 1982, p. 19.

(c) N. J. Turro and B. Kraeutler, in W. T. Borden, Ed., *Diradicals*, Wiley-Interscience, New York, 1982, p. 259.

113.* (a) V. F. Tarasov, E. M. Galimov, V. I. Mal'tsev, and A. L. Buchachenko, *Izv. Akad. Nauk SSSR*, 2156 (1982).

(b) V. F. Tarasov, A. D. Pershin, and A. L. Buchachenko, *Izv. Akad. Nauk SSSR*, 1927 (1980).

(c) See also Ref. 113d.

(d) V. F. Tarasov, D. B. Askerov, and A. L. Buchachenko, *Izv. Akad. Nauk SSSR*, 2023 (1982).

114.   N. J. Turro, *Pure Appl. Chem.*, **53**, 259 (1981).

115.   (a) F. S. Sarvarov and K. M. Salikhov, *Reaction Kinet. Catal. Lett.*, **4**, 33 (1976).
       (b) R. Z. Sagdeev, K. M. Salikhov, and Yu. N. Molin, *Usp. Khim.*, **46**, 569 (1977).
       (c) G. P. Zientara and J. H. Freed, *J. Chem. Phys.*, **70**, 1359 (1979).
       (d) P. A. Purtov and K. M. Salikhov, *Teor. Eksp. Khim.*, **16**, 579 (1980).
       (e) P. A. Purtov and K. M. Salikhov, *Teor Eksp. Khim.*, **16**, 737 (1980).

116.* N. J. Turro, M.-F. Chow, C.-J. Chung, and C.-H. Tung, *J. Am. Chem. Soc.*, **102**, 7391 (1980).

117.   See also ref. 6w.

118.* A. L. Buchachenko, V. F. Tarasov, and V. I. Mal'tsev, *Zh. Fiz. Khim.*, **55**, 1649 (1981).

119.* (a) A. L. Buchachenko, V. A. Belyakov, and V. I. Mal'tsev, *Izv. Akad. Nauk SSSR*, 1016 (1982).
       (b) See also ref. 108.

120.* (a) A. V. Podoplelov, T. V. Leshina, R. Z. Sagdeev, Yu. N. Molin, and V. I. Gol'danskii, *Pis'ma Zh. Eksp. Teor. Fiz.*, **29**, 419 (1979).
       (b) This suggestion has been attributed to Gol'danskii.[65g,o]

121.   (a) V. F. Tarasov, "Dynamics of the Enrichment of Magnetic Isotopes in Chemical Reactions," *Zh. Fiz. Khim.*, **54**, 2438 (1980).
       (b) V. F. Tarasov, A. L. Buchachenko, and V. I. Mal'tsev, "The Magnetic Isotope Effect and the Separation of Isotopes in 'Microreactors'," *Zh. Fiz. Khim.*, **55**, 1921 (1981).
       (c) Y. Sakaguchi, S. Nagakura, and H. Hayashi, "External Magnetic Field Effect on the Decay Rate of Benzophenone Ketyl Radical in a Micelle," *Chem. Phys. Lett.*, **72**, 420 (1980).
       (d) Y. Sakaguchi, S. Nagakura, A. Minoh, and H. Hayashi, "Magnetic Isotope Effect upon the Decay Rate of the Benzophenone Ketyl Radical in a Micelle," *Chem. Phys. Lett.*, **82**, 213 (1981).
       (e) Y. Sakaguchi, H. Hayashi, and S. Nagakura, "Laser-Photolysis Study of the External Magnetic Field Effect upon the Photochemical Processes of Carbonyl Compounds in Micelles," *J. Phys. Chem.*, **86**, 3177 (1982).
       (f) Y. Sakaguchi and H. Hayashi, "Laser-Photolysis Study of the Photochemical Processes of Carbonyl Compounds in Micelles under High Magnetic Fields," *Chem. Phys. Lett.*, **87**, 539 (1982).
       (g) J. C. Scaiano and E. B. Abuin, "Absolute Measurement of the Rates of Radical Exit and of Radical-Pair Intersystem Crossing in Anionic Micelles," *Chem. Phys. Lett.*, **81**, 209 (1981).
       (h) C. Doubleday, Jr., "Deuterium Isotope Effect on the Magnetic Field Dependence of CIDNP Derived from Biradicals," *Chem. Phys. Lett.*, **81**, 164 (1981).
       (i) W. Nakanishi, T. Jo, K. Miura, Y. Ikeda, T. Sugawara, Y. Kawada, and H. Iwamura, "Thermal Decomposition of ^{17}O-Labeled *t*-Butyl *o*-Methylthio- and *o*-Phenylthiobenzoates Studied by ^{17}O NMR. The sulfuranyl Radical Structure of the *o*-Thiobenzoyloxy Radicals," *Chem. Lett.*, 387 (1981).
       (j) N. J. Turro and J. Mattay, "Photochemistry of 1,2-Diphenyl-2,2-dimethyl-propanone-l in Micellar Solutions. Cage Effects, Isotope Effects and Magnetic Field Effects," *Tetrahedron Lett.*, **21**, 1799 (1980).
       (k) N. J. Turro and J. Mattay, "Photochemistry of Some Deoxybenzoins in

Micellar Solutions. Cage Effects, Isotope Effects, and Magnetic Field Effects," *J. Am. Chem. Soc.*, **103**, 4200 (1981).

(l) N. J. Turro, M.-F. Chow, C.-J. Chung, Y. Tanimoto, and G. C. Weed, "Time-Resolved Laser Flash Spectroscopic Study of Benzyl Radical Pairs in Micelle Cages," *J. Am. Chem. Soc.*, **103**, 4574 (1981).

122. References cited in ref. 77c.

123. See also refs. 5e and 6y.

124. (a) We thank Dr. Johnson for informing us of some of his results prior to publication.
     (b) See also refs. 124c and 124d.
     (c) M. Veber, K. N.-V-Duong, A. Gaudener, and M. D. Johnson, *J. Organometal. Chem.*, **209**, 393 (1981).
     (d) A. Bury, P. Bougeard, S. J. Corker, M. D. Johnson, and M. Perlmann, *J. Chem. Soc. Perkin II*, 1367 (1982).

125.* (a) M. R. Ashcroft, A. Bury, C. J. Cooksey, A. G. Davies, B. Dass Gupta, M. D. Johnson, and H. Morris, *J. Organometal. Chem.*, **195**, 89 (1980).
     (b) A. Bury, S. T. Corker, and M. D. Johnson, *J. Chem. Soc. Perkin I*, 645 (1982).
     (c) A. Bury, M. D. Johnson, and M. J. Stewart, *J. Chem. Soc., Chem. Commun.*, 622 (1980).
     (d) P. Bougeard and M. D. Johnson, *J. Organometal. Chem.*, **206**, 221 (1981).
     (e) P. Bougeard, A. Bury, C. J. Cooksey, M. D. Johnson, J. M. Hungerford, and G. M. Lampman, *J. Am. Chem. Soc.*, **104**, 5230 (1982).

126. In view of the statement that "cyclopentanes are formed in very low yield [8%] in some free-radical reactions of 1,5-diiodopentanes,"[125e] we do not understand the subsequent statement that the reactions being reported "are the first examples of homolytic displacement by attack of a remote radical at a saturated carbon of an alkyl chain . . . ."[125e] Note also that the label "very low yield" is applied to a reaction that proceeded in 8% yield, whereas one that proceeded in approximately 1% yield is described simply as one of "two definitive examples" of the characterization of a "bimolecular homolytic displacement at saturated carbon."[125e]

127.* (a) R. A. Jackson and M. Townson, *Tetrahedron Lett.*, 193 (1973).
     (b) R. A. Jackson and M. Townson, *J. Chem. Soc. Perkin II*, 1452 (1980).

128. (a) Several experiments are described in the Experimental section.[127b] However, this one, the only one for which a yield is reported, is not.
     (b) Elsewhere, it is stated that the same plot "*indicates that* [emphasis added] isobutane is formed as a primary product."[127b]
     (c) We estimate[128d] a conversion to isobutane of approximately 0.04% at that point. Also, we suggest that the reader construct a complete plot using *all* the data, as provided in the Experimental section.[127b]
     (d) An estimate is necessary because the amount of isobutane is reported in "arbitrary units," rather than as the actual amount.
     (e) The individual values were 31, 35, 42, and 61.

129. For example, see the following:
     (a) G. O. Pritchard and J. R. Dacey, *Can. J. Chem.*, **38**, 182 (1960).
     (b) G. O. Pritchard and M. J. Perona, *Int. J. Chem. Kinet.*, **2**, 281 (1970).
     (c) P. C. Kobrinsky, G. O. Pritchard, and S. Toby, *J. Phys. Chem.*, **75**, 2225 (1971).
     (d) R. D. Giles and E. Whittle, *Trans. Faraday Soc.*, **61**, 1425 (1965).
     (e) T. N. Bell and A. E. Platt, *Int. J. Chem. Kinet.*, **2**, 299 (1970).

(f) R. Hiatt and S. W. Benson, *Int. J. Chem. Kinet.*, **4**, 479 (1972).

(g) L. F. Loucks, M. T. H. Liu, and D. G. Hooper, *Can. J. Chem.*, **57**, 2201 (1979).

130. (a) P. B. Ayscough, J. C. Polanyi, and E. W. R. Steacie, *Can. J. Chem.*, **33**, 743 (1955).

(b) R. D. Giles, L. M. Quick, and E. Whittle, *Trans. Faraday Soc.*, **63**, 662 (1967).

(c) L. M. Quick and E. Whittle, *Trans. Faraday Soc.*, **67**, 1727 (1971).

131. (a) It has been customary to write reactions such as $CF_3 \cdot + CF_3I \rightarrow C_2F_6 + I \cdot$, $CH_3 \cdot + CH_3I \rightarrow C_2H_6 + I \cdot$, $CH_3 \cdot + CF_3I \rightarrow CF_3CH_3 + I \cdot$, $CF_3 \cdot + CH_3I \rightarrow CF_3CH_3 + I \cdot$, $CF_3 \cdot + C_2F_5I \rightarrow C_3F_8 + I \cdot$, and $C_2F_5 \cdot + C_2F_5I \rightarrow C_4F_{10} + I \cdot$, apparently without serious consideration.
Examples:

(b) R. Srinivasan and J. R. Lankard, *J. Phys. Chem.*, **78**, 951 (1974).

(c) D. Teclemariam and R. J. Hanrahan, *Radiat. Phys. Chem.*, **19**, 443 (1982).

(d) T. Hsieh and R. J. Hanrahan, *Radiat. Phys. Chem.*, **12**, 51 (1978).

(e) T. Hsieh and R. J. Hanrahan, *Radiat. Phys. Chem.*, **12**, 153 (1978).

(f) P. Gensel, K. Hohla, and K. L. Kompa, *Appl. Phys. Lett.*, **18**, 48 (1971).
Contrasts:

(g) T. L. Andreeva, V. I. Malyshev, A. I. Maslov, I. I. Sobel'man, and V. N. Sorokin, *Zh. Eksp. Teor. Fiz. Pis'ma Red.*, **10**, 423 (1969).

(h) V. Yu. Zalesskii, *Zh. Eksp. Teor. Fiz.*, **61**, 892 (1971).

(i) V. Yu. Zaleskii and T. I. Krupenikova, *Opt. Spektrosk.*, **30**, 813 (1971).

132. This possibility is strengthened by the observation that $CF_3CH_2C(CF_3)(CH_3)_2$ and $CF_3CH_2C(CH_3)_3$ are produced.

133. R. A. Jackson, F. Malek, and N. Özaslan, *J. Chem. Soc., Chem. Commun.*, 956 (1981).

134. See also ref. 10i.

135.* L. Vasáros, Yu. V. Norseev, and V. A. Khalkin, *Dokl. Akad. Nauk SSSR*, **263**, 119 (1982); Prepr. Joint Institute Nucl. Res. Dubna, P12-81-509 (1981).

136.* (a) V. V. Lipes, E. A. Bakova, and V. I. Morozova, *Dokl. Akad. Nauk SSSR*, **258**, 943 (1981).

(b) V. V. Lipes, E. A. Bakova, and V. I. Morozova, *Kinet. Katal.*, **23**, 568 (1982).

137.* (a) A. H. Laufer, *J. Phys. Chem.*, **85**, 3828 (1981).

(b) A. M. Renlund, F. Shokoohi, H. Reisler, and C. Wittig, *Chem. Phys. Lett.*, **84**, 293 (1981).

138. (a) N. L. Arthur and L. F. David, "Reactions of Trifluoromethyl Radicals. VI. Hydrogen Abstraction from Trifluoroacetic Acid and Trifluoroacetic($^2H_1$) Acid," *Aust. J. Chem.*, **34**, 1535 (1981).

(b) See also refs. 138c–h.

(c) B. C. Gilbert, R. O. C. Norman, G. Placucci, and R. C. Sealy, "Electron Spin Resonance Studies. Part XLV. Reactions of the Methyl Radical with some Aliphatic Compounds in Aqueous Solution," *J. Chem. Soc. Perkin II*, 885 (1975).

(d) Z. Y. Al-Saigh and J. R. Majer, "Hydrogen Abstraction from Fluorinated Aliphatic Acids," *J. Fluorine Chem.*, **7**, 589 (1976).

(e) N. L. Budeiko, V. E. Agabekov, N. I. Mitsketich, and E. T. Denisov, "Reaction of Undecyl Radicals with Monocarboxylic Acids," *React. Kinet. Catal. Lett.*, **8**, 71 (1978).

(f) I. Nemes, N. N. Ugarova, and O. Dobis, "Effects of Molecular Structure and of Solvent on the Relative Reactivities of Carboxylic Acids towards Free Methyl Radicals in the Liquid Phase," *Zh. Fiz. Khim.,* **40**, 466 (1966).

(g) O. Dobis, I. Nemes, and R. Kerepes, "Investigation of the Reactivity of Organic Compounds in Liquid Phase Free-Radical Reactions, II. The Relative Reactivities of Organic Acids with Methyl Radicals in Liquid Phase," *Acta Chim. Acad. Scient. Hung.,* **55**, 215 (1968).

(h) With regard to references 138f and 138g, note the possibility of $CH_3COOT \rightarrow TCH_2COOH$ by way of the enol.

139.* T. M. Lenhardt, C. E. McDade, and K. D. Bayes, *J. Chem. Phys.,* **72**, 304 (1980).

140.   (a) See also refs. 140b and 140c.

(b) G. M. Atri, R. R. Baldwin, G. A. Evans, and R. W. Walker, *J. Chem. Soc. Faraday I*, 366 (1978).

(c) G. A. Evans and R. W. Walker, *J. Chem. Soc. Faraday I*, 1458 (1979).

141.*   (a) H. H. Nelson and J. R. McDonald, *J. Phys. Chem.,* **86**, 1242 (1982).

(b) T. Ebata, K.-i. Obi, and I. Tanaka, *Chem. Phys. Lett.,* **77**, 480 (1981).

(c) T. Ebata, M. Kanamaya, K.-i. Obi, and I. Tanaka, *Kosoku Hanno Toronkai Koen Yokoshu, 14th (Proc. 14th Intl. Symp. Free Radicals*, Sanda, Japan), 64 (1979).

142.* C. E. McDade, T. M. Lenhardt, and K. D. Bayes, *J. Photochem.,* **20**, 1 (1982).

143.   (a) M. B. Neiman and G. I. Feklisov, *Zh. Fiz. Khim.,* **35**, 1064 (1961).

(b) D. E. Hoare and D. A. Whytock, *Can. J. Chem.,* **45**, 865 (1967).

(c) D. E. Hoare and D. A. Whytock, *Can. J. Chem.,* **45**, 2741 (1967).

144.   K. W. Watkins and W. W. Word, *Int. J. Chem. Kinet.,* **6**, 855 (1974).

145.   Plots of

$$\frac{(RCO_2Br)_{initial}}{(C_4H_9Br)_{final}} - 1 \quad \text{vs.} \quad \frac{1}{[C_4H_{10}]}$$

at three temperatures comprise the illustration. $[C_4H_{10}]$, which varied from two- to six-fold in the illustration, and $(C_4H_9Br)_{final}$ are the independent and dependent variables, respectively. $(RCO_2Br)_{initial}$ was constant. Superimposed on the plots of the data are three straight lines drawn through a (0,0) origin. From an examination of the points and the lines we conclude that the results are not inconsistent with the relationship.

146.   (a) S. A. Dombchik, Ph.D. thesis, University of Illinois (Urbana), 1969.

(b) R. A. Cooper, R. G. Lawler, and H. R. Ward, *J. Am. Chem. Soc.,* **94**, 545 (1972).

(c) E. A. Turetskaya, E. D. Skakovskii, S. V. Rykov, and Yu. V. Glazkov, *Dokl. Akad. Nauk BSSR*, **24**, 57 (1980).

(d) E. D. Skakovskii, S.V. Rykov, E. A. Turetskaya, N. A. Gurinovich, and Yu. A. Ol'dekop, *Vestsi Akad. Navuk BSSR,* (5), 34 (1982).

147.   (a) D. Griller and B. P. Roberts, *J. Chem. Soc. Perkin II*, 747 (1972).

(b) See also refs. 147c and 147d.

(c) P. Cadman, A. J. White, and A. F. Trotman-Dickenson, *J. Chem. Soc. Faraday I*, 506 (1972).

(d) E. A. Lissi, G. Massiff, and A. E. Villa, *J. Chem. Soc. Faraday I*, 346 (1973).

148.   (a) D. L. Rachmankylov, S. S. Zlotskii, V. Ch. Chamaev, V. N. Uzykova, S. A. Agisheva, V. A. Kas'yanov, and V. V. Zorin, *USSR Patent* 486013 (1976).

(b) M. Oyama, *J. Org. Chem.*, **30**, 2429 (1965).

(c) G. Fuller and F. F. Rust, *J. Am. Chem. Soc.*, **80**, 6148 (1958).

(d) E. P. Kalyazin, E. P. Petryaev, and O. I. Shadyro, *Zh. Org. Khim.*, **13**, 293 (1977).

(e) E. P. Petryaev, O. I. Shadyro, and P. N. Davidovich, *Zh. Org. Khim.*, **14**, 2488 (1978).

(f) E. P. Petryaev, O. I. Shadyro, and P. N. Davidovich, *Zh. Org. Khim.*, **14**, 2222 (1978).

(g) G. I. Nikishin, D. Lefort, and V. D. Vorob'ev, *Izv. Akad. Nauk SSSR*, 1271 (1966).

(h) G. I. Nikishin, E. K. Starostin, and S. I. Moryasheva, *Izv. Akad. Nauk SSSR*, 2630 (1976).

(i) E. P. Kalyazin and G. V. Kovalev, *Khim. Vys. Energ.*, **12**, 371 (1978).

(j) S. L. Kil'chitskaya, E. P. Petryaev, and E. P. Kalyazin, *Khim. Vys. Energ.*, **13**, 213 (1979).

(k) A. L. Karasev, E. P. Petryaev, and A. N. Rukhlya, *Khim. Vys. Energ.*, **16**, 284 (1982).

(l) J. Kollar, *U.S. Patent* 4337371 (1982).

(m) J. Kollar, *British Patent Application*, 2083037A (1982).

(n) This reaction deserves more thought because "formaldehyde" is not H_2CO to a significant extent under typical reaction conditions, a fact that has been considered[148o] only rarely.

(o) G. V. Kovalev, I. A. Abramenkova, A. V. Rudnev, and E. P. Kalyazin, *Khim. Vys. Energ.*, **16**, 331 (1982).

(p) A. M. Afanas'ev, Dissertation, Moscow State University, 1972; cited in ref. 148d.

(q) F. F. Rust, F. H. Seubold, and W. E. Vaughan, *J. Am. Chem. Soc.*, **70**, 4253 (1948).

(r) G. Cauquis, B. Sillion, and L. Verdet, *Tetrahedron Lett.*, 27 (1977).

(s) K. Maruyama, M. Taniuchi, and S. Oka, *Bull. Chem. Soc. Jpn.*, **47**, 712 (1974).

(t) F. S. Dainton, K. J. Ivin, and F. Wilkinson, *Trans. Faraday Soc.*, **55**, 929 (1959).

(u) Citations are extensive because we have tried to assemble a body of work not previously brought together.

149.* (a) H. Knoll, Á. Nacsa, S. Förgeteg, and T. Bérces, *React. Kinet. Catal. Lett.*, **15**, 481 (1980).

      (b) H. Knoll, G. Richter, and R. Schliebs, *Int. J. Chem. Kinet.*, **12**, 623 (1980).

      (c) H. Knoll, *React. Kinet. Catal. Lett.*, **15**, 431 (1980).

150. (a) Similar to a quantum yield in a photochemical reaction.

      (b) See ref. 150c.

      (c) A. V. Rudnev, G. V. Kovalev, K. S. Kalugin, and E. P. Kalyazin, *Zh. Fiz. Khim.*, **51**, 2031 (1977).

151.* A. Effio, D. Griller, K. U. Ingold, A. L. J. Beckwith, and A. K. Serelis, *J. Am. Chem. Soc.*, **102**, 1734 (1980).

152. Original data and results sufficient to permit recalculation of these numbers in response to a new value of $k(1°\ \text{alkyl} \cdot + Bu_3SnH \rightarrow)$[10j] are not reported in this Communication. Also, might there be a typographical error in its footnote 16?

153. For many of the reactions to be described questions remain as to whether the reactions proceed by way of a free radical mechanism and whether the observed

rotations of the products are the result of the presence of an impurity having a high specific rotation.

154.   H. Fujihara, M. Yoshihara, and T. Maeshima, *J. Polym. Sci., Polym. Lett. Ed.,* **18**, 287 (1980).

155.*   (a)   M. Yoshihara, H. Fujihara, A. Yoneda, and T. Maeshima, *Chem. Lett.,* 39 (1980).

   (b)   M. Yoshihara, K. Nozaki, H. Fujihara, A. Yoneda, and T. Maeshima, *Kinki Daigaku Rikogakubu Konkyu Hokoku,* No. 16, 63 (1981).

   (c)   J. H. Liu, K. Kondo, and K. Takemoto, *Angew. Makromol. Chem.,* **104**, 209 (1982).

   (d)   S. Murai, R. Sugise, and N. Sonoda, *Angew. Chem. Int. Ed. Engl.,* **20**, 475 (1981).

   (e)   M. Yoshihara, K. Nozaki, H. Fujihara, and T. Maeshima, *J. Macromol. Sci.-Chem.,* **A18**, 411 (1982).

   (f)   M. Yoshihara, K. Nozaki, H. Fujihara, and T. Maeshima, *J. Polym. Sci.: Polym. Lett. Ed.,* **19**, 49 (1981).

   (g)   M. Yoshihara, H. Fujihara, and T. Maeshima, *Chem. Ind.,* 201 (1980).

   (h)   M. Yoshihara, H. Fujihara, and T. Maeshima, *Chem. Lett.,* 195 (1980).

156.   (a)   Purified by use of vpc.

   (b)   "The reactions were found to be inhibited completely by the addition of a small amount of hydroquinone . . . ."[155a]

   (c)   A chemical yield of 66% and an optical yield of 4.1% were obtained in the absence of AIBN.

   (d) $(-)\text{-diop} \equiv$

   (e)   For $(R_1 = H, R_2 = CH_3, R_3 = C_6H_5)$ and $(R_1 = C_2H_5, R_2 = R_3 = CH_3)$ refs. 155e–g are inconsistent with respect to whether the reaction was run neat or in benzene as solvent.

   (f)   ". . . the reactions were completely inhibited with an addition of a small amount of hydroquinone, implying these reactions to proceed via radical process."

   (g)   Reported as 24% in ref. 155e.

   (h)   Reported as 30% in ref. 155e.

157.   M. Yoshihara and T. Maeshima, *J. Polym. Sci.: Polym. Lett. Ed.,* **19**, 1269 (1981).

158.   A. Rahm, M. Degueil-Castaing, and M. Pereyre, "Asymmetric Hydrostannation of Chiral α,β Unsaturated Esters, A New Exception to Prelog's Generalization," *Tetrahedron Lett.,* **21**, 4649 (1980).

159.*   A. Rahm and M. Pereyre, *Bull. Soc. Chim. Belg.,* **89**, 843 (1980).

160.   (a)   Reduction of $PhCCO_2(l\text{-menthyl})$ by $LiAlH_4$ in a different chemical environment gave a 7–8% optical yield of $(R)–PhCH(OH)CH_2OH$.[160b–d]

   (b)   M. J. Kubitscheck and W. A. Bonner, *J. Org. Chem.,* **26**, 2194 (1961).

   (c)   V. Prelog, M. Wilhelm, and D. B. Bright, *Helv. Chim. Acta,* **37**, 221 (1954).

   (d)   S. P. Bakshi and E. E. Turner, *J. Chem. Soc.,* 168 (1961).

161.   (a) See also refs. 161b and 161c.
        (b) T. H. Haskell, P. W. K. Woo, and D. R. Watson, *J. Org. Chem.*, **42**, 1302 (1977).
        (c) J. M. McIntosh and C. K. Schram, *Can. J. Chem.*, **55**, 3755 (1977).

162.   (a) See also ref. 162b.
        (b) E.-M. Acton, R. N. Goerner, H. S. Uh, K. J. Ryan, D. W. Henry, C. E. Cass, and G. A. Le Page, *J. Med. Chem.*, **22**, 518 (1979).

163.   (a) See also refs. 162b, 163b–f.
        (b) K. Tatsuta, K. Akimoto, and M. Kinoshita, *J. Am. Chem. Soc.*, **101**, 6116 (1979).
        (c) C. Copeland and R. V. Stick, *Aust. J. Chem.*, **30**, 1269 (1977).
        (d) J. J. Patroni and R. V. Stick, *Aust. J. Chem.*, **32**, 411 (1979).
        (e) R. H. Bell, D. Horton, D. M. Williams, and E. Winter-Mihaly, *Carbohy. Res.*, **58**, 109 (1977).
        (f) T. Hayashi, T. Dwaoka, N. Takeda, and E. Ohki, *Chem. Pharm. Bull.*, **26**, 1786 (1978).

164.   (a) See also ref. 164b.
        (b) R. E. Carney, J. B. McAlpine, M. Jackson, R. S. Stanaszek, W. H. Washburn, M. Cirovic, and S. L. Mueller, *J. Antibiot.*, **31**, 441 (1978).

165.   (a) See also ref. 165b.
        (b) P. J. L. Daniels and S. W. McCombie, *U.S. Patent*, 4053591 (1977).

166.*   (a) D. H. R. Barton, W. Hartwig, R. S. H. Motherwell, W. B. Motherwell, and A. Stange, *Tetrahedron Lett.*, **23**, 2019 (1982).
        (b) J.-K. Choi, D. J. Hart, and Y.-M. Tsai, *Tetrahedron Lett.*, 4765 (1982).
        (c) M. J. Robins and J. S. Wilson, *J. Am. Chem. Soc.*, **103**, 932 (1981).
        (d) *British Patent* 1561043 (1980).
        (e) L. L. Kostyukevich, U. B. Imashev, S. S. Zlotskii, R. A. Karakhanov, and D. L. Rakhmankulov, *Dokl. Akad. Nauk SSSR*, **250**, 391 (1980).

167.   (a) This work is referred to as "the first investigation into the kinetics and mechanism of the homolytic decomposition of formic acid esters, during which the corresponding alkyl radicals are formed."[166e] However, see refs. 167b–f.
        (b) J. C. J. Thynne and P. Gray, *Proc. Chem. Soc.*, 295 (1962).
        (c) J. C. J. Thynne, *Trans. Faraday Soc.*, **58**, 676 (1962).
        (d) J. C. J. Thynne, *Trans. Faraday Soc.*, **58**, 1394 (1962).
        (e) J. C. J. Thynne, *Trans. Faraday Soc.*, **58**, 1533 (1962).
        (f) J. C. J. Thynne and P. Gray, *Trans. Faraday Soc.*, **59**, 1149 (1963).

168.   (a) R. A. Jackson and F. Malek, *J. Chem. Soc. Perkin I*, 1207 (1980).
        (b) H. Redlich, H.-J. Neumann, and H. Paulsen, *J. Chem. Res.(M)*, 0352 (1982).
        (c) D. J. Hart and Y.-M. Tsai, *J. Am. Chem. Soc.*, **104**, 1430 (1982).
        (d) M. D. Bachi and C. Hoornaert, *Tetrahedron Lett.*, **22**, 2693 (1981).
        (e) D. H. R. Barton, W. B. Motherwell, and A. Stange, *Synthesis*, 743 (1981).
        (f) M. H. Beale, P. Gaskin, P. S. Kirkwood, and J. MacMillan, *J. Chem. Soc. Perkin I*, 885 (1980).
        (g) H. Redlich and W. Francke, *Angew. Chem. Int. Ed. Engl.*, **19**, 630 (1980).
        (h) J. Defaye, H. Driguez, B. Henrissat, and E. Bar-Guilloux, *Nouv. J. Chim.*, **4**, 59 (1980).
        (i) D. J. Hart, *J. Org. Chem.*, **46**, 367 (1981).
        (j) G. Just and C. Luthe, *Can. J. Chem.*, **58**, 1799 (1980).
        (k) J. Thiem and H. Karl, *Chem. Ber.*, **113**, 3039 (1980).

(l)   S. J. Cristol, M. W. Klein, M. H. Hendewerk, and R. D. Daussin, *J. Org. Chem.,* **46**, 4992 (1981).

(m)   T. S. Fuller and R. V. Stick, *Aust. J. Chem.,* **33**, 2509 (1980).

(n)   V. Pozsgay and A. Neszmélyi, *Carbohydr. Res.,* **85**, 143 (1980).

(o)   J. R. Rasmussen, C. J. Slinger, R. J. Kordish, and D. D. Newman-Evans, *J. Org. Chem.,* **46**, 4843 (1981).

(p)   J. R. Rasmussen, *J. Org. Chem.,* **45**, 2725 (1980).

(q)   T. Jikihara, T. Tsuchiya, H. Umezawa, and S. Umezawa, *Carbohydr. Res.,* **89**, 91 (1981).

(r)   D. H. R. Barton, W. Hartwig, and W. B. Motherwell, *J. Chem. Soc., Chem. Commun.,* 447 (1982).

169.*   G. E. Keck and J. B. Yates, "Carbon–Carbon Bond Formation via the Reaction of Trialkylallylstannanes with Organic Halides," *J. Am. Chem. Soc.,* **104**, 5829 (1982).

170.   See ref. 77g, note 7.

171.   G. W. Burton and K. U. Ingold, *J. Am. Chem. Soc.,* **103**, 6472 (1981).

172.   M. C. R. Symons and J. M. Stephenson, *J. Chem. Soc. Faraday I*, 1579 (1981).

173.   G. Leroy, Part I, in I. G. Csizmadia and R. Daudel, Eds., *Computational Theoretical Organic Chemistry,* Reidel, Hingham, Mass., 1981, p. 253.

174.   M. Ya. Mel'nikov and N. V. Fok, "Photochemical Reactions of Free Radicals in the Solid Phase," *Usp. Khim.,* **49**, 252 (1980).

175.   (a)   H. Adachi and N. Basco, *Int. J. Chem. Kinet.,* **13**, 367 (1981).

       (b)   P. Arrowsmith and L. J. Kirsch, *J. Chem. Soc. Faraday I,* 3016 (1978).

176.   M. C. R. Symons and P. J. Boon, *Chem. Phys. Lett.,* **89**, 516 (1982).

177.   D. J. DeFrees, R. T. McIver, Jr., and W. J. Hehre, *J. Am. Chem. Soc.,* **102**, 3334 (1980).

178.   L. Szirovicza and I. Szilágyi, *Int. J. Chem. Kinet.,* **12**, 113 (1980).

179.   (a)   P. D. Taylor, R. T. Tuckerman, and E. Whittle, *J. Photochem.,* **19**, 277 (1982).

       (b)   W. B. Nilsson and G. O. Pritchard, *Int. J. Chem. Kinet.,* **14**, 299 (1982).

180.   P. D. Pacey and J. H. Wimalasena, *Chem. Phys. Lett.,* **76**, 433 (1980).

181.   G. E. Keck and J. B. Yates, "A Novel Synthesis of (±)-Perhydrohistrionicotoxin," *J. Org. Chem.,* **47**, 3590 (1982).

182.   (a)   J. Pacansky and H. Coufal, *J. Chem. Phys.,* **72**, 3298 (1980).

       (b)   See also refs. 13b and 13d.

183.   (a)   M. C. R. Symons and D. Griller, *J. Magn. Res.,* **39**, 355 (1980).

       (b)   M. C. R. Symons, *Chem. Phys. Lett.,* **69**, 198 (1980).

       (c)   M. C. R. Symons, in "Annual Reports on the Progress of Chemistry," **78**, section C (Physical Chemistry), 1981 (Publ. 1982), p. 153.

184.   (a)   R, Mehnert, O. Brede, and Gy. Cserép, *Radiochem. Radioanal. Lett.,* **47**, 173 (1981).

       (b)   T. Shida, Y. Egawa, H. Kubodera, and T. Kato, *J. Chem. Phys.,* **73**, 5963 (1980).

       (c)   F. Gerson, J. Lopez, A. Krebs and W. Rüger, *Angew. Chem. Int. Ed. Engl.,* **20**, 95 (1981).

       (d)   F. Gerson, J. Lopez, R. Akaba, and S. F. Nelsen, *J. Am. Chem. Soc.,* **103**, 6716 (1981).

       (e)   H. Klusik and A. Berndt, *Angew. Chem. Int. Ed. Engl.,* **20**, 870 (1981).

       (f)   H. Eierdanz and A. Berndt, *Angew. Chem. Int. Ed. Engl.,* **21**, 690 (1982).

(g) K. Toriyama, K. Nunome, and M. Iwasaki, *J. Chem. Phys.*, **77**, 5891 (1982).

(h) O. Brede, R. Mehnert, W. Naumann, and G. Cserép, *Radiat. Phys. Chem.*, **20**, 155 (1982).

(i) R. Mehnert, O. Brede, and W. Naumann, *Ber. Bunsenges. Phys. Chem.*, **86**, 525 (1982).

(j) A. V. Kucherov and A. A. Slinkin, *Kinet. Katal.*, **23**, 1172 (1982).

(k) S. Shih, *J. Catal.*, **36**, 238 (1975).

185. (a) A. Berndt, H. Klusik, and K. Schlüter, *J. Organomet. Chem.*, **222**, C25 (1981).

       (b) H. Klusik and A. Berndt, *J. Organomet. Chem.*, **232**, C21 (1982).

186. M. Meot-Ner, *J. Am. Chem. Soc.*, **104**, 5 (1982).

187. (a) K. Glänzer, M. Quack, and J. Troe, *Chem. Phys. Lett.*, **39**, 304 (1976).

       (b) K. Glänzer, M. Quack, and J. Troe, *16th Intl. Symp. on Combustion*, The Combustion Institute, 1976, p. 949.

188. (a) K. Mukai, N. Iida, Y. Kumamoto, H. Kohama, and K. Ishizu, *Chem. Lett.*, 613 (1980).

       (b) K. Mukai, T. Yano, and K. Ishizu, *Tetrahedron Lett.*, **22**, 4661 (1981).

       (c) K. Mukai, N. Iida, and K. Ishizu, *Bull. Chem. Soc. Jpn.*, **55**, 1362 (1982).

       (d) H. Dugas and M. Ptak, *J. Chem. Soc., Chem. Commun.*, 710 (1982).

       (e) M. P. Eastman, D. E. Patterson, R. A. Bartsch, Y. Liu, and P. G. Eller, *J. Phys. Chem.*, **86**, 2052 (1982).

       (f) J. F. W. Keana, M. J. Acarregui, and S. L. M. Boyle, "2,2-Disubstituted-4,4-dimethylimidazolidinyl-3-oxy Nitroxides: Indicators of Aqueous Acidity through Variations of α_N with pH," *J. Am. Chem. Soc.*, **104**, 827 (1982).

189. D. Griller and K. U. Ingold, "Free-Radical Clocks," *Acc. Chem. Res.*, **13**, 317 (1980).

190. (a) C. L. Greenstock and R. H. Wiebe, *Can. J. Chem.*, **60**, 1560 (1982).

       (b) V. N. Belevskii, *Khim. Vys. Energ.*, **15**, 3 (1981).

       (c) Rate constants for the reactions ($\cdot CH_2OH + Bu^tNO$) and ($Me_2\dot{C}OH + PhCH=N(O)Bu^t$) are reported and attributed to ref. 190d. Although it is stated in that abstract[190d] that such rate constants were determined, numbers are not reported. In a subsequent publication,[190a] a value for the second reaction at an undisclosed temperature is reported, but it differs from that given[190b] for 20°.

       (d) C. L. Greenstock, R. H. Wiebe, and E. M. Gardy, *Proc. 61st Canadian Chem. Conf.*, Winnipeg, OR-5, 65 (1978).

       (e) T. Doba and H. Yoshida, *Bull. Chem. Soc. Jpn.*, **55**, 1753 (1982).

191. (a) V. R. Polishchuk, M. Yu. Antipin, V. I. Bakhmutov, N. N. Bubnov, S. P. Solodovnikov, T. V. Timofeeva, Yu. T. Struchkov, B. L. Tumanskii, and I. L. Knunyants, "2,3-Bis(Trifluoromethyl)-2,3-Bis(*p*-Fluorophenyl)hexafluorobutane. Length of a Central C–C Bond Equal to 1.671 Å," *Dokl. Akad. Nauk SSSR*, **249**, 1125 (1979).

       (b) H.-D. Beckhaus, G. Kratt, K. Lay, J. Geiselmann, C. Rüchardt, B. Kitschke, and H. J. Lindner, "Thermolabile Hydrocarbons, XIII. 3,4-Dicyclohexyl-3,4-dimethylhexane – Standard Heat of Formation, Thermal Stability, and Structure," *Chem. Ber.*, **113**, 3441 (1980).

       (c) R. Winiker, H.-D. Beckhaus, and C. Rüchardt, "Thermolabile Hydrocarbons, XIV. Thermal Stability, Strain Enthalpy, and Structure of sym. Hexaalkyl-substituted Ethanes," *Chem. Ber.*, **113**, 3456 (1980).

       (d) C. Rüchardt, "Steric Effects in Free Radical Chemistry," in *Topics in Current Chemistry*, vol. 88, Springer-Verlag, 1980, p. 1.

(e) C. Rüchardt and H.-D. Beckhaus, "Towards an Understanding of the Carbon—Carbon Bond," *Angew. Chem. Int. Ed. Engl.,* **19**, 429 (1980).

(f) G. Hellman, S. Hellman, H.-D. Beckhaus, and C. Rüchardt, "Thermolabile Hydrocarbons, XVI. Thermal Stability, Strain Enthalpy, and Structure of sym. Tetrasubstituted Ethanes," *Chem. Ber.,* **115**, 3364 (1982).

192. (a) P. C. Wong, D. Griller, and J. C. Scaiano, *J. Am. Chem. Soc.,* **104**, 5106 (1982).

(b) C. Chatgilialoglu, J. C. Scaiano, and K. U. Ingold, *Organometallics,* **1**, 466 (1982).

(c) G. D. Mendenhall, L. C. Stewart, and J. C. Scaiano, *J. Am. Chem. Soc.,* **104**, 5109 (1982).

(d) V. Malatesta and J. C. Scaiano, *J. Org. Chem.,* **47**, 1455 (1982).

(e) M. V. Encinas and J. C. Scaiano, *J. Am. Chem. Soc.,* **103**, 6393 (1981).

(f) B. P. Roberts and J. C. Scaiano, *J. Chem. Soc. Perkin II,* 905 (1981).

(g) P. K. Das, M. V. Encinas, S. Steenken, and J. C. Scaiano, *J. Am. Chem. Soc.,* **103**, 4162 (1981).

(h) A. Effio, D. Griller, K. U. Ingold, J. C. Scaiano, and S. J. Sheng, *J. Am. Chem. Soc.,* **102**, 6063 (1980).

(i) D. Griller, J. A. Howard, P. R. Marriott, and J. C. Scaiano, *J. Am. Chem. Soc.,* **103**, 619 (1981).

(j) J. C. Scaiano, *J. Am. Chem. Soc.,* **102**, 5399 (1980).

(k) S. K. Wong, *Int. J. Chem. Kinet.,* **13**, 433 (1981).

(l) K. Fuke, A. Hasegawa, M. Ueda, and M. Itoh, *Chem. Phys. Lett.,* **84**, 176 (1981).

(m) C. R. Park, S. A. Song, Y. E. Lee, and K. Y. Choo, *J. Am. Chem. Soc.,* **104**, 6445 (1982).

193. (a) P. Labbe, R. Le Goaller, H. Handel, G. Pierre, and J.-L. Pierre, *Electrochim. Acta,* **27**, 257 (1982).

(b) V. V. Zorin, N. A. Batyrhaev, S. S. Zlotskii, D. L. Rakhmankulov, and R. A. Karakhanov, *Dokl. Akad. Nauk SSSR,* **255**, 626 (1980).

194. G. B. Ellison, P. C. Engelking, and W. C. Lineberger, *J. Phys. Chem.,* **86**, 4873 (1982).

195. T. F. Hunter and K. S. Krisjansson, *J. Chem. Soc. Faraday II,* 2067 (1982).

196.* (a) K. Toriyama, K. Nunome, and M. Iwasaki, *21st ESR Symp.,* Tsukuba, 10/26/82, Abstr. 26P07, p. 48.

(b) K. Toriyama and M. Iwasaki, *46th Annual Meeting Chem. Soc. Japan,* Niigata, 10/3/82, Abstr. 2P12, p. 13.

197.* (a) Y. Endo, S. Saito, and E. Hirota, *IMS Ann. Rev.,* II-A-4 (1982).

(b) Y. Endo, S. Saito, and E. Hirota, *Symp. Struct. Chem.,* Tokyo, 1982, Abstr. 2D21, p. 170.

8

NITRENES

WALTER LWOWSKI

Department of Chemistry, New Mexico State University,
Las Cruces, New Mexico 88003

I. INTRODUCTION

During the last few years, nitrene chemistry has been active on several fronts. More physical data on nitrenes have been obtained and synthetic applications of intramolecular nitrene reactions have multiplied. Rather exotic molecules have been made, and new lines of investigation are beginning to bear fruit, promising more progress in years to come.

II. ALKYLNITRENES

The vacuum gas-phase pyrolysis of methyl azide was monitored by photo-electron spectroscopy.[1] Methyl azide decomposed only at 800 K (527°C), producing methyleneimine, $H_2C=NH$. MNDO calculations predict the activation energy for the singlet H_3C-N to $H_2C=NH$ conversion to be only 2.4 kcal/mol, leaving little chance for nitrene reactions.[1] Calculations of triplet methylnitrene show it to be 18 kcal/mol lower in energy than triplet $H_2C=NH$.[2] Singlet $H_2C=NH$ was calculated to be 46 kcal/mol below the triplet nitrene, and no substantial barrier was found for the conversion of singlet nitrene to $H_2C=NH$. Thus the results of the two studies agree well.

α-Substituted alkyl azides have been shown to decompose, in some systems, without nitrene formation. Neighboring group participation can occur (as in the well-known Curtius rearrangement), or radical mechanisms can lead to fragmentation prior to the loss of nitrogen.[3] Thus benzyl azides with sulfide, sulfoxide, or sulfone groups on the benzyl carbon decomposed with neighboring group participation (low decomposition temperature), radical fragmentation (CIDNP), or nitrene formation, respectively (Eq. 8.1, 8.2, 8.3).

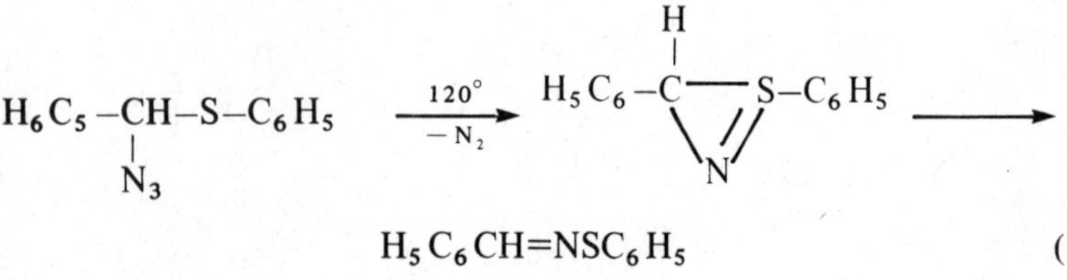

$$H_6C_5-CH-S-C_6H_5 \xrightarrow[-N_2]{120°} H_5C_6-\overset{\overset{H}{|}}{C}\diagdown\!\!\!\diagup S-C_6H_5 \longrightarrow$$
$$\underset{N_3}{|}$$

$$H_5C_6CH=NSC_6H_5 \qquad\qquad (8.1)$$

$$H_5C_6-\underset{\underset{N_3}{|}}{C}H-\overset{\overset{O}{||}}{S}-C_6H_5 \xrightarrow{70°} [H_5C_6CHN_3]^{\cdot} \; [H_5C_6SO]^{\cdot} \longrightarrow$$

$$H_5C_6\overset{\overset{O}{||}}{S}N=CHC_6H_5 + \text{other products} \qquad (8.2)$$

$$33\%$$

$$H_5C_6-\underset{\underset{N_3}{|}}{C}H-SO_2C_6H_5 \xrightarrow{150°} H_5C_6-\underset{\underset{:N:}{|}}{C}H-SO_2C_6H_5 \longrightarrow \text{products}$$

$$(8.3)$$

Photolysis of N-alkylbenzoquinoneimine-N-oxides gives alkylnitrenes, apparently in their triplet states, as indicated by their hydrogen abstraction to give alkylamines, and by addition to benzoquinone. Some cyclization to the pyrrolidine was observed when *tert*-octyl was the alkyl group (see Eq. 8.4).[4]

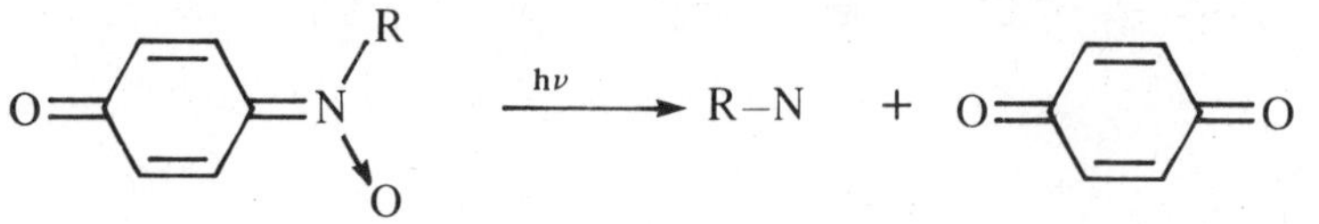

$$(8.4)$$

$$R = -C(CH_3)_3, \qquad -C(CH_3)_2-CH_2-C(CH_3)_3$$

The alkylnitrene–imine rearrangement continues to be studied and utilized for the construction of highly strained imines. It is not always clear whether a given reaction involves a free nitrene or is concerted.[5] The photolysis of 1-azidobicyclo[2.2.2]octane in inert solvents gave a polymer, but in alcohol solvents addition to the intermediate bridgehead imine gave 1-alkoxy-2-azabicyclo[3.2.2]nonane in 73% yield (Eq. 8.5).[6] Similarly, photolysis in methanol of 1-azidobicyclo[3.2.1]octane gave 1-methoxy-2-azabicyclo[3.2.2]nonane.[7] 1-Azidoadamantane reacted analogously with cyanide ion in a two-phase system,[8] as did 1-azidohomoadamantane.[9]

$$(8.5)$$

Ring expansion of monocyclic systems is also of synthetic interest. Isochromenes were converted to benzo-1,3-oxazepines in 30–50% yields (Eq. 8.6).[10] Annulenes of ring size 14 and 18, containing a nitrogen atom in the annulene system, were made by photochemically rearranging various condensed azidocyclopropanes;[11,12] Equation 8.7 gives one example. A ring expansion of the azidocyclopropane is followed — or is concerted with — valence isomerizations to aromatic aza[14]- and aza[18]annulenes. The enlargement of a three- to a four-membered ring, the azidocyclopropane to azetine conversion, has been known for a number of years.[13] It has recently been used with 1-azidocyclopropane-1-thio ethers (Eq. 8.8).[14] The presence of hetero atoms in an C-azido-substituted ring need not interfere with ring-expanding alkylnitrene rearrangements. A benzo-1,3-dithiolane was converted to a mixture of isomeric benzodithiazines (Eq. 8.9).[15]

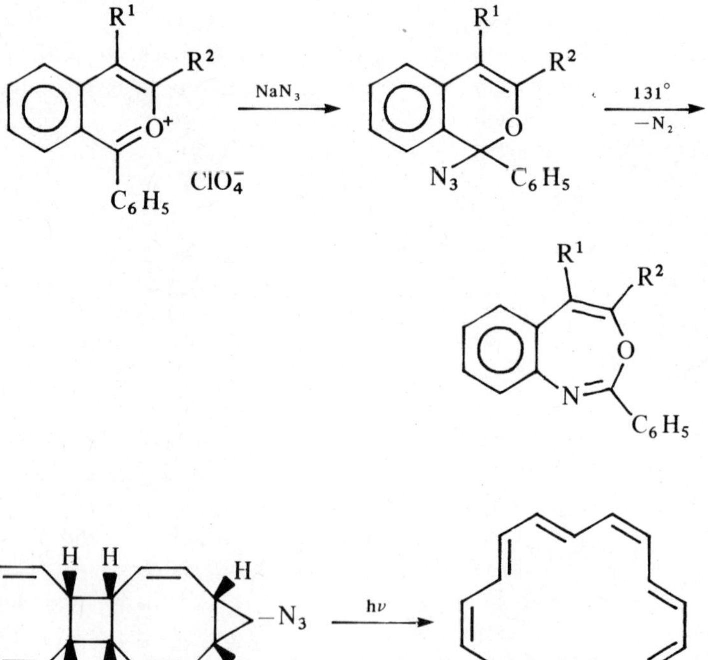

$$(8.6)$$

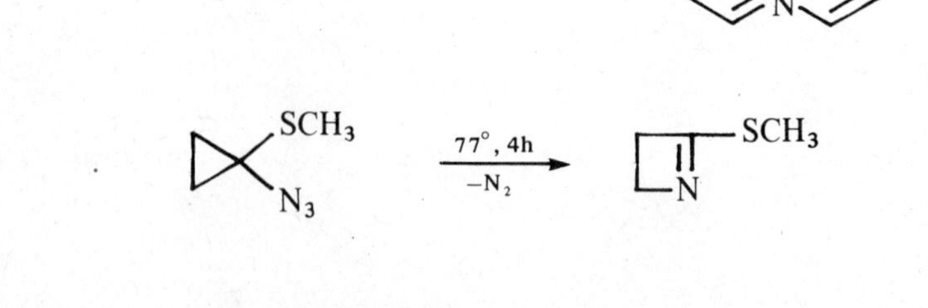

$$(8.7)$$

$$(8.8)$$

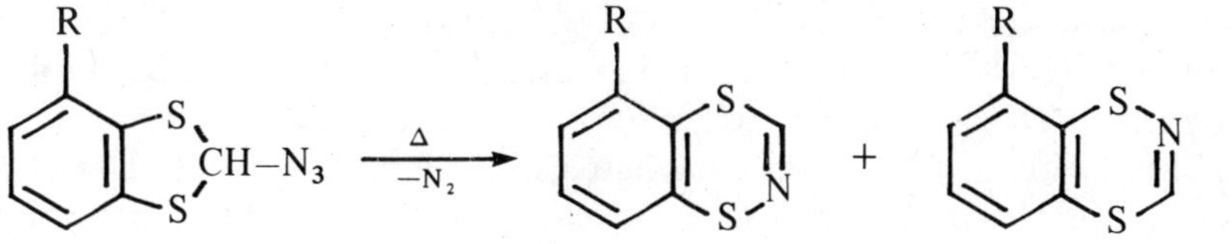

$$(8.9)$$

III.   VINYLNITRENES

Vinylnitrenes continue to attract interest, both because of their synthetic potential and because of their valence tautomerisms. For reviews, see References 16 and 17.

Thermolysis and photolysis of vinyl azides give azirines, and the latter readily ring-open to vinylnitrenes, which either revert to azirine or react further. Photolysis of the azirines, however, yields nitrile ylides, which then react. A general scheme for the decomposition of β-styryl azide is given, simplified, in (Eq. 8.10), and examples and references are found below.

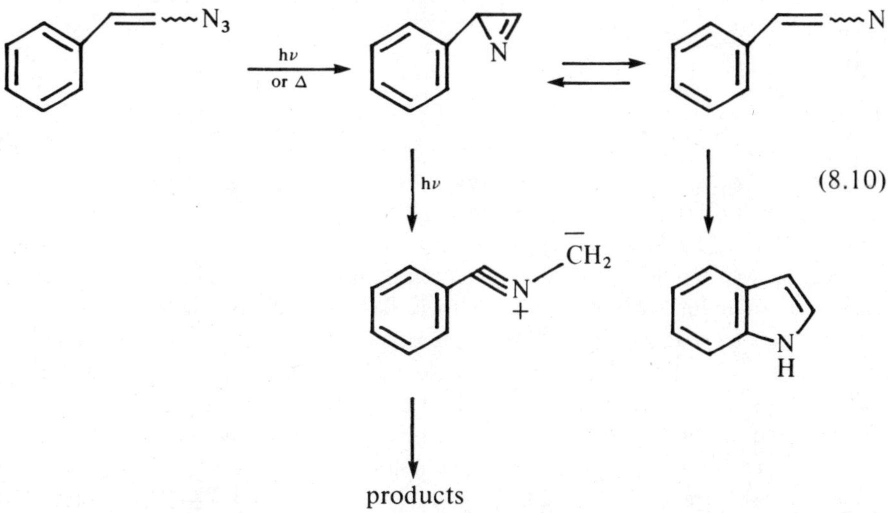

$$(8.10)$$

The kinetic parameters of vinyl azide thermolyses suggest that the azirines are formed from the azides directly,[18] and not through nitrene intermediates. This contention rests on the similarity of the vinyl azide thermolysis parameters to those of the Curtius rearrangement, whereas the parameters for the decomposition of, for example, phenyl azide are drastically different. For the thermolysis of vinyl azides, mostly aryl substituted, the activation energies were between 26 and 30 kcal/mol, and $\Delta S^{\ddagger}$ was between -3 and $+5$ e.u.[18] The Curtius rearrangement has E_a around 28 kcal/mol and $\Delta S^{\ddagger}$ around 4.2 e.u.,[19] whereas phenyl azide thermolysis has $E_a = 39$ kcal/mol and $\Delta S^{\ddagger} = 18.7$ e.u.

The question of an azirine–vinylnitrene equilibrium, and the rates of its forward and back reactions, is not answered by the kinetic studies. Older work[20] has been interpreted in favor of a rapid equilibrium. Recent rate data show the equilibrium to exist and the rate of reversion of the nitrene to the azirine to be quite high. Optically active 2-methyl-3-phenylazirine, when heated, racemized much faster than it rearranged to 2-methylindole, with $k_{rac.} = 2.57 \times 10^{-4}$/sec at $120°$ (Eq. 8.11).[21]

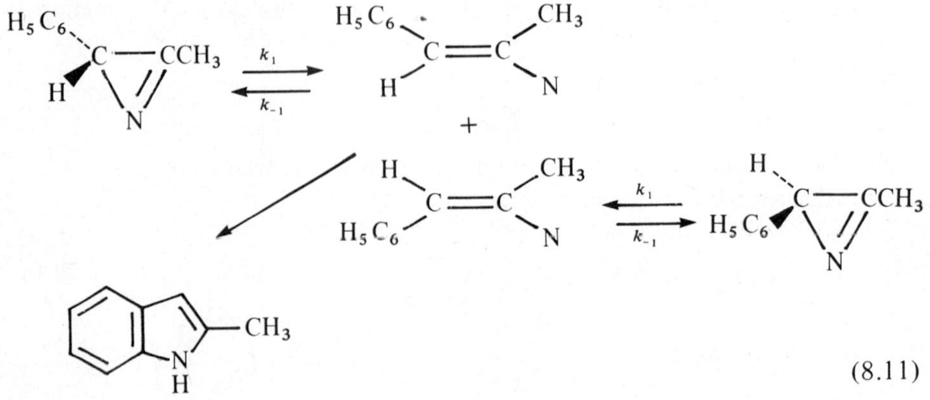

(8.11)

The cyclization of ortho-substituted β-styryl azides can be modified by catalytic quantities of iodine.[22] For example, ethyl o-benzyl-l-azidocinnamate upon thermolysis gave a 40% yield of the corresponding indole and a 25% yield of 3-ethoxycarbonyl-1,2-dihydro-1-phenylisoquinoline. Addition of catalytic quantities of iodine changed these yields in favor of the isoquinoline, to <10 and 40%, respectively (Eq. 8.12).[22] Apparently, the presence of the iodine causes the nitrene to go to its triplet state, which then reacts with the benzyl CH_2 by abstraction-recombination. This is not a mere heavy atom effect. The authors conducted experiments in 2-iodopropane solution without observing favored formation of the dihydroisoquinoline. One may thus assume the formation of nitrene–iodine adduct, which loses iodine to give –ṄI or another species capable of abstracting H· from the benzyl CH_2. The first step of the azide thermolysis is likely to be azirine formation, followed by thermal opening to the nitrene. Photolysis of 1-azidocinnamate gave an azirine trimer[23] rather than a nitrene cyclization product, and the structure of the trimer, proved by X-ray analysis, agrees with a nitrile ylide as the reactive intermediate. Also, the

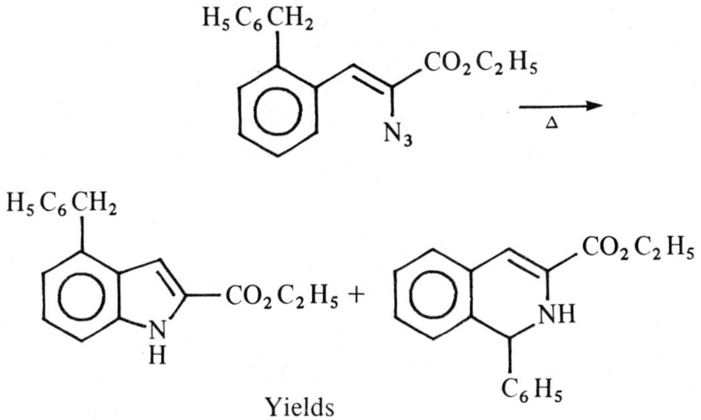

$$(8.12)$$

Yields		
without I_2	40%	25%
with I_2	< 10%	40%

pyridine analogue of azidocinnamates, ethyl α-azido-β-(2-pyridyl)acrylate, cyclized upon thermolysis to give a 95% yield of a pyrazolopyridine (**1**). However, upon photolysis a 55% yield of the imidazopyridine (**2**) was formed (Eq. 8.13).[23]

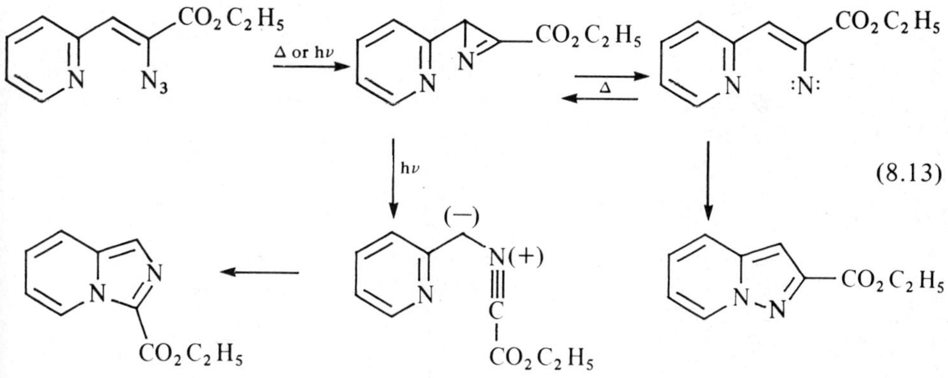

$$(8.13)$$

α-Azidocinnamates containing unsaturated orthosubstituents react, upon heating, both through azide cycloaddition and by the azirine–nitrene mechanism.[24]

α-Azido-β-(3-methyl-2-thienyl)acrylates reacted preferentially at the thiophene methyl group (Eq. 8.14).[25] In this latter reaction, the product is formed

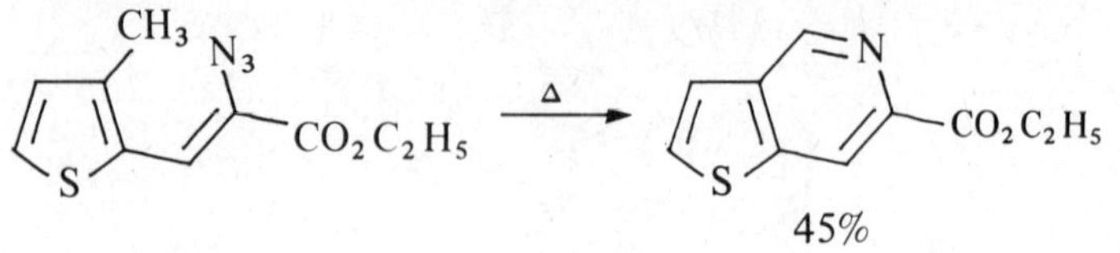

$$\text{(8.14)}$$

45%

by dehydrogenation (possibly by more nitrene) of an intermediate dihydro product.[25]

IV. ARYL AND HETEROARYLNITRENES

A. Spectra and Theory

Electronic absorption spectra of some arylnitrenes have been measured in ethanol glass matrices at 77 K. With electron-donating or electroneutral substituents on the aryl group, the spectra were similar to those of benzyl radicals. This is expected, inasmuch as the triplet nitrene absorptions should be largely determined by the interactions of the nitrene p_y orbital with the π-system of the ring. Absorption bands at 400–450, 320, and 250 nm were observed.[26] Examples are given in Table 8.1. Those of the arylnitrenes which did not contain electron-withdrawing substitutents showed T_1 to T_0 luminescence.[26]

The infrared spectrum of pentafluorophenylnitrene has been observed in nitrogen or argon matrices at 12 K.[27] This was possible because pentafluorophenylnitrene does not, under these conditions, ring-expand spontaneously.

Electron-spin resonance spectra were obtained by the photolysis of matrix-isolated 8-substituted 1-azidonaphthalenes at 4 K. Mixtures of quinomethane

TABLE 8.1 Electronic absorption of some *para*-Substituted Phenylnitrenes in ethanol glasses at 77 K[a]

$p-R-C_6H_4-N$	Major Absorption Bands (λ in nm, ϵ in $mol^{-1}\,cm^{-1}$)					
(R)	λ	ϵ	λ	ϵ	λ	ϵ
H	243	8,500	304	2200	383	860
CH_3	250	11,500	315	5600	385	1100
NH_2	309	10,800	356	2100	437	3000
OCH_3	271	16,200	330	7600	388	2200
NO_2	312	13,800	380	700	500	370
$COCH_3$	276	16,200	410	870	508	420

[a] Data from Reference 26.

TABLE 8.2. Zero Field esr. parameters D/hc for 8-substituted 1-nitrenona-phthalenes at 4 K in argon matrices[a]

8-Substituent	D/hc	8-Substituent	D/hc
H	0.79	OH	0.75
CH$_3$	0.79	Cl	0.72
CO$_2$H	0.78	NH$_2$	0.62
NO$_2$	0.78	Br	0.57
N$_3$	0.75		

[a] Data from Reference 28.

diradicals and triplet nitrenes were obtained. Fortunately, the two have widely separated esr spectra. Some zero-field D/hc values are given in Table 8.2.[28] On warming of the matrix to 98 K, the quinomethane diradical esr signals disappeared, whereas the nitrene signals persisted.

The decreasing spin-spin interactions, reflected in Table 8.2 as decreasing D/hc values, indicate increasing interaction of a nitrene spin with the rest of the system in the substituent order: $H \sim CH_3 < CO_2H < OH < Cl < Br$.[28]

B. Unimolecular Arylnitrene Reactions

Before discussing nitrene reactions which (at least formally) do not involve other species, it seems appropriate to point to the possibility of chain decomposition of aryl azides, a process that could interfere with the investigation of such systems. The photolyses of substituted phenyl azides in several deaerated solvents proceeded with quantum yields as high as 300. Even in 10^{-5} molar solutions, quantum yields larger than unity were found.[29] A possible mechanism is shown in (Eq. 8.15).[30]

$$Ar-N_3 + N-Ar \longrightarrow Ar-N=N-N=N-Ar \longrightarrow$$
$$Ar-N + N_2 + N-Ar \qquad (8.15)$$
$$\downarrow \qquad\qquad \downarrow$$
$$\text{chain} \qquad\quad \text{chain}$$

Unimolecular reactions of the singlet phenylnitrene system have been discussed for a considerable time.[5,16,19,31] The available evidence agrees with the scheme of (Eq. 8.16), in which significant intermediates include both an azanorcaradiene and an azacycloheptatetraene, in addition to a singlet phenyl-

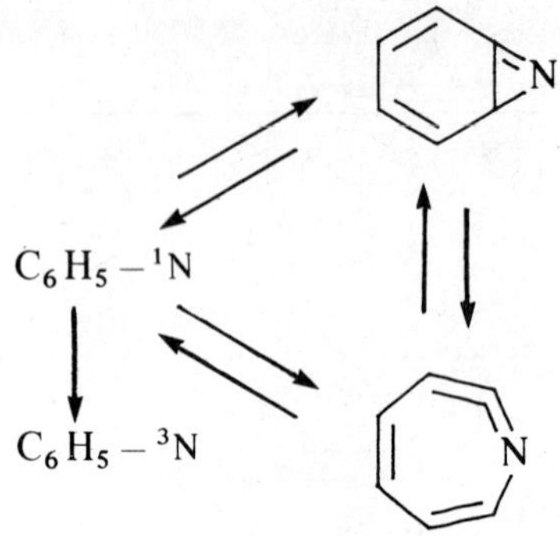

(8.16)

nitrene. Products may be derived from any of these intermediates, although the azacycloheptatetraene is not strictly required to explain the products. It has, however, been observed in matrix isolation spectra, so that there is no doubt of its physical existence. The ring-expansion sequence has been discussed lately.[32] The analogous carbon system, containing 1,2,4,6-cycloheptatetraene, has recently been demonstrated.[33]

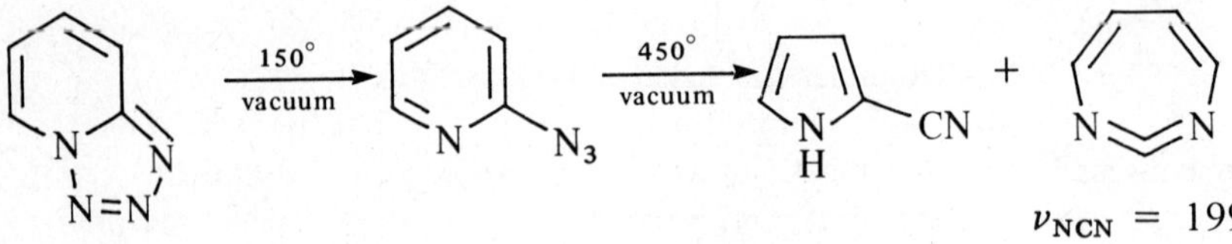

(8.17)

$$\nu_{NCN} = 1995 \text{ cm}^{-1}$$

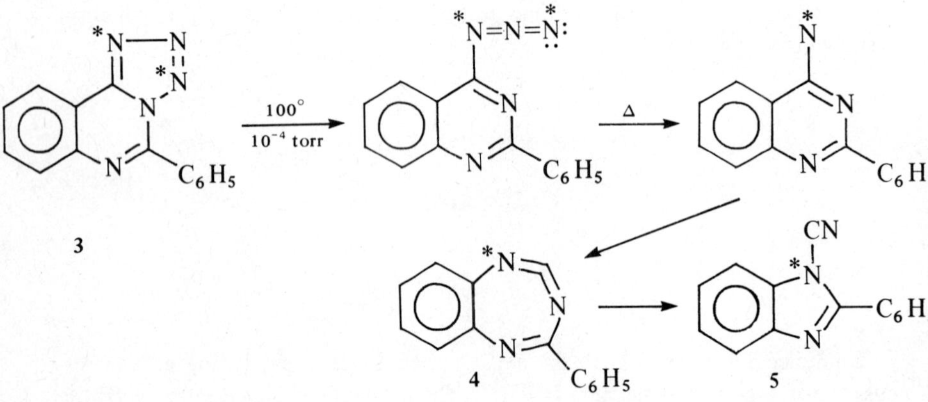

$$\nu_{NCN} = 2010, 1610 \text{ cm}^{-1}$$

(8.18)

Its seems that perfluorophenyl azides have a lesser tendency to involve ring expansion in their reaction paths. No ring-expansion product was observed by spectrometry in matrix-isolated perfluorophenylnitrene,[27] and ring-expansion products were formed only in modest yields in the flash vacuum pyrolysis of $C_6F_5-N_3$.[34] 2-Pyridylnitrene, produced by flash vacuum pyrolysis, gave 1,3-diaza-1,2,4,6-cycloheptatetraene, which was trapped at $-170°C$ and displayed an infrared absorption at $1995\ cm^{-1}$ (Eq. 8.17).[35] Similarly, the benzotetrazolopyrimidine (3) in (Eq. 8.18), labeled with ^{15}N, gave 4 and 5.[36]

C.   Aryl and Heteroarylnitrene Reactions in Solution

Solution photolysis and thermolysis of aryl and heteroaryl azides has become a valuable synthetic tool.[5,31,37] Aromatic amines are accessible by opening of intermediate azanorcaradienes, and azepines by stabilizing ring-expanded intermediates or azanorcaradienes. Thus bicyclic azepines, diazepines, and quinolyl- and isoquinolyldiamines have been prepared from quinolyl- and isoquinolyl azides. Solvents such as tetrahydrofuran, tetramethylethylenediamine, and dichloromethane stabilize the singlet nitrenes in these systems, suppressing hydrogen abstraction reactions and increasing the yields of products from nucleophilic addition to intermediates.[38,39] The photolysis of 4-azido-7-chloroquinoline in KOH/CH₃OH/dioxane, for example, gave the methoxybenzo-1,4-diazepine and its hydrolysis product (Eq. 8.19).[39]

20% yield   20% yield

(8.19)

Irradiation of 4-azidopyridines, -quinolines, and -isoquinolines in $6\,N$ HCl or HBr gave mixtures of the ring-opening products of the intermediate azirines, according to the example in (Eq. 8.20). The corresponding N-oxides behave differently; 3-halo-4-aminopyridine-N-oxides are formed in high yields (Eq. 8.20). The reason for the absence of isomers with the nitrogen function in the 3-position is, presumably, that the N-oxides react by way of a nitrenium ion, rather than an azirine.[40]

2-Azidophenazines, irradiated under neutral conditions in alcohols, give 2-

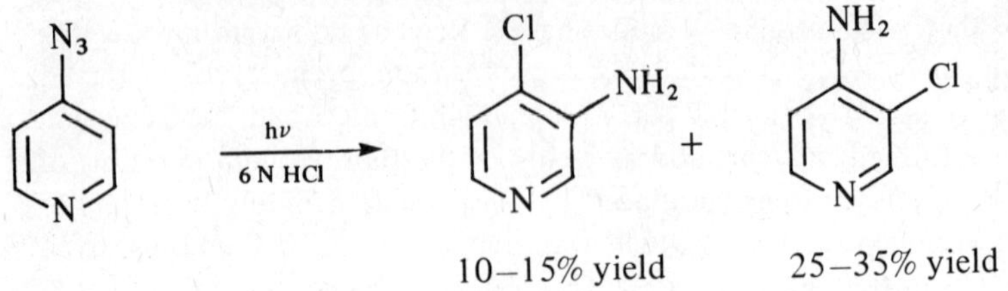

$$(8.20)$$

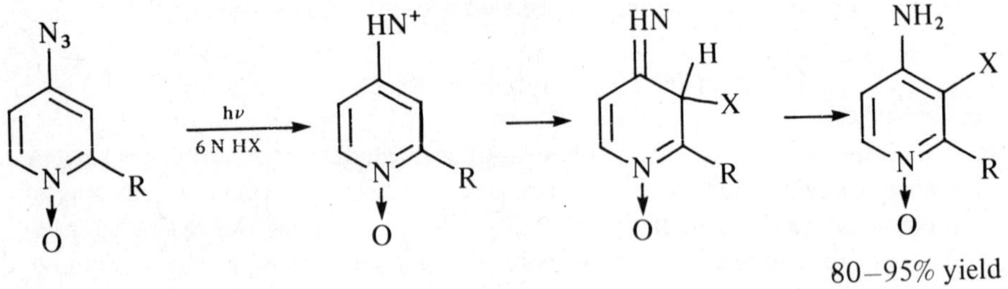

10–15% yield   25–35% yield

80–95% yield

amino-1-alkoxyphenazines in low yields. In acidic solution, the 1-methoxy-2-aminophenazine was formed in 30% yield, and its 2-alkoxy-1-amino isomer in 36% yield. In basic solution, ring-expansion product was isolated in 15% yield, whereas, in the presence of cyanide ion, a 46% yield of 1-amino-2-cyano-phenazine was obtained (Eq. 8.21).[41]

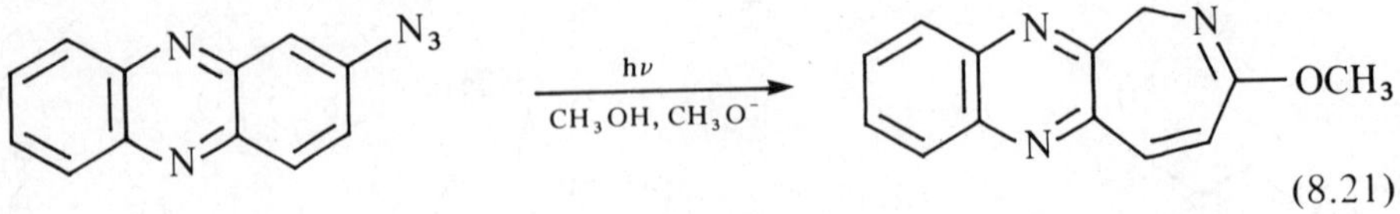

$$(8.21)$$

In acetic acid, the thermolysis of phenyl azide proceeds at the same rate as it does in dioxane solution.[42] From this reaction, and from the corresponding photolysis, a product mixture is obtained which contains azepinone (and, in the presence of phenol, 2-phenoxy-3H-azepine) and products of nucleophilic attack on the phenyl ring, such as 2-acetoxyaniline.[42] Apparently, the primary azide decomposition product is phenylnitrene, which either ring expands and is stabilized by solvent, or forms a nitrene–acetic acid complex, which is attacked on the ring (Eq. 8.22).

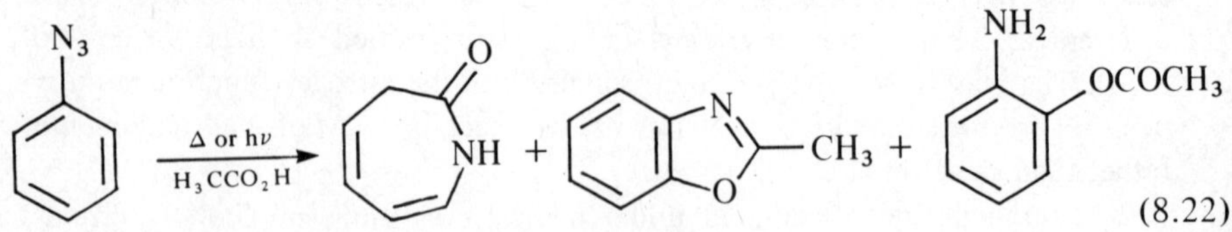

$$(8.22)$$

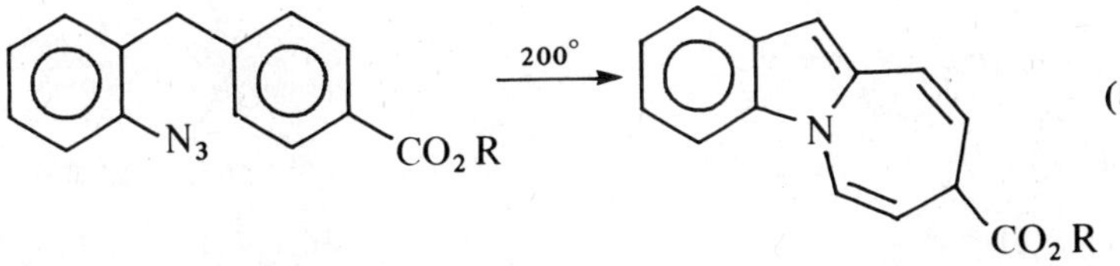

$$(8.23)$$

Phenyl rings bearing electron-withdrawing substitutents are normally not attacked by arylnitrenes. However, at 200° an intramolecular attack on an electronically isolated benzoate has been reported (Eq. 8.23).[43]

Singlet nitrenes react preferentially in single-step, concerted processes, during which the electrons on the nitrene and the reactant all remain paired. C–H insertion, concerted addition to multiple bonds, and addition to unshared electron pairs on hetero atoms are typical examples. Triplet nitrenes, in contrast, often behave like 1,1-diradicals, performing sequentially two radical reactions, such as hydrogen abstraction, sometimes followed by combination of the resulting radical pair – NH$^{\bullet}$ $^{\bullet}$X, or by further hydrogen abstraction to give – NH$_2$ functions. Examples were mentioned in the previous volume of this series.[31] Suschitzky's group at Salford has provided many more examples and synthetic applications.[37] A simple scheme for generalized arylnitrene reactions with different reaction courses for the singlet and the triplet is given in (Eq. 8.24). A specific example is the decomposition of 1-isopropyl-2-(o-azidophenyl) benzimidazole.[44] Thermolysis gave cyclization to nitrogen, whereas triplet-sensitized photolysis resulted in hydrogen abstraction and radical recombination

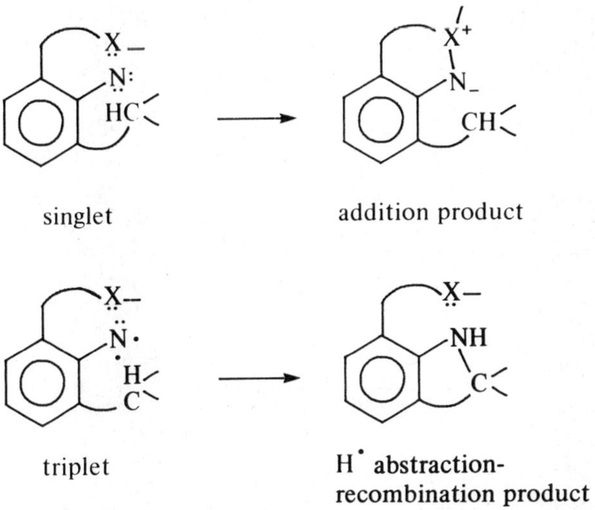

singlet   addition product

triplet   H$^{\bullet}$ abstraction-
                       recombination product

$$(8.24)$$

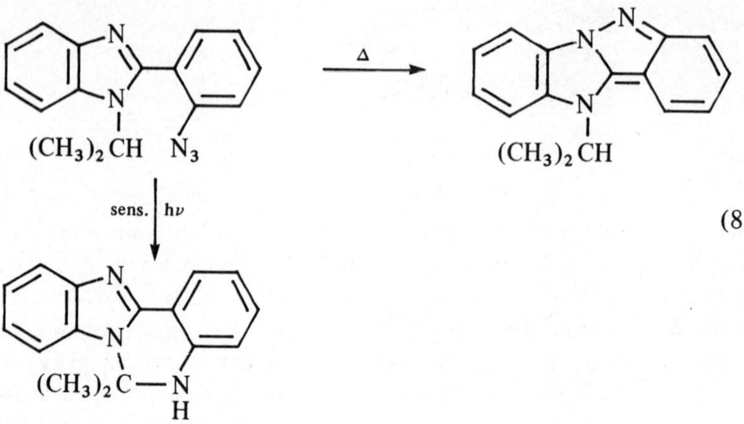

(8.25)

at the isopropyl group (Eq. 8.25). The corresponding 2-(*o*-azidophenyl)benzo-1, 3-thiazole, however, gave only one product, regardless of the mode of azide decomposition.[44] No cyclization to sulfur was found in this and many other cases, although nitrene addition to electron pairs on sulfur is usually facile. Thermodynamic control due to reversibility of the addition to sulfur might be the reason. In a system in which a potentially more stable product does not compete, such cyclization to sulfur was indeed observed (Eq. 8.26).[45] The reasons for obtaining one and the same product, under singlet and under triplet conditions, in the benzothiazole reaction are not yet clear.[44] Some other systems behave analogously.[46,47] Electron-transfer or rapid triplet-singlet inter-conversion, together with slow forward reaction of the triplet, might be the causes (Eq. 8.27).[46,47]

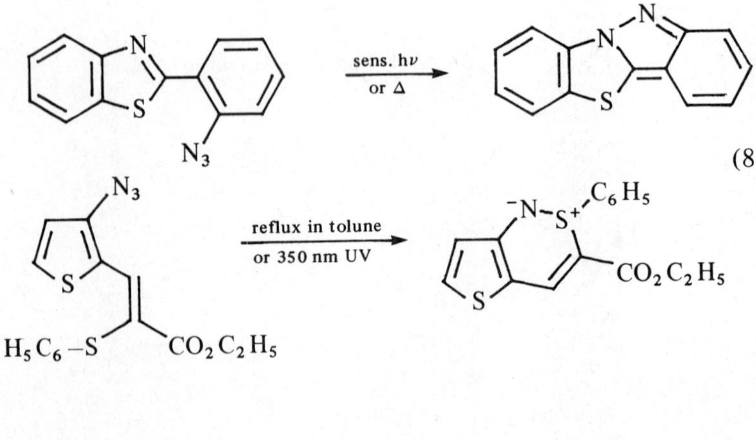

(8.26)

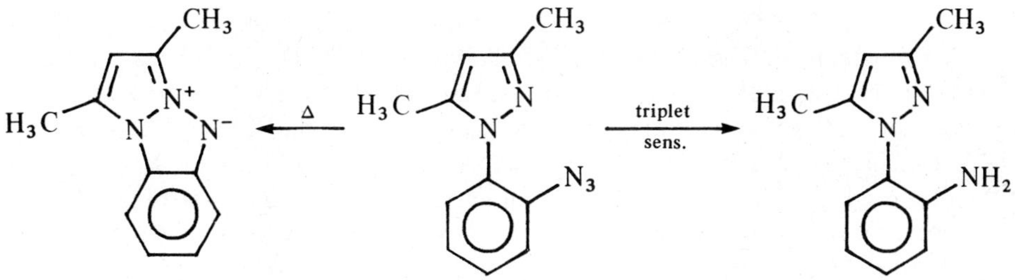

(8.27)

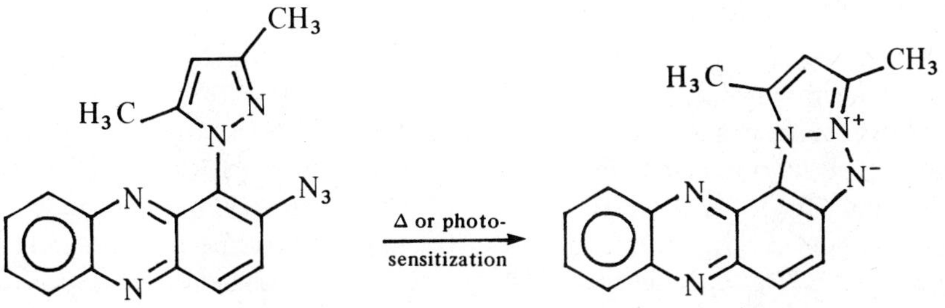

Intermolecular attack of phenylnitrene has been demonstrated in a case in which rearrangement of the adduct competes with the potential reversion to starting materials (Eq. 8.28).[48]

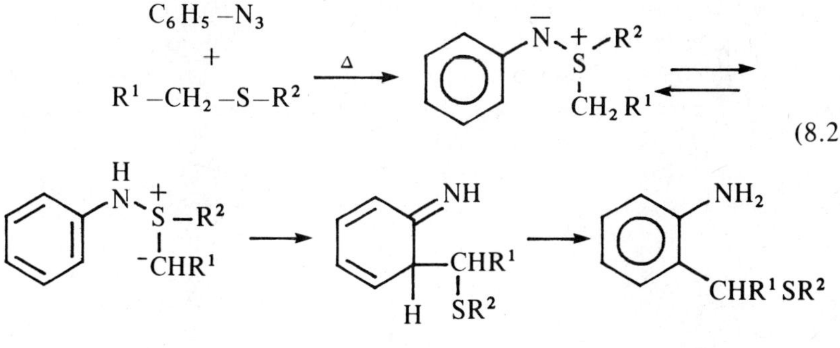

(8.28)

30% yield

D.   Photoaffinity Labeling

A large and fast-growing application of aryl and heteroaryl nitrenes is photoaffinity labeling. In a typical mode of application, molecules that are reversibly bound by biomolecules, such as enzymes, are endowed with an azido group. After binding to the biomolecule, irradiation converts the azide to a nitrene function. This is then supposed to react rapidly and irreversibly with a nearby part of the biomolecule, making a permanent bond and labeling the site of attachment. Later, degradative or other studies pinpoint this location. It is often assumed that the arylnitrenes used insert into C–H bonds of the bio-molecule. In view of the known reactivity of arylnitrenes, this is unlikely in most cases, at least as long as singlet insertion is assumed. Triplet arylnitrenes can abstract a hydrogen atom, and the radical pair so formed can combine (after spin inversion) to make a covalent bond. However, in view of the three steps involved (one being spin inversion), such a reaction is likely to lead to loss of much of the labeling group to solvent reactions. Substituting an aryl group with electron-withdrawing groups (F, NO_2, etc.) increases the nitrene's reactivity (and often shifts the azide's light absorption to longer wavelengths), improving the labeling yield. Presumably, reactions with amino groups or sulfide links are more important than those with C–H bonds. Thus there seems to be room for improvement of the technique by constructing better azide precursors. These should be readily introduced into the molecules of biochemical interest, should give reactive nitrenes in high quantum yields, using light of a wavelength not harmful to the biosystem, and should not be toxic to the system of interest. Some such new label precursors have been proposed.[49]

Whatever the detailed chemistry of the labels presently in use, the method is successful enough to be applied widely. A very few recent examples follow, to illustrate the diversity of the technique. Phenyl azide groups, with no special activation by electron-withdrawing substituents, were found to be useful as part of peptide chains (p-azidophenylalanine),[50] or attached to phloretin,[51] or puromycin.[52] The intervention of azacycloheptatetraenes (see above) has been postulated for such applications.[53] Phenyl azide groups bearing activating substituents were used in peptide chains attached to glutathione[54] and phenyl-glycine,[55] or on carazolol,[56] atractyloside,[57] glutathione,[58] or as p-azidobenzoyl coenzyme A.[59] 8-Azidoadenine has been made[60] and applied,[61,62] as has azido-cytosine.[63]

V.   CARBONYLNITRENES

A.   Theory

Ab initio calculations of formylnitrene show the ground state to be triplet $^3A''$, and the first two excited states to be the $^1A'$ and $^1A''$ singlets, 36.8 and 39 kcal/mol above the ground state, respectively. Little conjugation between nitrogen and carbonyl was found. The ground state (triplet) C=O bond length was calculated to be 1.214 Å, longer than the 1.193 Å measured in formamide, but very close to the C=O bond length calculated by the SCF method for formamide, 1.212 Å. The C—N bondlength for the triplet formylnitrene was calculated as 1.447 Å, longer than the experimental (1.376 Å) and calculated (1.363 Å) values for formamide.[64]

B.   Intramolecular Reactions

Phenyl azidoformates, upon spray flash pyrolysis, cyclize to benzoxazolones in good yields.[65] α-Naphthyl azidoformate gave, under these conditions, only the naphthoxazolone formed by attack on the β-position. Biphen-2-yl azidoformate gave both 7-phenylbenzoxazolone (24% yield) and an azepine from attack on a β-position of the adjacent phenyl ring (Eq. 8.29). Blocking both ortho positions by using 2,6-dimethylphenyl azidoformate gave more complex products, derived, however, from initial attack on an ortho position.[65]

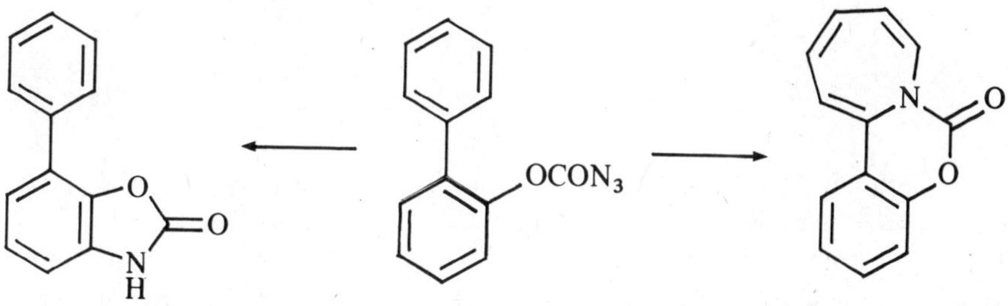

(8.29)

The spray flash pyrolysis of benzyl azidoformate at 350° gave the dimer of an azepine formed by attack on the ortho position (Eq. 8.30).[66]

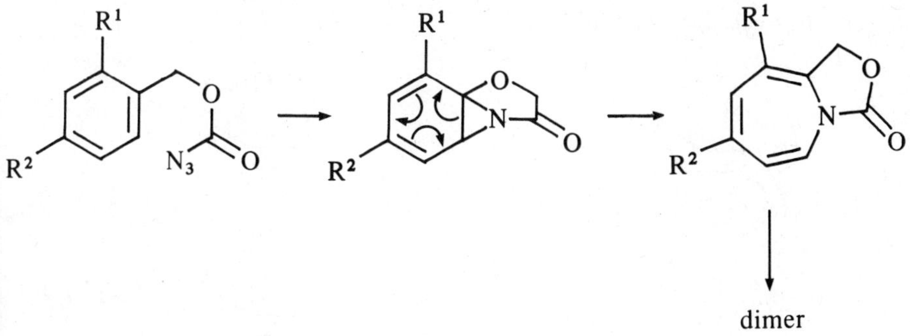

dimer

(8.30)

C.   Intermolecular Reactions

Ethoxycarbonylnitrene proved reactive enough to convert hexafluorobenzene to hexafluoro-*N*-carbethoxyazepin in 21% yield.[67] Enamines add ethoxycarbonylnitrene on the nitrogen, rather than across the double bond, in accord with the general experience that unshared electron pairs — if able to form a stable bond — are preferred as targets over π-systems (Eq. 8.31).[68]

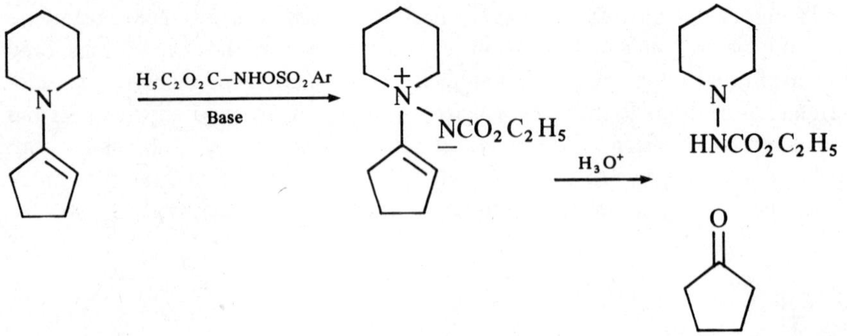

$$(8.31)$$

The sulfur in 4-*tert*-butylthiane is attacked indiscriminantly from the axial or equatorial sides by ethoxycarbonylnitrene, to give a 1–1 mixture of the cis and the trans sulfimines (Eq. 8.32).[69] Tosylnitrene behaves similarly, but bis(ethoxycarbonyl)carbene and bis(acetyl)carbene add exclusively in the equatorial position. After considering equilibration and *S*-inversion, the authors[69] conclude that the nitrene addition is kinetically controlled. The stereochemistry of the carbene products is governed by the greater steric demand of the carbenes, in comparison with ethoxycarbonylnitrene. This agrees with the finding that dimethyl- and diisopropylsulfide are equally reactive towards ethoxycarbonylnitrene.[69]

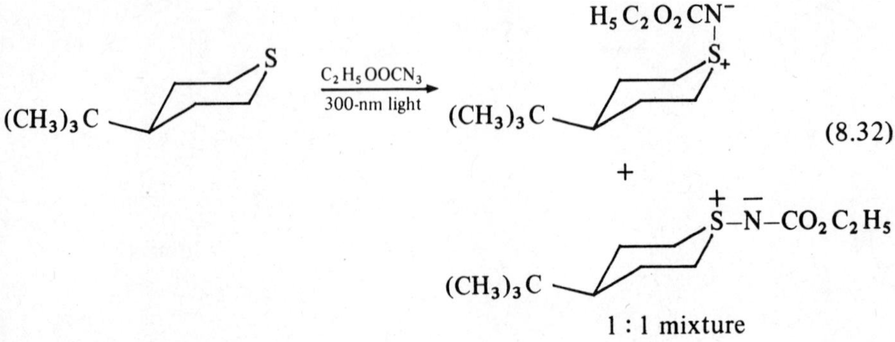

$$(8.32)$$

1 : 1 mixture

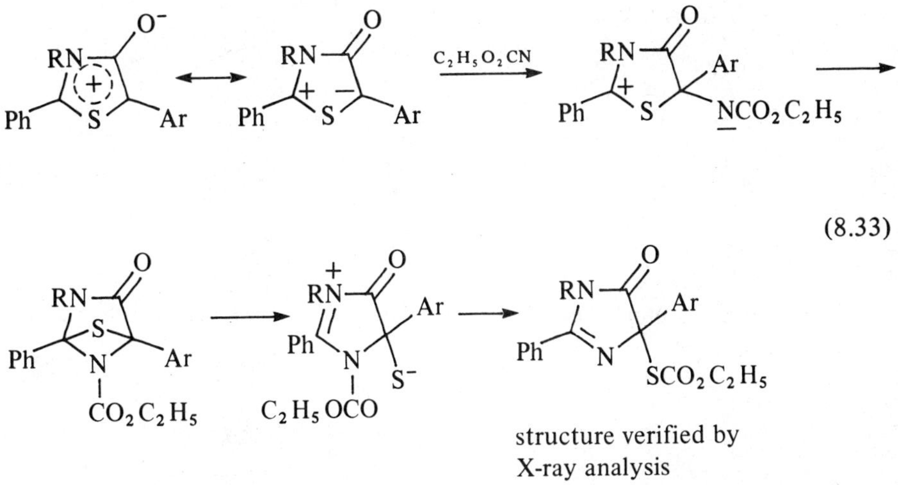

(8.33)

structure verified by
X-ray analysis

A mesoionic thiazolone adds ethoxycarbonylnitrene, resulting in a ring transformation (Eq. 8.33).[70]

Ethoxycarbonylnitrene adds stereospecifically to E and Z-1,2-diisopropyl-diimine (Eq. 8.34). The diphenyl- and ditolyldiimines gave the same azimine from both isomers.[71] The azimines show the carbonyl absorption at 1670–1680 cm^{-1}, a frequency lower than that of carbamates. Fairly strong absorption is observed between 280 and 290 nm (ϵ 7000–8000).[71]

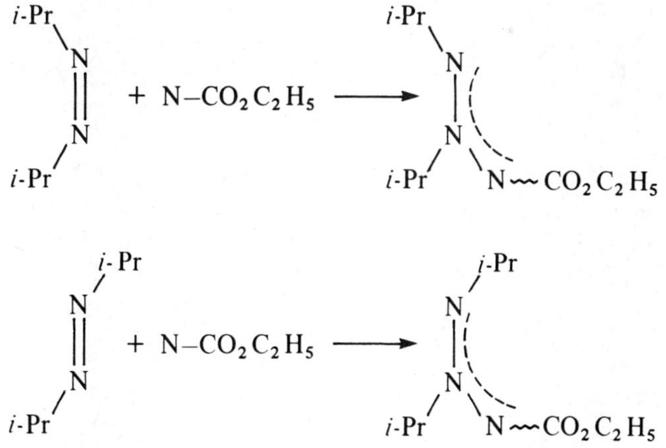

(8.34)

Irradiation of either stereoisomer of 1-ethoxycarbonyl-2,3-diisopropyl-azimine gave a 20% yield of 1-ethoxycarbonyl-2,3-diisopropyltriaziridine, in the first synthesis of a triaziridine (Eq. 8.35).[72]

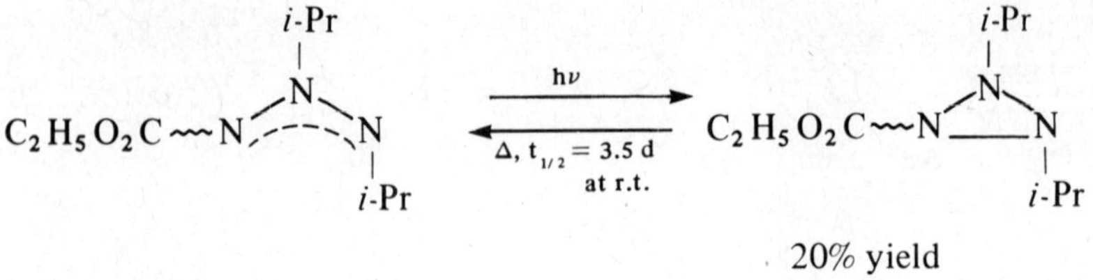

$$\text{C}_2\text{H}_5\text{O}_2\text{C}\text{---}\text{N} \underset{i\text{-Pr}}{\overset{i\text{-Pr}}{\text{N}\cdots\text{N}}} \quad \underset{\underset{\text{at r.t.}}{\Delta,\, t_{1/2} = 3.5\,\text{d}}}{\overset{h\nu}{\rightleftharpoons}} \quad \text{C}_2\text{H}_5\text{O}_2\text{C}\text{---}\text{N}\underset{i\text{-Pr}}{\overset{i\text{-Pr}}{\text{N}\text{---}\text{N}}}$$

20% yield

(8.35)

Phosphatidyl cholines bearing azidoformate groups have been developed as photoaffinity labels.[73]

VI.   SULFENYLNITRENES

2,4-Dinitrophenylsulfenylnitrene adds readily to electron-rich olefins, such as α- and β-methylstyrene (61% and 64% yields, respectively), but does not add to methyl acrylate, ethyl cinnamate, or 2-acetylbenzofuran.[74]

VII.   SULFONYLNITRENES

Intramolecular attack on phenyl in ω-azidosulfonylalkylbenzenes gave the corresponding sultams when the carbon chain contained 2, 3, or 4 CH_2 groups, either upon flash thermolysis of the azides,[75] or by thermolysis in solution.[76] With tetradecane as solvent, attack on solvent was the main reaction, and with two CH_2 groups in the chain, a 47% yield of insertion into tetradecane C–H bonds was obtained. The use of 1,1,2-trichloro-1,2,2-trifluoroethane as solvent increased the sultam yields.

Using nitrogen-labeled tosyl azide, it was found that a ^{15}N label in the sulfonyl azide group can be scrambled by the well-known diazo transfer reaction.[77] In the presence of base and tosyl amide, for example, diazo transfer between azide and amide anion occurs, giving azide derived from the amide, which contains the center nitrogen of the original azide at the end of the nitrogen chain of the new (amide derived) azide.

VIII.   CYANONITRENE

The high reactivity of cyanonitrene continues to stimulate attempts to generate it by routes other than through cyanogen azide. Swern found that the reaction of sodium cyanamide, tert-butyl hypochlorite, and tertiary amines produces N-cyanoaminimides, $R_3 N^+-N^--CN$.[78] The N–CN moiety is thus transferred, as a "cyanonitrene-like species." In the absence of tertiary amine, 1,2-dicyanodiimine, NC–N=N–CN is formed.[78] Methanol is used as the solvent and is apparently not attacked, indicating a nitrenoid intermediate which is less reactive than free cyanonitrene.

IX. PHOSPHINONITRENE

Phosphinonitrene and its isomers have been investigated by an *ab initio* calculation.[79] The singlet states of $H_2 P-N$, HP=NH, and $P-NH_2$, were all found to be planar, whereas triplet $H_2 P-N$ and $P-NH_2$ have pyramidal phosphino and amino groups, respectively. Singlet and triplet $H_2 P-N$ are high in energy relative to the other isomers – the singlet is 42.6 and the triplet 6.0 kcal/mol above triplet $P-NH_2$.

X. AMINONITRENES

Recent calculations show the energy of singlet $H_2 N-N$ to be 24.5 kcal/mol above that of *trans*-diimide,[80] with a very high barrier[80,81] for intramolecular conversion of *trans*-diimide to aminonitrene.

More detailed physical data on the "persistent" aminonitrenes (bearing two tertiary alkyls on the amino nitrogen) are now available. In agreement with theoretical studies done after 1976, the ground state is a singlet, stabilized by high double-bond character of the N–N bond.[82] N-nitreno-2,2,5,5-tetramethylpyrrolidine shows a n-π^* transition close to the predicted one at $\lambda_{max} = 497$ nm ($\epsilon = 20$) and $\lambda_{0,0} = 572$ nm (in dichloromethane). The N–N stretching frequency is 1638 cm^{-1} and is shifted to 1612 cm^{-1} when the terminal nitrogen is labeled with ^{15}N, in accord with Hooke's law.[83] The activation parameters for the dimerization of N-nitreno-2,2,5,5-tetramethylpyrrolidine to the tetrazene are $E_a =$

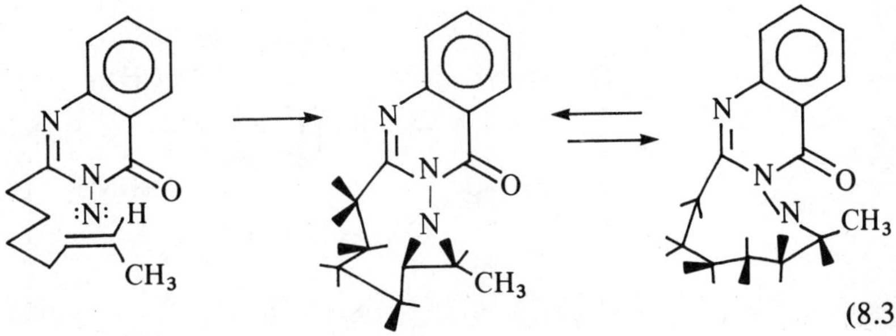

$$(8.36)$$

6.4 ± 0.9 kcal/mol and $\Delta S^{\ddagger} = -43.2$ e.u. 1,1-di-*tert*-butylaminonitrene is less stable, and its spectra had to be measured at $-127°$.[84]

The feasibility of intramolecular cyclization of aminonitrenes with electron-rich aryl groups requires not only proper electronic activation, but must also

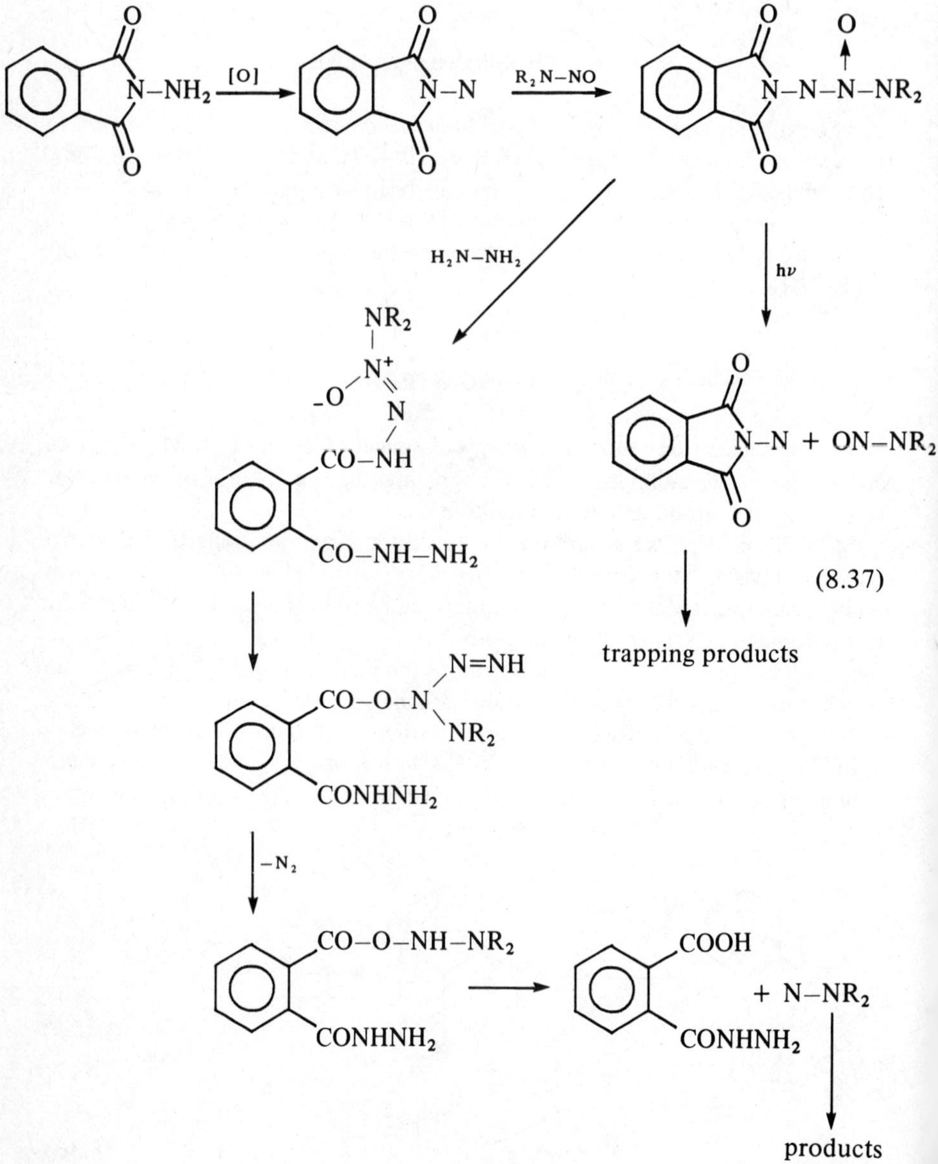

(8.37)

meet the steric requirements for a seven-membered or larger transition state.[85] Where steric factors are nonselective in such reactions, regioselectivity comes from electronic factors.[86] Such reactions produced flexible, N-ring-fused aziridines, in which cis and transfused aziridines were found to be in equilibrium (Eq. 8.36).[87]

Phthalimidonitrene displayed full regioselectivity in reactions with 1-alkyl-2-aryldiimines, always attaching itself to the nitrogen bearing the alkyl group.[88] The addition of phthalimidonitrene to aliphatic azo compounds is stereospecific and photoreversible, just as is the addition of ethoxycarbonylnitrene (see above).[89]

Nitrosamines add phthalimidonitrene to give 1,1-dialkyl-4,4-phthaloyl-2-tetrazene-2-oxides.[90] The reaction again is photoreversible. Nucleophilic attack on one of the phthaloyl carbonyls causes an intramolecular nucleophilic rearrangement, followed by fragmentation to give a new aminonitrene, derived from the original nitrosamine (Eq. 8.37).[90]

One of the resonance structures of diazoalkanes is formally an aminonitrene, a N-nitrenoketimine. 1,1-Cycloadditions such as one might predict for these species have indeed been reported, although it is by no means clear that the reaction mechanisms or transition states in these processes are like those of aminonitrenes, R_2N-N.[91,92] (Equation 8.38) gives an example.[92]

$$R_2C=\overset{+}{N}=\overset{..}{\underset{..}{N}} \longleftrightarrow R_2C=\overset{..}{N}-\overset{..}{\underset{..}{N}}$$

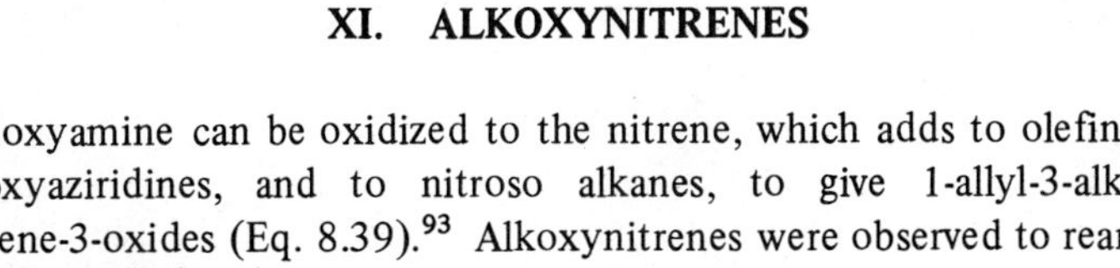

$$(8.38)$$

XI.   ALKOXYNITRENES

Allyloxyamine can be oxidized to the nitrene, which adds to olefins to give N-allyloxyaziridines, and to nitroso alkanes, to give 1-allyl-3-alkyl-1,2,3-oxadiazene-3-oxides (Eq. 8.39).[93] Alkoxynitrenes were observed to rearrange to nitrosoalkanes.[94,95]

$$(CH_3)_2CR-N=O \;+\; N-O-CH_2-CH=CH_2 \longrightarrow$$

$$(CH_3)_2CR-N=N-O-CH_2CH=CH_2 \qquad (8.39)$$
$$\underset{O}{\downarrow}$$

XII. NITRENIUM IONS

Nitrenium ions, R_2N^+, are expected to have triplet ground states, unless strong electronic interactions with the groups R are possible. Aryl- and acyl-arylnitrenium ions are predicted to have singlet ground states on the basis of semiempirical MNDO calculations.[96] In arylnitrenium ions only 20% of the positive charge is localized on the nitrogen, and nucleophiles are therefore predicted to attack the aryl group. In the case of Ar = phenyl, attack in the ortho and para positions is expected to form intermediates of the kind known in nucleophilic aromatic substitution. The azepinium ion was calculated to be 16 kcal/mol more stable than the phenylnitrenium ion, but a 39 kcal/mol activation energy for the ring expansion seems to prohibit this reaction.

The predicted attack on the aromatic ring of arylnitrenium ions has been demonstrated.   1-(3-Azidophenyl)-2-phenylethane, when decomposed in mixtures of $F_3C–SO_3H$ and $F_3C–CO_2H$, gives the product of attack on the para position of the inferred phenylnitrenium ion in 72% yield, (Eq. 8.40).[97]

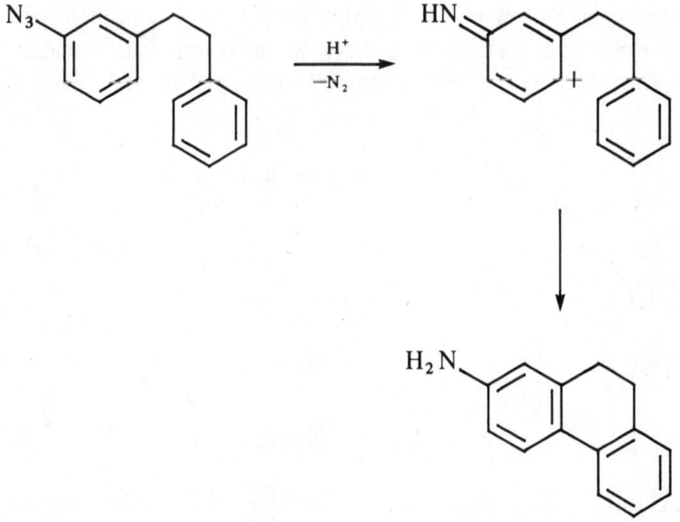

$$(8.40)$$

Strong acids, such as $F_3C–CO_2H$, greatly accelerate the decomposition of phenyl azide. In hydrocarbon solution, temperatures of 130° are required, but in trifluoroacetic acid decomposition is complete in 20 min at 25°. Thus the decomposing species is likely to be the protonated azide, and the nitrenium ion, rather than a nitrene, is the product of the first step of the reaction.[98] Protonated ethoxycarbonylnitrene attacks substituted benzenes, to give N-arylcarbamates.[31,99]

XIII. REFERENCES

1. H. Bock, R. Dammel, and L. Horner, *Chem. Ber.* **114**, 220 (1981).

2. J. J. Demuynck, J. Fox, Y. Yamaguchi, and H. F. Schaefer, *J. Am. Chem. Soc.* **102**, 6204 (1980).

3. B. B. Jarvis, P. E. Nicholas, and J. O. Midiwo, *J. Am. Chem. Soc.* **103**, 3878 (1981).

4. P. J. Baldry, A. R. Forrester, M. M. Ogilvy, and R. H. Thomson, *J. Chem. Soc. Perkin Trans. 1*, 2027 (1982).

5. W. Lwowski in M. Jones, Jr. and R. A. Moss, Ed., *Reactive Intermediates*, Vol. 2, Wiley, New York, 1981, p. 315.

6. H. Quast and B. Seiferling, *Annalen*, 1553 (1982).

7. K. B. Becker and C. A. Gabutti, *Tetrahedron Lett.*, **23**, 1883 (1982).

8. T. Sasaki, S. Eguchi, and T. Okano, *J. Org. Chem.* **46**, 4474 (1981).

9. T. Sasaki, S. Eguchi, S. Hattori, and T. Okano, *J. Chem. Soc. Chem. Commun.*, 1193 (1981).

10. J. P. LeRoux, P. L. Desbene, and J. C. Cherton, *J. Heterocycl. Chem.*, **18**, 847 (1981).

11. W. Gilb and G. Schroeder, *Chem. Ber.*, **115**, 240 (1982).

12. H. Roettele and G. Schroeder, *Chem. Ber.*, **115**, 248 (1982).

13. G. Szeimies, U. Siefken, and R. Rink, *Angew. Chem. Int. Ed. Engl.*, **12**, 161 (1973).

14. R. Jorritsma, H. Steinberg, and T. J. DeBoer, *Rec. Trav. Chim.* **100**, 307 (1981).

15. J. Nakayama, H. Fukushima, R. Hashimoto, and M. Hoshino, *J. Chem. Soc. Chem. Commun.* 612 (1982).

16. C. Wentrup in A. R. Katritzky and A. J. Boulton, Eds., *Advances in Heterocyclic Chemistry*, Vol. 28, Academic, New York, 1981, p. 232.

17. G. L'abbé, *Angew. Chem. Int. Ed. Engl.*, **14**, 775 (1974).

18. G. L'abbé, *J. Org. Chem.*, **39**, 1778 (1974).

19. W. Lwowski in W. Lwowski, Ed., *Nitrenes*, Wiley, New York, 1970, p. 185.

20. T. Nishiwaki, *J. Chem. Soc. Chem. Commun.*, 565 (1972).

21. K. Isomura, G.-I. Ayabe, S. Hatano, and H. Taniguchi, *J. Chem. Soc. Chem. Commun.*, 1252 (1980).

22. D. M. B. Hickey, C. J. Moody, J. Christopher, and C. W. Rees, *J. Chem. Soc. Chem. Commun.*, 3 (1982).

23. D. M. B. Hickey, C. J. Moody, C. W. Rees, and D. J. William, *J. Chem. Soc. Chem. Commun.*, 4 (1982).

24. D. M. B. Hickey, C. J. Moody, and C. W. Rees, *J. Chem. Soc. Chem. Commun.*, 1419 (1982).

25. T. L. Gilchrist, C. W. Rees, and J. A. R. Rodriguez, *J. Chem. Soc. Chem. Commun.*, 627 (1979).

26. V. A. Smirnoff and S. B. Brichkin, *Chem. Phys. Lett.* **87**, 548 (1982).

27. I. R. Dunkin and P. C. P. Thomson, *J. Chem Soc. Chem. Commun.*, 1192 (1982).

28. M. S. Platz, G. Carrol, F. Pierrat, J. Zayaz, and S. Auster, *Tetrahedron*, **38**, 777 (1982).

29. W. H. Waddell and C. L. Go, *J. Am. Chem. Soc., 104,* 5804 (1982).

30. P. A. S. Smith in W. Lwowski, Ed., *Nitrenes,* Wiley, New York, 1970, p. 115.

31. W. Lwowski in M. Jones, Jr. and R. A. Moss, Eds., *Reactive Intermediates,* Vol. 1, Wiley, New York, 1978, p. 197.

32. O. L. Chapman, R. S. Sheridan, and J.-P. LeRoux, *Rec. Trav. Chim., 98,* 334 (1979).

33. P.R. West, O. L. Chapman, and J.-P. LeRoux, J. Am. Chem. Soc., *104,* 1779 (1982).

34. R. E. Banks, N. D. Venayak, and T. A. Hamor, *J. Chem. Soc. Chem. Commun.,* 900 (1980).

35. C. Wentrup and H.-H. Winter, *J. Am. Chem. Soc., 102,* 6159 (1980).

36. C. Wentrup, C. Thetaz, E. Tagliaferri, H. J. Lindner, B. Kitschke, H. Wilhelm, and H. P. Reisenauer, *Angew. Chem., 92,* 556 (1980).

37.* B. Iddon, O. Meth-Cohn, E. F. V. Scriven, H. Suschitzky, and P. T. Gallagher, *Angew. Chem. Int. Ed. Engl., 18,* 900 (1979).

38. F. Hollywood, B. Nay, E. F. V. Scriven, H. Suschitzky, Z. U. Khan, and R. Hull, *J. Chem. Soc. Perkin Trans. 1,* 421 (1982).

39. F. Hollywood, Z. U. Khan, E. F. V. Scriven, R. K. Smalley, H. Su, D. R. Thomas, and R. Hull, *J. Chem. Soc. Perkin Trans. 1,* 431 (1982).

40.* H. Sawanishi, T. Hirai, and T. Tsuchiya, *Heterocycles, 19,* 1043 (1982).

41. A. Albini, G. F. Bettinetti, E. Fasani, and S. Pietra, *Gazz. Chim. Ital., 112,* 13 (1982).

42. H. Takeuchi and K. Koyama, *J. Chem. Soc. Perkin Trans. 1,* 1269 (1982).

43. R. N. Cade, P. C. Hayes, G. Jones, and C. J. Cliff, *J. Chem. Soc. Perkin Trans. 1,* 1132 (1981).

44.* D. Hawkins, J. M. Lindley, I. McRobbie, and O. Meth-Cohn, *J. Chem. Soc. Perkin Trans. 1,* 2387 (1980).

45. C. J. Moody, C. W. Rees, S. C. Tsoi, and D. J. Williams, *J. Chem. Soc. Chem. Commun.,* 927 (1981).

46. A. Albini, G. F. Bettinetti, and G. Minoli, *Chem. Lett.,* 331 (1981).

47. A. Albini, *J. Chem. Soc. Perkin Trans. 1,* 4 (1981).

48. L. Benati, M. Grossi, P. C. Montevecchi, and S. P. Spagnolo, *J. Chem. Soc. Chem. Commun.,* 763 (1982).

49.* W. Lwowski, *Ann. N. Y. Acad. Sci., 346,* 491 (1980).

50. E. Escher, R. Coutoure, G. Champagne, J. Mizrahi, and D. Regoli, *J. Med. Chem., 25,* 470 (1982).

51. E. M. Gibbs, M. Hosang, B. F. X. Reber, G. Stemenza, and D. D. Diederich, *Biochem. Biophys. Acta, 688,* 547 (1982).

52. A. W. Nicholson, C. S. Hall, W. A. Strycharz, and B. S. Cooperman, *Biochemistry, 21,* 3809 (1982).

53. P. E. Nielsen and O. Burchardt, *Photochem. Photobiol., 35,* 317 (1982).

54. C. D'Silva, A. P. Seddon, and K. D. T. Douglas, *J. Chem. Soc. Perkin Trans. 1,* 3029 (1981).

55. I. Schwartz and J. Offengand, *Biochem. Biophys. Acta, 697,* 330 (1982).

56.   T. N. Lavin, P. Nambi, S. L. Heald, P. W. Jeffes, R. J. Lefkowitz, and M. G. Caron, *J. Biol. Chem.*, **257**, 12, 332 (1982).

57.   F. Boulay, G. J. M. Lauquin, and V. P. Vignais, *FEBS Lett.*, **143**, 268 (1982).

58.   J. O. Hoeg, K. T. Douglas, C. D'Silva, and A. Holmgren, *Biochem. Biophys. Res. Commun.*, **107**, 1475 (1982).

59.   A. Basu, S. Subramanian, and C. SivaRaman, *Biochemistry*, **21**, 4434 (1982).

60.   S. Marburg, D. Jorn, and R. L. Tolman, *J. Heteroycyl. Chem.*, **19**, 671 (1982).

61.   M. M. King, G. M. Carlson, B. E. Haley, *J. Biol. Chem.*, **257**, 14, 508 (1982).

62.   R. J. Jackson and G. Sachs, *Biochem. Biophys. Acta*, **717**, 453 (1981).

63.   V. P. Demushkin and Yu. V. Kotelevtzev, *Bioorg. Khim.*, 8, 6219 (1982); *Chem. Abstr.*, **97**, 50236e (1982).

64.   A. Mavridis and J. F. Harrison, *J. Am. Chem. Soc.* **102**, 7651 (1980).

65.*   O. Meth-Cohn and S. Rhougat, *J. Chem. Soc. Chem. Commun.*, 241 (1981).

66.   O. Meth-Cohn and S. Rhougat, *J. Chem. Soc. Chem. Commun.*, 1161 (1980).

67.   M. G. Barlow, G. M. Harrison, R. N. Haszeldine, W. D. Morton, P. Shaw-Luckman, and M. D. Ward, *J. Chem. Soc. Perkin Trans. 1*, 2101 (1982).

68.   L. Pellacani, P. Pulcini, and P. A. Tardella, *J. Org. Chem.*, **47**, 5023 (1982).

69.   D. C. Appleton, D. C. Bull, J. McKenna, and A. R. Walley, *J. Chem. Soc. Perkin Trans 2.*, 385 (1980).

70.*   T. Sheradsky and D. Zbaida, *Tetrahedron Lett.*, **22**, 1639 (1981).

71.   C. Leuenberger, L. Hoesch, and A. S. Dreiding, *Helv. Chim. Acta*, **65**, 217 (1982).

72.   C. Leuenberger, L. Hoesch, and A. S. Dreiding, *J. Chem. Soc. Chem. Commun.*, 1197 (1980).

73.   M. L. Tsrenina, T. N. Simonova, N. A. Koltovaya, E. E. Golubeva, and A. N. Ushakov, *Bioorg. Khim.*, 7, 1283 (1981); *Chem. Abstr.*, **95**, 219747 n (1982).

74.   R. S. Atkinson and D. Judkins, *J. Chem. Soc. Perkin Trans. 1*, 2615 (1981).

75.   R. A. Abramovitch, W. D. Holcomb, and S. Wake, *J. Am. Chem. Soc.*, **103**, 1525 (1981).

76.   R. A. Abramovitch, S. B. Hendi, and A. D. Kress, *J. Chem. Soc. Chem. Commun.*, 1087 (1981).

77.   C. Casewit and J. D. Roberts, *J. Am. Chem. Soc.*, **102**, 2364 (1980).

78.   M. G. K. Hutchins and D. Swern, *J. Org. Chem.*, **47**, 4847 (1982).

79.   G. Trinquier, *J. Am. Chem. Soc.*, **104**, 6969 (1982).

80.   C. A. Parsons and D. E. Dykstra, *J. Phys. Chem.*, **71**, 3025 (1979).

81.   M. H. J. Kemper and H. M. Buch, *Can. J. Chem.*, **59**, 3044 (1981).

82.   J. H. Davies and W. A. Goddard, *J. Am. Chem. Soc.*, **99**, 711 (1977).

83.   W. D. Hinsbergs, III, P. G. Schultz, and P. B. Dervan, *J. Am. Chem. Soc.*, **104**, 766 (1982).

84.   D. K. McIntyre and P. B. Dervan, *J. Am. Chem. Soc.*, **104**, 6466 (1982).

85.   R. S. Atkinson, J. R. Malpass, and K. L. Woodthorpe, *J. Chem. Soc. Perkin Trans. 1*, 2407 (1982); *J. Chem. Soc. Chem. Commun.*, 161 (1981).

86.   R. S. Atkinson, J. R. Malpass, K. L. Skinner, and K. L. Wood, *J. Chem. Soc. Chem. Commun.*, 549 (1981).

87.   R. S. Atkinson and K. L. Skinner, *J. Chem. Soc. Chem. Commun.* 22 (1983).

88.   M. A. Kuznetzov and A. A. Suvorov, *Zh. Org. Khim.*, **17**, 1122 (1981).

89.[*]   C. Leuenberger, L. Hoesch, and A. S. Dreiding, *Helv. Chim. Acta*, **64**, 1219 (1981).

90.[*]   L. Hoesch, *Helv. Chim. Acta*, **64**, 890 (1981).

91.   A. Padwa and K. Ku, *Tetrahedron Lett.*, **21**, 1009 (1980).

92.   Y. Nishizawa, T. Miyashi, and T. Mukai, *J. Am. Chem. Soc.*, **102**, 1176 (1980).

93.   Yu. P. Artsybasheva, I. V. Suvorova, and B. V. Ioffe, *Zh. Org. Khim.*, **17**, 435 (1981).

94.   B. V. Ioffe and Yu. P. Artsybasheva, *Zh. Org. Khim.*, **17**, 911 (1981).

95.   P. J. Baldry, A. R. Forrester, M. M. Ogilvy, and R. H. Thomson, *J. Chem. Soc. Perkin Trans. 1*, 2035 (1982).

96.   G. P. Ford and J. D. Scribner, *J. Am. Chem. Soc.*, **103**, 4281 (1981).

97.   R. A. Abramovitch, M. Cooper, S. Iyer, R. Jeyaraman, and J. A. R. Rodriguez, *J. Org. Chem.*, **47**, 4819 (1982).

98.   H. Takeuchi, K. Takano, and K. Koyama, *J. Chem. Soc. Chem. Commun.*, 1155 (1982).

99.   H. Takeuchi and K. Koyama, *J. Chem. Soc. Chem. Commun.*, 226 (1982).

9

SILYLENES

PETER P. GASPAR

Department of Chemistry, Washington University, St. Louis, Missouri 63130

I.   INTRODUCTION

Under the heading "Future Developments" in the last installment of this ongoing effort to keep abreast of the evolution of silylene chemistry,[1] it was predicted (guessed) that the intramolecular reactions of substituted silylenes would receive more attention, and this has proven correct. That the detailed elucidation of silylene reaction mechanisms would move to the center of the stage was also forecast, but although mechanistic studies have certainly increased, their need is even more sharply felt in 1983 than in 1980. The phenomena of silylene chemistry are now so diverse as to appear occasionally inconsistent. The "gee whiz" stage in which every attempt to understand an old reaction uncovers two new ones and has been fun for the participants, giving a taste of the euphoria of organic chemistry a century ago. It is now clear however that a foundation of mechanistic knowledge is required if the lessons that silylene chemistry could teach us about structure-reactivity relationships and the control of chemical transformations are really to be learned. It was a century before the elucidation of organic reaction mechanisms attained an esteem even vaguely comparable to that accorded the exercise of making new compounds and finding new reactions. The health and the usefulness of silylene chemistry requires a foreshortening of that history.

A major survey of the reactions of silicon atoms and silylenes by Tang has appeared.[2] Somewhat more specialized but no less useful is Walsh's summary of a seven-year effort to establish a consistent and reliable set of bond-dissociation energies for silicon-containing compounds.[3] The painstaking kinetic studies by the Walsh group are a model of the needed mechanistic work mentioned before, and of its rewards. Thanks to Walsh we now know that silicon—hydrogen bonds have a consistent strength near 90 kcal/mol (pace SI units), not much weaker than their C—H analogues. Walsh's bond dissociation energy values place the extra stability of silylenes, compared to carbenes, on a firm experimental basis and thus lend credence to various rearrangement mechanisms involving silylene intermediates discussed later.

The important work of Kumada and Ishikawa on the light-induced generation of silylenes has been reviewed in the more general context of polysilane photochemistry.[4] Hungarian readers have been specially served.[5]

This chapter records the progress of silylene chemistry made in 1980 to 1982, omitting 1980 work included in our last review[1] and including some 1983 references. Only a few earlier citations are given as needed, the remainder being included in Volumes 1 and 2 of this series.[1,6] Perhaps the most spectacular recent successes have been the synthesis of a stable disilene by West, Fink, and Michl,[7] who employed the silylene dimerization route pioneered by Conlin and Sakurai,[1] and the fantastic range of silylene rearrangements found by the Barton group and discussed in Section III.E.

Again, the author must confess that omissions reflect not value judgements but deficiencies in his method.

II.   THE GENERATION OF SILYLENES

A.   Thermolysis of Polysilanes

Pride of place must be given to the extrusion of silylenes from polysilanes, the reaction that played such an important role in opening up the study of silylenes,[1,6] and, as indicated in subsequent sections of this review, this is still the bread and butter method for generating silylenes in the gas phase. Important new facets of polysilane pyrolysis continue to appear, indicating that there is more going on than the classic silicon–silicon bond cleavage with simultaneous migration of a group from one silicon to the other:

$$XYZSi\text{–}SiXYZ \quad \xrightarrow{\Delta} \quad XYSi\cdot \; + \; XYSiZ_2$$

Study of the homogeneous gas-phase decomposition kinetics of disilane by a single-pulse shock-tube technique in the temperature range 850–1000 K has revealed that loss of a hydrogen molecule, formulated as an α-elimination on the basis of its activation energy, is a primary process, with $k_1 : k_2 \simeq 4.26 \pm 1.0$.[8]

$$Si_2D_6 \quad \xrightarrow{k_1} \quad SiD_4 \; + \; \ddot{S}iD_2$$

$$Si_2D_6 \quad \xrightarrow{k_2} \quad SiD_3\ddot{S}iD \; + \; D_2$$

The activation parameters for SiH_2 formation by disilane pyrolysis have been refined to $\log k_1 \; (s^{-1}) = 14.36 \pm 0.09 - (48810 \pm 300\,cal)/2.303\,RT$. The higher activation energy for hydrogen loss, $\log k_2 \; (s^{-1}) = 15.3 - (55300 \pm 2200\,cal)/2.303\,RT$ accounts for its decreased contribution at lower temperatures.

A new wrinkle has been provided by the pyrolysis of trisilanes capable of the sequential pairwise shedding of all four substituents from the central silicon and thus functioning as silicon atom synthons.[9] When comparison of this process, involving serial formation of silylenes, with the reactions of atomic silicon has been possible, the product yield from the synthon has been much higher.[10]

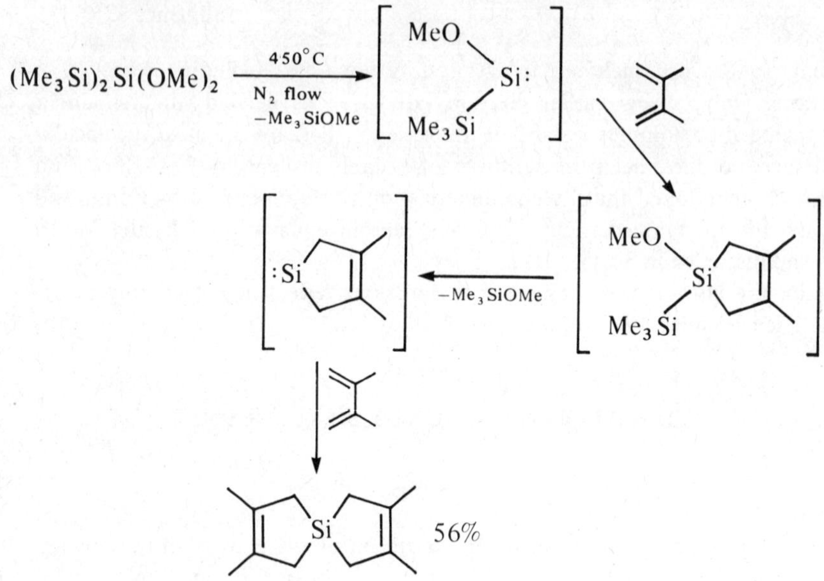
(Me3Si)2Si(OMe)2
450°C
N2 flow
−Me3SiOMe
MeO
Si:
Me3Si
MeO
Si
Me3Si
:Si
−Me3SiOMe
Si
56%

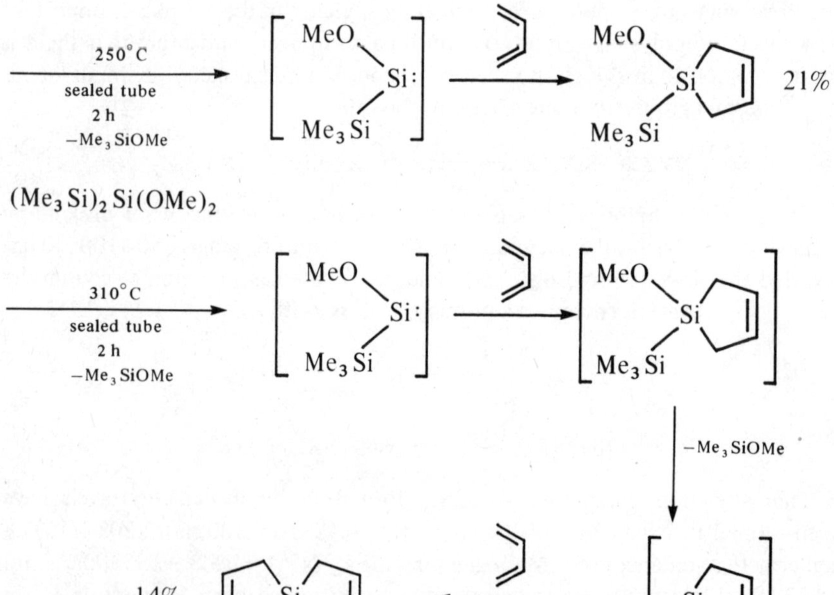
250°C
sealed tube
2 h
−Me3SiOMe
MeO
Si:
Me3Si
MeO
Si
Me3Si
21%
(Me3Si)2Si(OMe)2
310°C
sealed tube
2 h
−Me3SiOMe
MeO
Si:
Me3Si
MeO
Si
Me3Si
−Me3SiOMe
14%
Si
:Si

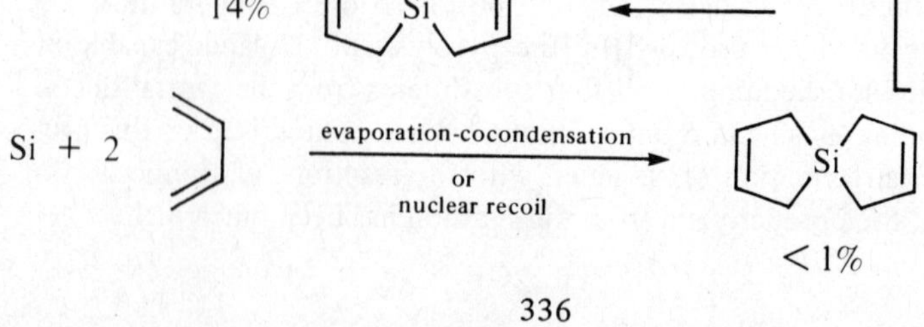
Si + 2
evaporation-cocondensation
or
nuclear recoil
Si
< 1%

A newly discovered, more circuitous pyrolysis of polysilanes to silylenes appears to involve initial homolysis to a silyl radical that is converted by disproportionation to a silene that in turn rearranges (see below) to a silylene.[11,12]

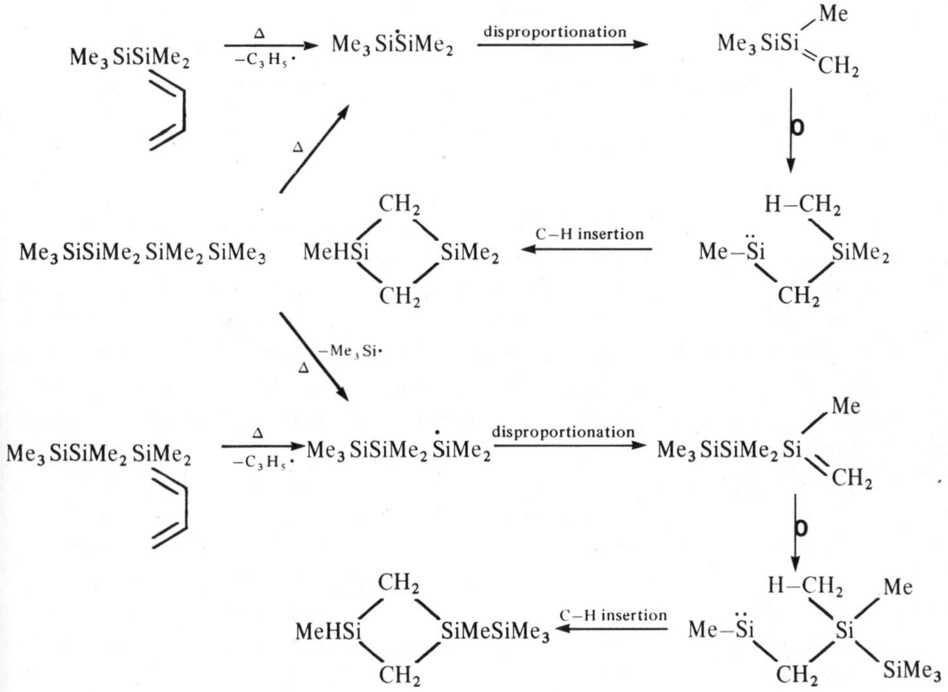

$$Me_3SiSiMe_2 \xrightarrow[-C_3H_5\cdot]{\Delta} Me_3Si\dot{S}iMe_2 \xrightarrow{\text{disproportionation}} Me_3SiSi\overset{Me}{\underset{CH_2}{\diagdown}}$$

$$Me_3SiSiMe_2SiMe_2SiMe_3 \quad MeH\dot{S}i\overset{CH_2}{\underset{CH_2}{\diamond}}SiMe_2 \xleftarrow{\text{C-H insertion}} Me-\ddot{S}i\overset{H-CH_2}{\underset{CH_2}{\diamond}}SiMe_2$$

$$Me_3SiSiMe_2SiMe_2 \xrightarrow[-C_3H_5\cdot]{\Delta} Me_3SiSiMe_2\dot{S}iMe_2 \xrightarrow{\text{disproportionation}} Me_3SiSiMe_2Si\overset{Me}{\underset{CH_2}{\diagdown}}$$

$$MeH\dot{S}i\overset{CH_2}{\underset{CH_2}{\diamond}}SiMeSiMe_3 \xleftarrow{\text{C-H insertion}} Me-\ddot{S}i\overset{H-CH_2}{\underset{CH_2}{\diamond}}Si\overset{Me}{\underset{SiMe_3}{\diagdown}}$$

(−Me₃Si·)

B.   Silicon Atom Reactions

As long ago as 1966 it was suggested that the reactions of high-energy silicon atoms recoiling from the $^{31}P(n, p)^{31}Si$ nuclear transformation lead to the formation of $^{31}SiH_2$,[13] whose insertions and additions produced the end products observed. There has been qualitative agreement between the results of product studies, the outcome of experiments employing various scavengers and moderators, and the expectations based on the intermediacy of silylene. Nevertheless, apparent differences in reactivity were observed between the intermediates in hot atom studies and thermally generated SiH_2.[1,6] These have turned out to be due to differences in reaction conditions.

It was recently possible to compare quantitatively the selectivity of the nucleogenic species presumed to be $^{31}SiH_2$ and SiH_2 from the pyrolysis of SiH_3SiMe_3.[14] The following thermal reactions were studied in a low-pressure (15–25 torr) pyrolysis flow system:

$$\text{SiH}_3\text{SiMe}_3 \xrightarrow{\Delta} \text{SiH}_2 + \text{HSiMe}_3$$

$$\text{SiH}_2 + \text{SiH}_3\text{Me} \xrightarrow{k_1} \text{SiH}_3\text{SiH}_2\text{Me}$$

$$\text{SiH}_2 + \text{SiH}_4 \xrightarrow{k_2} \text{SiH}_3\text{SiH}_3$$

$$\text{SiH}_2 + \; \raisebox{0pt}{\includegraphics{}} \; \xrightarrow{k_3} \; \text{H}_2\text{Si}\raisebox{0pt}{\includegraphics{}}$$

The ratio $k_1:k_2 = 1.38 \pm 0.08$ was found not to vary in the temperature range 385–460°C and to agree closely with the selectivity ratio 1.36 ± 0.09 found from 27 to 202°C for the intermediates in the recoil studies giving rise to the same products. This agreement in relative reactivity proved conclusively the intermediacy of ground state singlet silylene in the recoil chemistry. That the nucleogenic SiH_2 is thermally relaxed was indicated by the similarity of the ratio $k_2:k_3 = 9 \pm 1$ measured in a recoil experiment and the value 13 ± 4 obtained by extrapolating to room temperature the temperature-dependent ratio obtained for thermally generated SiH_2 in the range 377–470°C.[14]

The use of hot atom techniques in kinetic and mechanistic studies has received recent attention with some emphasis on silicon and the study of silylenes.[15,16] The measurement of accurate relative rate constants for SiH_2 at room temperature in the gas phase can now be conveniently accomplished. Evidence has been presented that nucleogenic SiH_2 is formed by way of stepwise acquisition of hydrogen atoms,[17] and thus silyne ^{31}SiH reactions may also be clarified by recoil studies.

The reactions of monomeric difluorosilylene are also conveniently examined by means of hot atom experiments, and several examples are discussed below. $^{31}\text{SiF}_2$ has been shown to result from stepwise fluorine acquisition.[17] Use of the term "acquisition" purposely begs the unanswered question whether hydrogen- or fluorine-abstraction occurs, or instead does an insertion-decomposition mechanism operate:

$$\text{Si} + \text{Z–Q} \xrightarrow[-\text{Q}]{\text{abstraction}} \text{Si–Z} \xrightarrow[\text{Z–Q}]{\text{abstraction}} :\text{SiZ}_2$$

$$\text{or} \quad \text{Si} + \text{Z–Q} \xrightarrow{\text{insertion}} \text{Z–}\ddot{\text{Si}}\text{–Q} \xrightarrow{\text{decomposition}} \text{Si–Z}$$

$$\text{Si–Z} + \text{Z–Q} \xrightarrow{\text{insertion}} \text{Z}_2\dot{\text{Si}}\text{–Q} \xrightarrow[-\text{Q}]{\text{decomposition}} :\text{SiZ}_2$$

$$\text{Z} = \text{H, F} \qquad \text{Q} = \text{Si, P}$$

Silylsilylene formation by insertion of a silicon atom into an Si–H bond has long been known,[6] and has been demonstrated in a recoil experiment by studying the competing insertion and rearrangement (see Section III.E) of trimethylsilylsilylene.[18]

$$^{31}Si + HSiMe_3 \xrightarrow{\text{insertion}} H-^{31}Si-SiMe_3 \xrightarrow[HSiMe_3]{\text{insertion}} Me_3Si^{31}SiH_2SiMe_3$$

$$\downarrow \text{rearrangement}$$

$$Me_2HSiCH_2{}^{31}SiH \xrightarrow[HSiMe_3]{\text{insertion}} Me_2HSiCH_2{}^{31}SiH_2SiMe_3$$

An interesting new silylene has been observed upon matrix isolation in solid argon of thermally evaporated silicon atoms together with water and its isotopic variants.[19] An initial silicon–water complex is believed to rearrange thermally and photochemically to hydroxysilylene, whose irradiation produces silicon monoxide and hydrogen.

$$Si + H_2O \longrightarrow Si \cdots OH_2 \longrightarrow H-\overset{..}{Si}-OH \xrightarrow{h\nu} SiO + H_2$$

A normal coordinate analysis for hydroxysilylene was made possible by measured infrared frequencies for the six isotomers HSiOH, HSiOD, DSiOD, HSi^{18}OH, HSi^{18}OD, and DSi^{18}OD, giving bond lengths H–Si = 1.52 ± 0.030, Si–O = 1.591 ± 0.100, and O–H = 0.958 ± 0.005 Å, and bond angles H–Si–O = $96.6 \pm 4.0°$ and Si–O–H = $114.5 \pm 6°$ in a trans geometry.

The fluorosilylene molecule HSiF was observed for the first time in a similar experiment by allowing silicon atoms to react with HF after deposition in solid argon.[20]

C.   Photolysis of Silanes

a.   Photolysis of Monosilanes

It has been proposed that triplet benzene sensitizes the dissociation of 1,1-dimethyl-1-silacyclobutane to cyclopropane and dimethylsilylene.[21]

$$C_6H_6(^3B_{1u}) + Me_2Si(CH_2)_3 \rightarrow Me_2Si + (CH_2)_3 + C_6H_6(^1A_{1g})$$

If spin is conserved in this transformation, then a triplet silylene may, for the first time, be available for study. As yet no products from photosensitized silylene generation have been reported.

Formation of an excited $^1B_1$ singlet state of SiH$_2$ has been reported in the infrared multiphoton decomposition of SiH$_4$.[22] Irradiation of pure silane, or an SiH$_4$–SiF$_4$ mixture, with an unfocused CO$_2$ laser led to the observation of two emission bands at 6323–6155 and 5945–5864 Å respectively, corresponding to the transitions $^1A_1 (v'' = 1) \leftarrow {}^1B_1 (v = 2, 3)$ of SiH_2. The role of SiF_4 is that of a photosensitizer:

$$SiF_4 + n\, h\nu\, (1025.3\, cm^{-1}) \rightarrow SiF_4^{**}$$

$$SiF_4^{**} + SiH_4 \rightarrow SiH_4^* + SiF_4^*$$

$$SiH_4^* \rightarrow SiH_2 + H_2$$

Decomposition of SiH_4 to SiH_2 is believed to require 2.46 eV excitation energy, whereas 4.57 eV is required for the formation of $^1B_1$ excited singlet silylene. The lifetime of the $^1B_1$ excited state of SiH_2 could not be determined, but the emission intensity reached its maximum only some 200 μsec after the 1-μsec laser flash and decayed with a halftime of about 200 μsec. The form of this time dependence for emission is believed to be determined by the collisional energy pooling processes that produce highly excited silane molecules.

For the decomposition of SiH_4, energy absorption from the laser beam can be increased by raising the pressure of SiH_4 or helium, indicating that collisions are necessary to pump molecules into the quasicontinuum from which resonant absorption of the laser photon readily occurs.[23] The increase in energy absorption is due to pressure broadening of the rotational fine structure. The number of infrared photons absorbed per molecule increases from two at 10 torr to more than four at 26 torr. The amount of decomposition per pulse also increases with increasing helium pressure, indicating that the reaction is not a thermal one.

b.   Photolysis of Polysilanes

The photolysis of ethynylpolysilanes produces a wealth of interesting organosilicon species, including silylenes. Dimethylsilylene has been trapped by Si–H bond insertion in the irradiation of 1-ethynylheptamethyltrisilane.[24] In the presence of methanol, a silacyclopropene methanolysis product was obtained, and thus silylene extrusion and silacyclopropene formation were suggested to be competing processes.

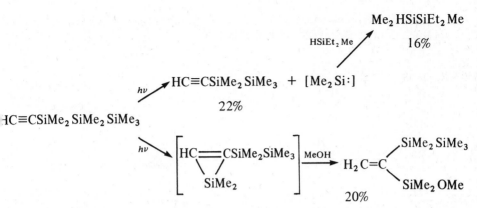

In the absence of trapping reagent the only volatile product arose from rearrangement of the silacyclopropene, which probably proceeded by way of a diradical intermediate.

$$HC{=\!=\!=}CSiMe_2 SiMe_3 \longrightarrow HC{=}\overset{\bullet}{C}SiMe_2 SiMe_3 \longrightarrow$$
with SiMe₂ below and ·SiMe₂ below

$$Me_2 HSiC{\equiv}CSiMe_2 SiMe_3$$

Photolysis of tris(trimethylsilyl)ethynylsilane produced hexamethyldisilane, in addition to a silacyclopropene detected by means of its methanolysis product. Presumably a silylene extrusion reaction is responsible. The silylene itself was not trapped by Si–H insertion, and it is curious that no methanol trapping product of a silylene was reported.

$$C{\equiv}CSi(SiMe_3)_3$$

$$\xrightarrow{h\nu} Me_3 SiSiMe_3 + [HC{\equiv}C{-}\overset{..}{Si}{-}SiMe_3]$$
16%   ?

HSiEt₂Me → HC≡CSiHSiMe₃ , SiEt₂Me

$$\xrightarrow{h\nu} \left[HC{=\!=\!=}CSiMe_3 \right] \xrightarrow{MeOH} H_2 C{=}C \begin{smallmatrix} SiMe_3 \\ Si(SiMe_3)_2 \\ | \\ OMe \end{smallmatrix}$$ 5%
with Si(SiMe₃)₂ below

Other ethynylpolysilanes gave evidence for the formation of silapropadienes in addition to silylenes and silacyclopropenes.[24]

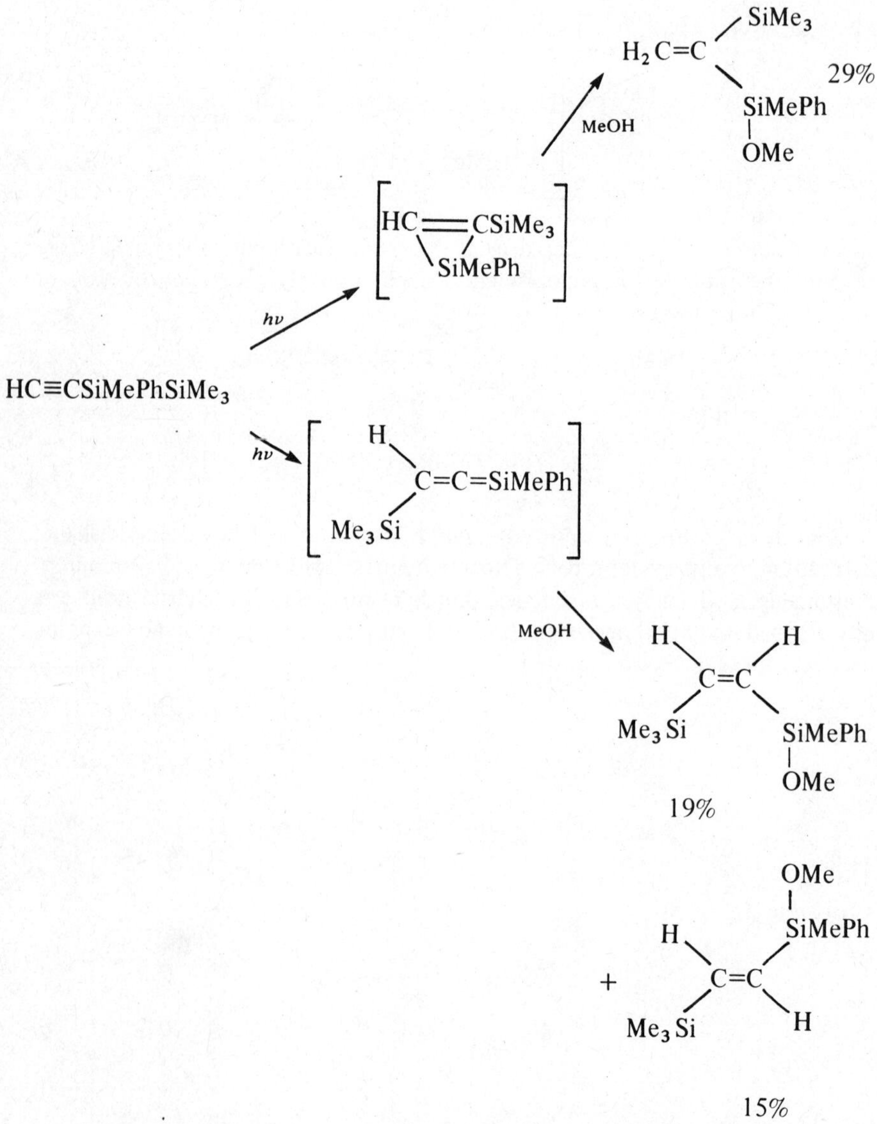

$$HC\equiv CSiMePhSiMe_3$$

Reaction scheme:

$$\left[\underset{SiMePh}{HC}=CSiMe_3 \right] \xrightarrow{MeOH} H_2C=C\underset{\underset{OMe}{\overset{|}{SiMePh}}}{\overset{SiMe_3}{\diagup}} \quad 29\%$$

$$\left[\underset{Me_3Si}{\overset{H}{\diagdown}}C=C=SiMePh \right] \xrightarrow{MeOH} \underset{Me_3Si}{\overset{H}{\diagdown}}C=C\underset{\underset{OMe}{\overset{|}{SiMePh}}}{\overset{H}{\diagup}} \quad 19\%$$

$$+ \quad \underset{Me_3Si}{\overset{H}{\diagdown}}C=C\underset{H}{\overset{\overset{OMe}{\overset{|}{SiMePh}}}{\diagup}} \quad 15\%$$

Similar results were obtained with 1,1-diphenyl-1-ethynyltrimethyldisilane, but when a phenyl group was attached to the other silicon yet another reaction, silene formation, was detected:[24]

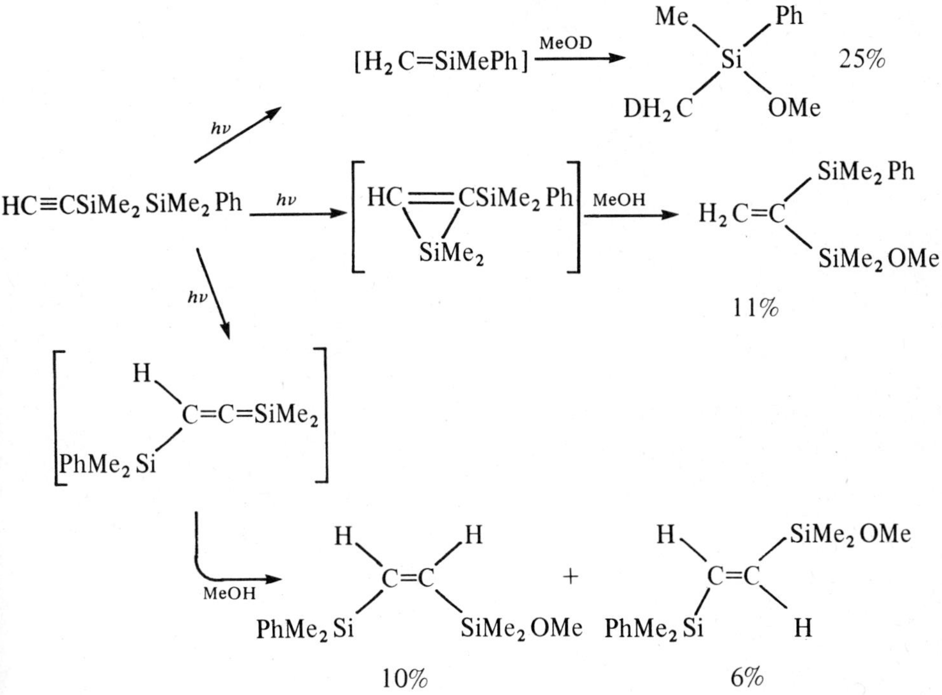

The photochemical decomposition of 1-(trimethylsilyl)-1,1-diphenyl-2,2,3-trimethyldisilane gave evidence for the formation of diphenylsilylene, but whether it arose directly from the starting material, or instead from silacyclopropene photolysis was not determined.[25] Silacyclopropenes have previously been known to be photochemical sources of silylenes.[1]

The Kumada group has employed *ab initio* self-consistent field molecular orbital calculations to find the orbitals most likely to be intimately involved in the photochemical transformations of ethynyldisilanes.[26]

Silapropadiene formation is envisioned as being due to rebonding in a molecular orbital to which an electron is promoted upon absorption of a photon by the ethynyldisilane. For the simple model ethynyldisilane, the orbital is the ground state LUMO, while excitation to the (LU + 2) MO leads to a 1,2-silyl shift producing a diradical that can close to a silacyclopropene.

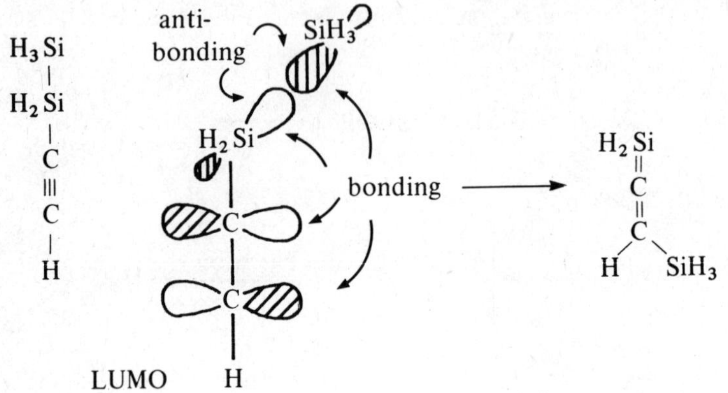

anti-
bonding
bonding
LUMO

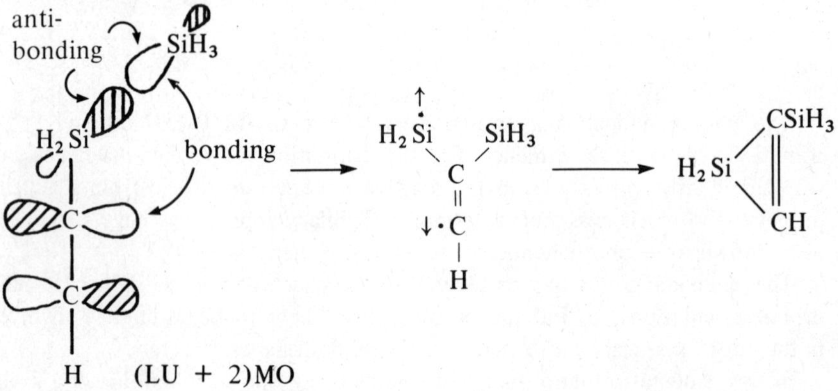

anti-
bonding
bonding
(LU + 2)MO

For the more complex model phenylethynyldisilane, there are corresponding orbitals, (LU + 3)MO and (LU + 4)MO, that can be used to explain the formation of a propadiene and a diradical intermediate leading to a silacyclopropene, respectively. The intermediacy of a silenylcarbene in silacyclopropene formation, for example,

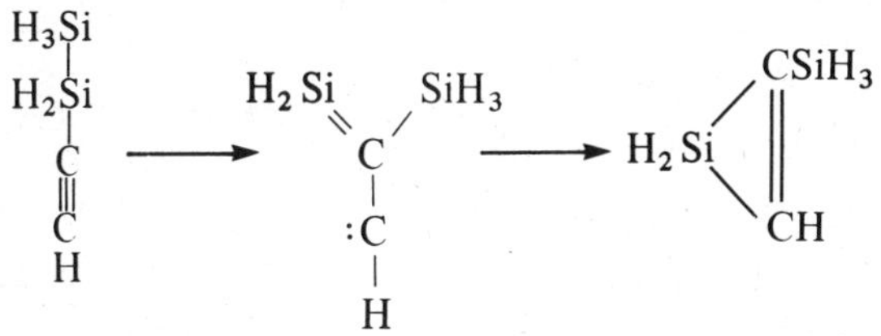

was considered less likely, owing to a calculated energy difference of 18 kcal/mol between singlet $H_2Si{=}CH{-}CH{:}$ and planar singlet $H_2\dot{S}i{-}CH{=}\dot{C}H$. The energy difference is calculated to be less (about 6 kcal/mol) for the corresponding triplet species, but the intermediacy of both triplet species and free radicals was believed to be excluded by scavenger experiments; for example, the absence of hydrogen abstraction products from irradiation of phenylethynylpentamethyldisilane in the presence of $HSiEt_2Me$. In this case the silacyclopropene could be isolated as well as a silylene insertion product.

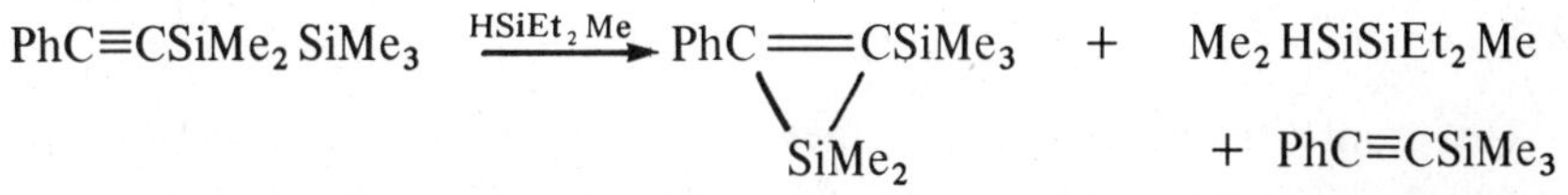

The use of benzene as solvent for many of the photochemical transformations of phenyl-substituted polysilanes by the Kumada group does however raise the possibility of photosensitization and triplet or free radical intermediates.

The formation of a silene, for instance, can be rationalized by a disproportionation of silyl radicals, for which precedent exists.[1]

$$HC{\equiv}CSiMe_2SiMe_2Ph \xrightarrow{h\nu} [HC{\equiv}C\dot{S}iMe_2 + \dot{S}iMe_2Ph] \longrightarrow$$

$$HC{\equiv}CSiHMe_2 + CH_2{=}SiMePh$$

Indeed, the first photochemical generation of a silene by Boudjouk, Sommer, et al. may have followed this mechanism:[27]

$$Ph_3SiSiMePh_2 \xrightarrow{h\nu} [Ph_3Si\cdot + \cdot SiMePh_2]$$

$$Ph_3SiH + [CH_2=SiPh_2]$$

$$\downarrow MeOD$$

$$CH_2DSiPh_2OMe$$

Silylene formation in ethynylpolysilane photolysis may be due to silylene loss from a diradical intermediate that could itself act as a silylene transfer agent, a "silylenoid".

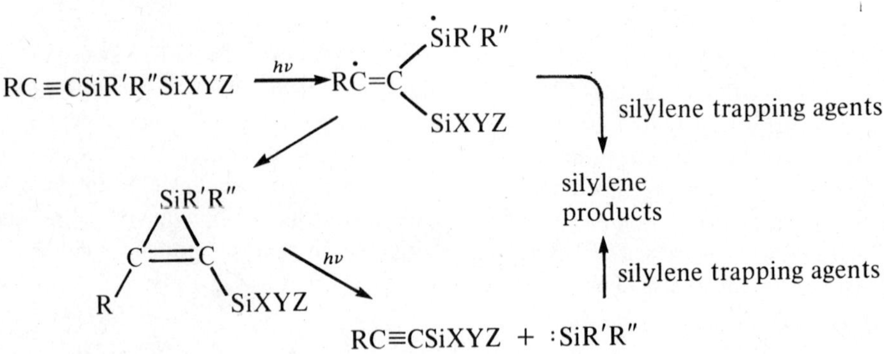

The primary photochemistry in these transformations has begun to be explored. The Kumada group reports that both the fluorescence and reactions of phenyldisilanes originate with a singlet $2p\pi$, $3d\pi$ intramolecular charge-transfer state.[28] Molecular orbital calculations indicate that either $2p\pi^*(C) \rightarrow \sigma^*(Si-Si)$ or $2p\pi^*(C) \rightarrow 3d\pi(Si)$ charge transfer may be responsible for the photochemistry of the ethynyldisilanes.[26]

A fascinating new photochemical synthesis of a stable disilene may produce a silylene as well.[29] Masamume and co-workers found that irradiation of hexakis-(2,6-dimethylphenyl)cyclotrisilane, itself the unexpected stable first member of a new class of compounds, led to the formation of the tetraaryldisilene. The primary coproduct may well be di(2,6-dimethylphenyl)silylene, whose dimerization produces further disilene. Total rupture of the cyclotrisilane, while less likely, could give the same end product.

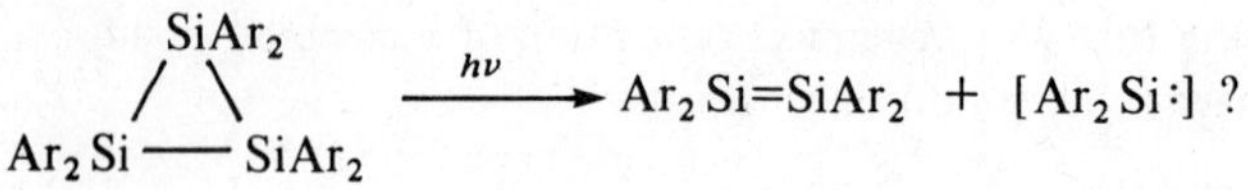

$$Ar_2Si \underset{\displaystyle SiAr_2}{\overset{\displaystyle SiAr_2}{\diagup \diagdown}} \xrightarrow{h\nu} Ar_2Si=SiAr_2 + [Ar_2Si\colon]\ ?$$

D.   Pyrolysis of Monosilanes

It will be a long time before the dust settles from the controversy surrounding Conlin's discovery that pyrolysis of 1-methyl-1-silacyclobutane leads to the formation of dimethylsilylene. This was evidenced by the high yields of products expected from addition, insertion, and dimerization followed by rearrangement:[30]

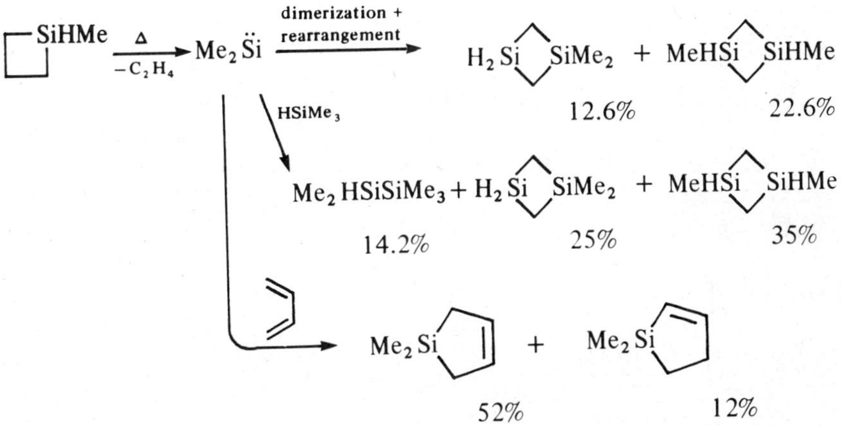

Conlin's proposed mechanism for silylene formation from silacyclobutane pyrolysis involves silene rearrangement and is discussed together with mechanistic alternatives and other examples in the next section.

Unsubstituted silacyclobutane was found to yield silylene, methylsilylene, and silaethylene.[31] Competing stepwise fragmentations followed by rearrangement were suggested:

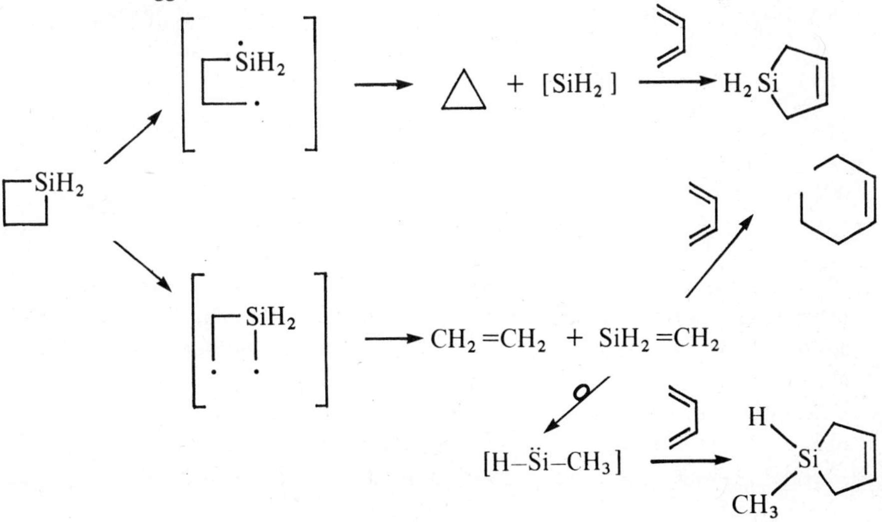

The formation of dichlorosilylene from flow pyrolysis of 1,1-dichloro-1-silacyclobutane has also been reported, and the analogous products found.[32]

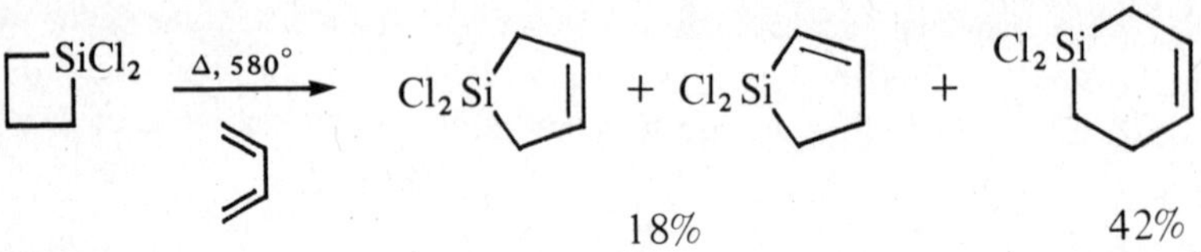

$$\boxed{}\text{-SiCl}_2 \xrightarrow{\Delta,\ 580°} \text{Cl}_2\text{Si} + \text{Cl}_2\text{Si} + \text{Cl}_2\text{Si}$$

18%   42%

The mild conditions under which thermolysis of hexamethylsilirane undergoes loss of a dimethylsilylene unit have made it a highly attractive source of dimethylsilylene.[1,6] Its synthesis is now fully described,[33] and important evidence has been presented for concerted extrusion of the silylene as the decomposition mechanism.[34]

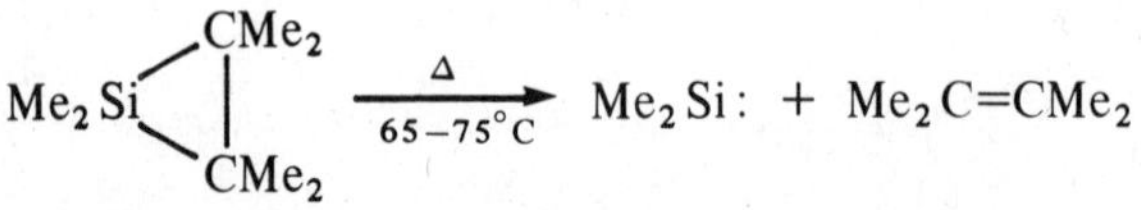

$$\text{Me}_2\text{Si} \overset{\text{CMe}_2}{\underset{\text{CMe}_2}{|}} \xrightarrow[65-75°C]{\Delta} \text{Me}_2\text{Si}: + \text{Me}_2\text{C}=\text{CMe}_2$$

In light of previous questions about the possibility of a diradical intermediate in silirane thermolysis, it is significant that the Seyferth group has demonstrated that dimethylsilylene transfer from hexamethylsilirane to an olfein is *stereospecific*,[34] as had previously been found for photogenic Me$_2$Si.[1]

If Me$_2$Si extrusion from hexamethylsilirane were stepwise, there would also be the possibility that the diradical intermediate was a silylene-transfer agent and that formation of free silylene was avoided. If silylene transfer from Me$_2\dot{\text{S}}$iCMe$_2$-$\dot{\text{C}}$Me$_2$ to olefins occurred, the immediate product would likely be another diradical:

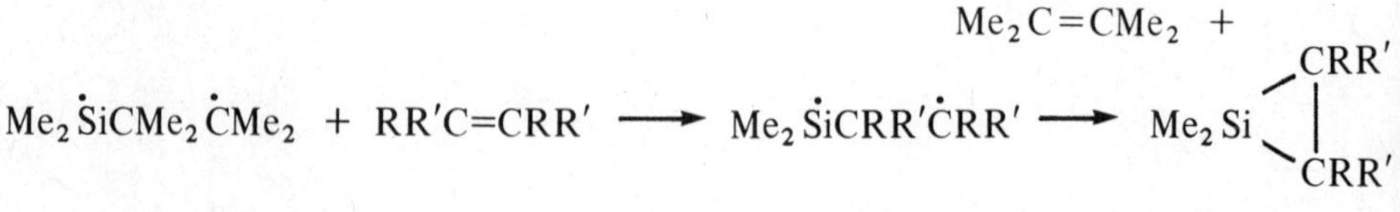

$$\text{Me}_2\dot{\text{S}}\text{iCMe}_2\dot{\text{C}}\text{Me}_2 + \text{RR}'\text{C}=\text{CRR}' \longrightarrow \text{Me}_2\dot{\text{S}}\text{iCRR}'\dot{\text{C}}\text{RR}' \longrightarrow \text{Me}_2\text{Si}\overset{\text{CRR}'}{\underset{\text{CRR}'}{|}}$$

$$\text{Me}_2\text{C}=\text{CMe}_2 +$$

For this reaction to be *stereospecific*, the lifetime of the final diradical intermediate must be shorter than its rotational period. Since this is unlikely, stereospecific silylene transfer is significant evidence for concerted silylene extrusion.

The thermolysis of twenty-two, 7-sila-2,3-benzo-norbornadienes, 18 of them new compounds, to silylenes has been reported by Mayer and Neuman![35] A stepwise cleavage was suggested, with a diradical capable of rebonding to form an epimer of the starting compound acting as intermediate.

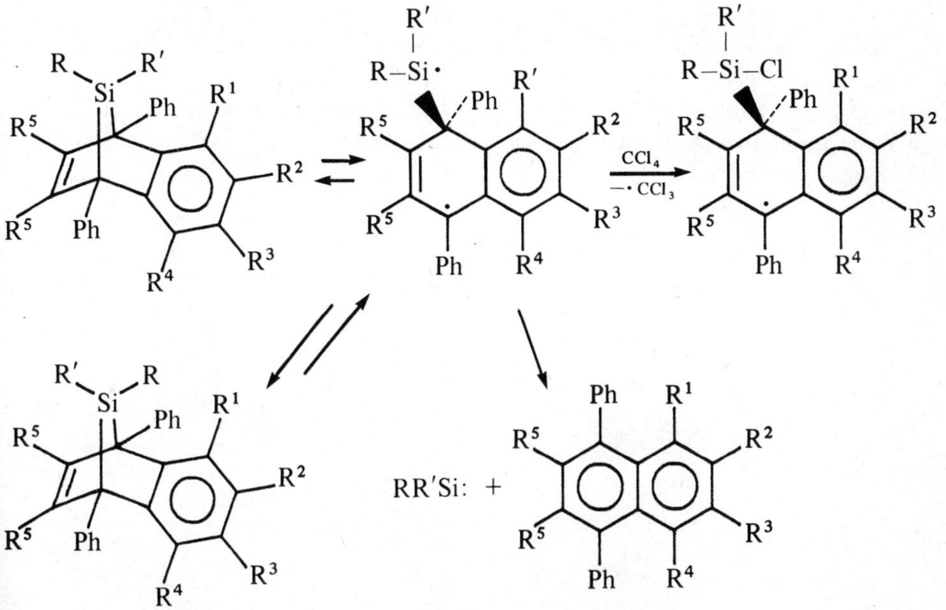

The thermal stability of the silanorbornadiene increased with the bulk of the substituents R^1, R^4, and R^5, presumably due to decreased conjugation of a carbon radical center with a 1- or 6-phenyl substituent.

Although diradical intermediates were confirmed by Barton and co-workers in the thermal decomposition of several 7-silanorbornadienes, the epimerization observed by Neumann could not be repeated.[36]

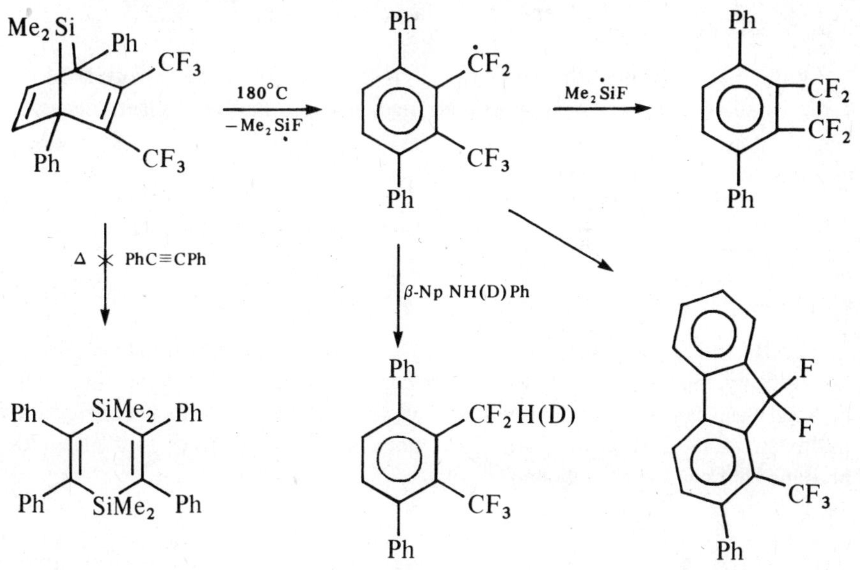

That no disilacyclohexadiene was formed in the presence of tolane seems to rule out the formation of dimethylsilylene. When the analogous tetraphenylsila-norbornadiene is pyrolyzed, or photolyzed in CCl_4, radical products are obtained, but photolysis in the presence of tolane does give the disilacyclohexa-diene, generally believed to arise from a silylene intermediate by way of a variety of possible mechanisms.[1]

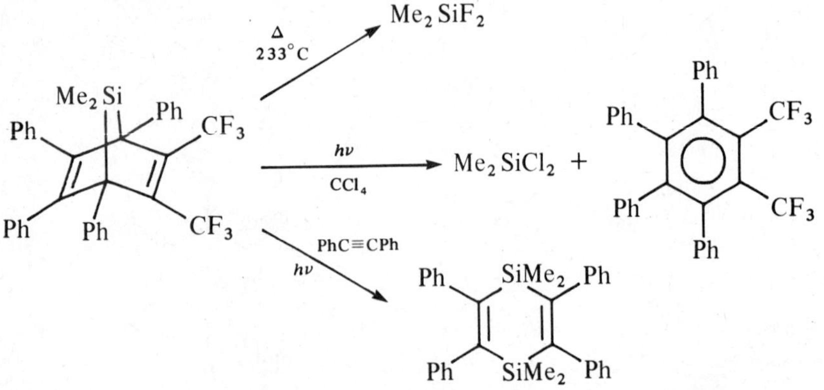

Since the diradical intermediates postulated by both Barton and Neumann for 7-silanorbornadiene fragmentation could act as silylene donors, the formation of free silylenes in this reaction remains to be unequivocally established.

E.   The Generation of Silylenes by the Rearrangement of Silenes, Silacyclo-propenes, and Silacyclopropanes

Conlin has proposed that dimethylsilylene arises from the pyrolysis of 1-methyl-1-silacyclobutane by way of rearrangement of 1-methyl-1-silaethylene:[30]

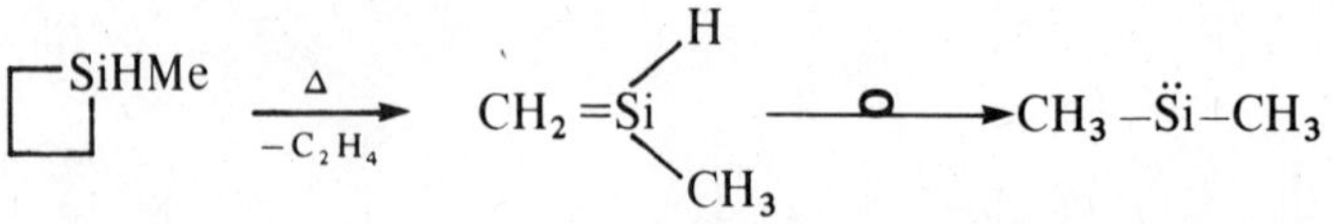

The Barton group has proposed an alternative mechanism for liberation of dimethylsilylene in which stepwise cleavage of the silacyclobutane ring commences with carbon–carbon bond homolysis to a diradical which undergoes a hydrogen atom shift prior to the final bond scission, or closes to a silacyclo-propane that can expel a silylene.[12]

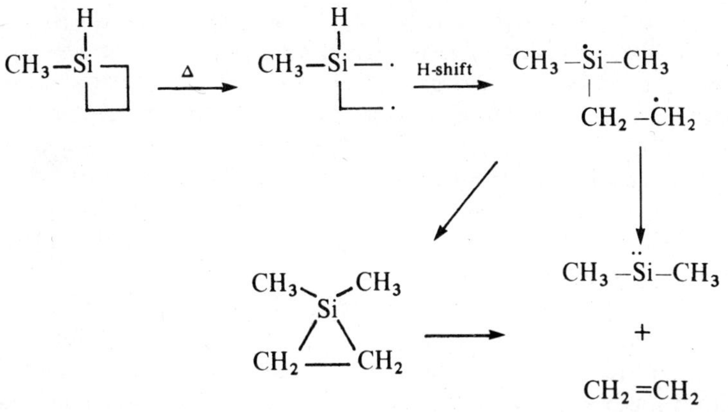

$$\underset{\text{H}}{\overset{\text{H}}{\text{CH}_3-\text{Si}}}\square \quad \xrightarrow{\Delta} \quad \overset{\text{H}}{\underset{}{\text{CH}_3-\text{Si}}}\,\llcorner\cdot \quad \xrightarrow{\text{H-shift}} \quad \underset{\text{CH}_2-\overset{\cdot}{\text{CH}_2}}{\overset{\cdot}{\text{CH}_3-\overset{}{\text{Si}}-\text{CH}_3}}$$

$$\underset{\text{CH}_2-\!-\!-\text{CH}_2}{\overset{\text{CH}_3\diagdown\,\diagup\text{CH}_3}{\text{Si}}} \quad \longrightarrow \quad \begin{array}{c} \text{CH}_3-\overset{..}{\text{Si}}-\text{CH}_3 \\ + \\ \text{CH}_2=\text{CH}_2 \end{array}$$

Barton added doubt concerning the involvement of free silene in the methyl-silacyclobutane pyrolysis by generating this silene from a precursor different from Conlin's and observing the products of silene dimerization and addition to butadiene without finding any products from dimethylsilylene.[12]

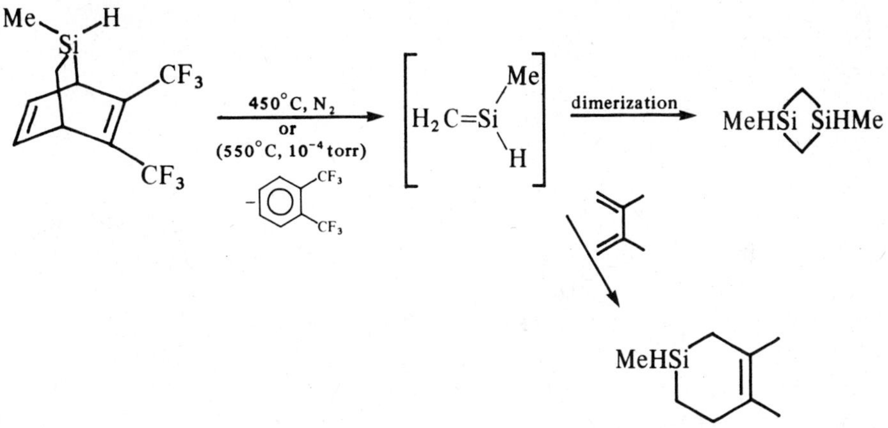

$$\xrightarrow[\substack{\text{or} \\ (550°\text{C}, 10^{-4}\text{torr})}]{450°\text{C}, \text{N}_2} \quad \left[\text{H}_2\text{C}=\underset{\text{H}}{\overset{\text{Me}}{\text{Si}}} \right] \quad \xrightarrow{\text{dimerization}} \quad \text{MeHSi}\;\square\;\text{SiHMe}$$

MeHSi⟨ring⟩

More recently, however, Conlin has reported the trapping of dimethylsilylene from a methylsilabicyclo[2.2.2]octadiene.[37]

Theoreticians were quick to savage Conlin. Schaefer reported in 1979 that a configuration interaction calculation employing a "double zeta plus d" basis set predicted a near degeneracy for silaethylene and methylsilylene, the latter only 0.08 kcal/mol more stable than the former.[38] In a 1980 paper this energy difference is given as 0.4 kcal/mol with an estimated uncertainty of 5 kcal/mol.[39] A large energy barrier of 41.0 kcal/mol was calculated for the $\text{CH}_2=\text{SiH}_2 \rightarrow \text{CH}_3\text{-SiH}$ interconversion, using a configuration interaction calculation corrected for the contribution of configurations higher than doubly excited. In an even heftier calculation adding p polarization functions for hydrogen as well as d functions for carbon and silicon, and combining 9003 configurations for sila-ethylene and 32,131 for the transition state, the barrier shrank by less than half a kilocalorie to 40.6 kcal/mol.[40]

The confidence of modern theory is such that Schaefer suggested that alternative explanations be sought for the observations of Conlin and others (see below),[40] since the large calculated barrier seemed incompatible with a facile silene-to-silylene rearrangement.

The Conlin mechanism has however received strong support from the temperature dependence of the yield ratio of products reasonably formulated as arising from reactions of butadiene with silaethylene and methylsilylene respectively in the copyrolysis of unsubstituted silacyclobutane and butadiene.[31] (See Section II.D for the additional formation of SiH_2 in this reaction.)

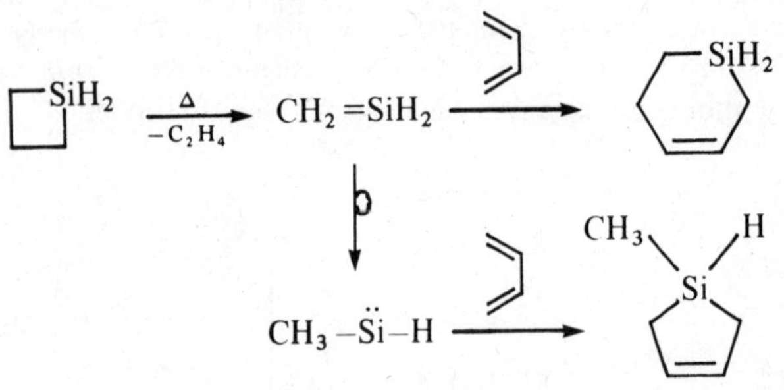

The ratio of silylene product to silene product increases from 1.2 at a pyrolysis temperature of 556°C to 5.1 at 697°C. This change is probably due to increased silene-to-silylene isomerization, which should be able to surmount even a 40 kcal/mol barrier at the high temperatures employed.

If the formation of methylsilylene depended on a competition between a 1,2-hydrogen shift and fragmentation of a diradical, the much larger activation entropy of the cleavage should lead to enhancement of silene formation with increasing temperature, just the opposite of the observed trend.[31]

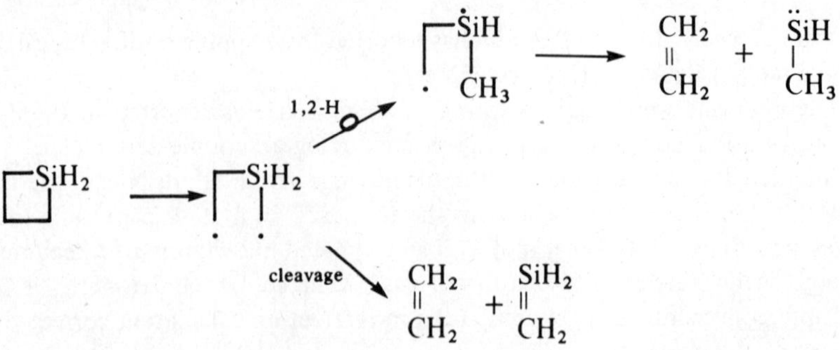

The only thing certain at this point (May, 1983) in the controversy about the silene-to-silylene rearrangement is that it will continue. There is a feeling among workers in the field that silylenes are more stable than their silene isomers, and thus there is less reluctance on Barton's part in postulating the migration of a labile trimethysilyl group in such a rearrangement.[11] Other rearrangements will be introduced as segments in silylene rearrangement sequences (Section III.E).

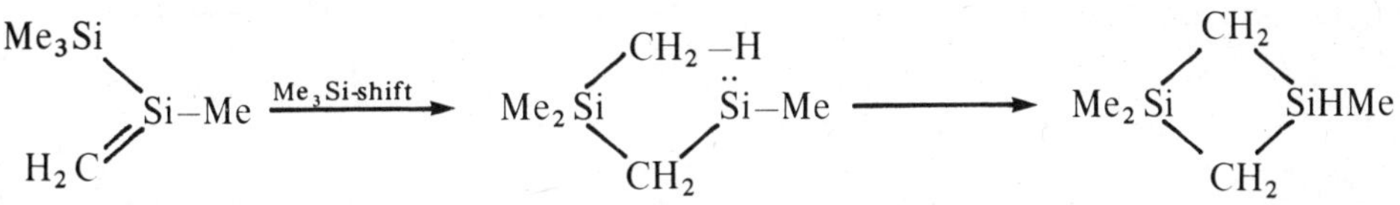

If silenes and silylenes are readily interconverted, then their rates of reaction as well as their energy difference will determine the ratio of products from the isomeric species.

Ring strain also seems to favor the silene-to-silylene rearrangement. Such a process would explain the formation of silacyclopentenes from pyrolysis of 4-sila[3.3]spiroheptane,[41] first reported by Gusel'nikov in 1973.[42]

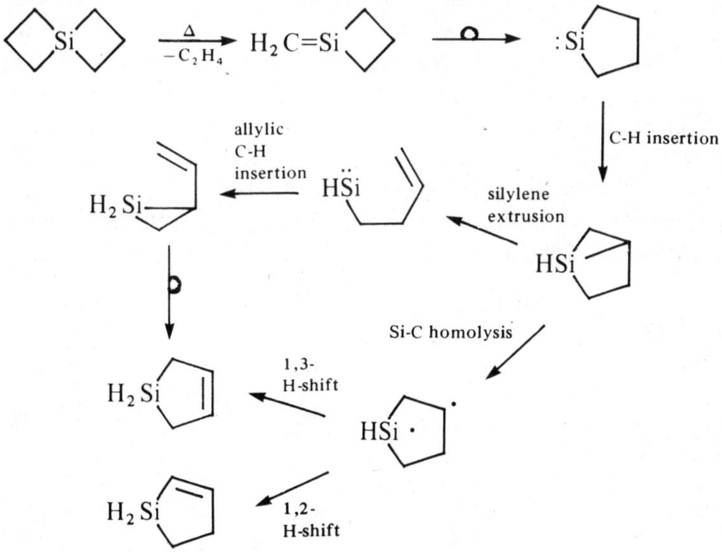

Drahnak, West, and Michl have reported an interesting low-temperature thermal isomerization of methylsilene to dimethylsilylene,[43] but the evidence presented should be carefully scrutinized. When dodecamethylcyclohexasilane is irradiated at 254 nm in a frozen argon matrix at 10 K, a yellow species is formed (λ_{max} 450 nm) that is believed to be dimethylsilylene. Irradiation of the frozen matrix at 450 nm causes the yellow color and infrared spectroscopic features assigned to Me_2Si to disappear, and they are replaced with ir bands assigned to methylsilene, whose cyclic dimer is formed when the matrix is annealed at 50 K.

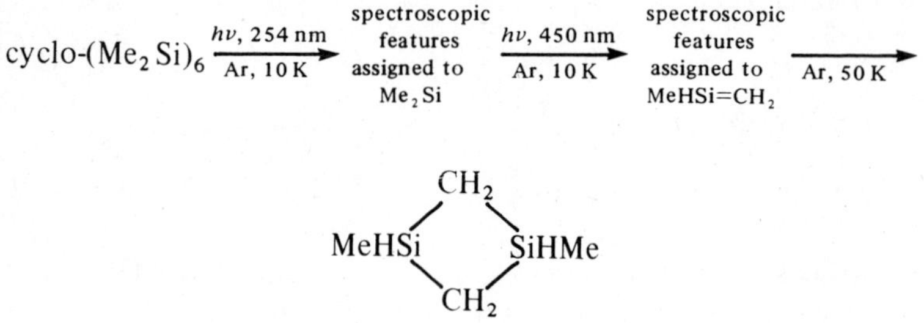

$$\text{cyclo-}(Me_2Si)_6 \xrightarrow[\text{Ar, 10 K}]{h\nu,\ 254\ nm} \begin{array}{c}\text{spectroscopic}\\\text{features}\\\text{assigned to}\\Me_2Si\end{array} \xrightarrow[\text{Ar, 10 K}]{h\nu,\ 450\ nm} \begin{array}{c}\text{spectroscopic}\\\text{features}\\\text{assigned to}\\MeHSi{=}CH_2\end{array} \xrightarrow{\text{Ar, 50 K}}$$

$$\begin{array}{ccc} & CH_2 & \\ & \diagup \quad \diagdown & \\ MeHSi & & SiHMe \\ & \diagdown \quad \diagup & \\ & CH_2 & \end{array}$$

The thermal reversion of methylsilene to dimethylsilylene was inferred from the observation that photodestruction of the yellow species in solid 3-methyl-pentane doped with trapping reagents at 77 K, followed by annealing of the matrix at 100 K, led to the products of silylene reactions. It was therefore deduced that $MeHSi{=}CH_2$ was formed, as shown, but rearranged at 100 K to Me_2Si. Although this is a reasonable interpretation, its validity depends on the correctness of the association of the yellow color and other spectral features with dimethylsilylene. An alternative explanation is that since bleaching of the yellow color did *not* lead to a diminution in the yield of silylene trapping products, the yellow color is *not* associated with Me_2Si.

Maier and co-workers report reversible photoisomerization between silenes and silylenes.[44] Silenes were generated by the Barton retro-Diels-Alder method

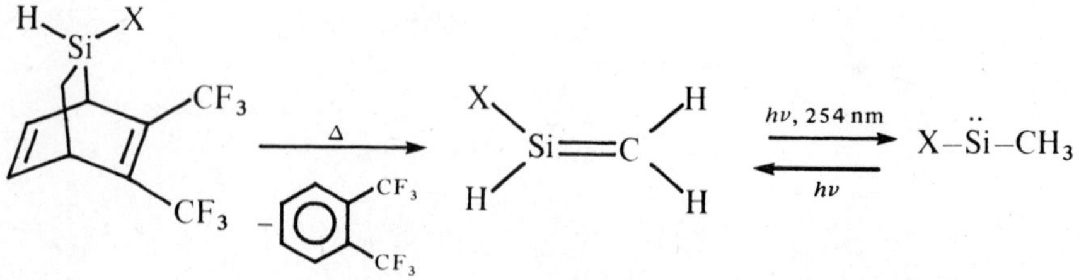

$$\begin{array}{ccc} & & \\ H\diagdown\ _{Si}\diagup X & & \\ \end{array} \xrightarrow{\Delta} \quad \begin{array}{c}X\diagdown\quad\diagup H\\ Si{=}C\\ H\diagup\quad\diagdown H\end{array} \underset{h\nu}{\overset{h\nu,\ 254\ nm}{\rightleftarrows}} X{-}\ddot{Si}{-}CH_3$$

	λ_{max} (nm)		λ_{max} (nm)
X = H	258 (Ar, N_2)	480	330
Cl	255 (Ar, N_2)	407 (Ar)	387 (N_2)
CH_3	260 (Ar)	460 (Ar)	425 (N_2)

and matrix isolated at 10 K in argon and nitrogen matrices. Short-wavelength (254 nm) irradiation of the silenes was reported to lead to quantitative conversion to the isomeric silylenes, which could in turn be reconverted to the silene by longer-wavelength irradiation at their absorption maxima.

The dramatic effect of a change of matrix on the absorption maxima for the silylene raises some doubts about their assignment to the same species. In light of Conlin's results it is also surprising that some thermal isomerization of CH_3-$HSi=CH_2$ did not occur in its initial preparation.

A new route to silylenes is the isomerization of silacyclopropenes.[45] This rearrangement to vinylsilylenes was inferred by the Barton group from a stable product itself arising from rearrangement of a silirene formed by addition of the vinylsilylene to an acetylene molecule. Since the original silirene was also formed by silylene addition, the following sequence contains two silylenes and three silirenes:

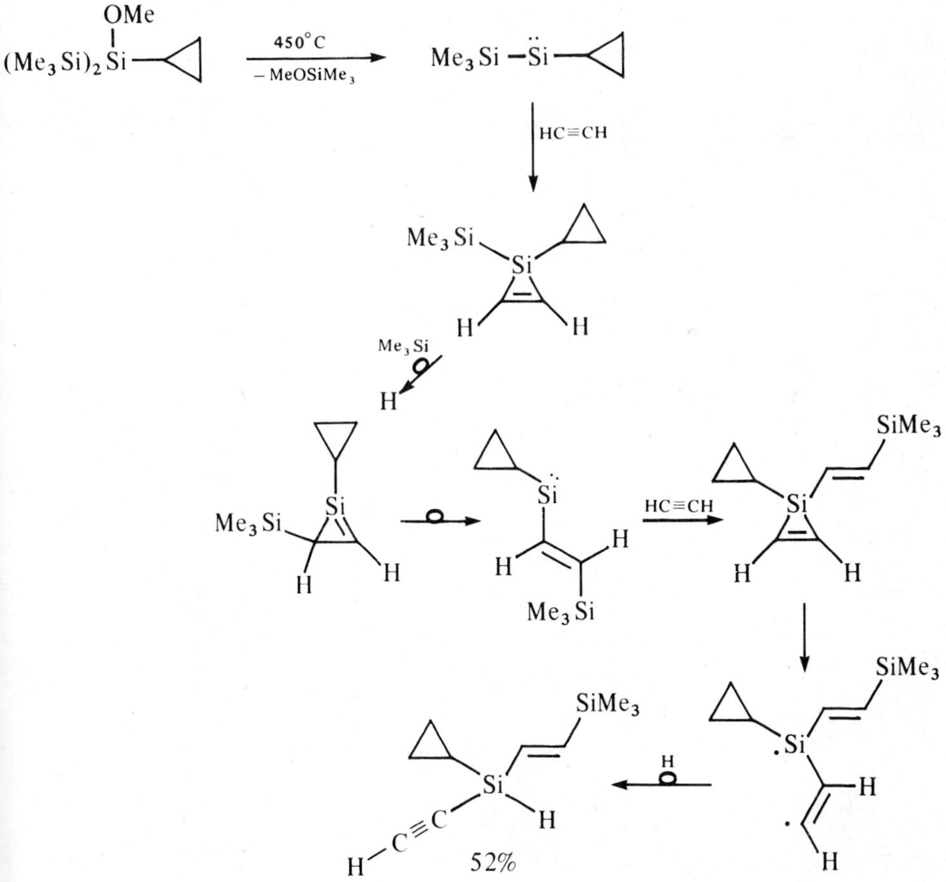

Similar mechanisms were proposed for the following three reactions, in which trimethylsilyl and hydrogen migration respectively are proposed for the silirene-to-vinylsilylene rearrangement. Use of an asymmetric acetylene gave an indication that this rearrangement is not regiospecific.

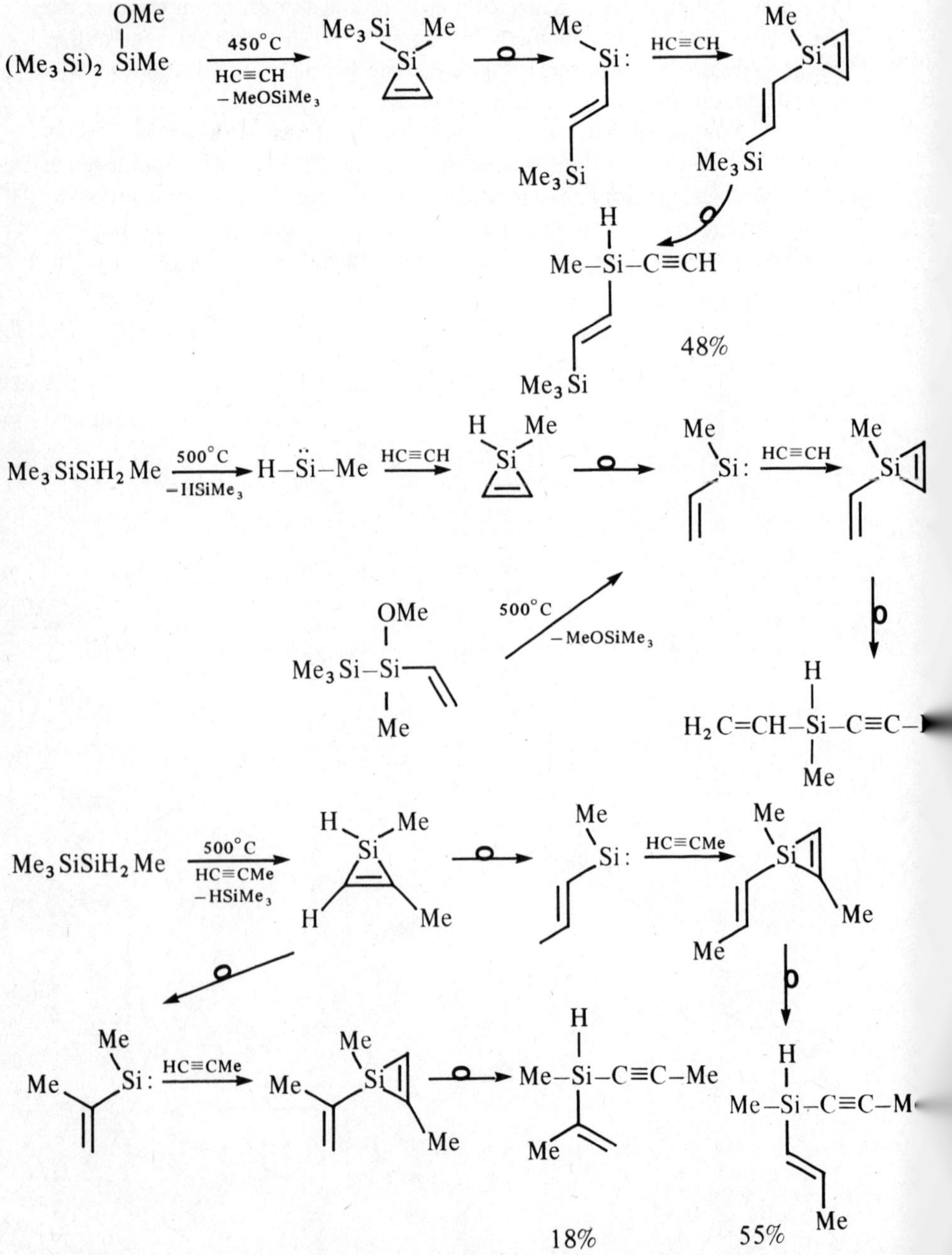

The problem with this mechanism is that it seems to require a rather high-energy[46] 1-silacycloprop-1-ene intermediate, one also invoked for the reverse process (see Section III.E). One wonders whether a direct rearrangement from silirene to silylene might not also operate, under conditions such that neither pathway is reversible.

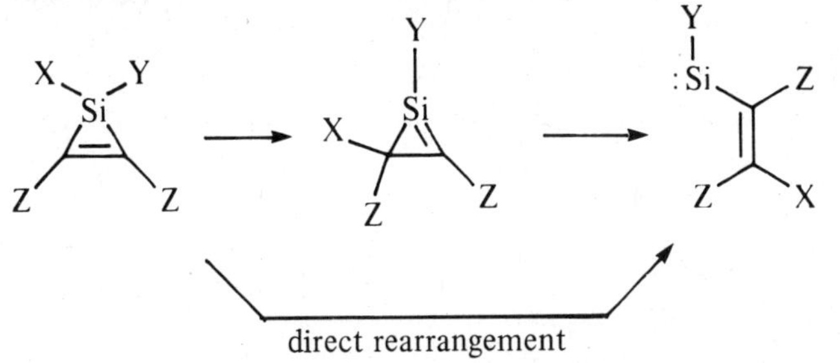

direct rearrangement

Such a direct process would explain a newly discovered reaction that competes with the usual vinylsilacyclopropane-to-silacyclopentene rearrangement[1,6] in the addition of silylene to 1,3-dienes.[47]

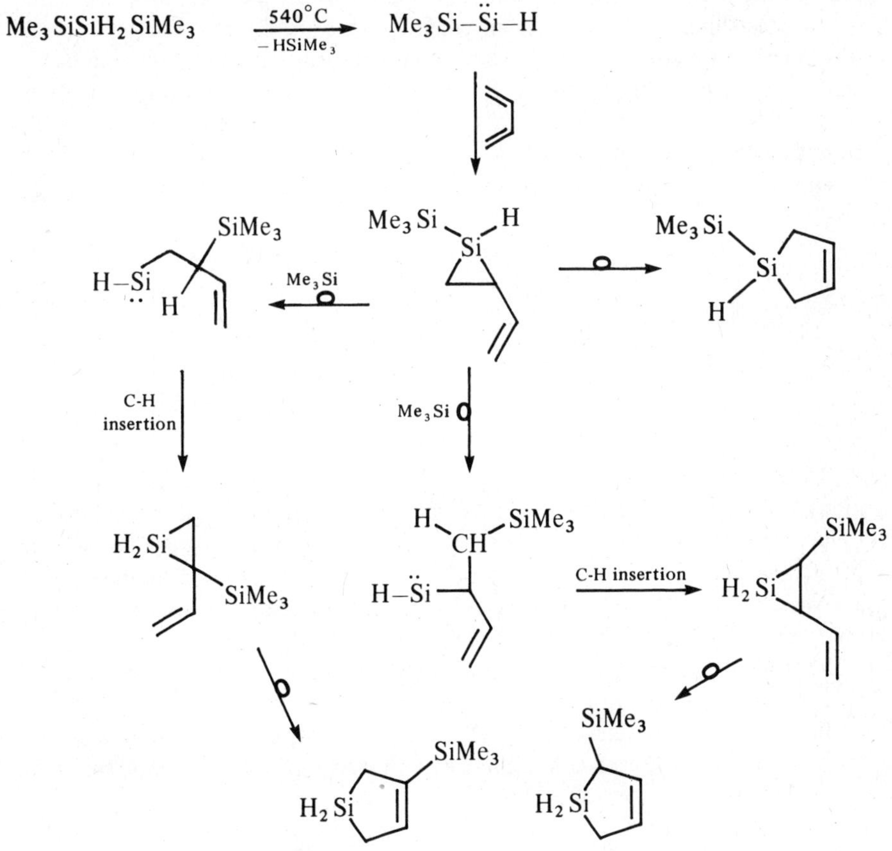

Here the presence of a labile trimethylsilyl group seems to permit another new process, a silirane-to-silylene intramolecular rearrangement, to compete with other modes of silirane loss.

F.   Miscellaneous Methods for the Generation of Silylenes

The elimination of $SiCl_2$ from a series of dichlorosilanes upon impact by 30-eV electrons has been investigated by measuring the intensity of the $(P-SiCl_2)^+$ ion in a mass spectrometer.[48]

$$RR'SiCl_2 \; + \; e \longrightarrow [RR'SiCl_2]^{+*} \; + \; 2e$$

$$\downarrow$$

$$R-R'^+ \; + \; :SiCl_2$$

When R and R' are both aryl, heteroaromatic, or vinyl groups the largest peak in the mass spectrum, accounting for 25–50% of the total ion current, is due to this elimination with rearrangement. Loss of $SiCl_2$ from the parent ions was in each case confirmed by observation of the appropriate metastable peak. If R' is hydrogen, elimination of $SiCl_2$ is still copious, but no $SiCl_2$ formation occurs if R' is alkyl, aryl, or chloro. Cyclic unsaturated dichlorosilanes, of which several examples are shown below, also eliminate $SiCl_2$ on electron impact. It is surprising that even the presence of an Si–O bond does not prevent $SiCl_2$ loss.

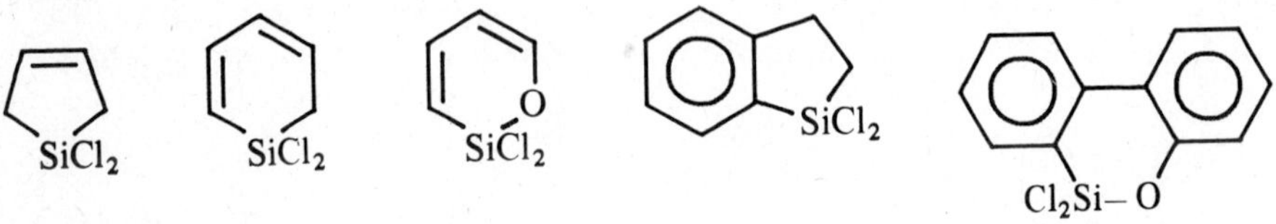

Silylenes have been implicated in two modern industrial processes of enormous importance: plasma etching of silicon surfaces in the manufacture of electronic devices and the deposition of silicon thin films by the decomposition of silane vapors. This is not the place for a complete review, since both subjects are vast,[49] but it is thought that a few recent leading references may be of interest to chemists concerned with silylenes.

Vasile and Stevie have reported mass spectrometric evidence for the formation of SiF_2 together with SiF_4 as gas-phase products from a low-density beam of fluorine atoms impinging on a silicon surface within the ion source of the mass spectrometer.[50] Activation energies of 0.09 ± 0.02 eV (2.1 ± 0.5 kcal/mol) and $0.15 \pm 0.02\,eV$ $(3.5 \pm 0.5\,kcal/mol)$ were deduced from formation of SiF_2 and SiF_4 respectively.

This is important evidence for the formation of SiF_2 as an intermediate, ultimately converted to SiF_4, in the plasma etching of silicon by atomic fluorine, the principal chemical reactant liberated in electric discharges through CF_4-O_2 mixtures.

The visible chemiluminescence observed during fluorine plasma etching of silicon has been postulated as being due to excited trifluorosilyl radicals produced from gas-phase reactions of SiF_2.[51]

$$SiF_2 + F(F_2) \rightarrow SiF_3^*$$

This interpretation has received recent support by observation of the same chemiluminescence spectrum from direct reaction of SiF_2 and fluorine.[51] Although the chemiluminescence experiments placed rather wide limits, 0.01–100%, on the yield of SiF_2 during plasma etching,[51] the mass spectrometric studies point to at least 50% of the desorbed species being silicon fluorides lower than SiF_4.[50]

It has been suggested that SiH_2 is the key species in the deposition of amorphous silicon films in the plasma decomposition of SiH_4, Si_2H_6, and Si_3H_8.[52] It was found that the amorphous silicon deposition rate was more than 10 times greater for the higher silanes than for SiH_4, and this was believed due to the greater ease of SiH_2 formation. No detailed mechanism for silicon deposition was suggested, and we leave this as an exercise for the reader.

The formation of silylenes by interaction of alkali metals with dihalosilanes has a long and somewhat ambiguous history.[6] Some of the metal-induced insertion and addition processes have received renewed attention. Nefedov and Skell have given several new explanations for the formation of higher homologues from reaction of Me_2Si with trimethylsilane.[53]

$$Me_2SiCl_2 \xrightarrow[\substack{gas\text{-}phase \\ -2\,MCl}]{K/Na} Me_2Si: \xrightarrow{HSiMe_3} \underset{n = 2\text{--}5}{Me(SiMe_2)_n H}$$

The occurence of Si–Si insertion was ruled out by the failure to observe octamethyltrisilane from reaction of Me_2Si with hexamethyldisilane. Activation of the Si–H bonds in a polysilane by intramolecular hydrogen bonding to a silicon atom is an interesting if unlikely explanation. Insertion into an Si–H bond of the silylene dimer, tetramethyldisilene, either directly, or by way of abstraction-recombination, is also unlikely, since one would also expect insertion products from isomeric silylenes formed by rearrangement of the disilene:[1] (see Section II.E).

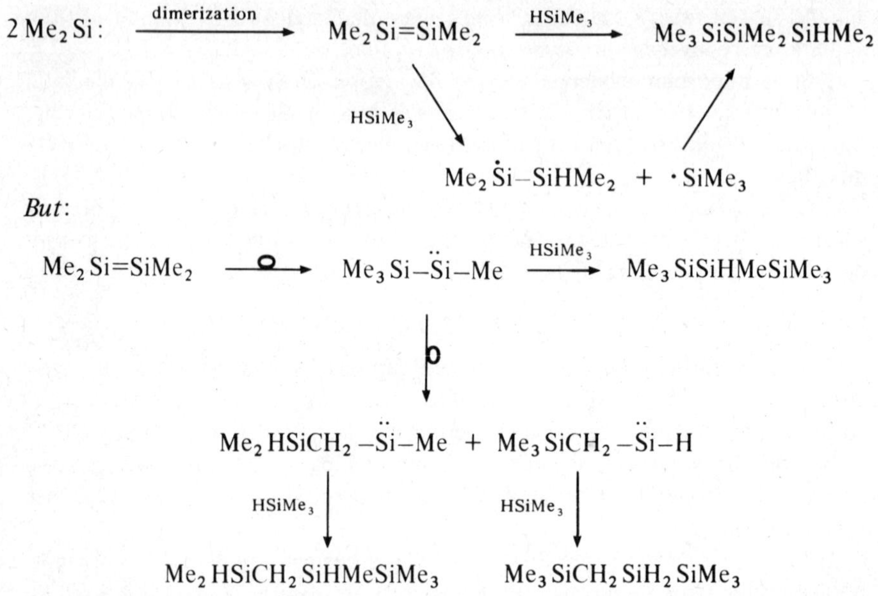

$$2 \, Me_2 Si: \quad \xrightarrow{\text{dimerization}} \quad Me_2 Si = SiMe_2 \quad \xrightarrow{HSiMe_3} \quad Me_3 SiSiMe_2 SiHMe_2$$

$$Me_2 Si = SiMe_2 \quad \xrightarrow{HSiMe_3} \quad Me_2 \overset{\cdot}{Si} - SiHMe_2 \; + \; \cdot SiMe_3$$

But:

$$Me_2 Si = SiMe_2 \quad \xrightarrow{a} \quad Me_3 Si - \overset{..}{Si} - Me \quad \xrightarrow{HSiMe_3} \quad Me_3 SiSiHMeSiMe_3$$

$$\downarrow b$$

$$Me_2 HSiCH_2 - \overset{..}{Si} - Me \; + \; Me_3 SiCH_2 - \overset{..}{Si} - H$$

$$\big\downarrow HSiMe_3 \qquad\qquad \big\downarrow HSiMe_3$$

$$Me_2 HSiCH_2 SiHMeSiMe_3 \qquad\qquad Me_3 SiCH_2 SiH_2 SiMe_3$$

The gas-phase generation of silylenes by alkali metal dehalogenation of substituted dichlorosilanes has been put to interesting use in the study of intramolecular reaction of silylenes.[54] In both cases shown here the intramolecular insertion into a C–H bond yielding a silacyclopropane was the major process. It was accompanied by insertion of the divalent silicon center into a C–H bond of the trimethylsilyl group.

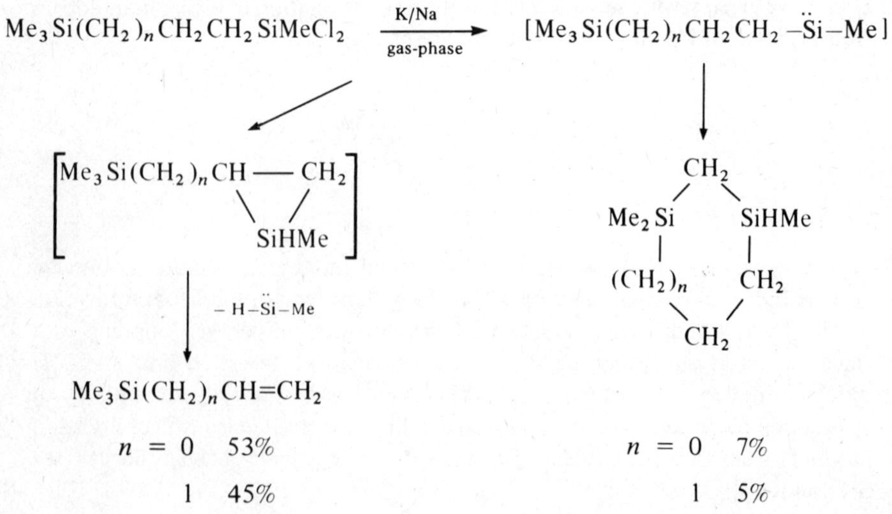

$$Me_3 Si(CH_2)_n CH_2 CH_2 SiMeCl_2 \quad \xrightarrow[\text{gas-phase}]{K/Na} \quad [Me_3 Si(CH_2)_n CH_2 CH_2 - \overset{..}{Si} - Me]$$

$$\left[Me_3 Si(CH_2)_n CH \overset{\textstyle \diagdown \; \diagup}{\underset{SiHMe}{-\!\!\!-\!\!\!-} CH_2} \right]$$

$$\downarrow \; -H - \overset{..}{Si} - Me$$

$$Me_3 Si(CH_2)_n CH = CH_2$$

$$n = 0 \quad 53\%$$
$$1 \quad 45\%$$

$$\begin{array}{c} CH_2 \\ \diagup \quad \diagdown \\ Me_2 Si \qquad SiHMe \\ | \qquad\qquad | \\ (CH_2)_n \qquad CH_2 \\ \diagdown \qquad \diagup \\ CH_2 \end{array}$$

$$n = 0 \quad 7\%$$
$$1 \quad 5\%$$

The other major products, $Me_3Si(CH_2)_nCH_2CH_2SiH_2Me$, resulted from reduction of the silicon–chlorine bonds and were formed in 25 and 35% yields.

Conlin has found that in solution sodium–potassium alloy promotes the transfer of a dimethylsilylene unit from dodecamethylcyclohexasilane to butadiene and silanes.[55]

$$\text{cyclo-}(Me_2Si)_6 \ + \ HSiEt_3 \ \xrightarrow[\text{THF, 25°C}]{\text{Na/K}} \ Me_2HSiSiEt_3 \ + \ \text{cyclo-}(Me_2Si)_5$$

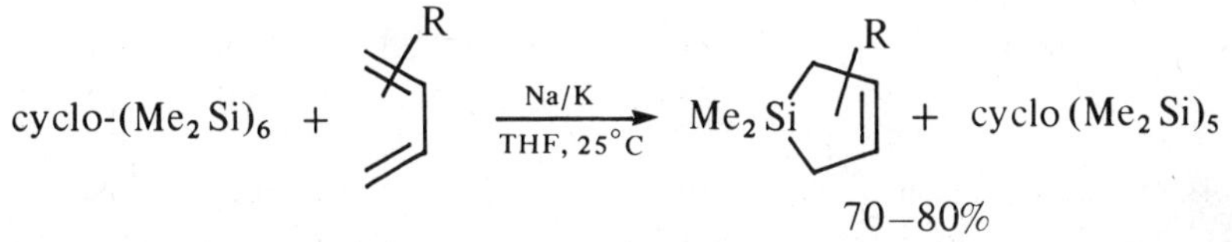

$$60\%$$

$$70-80\%$$

The occurrence of Si–H insertion suggests the intermediate formation of free dimethylsilylene or its radical anion $Me_2\overset{\cdot}{Si}{:}^-$, but the higher rates of the reactions with dienes (complete within a few minutes compared with 20% reaction in 12 h with $HSiEt_3$) and the consumption of excess diene point to a different mechanism.

III.   THE REACTIONS OF SILYLENES

A.   Silylene Insertion Reactions

Silylene reactions have begun to capture the attention of theoreticians. *Ab initio* self-consistent field calculations augmented by second-order perturbation corrections have been found to predict that the insertion of SiH_2 into the hydrogen molecule to form SiH_4 has a 36.0 kJ/mol (8.60 kcal/mol) barrier,[56] compared to an experimental value of 23 kJ/mol (5.50 kcal/mol).[57] A similar calculation for insertion of CH_2 into H_2 indicated that there was no barrier.

The geometry of the transition state calculated for SiH_2 insertion follows.

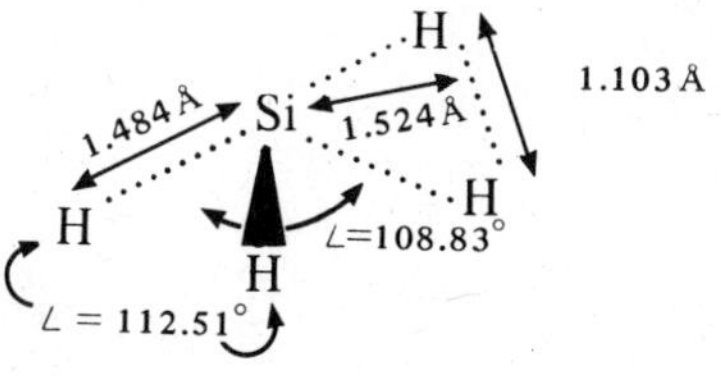

This transition state closely resembles the geometry predicted from simple frontier orbital considerations for reasonable HOMO-LUMO positive overlap:

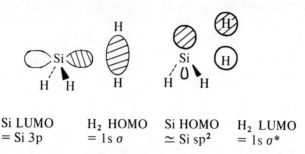

Si LUMO	H$_2$ HOMO	Si HOMO	H$_2$ LUMO
= Si 3p	= 1s σ	$\simeq$ Si sp^2	= 1s σ*

It is therefore surprising to read the suggestion by Bell that reaction of $^1A_1$ SiH$_2$ with H$_2$ is symmetry forbidden, and that the singlet-triplet energy difference must be supplied as activation energy for the reaction to take place.[58] This argument assumes a C$_{2V}$ symmetry for the insertion, but no justification is given for preferring this geometry over that shown before, except that its

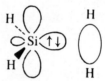

occurrence would account for a large difference in activation energy between insertion of silylenes into H–H and Si–H bonds.

Ab initio calculations on the insertion of SiH$_2$ into O–H bonds have led to a description of this reaction as consisting of initial coordination of an oxygen lone pair of electrons with the empty 3p orbital of the silylene leading to a stable zwitterionic complex ($\Delta E = -12.1$ kcal/mol) which has a significant barrier for rearrangement to silanol (23.4 kcal/mol).[59]

Insertion of a second SiH_2 into the remaining O–H bond of SiH_3OH is predicted to follow a similar pathway with an overall activation barrier (4.9 kcal/mol) and exothermicity (75.7 kcal/mol) each lower than the corresponding values of 11.3 and 70.1 for the first insertion. For comparison, the insertion of singlet CH_2 into water forming methanol is predicted to be exothermic by 94.9 kcal/mol.

Since the barrier to O–H insertion is predicted by calculation to be in the hydrogen migration step, an isotope effect is expected and has indeed been found by Weber and Steele for photochemically generated Me_2Si and MeSiPh in reactions with ethanol and *tert*-butanol,[60] as well as for insertion into the Si–H bond of *n*-butyldimethylsilane:

$$\text{cyclo-}(Me_2Si)_6 \xrightarrow[\text{-cyclo-}(Me_2Si)_5]{} Me_2Si: \xrightarrow[\text{solvent}]{\text{EtO(H/D)}} Me_2(H/D)SiOEt$$

$$k_H/k_D = 2.33 \pm 0.2 \ (Et_2O)$$
$$2.14 \pm 0.18 \ (\text{cyclohexane})$$
$$2.17 \pm 0.13 \ (\text{THF})$$

$$\xrightarrow[\text{solvent}]{t\text{-BuO (H/D)}} Me_2(H/D)SiO(t\text{-Bu})$$

$$k_H/k_D = 1.80 \pm 0.07 \ (Et_2O)$$

$$\xrightarrow[\text{solvent}]{n\text{-BuMe}_2Si(H/D)} Me_2(H/D)SiSiMe_2 \ (n\text{-Bu})$$

$$k_H/k_D = 1.37 \pm 0.09 \ (Et_2O)$$
$$1.32 \pm 0.19 \ (\text{cyclohexane})$$

$$Me_3Si(SiMePh)_2SiMe_3 \xrightarrow{h\nu} Me{-}\ddot{Si}{-}Ph \xrightarrow{t\text{-BuO(H/D)}} MePh(H/D)SiO(t\text{-Bu})$$

$$k_H/k_D = 1.76 \pm 0.12 \ (Et_2O)$$
$$2.09 \pm 0.06 \ (\text{cyclohexane})$$

The insensitivity of the measured isotope effect to change in solvent should be contrasted with the differences in selectivity observed for silylene insertions with change in solvent.

Steele and Weber have found that photochemically generated dimethyl-silylene is much more selective between pairs of alcohols in ethereal solvents than in cyclohexane, and that the selectivity parallels the polarity or the basicity of the solvent.[61] The more sterically hindered an alcohol, the less reactive.

	Relative Reactivity in		
Alcohol pairs	cyclohexane	Et_2O	THF
EtOH/MeOH	0.9		0.7
EtOH/i-PrOH	1.2	2.0	2.7
EtOH/t-BuOH	1.8	4.7	9.6
EtOH/neoPentOH	1.0	1.9	2.6
i-PrOH/t-BuOH	1.2	2.1	4.3

A greater selectivity was also observed for Si—H insertion. The less sterically congested n-BuMe$_2$SiH was found to react with Me$_2$Si 1.3 times as rapidly as $(i$-Pr$)_2$MeSiH in cyclohexane, but the reactivity ratio was 3.2 in THF.

The enhanced selectivity in ethereal solvents was ascribed to complexation of the silylene by the ether, for example,

$$Me_2Si: \; + \; O \longrightarrow Me_2Si \cdots O$$

Zwitterionic silylene—ether adducts had already been suggested as inter-mediates in reactions with strained cyclic ethers.[1] That these complexes should act as silylenoids is quite plausible. The insensitivity of the O—H/O—D isotope effect to the nature of the solvent suggests that the solvent is not intimately involved in the rate-determining transition state. This is consistent with a reversible transfer of the silylene unit from the ether complex to the alcohol as a fast step followed by a slower hydrogen shift.

$$R_2O \cdots SiMe_2 \; + \; R'O(H/D) \rightleftarrows R_2O \; + \; \overset{+}{\underset{-SiMe_2}{R'O(H/D)}} \longrightarrow R'OSi(H/D)Me_2$$

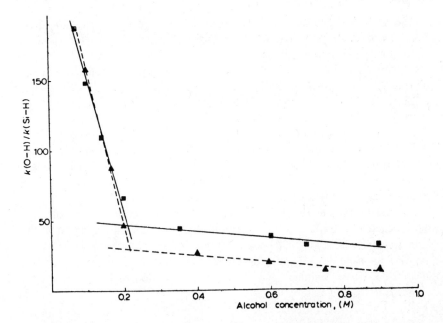

FIGURE 9.1   Variation of relative rates of dimethylsilylene insertion into oxygen—hydrogen and silicon—hydrogen bonds with alcohol concentration. ■, ethyl alcohol/n-butyldimethylsilane; ▲, t-butyl alcohol/n-butyldimethylsilane.

Life is, however, more complicated, and a concentration dependence has been found for the relative reactivity of Si—H and O—H bonds toward silylene insertion.[62]

As shown in Figure 9.1, both ethanol and *tert*-butanol are much more reactive toward photogenic dimethylsilyene than is n-BuMe$_2$SiH. The reactions are O—H and Si—H insertion respectively.

$$\text{cyclo-(Me}_2\text{Si)}_6 \quad \xrightarrow[\text{-cyclo-(Me}_2\text{Si)}_5]{h\nu} \quad \text{Me}_2\text{Si:}$$

$$\text{Me}_2\text{Si:} \quad \begin{array}{c} \xrightarrow{\text{ROH}} \quad \text{Me}_2\text{HSiOR} \\ \\ \xrightarrow{n\text{-BuMe}_2\text{SiH}} \quad \text{Me}_2\text{HSiSiMe}_2\ (n\text{-Bu}) \end{array}$$

The relative reactivities of both alcohols drop sharply with increasing alcohol concentration in cyclohexane up to about $0.2\,M$ and then continue to decrease more slowly. This is attributed to increasing self-association of the alcohols with increasing concentration; the aggregates being less reactive toward the silylene than monomeric alcohol.

When the O–H, Si–H competition is carried out with a high concentration of alcohol in THF rather than cyclohexane, the selectivity in favor of the alcohol is as great as in cyclohexane at the lowest alcohol concentrations employed. Weber explains this by suggesting that in THF the alcohol is hydrogen bonded to the solvent rather than self-associated and that the hydrogen bonding enhances the reactivity of the O–H bond. The increase in reactivity may, however, be due to complexation of the silylene as discussed before.

Gu and Weber have shown that Si–H and Si–OR bonds of polysilanes are more reactive toward silylene insertion than the corresponding substituted monosilanes.[63] Thus $Me_3SiSiHMe_2$ was found to be 1.5 times as reactive toward Me_2Si than $n\text{-}BuMe_2SiH$, and $Me_3SiSiMe_2OEt$ was seven times as reactive as Me_3SiOMe. These findings help explain the formation of higher silanes from monosilane starting materials:

$$\text{cyclo-}(Me_2Si)_6 + HSiMe_3 \quad \xrightarrow[\substack{[Me_2Si]:[H-Si]\\1:3}]{h\nu}$$

$$HSiMe_2SiMe_3 + H(SiMe_2)_2SiMe +$$
$$28.5\% \qquad\qquad 22\%$$
$$H(SiMe_2)_3SiMe_3 \quad 11\%$$

$$\text{cyclo-}(Me_2Si)_6 + EtOSiMe_3 \quad \xrightarrow[\substack{[Me_2Si]:[O-Si]\\1:2}]{h\nu}$$

$$EtOSiMe_2SiMe_3 + EtO(SiMe_2)_2SiMe_3 +$$
$$10\% \qquad\qquad 25\%$$
$$EtO(SiMe_2)_3SiMe_3 \quad 32\%$$

$$\text{cyclo-}(Me_2Si)_6 \quad + \quad \text{[} \overset{\displaystyle SiMe_2}{O} \text{]} \quad \xrightarrow[\substack{[Me_2Si]:([H-Si]\\+[O-Si])=1:4}]{h\nu}$$

$$22.5\% \qquad\qquad 27.5\% \qquad\qquad 34\%$$

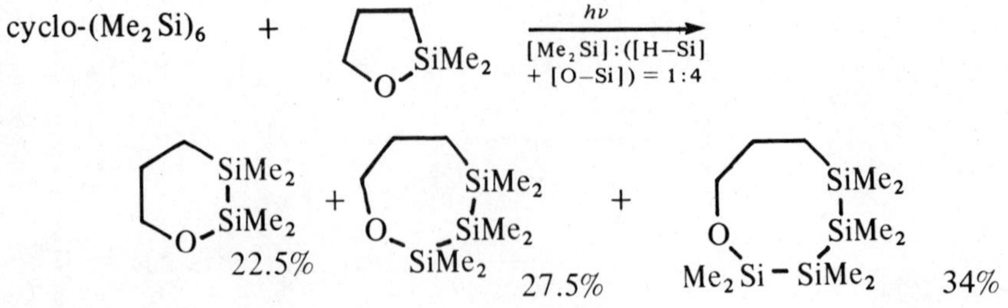

The Si—O bond of $Me_3SiSiMe_2OEt$ was found to be 1.14 times as reactive as the Si—H bond of n-$BuMe_2SiH$, but the relative reactivity could not be determined in an intramolecular competition:

$$cyclo\text{-}(Me_2Si)_6 + HSiMe_2OMe \xrightarrow[\substack{[Me_2Si]:([H-Si] \\ + [O-Si]) = 1:4}]{h\nu} \underset{9\%}{H(SiMe_2)_2OMe}$$

$$+ \underset{21.5\%}{H(SiMe_2)_3OMe} + \underset{41.5\%}{H(SiMe_2)_4OMe}$$

Davidson and co-workers studied the insertion of photochemically generated Me_2Si in the gas phase.[64] Pentamethyldisilane was found to be about three times as reactive toward Me_2Si as trimethylsilane, in agreement with the work of Weber. The relative reactivity was independent of temperature in the range 413–510 K, and the near equality of the activation energies was attributed to their smallness, both believed nearly zero. Reference was made in Section II.B to the observation that the relative reactivity of SiH_4 and SiH_3Me toward Si—H insertion by SiH_2 shows no temperature dependence in the range 658–733 K.[14] The relative reactivity toward Si—H insertion of Me_2Si of $Me_2HSiSiHMe_2$ and $Me_3SiSiHMe_2$ was 1.4, again independent of temperature. In contrast, the insertion of Me_2Si into HCl exhibited an activation energy of 27.7 ± 1 kJ/mol (6.6 kcal/mol) in the range 423–478 K.

In all of these reactions minor products resulted from consecutive insertions and the incursion of radical processes.

$$cyclo\text{-}(Me_2Si)_6 \xrightarrow{h\nu} Me_2Si: \xrightarrow{Me_3SiH} Me_3SiSiHMe_2$$

$$\xrightarrow{Me_3SiSiHMe_2} Me_3SiSiMe_2SiHMe_2$$

$$\xrightarrow{HCl} Me_2HSiCl$$

It is interesting to note that little insertion of Me_2Si into the Si—Cl bond of Me_3SiCl was observed in photochemical experiments at temperatures up to 488 K.[65] (See Section III.D for a thermally induced Si—Cl insertion.) The products of radical processes were observed, and it is intriguing to speculate as to the radical source:

$$Me_2 \overset{\cdot}{S}iH + Me_3 SiCl \longrightarrow Me_2 SiHCl + Me_3 Si \cdot$$

$$2 Me_3 Si \cdot \longrightarrow Me_3 SiSiMe_3$$

Davidson has suggested that Me_2SiH comes either directly from photolysis of $(Me_2Si)_6$ or from hydrogen abstraction by Me_2Si. The latter process would be interesting as a possible reaction of *triplet* Me_2Si, no hydrogen abstraction having been found for thermally generated Me_2Si in an experiment designed to search for hydrogen abstraction.[6]

The photolysis of $(Me_2Si)_6$ in the presence of Me_2HSiCl and $MeHSiCl_2$ led mostly to the formation of radical reaction products with photolysis of the chlorosilane as well as the cyclohexasilane initiating the radical reactions.[65]

An important result of Davidson's work is an explanation for the curious result that pyrolysis of pentamethyldisilane does *not* lead to the formation of detectable amounts of heptamethyltrisilane despite the formation of dimethyl-silylene, as evidenced by trapping with monosilanes.[6] It has been inferred that the activation energy for decomposition of heptamethyltrisilane is lower than for pentamethyldisilane, 150 kJ/mol (36 kcal/mol) versus 198 kJ/mol (47 kcal/mol) with nearly equal A values. Thus the insertion of dimethylsilylene into the Si–H bond of pentamethyldilsilane, which is believed to have no activation barrier and $\log A = 9.5$, is reversible under the reaction conditions, the silylene being siphoned off by polymer formation.

$$Me_3 SiSiHMe_2 \overset{\Delta}{\rightleftharpoons} Me_3 SiH + Me_2 Si:$$

$$Me_2 Si: + Me_3 SiSiHMe_2 \rightleftharpoons Me_3 SiSiMe_2 SiHMe_2$$

$$Me_2 Si: \longrightarrow polymer$$

Several new silylene insertions have been found recently. Blazejowski and Lampe report that SiH_2 formed by infrared multiphoton laser photolysis of SiH_4 inserts into a P–H bond of phosphine at room temperature with a rate constant 5.3×10^8 l/mol(sec).[66]

$$SiH_4 \xrightarrow[-H_2]{n h\nu} :SiH_2 \xrightarrow{PH_3} H_3 SiPH_2$$

An earlier absence of methylsilylphosphine from the high-temperature thermal generation of SiH_2 in the presence of methylphosphine may have been due to competition from even more reactive Si–H bonds, or, possibly, due to the decomposition of the silylphosphine under the reaction conditions.[67]

Chihi and Weber have found insertion by photochemically generated Me_2Si into silicon–sulfur and sulfur–sulfur single bonds.[68]

$$\text{cyclo-}(Me_2Si)_6 \xrightarrow{h\nu} Me_2Si: \xrightarrow{EtS-SiMe_3} EtS-SiMe_2SiMe_3$$

$$\xrightarrow{MeS-SMe} MeS-SiMe_2-SMe$$

It is interesting that dimethylsilylene will insert into an unstrained Si–S bond, in light of the earlier finding that only angle-strained siloxanes undergo Si–O insertion.[1] Insertion into Si–O bonds of alkoxysilanes does, however, occur in the absence of strain.[1,6]

The products of photochemical generation of trimethylsilylphenylsilylene in the presence of alkyl halides provide considerable insight into the mechanism of silylene insertion into carbon–chlorine bonds.[69] Ishikawa, Kumada, and co-workers observed only the products of formal C–Cl insertion from octyl chloride and cyclopropylcarbinyl chloride. From *sec*-butyl chloride both the products of formal C–Cl insertion and HCl abstraction were observed, whereas *tert*-butyl chloride gave only abstraction product.

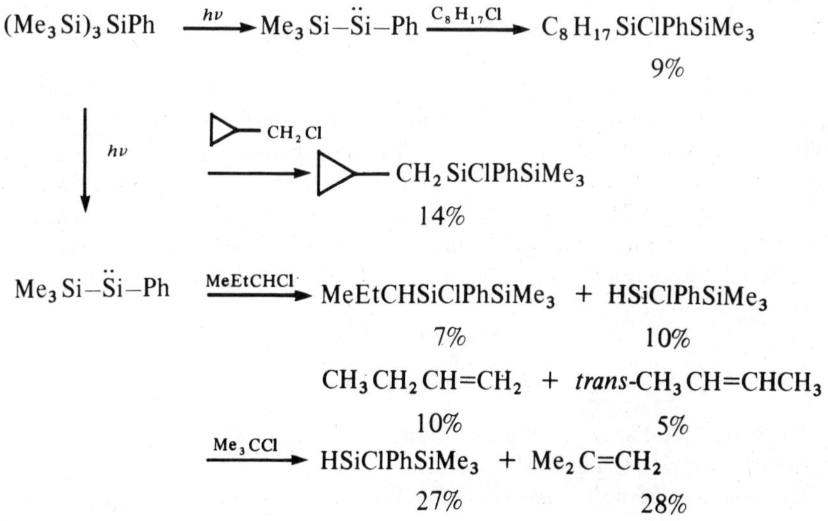

$$(Me_3Si)_3SiPh \xrightarrow{h\nu} Me_3Si-\overset{..}{Si}-Ph \xrightarrow{C_8H_{17}Cl} C_8H_{17}SiClPhSiMe_3$$
$$9\%$$

$$\xrightarrow{CH_2Cl} \quad CH_2SiClPhSiMe_3$$
$$14\%$$

$$Me_3Si-\overset{..}{Si}-Ph \xrightarrow{MeEtCHCl} MeEtCHSiClPhSiMe_3 + HSiClPhSiMe_3$$
$$ 7\% 10\%$$
$$CH_3CH_2CH=CH_2 + \textit{trans-}CH_3CH=CHCH_3$$
$$ 10\% 5\%$$
$$\xrightarrow{Me_3CCl} HSiClPhSiMe_3 + Me_2C=CH_2$$
$$ 27\% 28\%$$

Although these yields are all low, their mechanistic implication is clearly that C–Cl insertion proceeds by way of a zwitterionic ylid intermediate that can undergo competitive rearrangement and concerted elimination. The lack of ring-opened products from the cyclopropylcarbinyl chloride seems to rule out fragmentation to a carbocation.

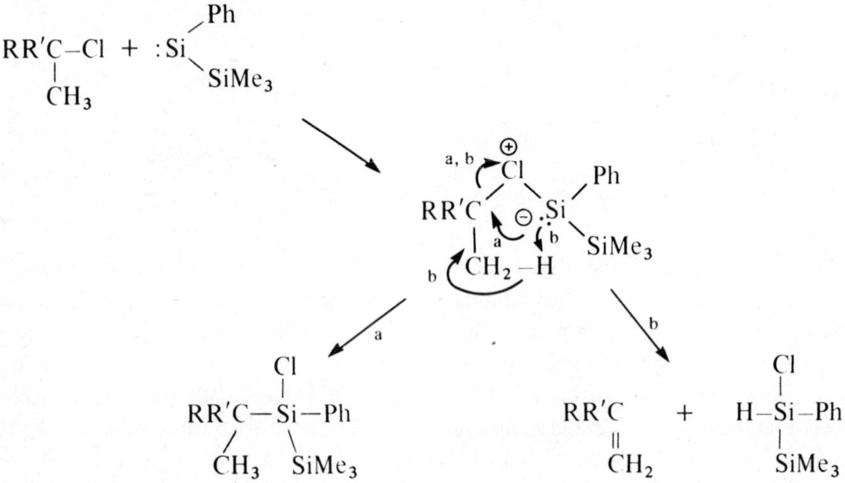

A most unusual product has been reported from photochemically induced insertion of matrix isolated SiO (at least formally a silylene) into the F–F bond.[70]

$$SiO + F_2 \xrightarrow{h\nu} OSiF_2$$

The force constant of the Si=O double bond was calculated as 9×10^2 N/m with the principal ir band at 1309.4 cm^{-1}. This wonderful molecule was previously reported in the gas phase, detected by mass spectrometry above the BeO-SiO$_2$-Na$_2$BeF$_4$ system.[71] This reader prefers to reserve judgement on the existence and stability of monomeric OSiF$_2$. The mass spectral data may be due to the decomposition of a more mundane compound, SiF$_3$OSiF$_3$.

B.   Silylene Addition Reactions

a.   Addition to Olefins and Dienes

An important paper by Wu and Jones extends their study of the stereochemistry of silylene additions to olefins.[1]

The addition of both dimethyl- and diphenylsilylene to the 2-butenes was demonstrated to be stereospecific (> 95%) by the finding of only a single stereoisomer upon methanolysis of the intermediate silacyclopropanes with MeOD.[72] The analysis was made possible by the formation of a *diastereotopic* pair of hydrogens, H_a and H_b.

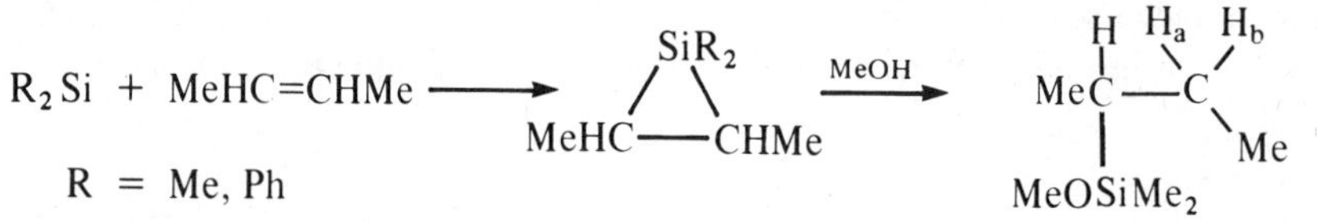

R_2Si + MeHC=CHMe
SiR_2
MeHC—CHMe
MeOH
H H_a H_b
MeC—C
Me
$MeOSiMe_2$
R = Me, Ph

In these experiments $(Me_2Si)_6$ was the source of dimethylsilylene, and 2,2-diphenylhexamethyltrisilane was photolyzed to yield diphenylsilylene, as demonstrated by diene trapping. The products were those expected from rearrangement of an intermediate vinylsilacyclopropane:[1,5]

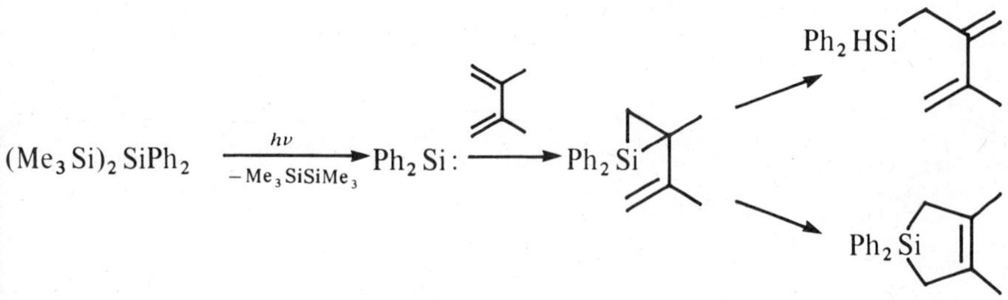

$(Me_3Si)_2SiPh_2$
$h\nu$
$-Me_3SiSiMe_3$
Ph_2Si:
Ph_2Si
Ph_2HSi
Ph_2Si

The silacyclopropanes formed by addition of diphenylsilylene to the 2-butenes were much less stable than the analogous dimethylsilylene adduct. To trap the diphenyldimethylsilirane with methanol it was necessary for the alcohol to be present together with the olefin when the silylene was generated.

Wu and Jones were able to establish beyond cavil that silylene addition occurred in a cis manner by methanolysis of the siliranes formed from addition to cyclopentene and cyclohexene. By use of MeOD stereospecific cis silirane ring opening was also demonstrated.

R_2Si: +
R_2Si
R = Me, Ph
MeOD
H
OMe
SiR_2
H
D

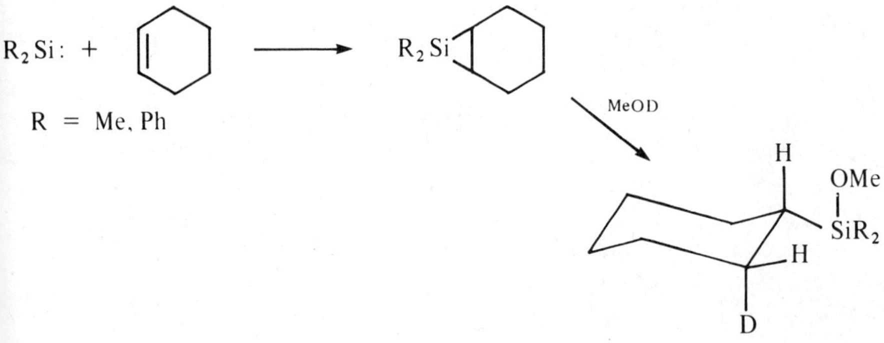

The stereochemistry of the silylene–olefin addition was approached in a more direct manner by the Seyferth group. They employed the thermolysis of hexamethylsilirane to generate dimethylsilylene under conditions so mild that the products of addition to *cis*- and *trans*-propenyltrimethylsilane could by analyzed by ^{1}H and ^{13}C NMR spectroscopy *prior* to methanolysis.[34] Although the arguments required for the interpretation of the spectra obtained are long and somewhat oblique, these results are consistent with the Princeton and Kyoto findings.

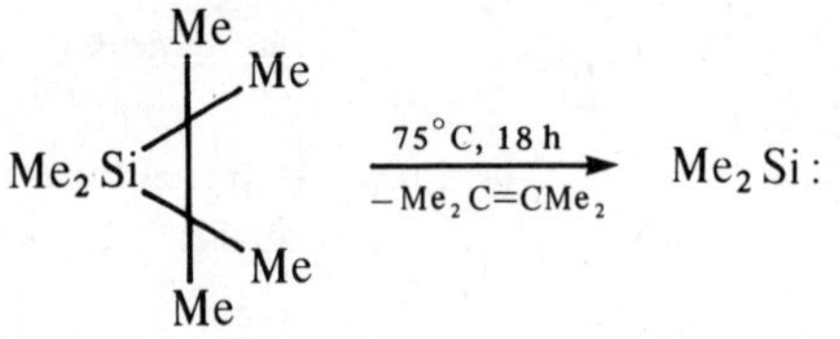

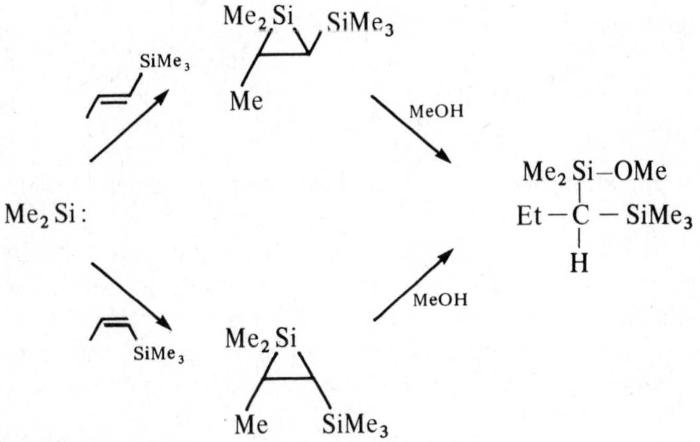

Ishikawa, Kumada, and co-workers have studied the addition of photochemically generated phenyltrimethylsilylsilylene to various substituted olefins.[73] Addition to vinyl chloride and to 1-bromo-2-methylpropene afforded the products of silirane ring opening with a 1,2-shift of a halogen atom.[73] Vinyl acetate reacts by a similar mechanism, but in this case ring opening and acetate migration are more likely to be concerted.

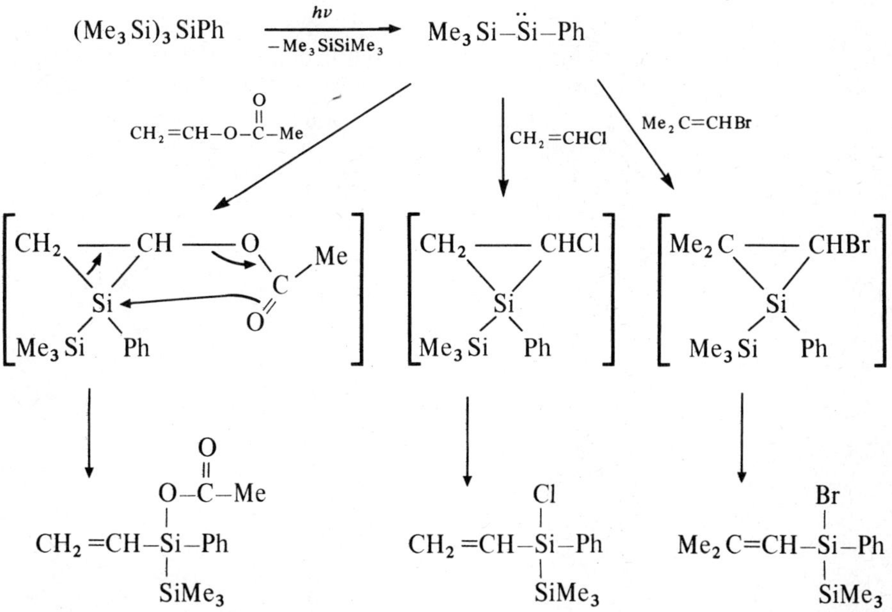

$$(Me_3Si)_3SiPh \xrightarrow[-Me_3SiSiMe_3]{h\nu} Me_3Si-\ddot{S}i-Ph$$

$$CH_2=CH-O-\overset{\overset{O}{\parallel}}{C}-Me$$

$$\begin{bmatrix} CH_2 \!-\!\! CH \!-\!\! O \\ \quad\diagdown\!\!\diagup Si \diagup\quad\quad C\!-\!Me \\ \quad\quad\quad\quad\overset{\diagdown}{O}\!\!\diagup \\ Me_3Si \quad Ph \end{bmatrix} \quad \begin{bmatrix} CH_2 \!-\!\! CHCl \\ \diagdown\!\!Si\!\!\diagup \\ Me_3Si \quad Ph \end{bmatrix} \quad \begin{bmatrix} Me_2C \!-\!\! CHBr \\ \diagdown\!\!Si\!\!\diagup \\ Me_3Si \quad Ph \end{bmatrix}$$

$$CH_2=CHCl \qquad Me_2C=CHBr$$

$$\underset{\displaystyle \underset{SiMe_3}{|}}{CH_2=CH-\overset{\overset{\displaystyle O-\overset{O}{\overset{\parallel}{C}}-Me}{|}}{Si}-Ph} \qquad \underset{\displaystyle \underset{SiMe_3}{|}}{CH_2=CH-\overset{\overset{\displaystyle Cl}{|}}{Si}-Ph} \qquad \underset{\displaystyle \underset{SiMe_3}{|}}{Me_2C=CH-\overset{\overset{\displaystyle Br}{|}}{Si}-Ph}$$

Addition of the same silylene to ethyl vinyl ether gave a silirane that could be observed by ^1H NMR and could be trapped by addition of methanol or HCl after irradiation.

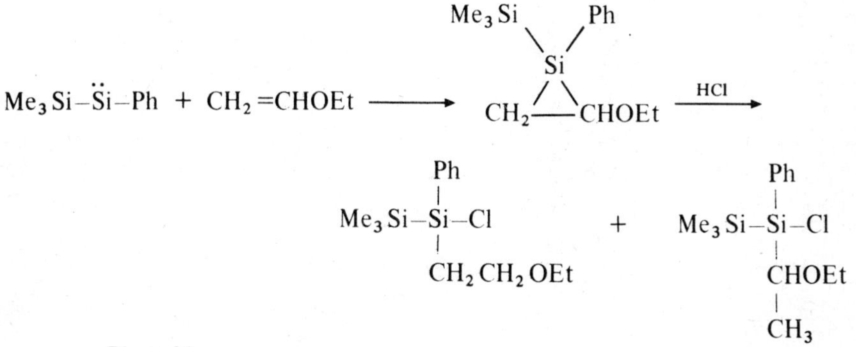

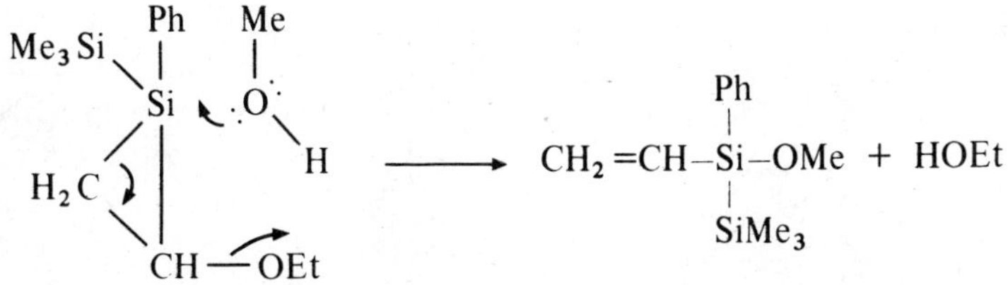

$$Me_3Si-\ddot{S}i-Ph + CH_2=CHOEt \longrightarrow \underset{\displaystyle CH_2\!-\!\!CHOEt}{\overset{\displaystyle \overset{Me_3Si \quad Ph}{\diagdown\!\!\diagup}}{\underset{\displaystyle Si}{}}} \xrightarrow{HCl}$$

$$\underset{\displaystyle CH_2CH_2OEt}{\overset{\displaystyle Ph}{\underset{\displaystyle |}{Me_3Si-\overset{\overset{|}{}}{Si}-Cl}}} \quad + \quad \underset{\displaystyle \underset{CH_3}{\overset{|}{CHOEt}}}{\overset{\displaystyle Ph}{\underset{\displaystyle |}{Me_3Si-\overset{\overset{|}{}}{Si}-Cl}}}$$

$$\underset{\displaystyle CH\!-\!OEt}{\overset{\displaystyle \overset{Me_3Si}{\diagdown}\ \overset{Ph}{\overset{|}{Si}}\ \overset{Me}{\overset{|}{O}}}{\underset{\displaystyle H_2C}{}}} \longrightarrow \underset{\displaystyle SiMe_3}{\overset{\displaystyle Ph}{\underset{\displaystyle |}{CH_2=CH-\overset{\overset{|}{}}{Si}-OMe}}} + HOEt$$

While the sheer fun of arrow-pushing does not prove that the ring opening with elimination induced by methanol occurred by a concerted cyclic process, this is a possibility. That the alcoholysis is concerted is strongly suggested by its stereospecificity in the following examples, which also confirm the stereo-specificity of the silylene addition:

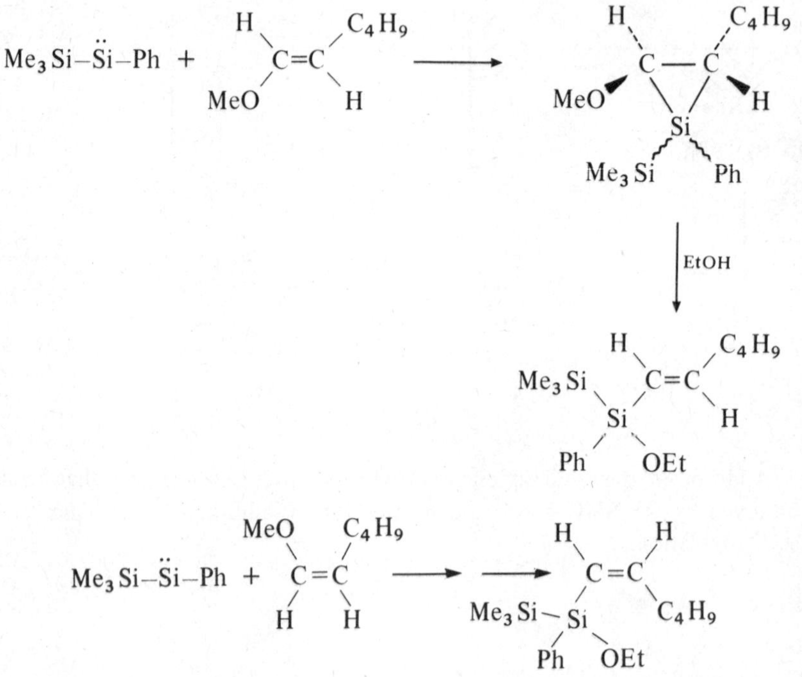

$$Me_3Si-\ddot{S}i-Ph \;+\; \underset{MeO}{\overset{H}{\diagdown}}C=C\underset{H}{\overset{C_4H_9}{\diagup}} \longrightarrow \underset{MeO}{\overset{H}{\diagdown}}C\!-\!C\underset{H}{\overset{C_4H_9}{\diagup}} \quad \underset{Me_3Si \qquad Ph}{\overset{Si}{\diagup}}$$

$$\Big\downarrow EtOH$$

$$Me_3Si\diagdown \underset{Si}{\diagup}C=C\underset{H}{\overset{C_4H_9}{\diagup}} \qquad Ph \diagup \diagdown OEt$$

$$Me_3Si-\ddot{S}i-Ph \;+\; \underset{H \quad H}{\overset{MeO \quad C_4H_9}{\diagdown \diagup}}C=C \longrightarrow \longrightarrow \underset{Me_3Si-Si}{\overset{H}{\diagdown}}C=C\underset{C_4H_9}{\overset{H}{\diagup}} \quad Ph \diagup \diagdown OEt$$

This elegant experiment would have been even more valuable had the chiral silicon center also been utilized to determine the steric course of the ring open-ing.

The addition to butadiene of the novel silylene 1-silacyclopent-3-en-1-ylidene, itself formed by α-elimination from a silylene–butadiene adduct has been men-tioned in Section II.A.

Ishikawa, Kumada, and co-workers have studied the photolysis of 2-phenyl-heptamethyltrisilane in the presence of various 1,3-dienes.[125] The observed products indicate that two different reactive intermediates are formed in com-parable quantities. One is methylphenylsilylene and the other a conjugated silene that is a rearrangement product of the starting trisilane. It may arise by a photoinduced, 1,3-trimethylsilyl shift, or instead by homolysis to a radical pair followed by rebonding. The silylene adds in 1,2-fashion to the butadiene, and the resulting vinylsilirane can be trapped with methanol, rearrange to a silacyclopentene, or undergo ring opening. The conjugated silene seems to undergo an ene reaction with 2,3-dimethylbutadiene.

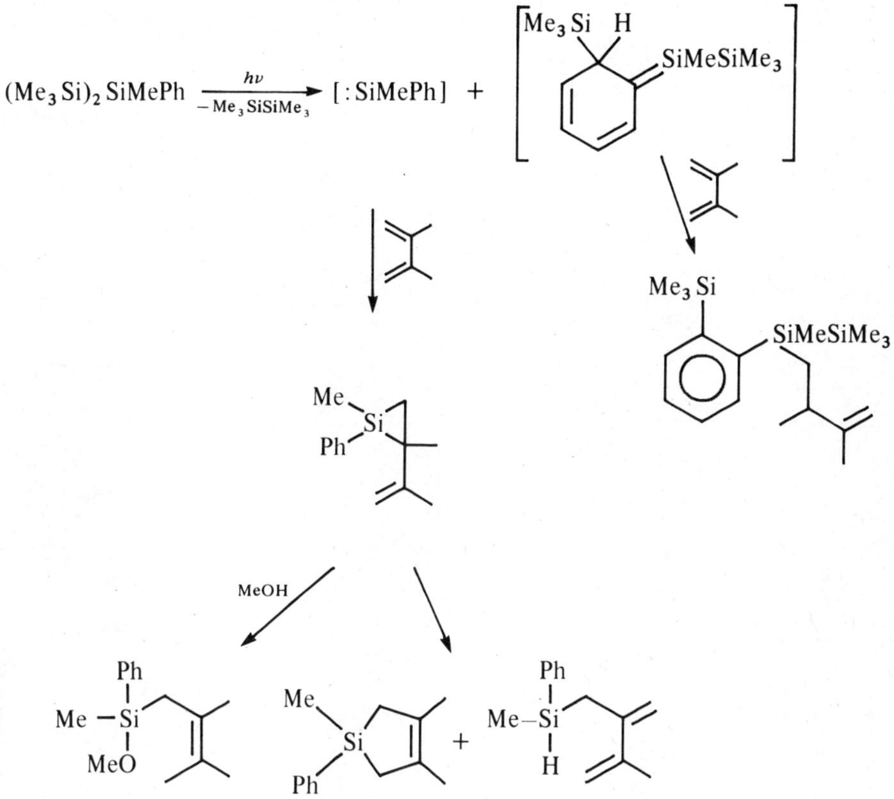

It should be noted that a different structure had previously been assigned to the product of the ene reaction.[6]

The photochemical reaction of 2-phenylheptamethyltrisilane with isoprene was found to give somewhat similar results, complicated by the participation of the two nonequivalent double bonds of isoprene in both silylene addition and the ene reaction. Only low yields of products were obtained when cyclopentadiene was the diene substrate, and none from 1,3-cyclohexadiene. 1,3-Cyclooctadiene, however, gave a variety of products from silylene addition, as well as an ene-reaction product:

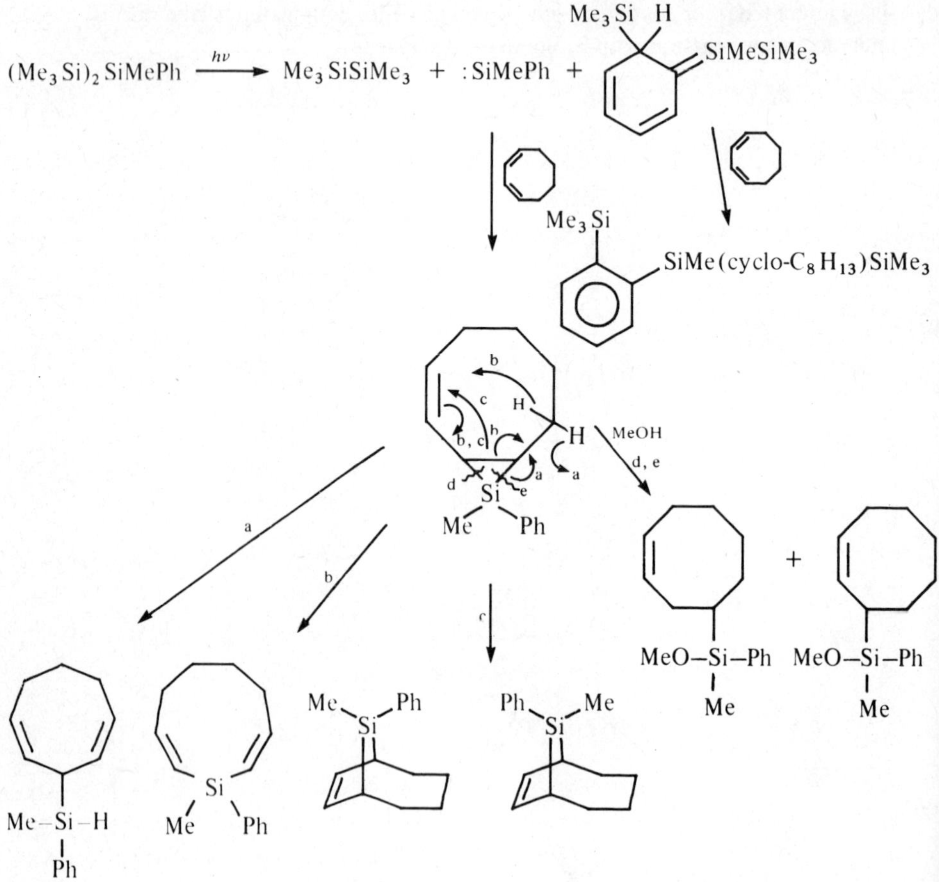

As unusual as these rearrangement products of the initial silylene adduct appear, they all have precedent in earlier studies of silylene addition to cyclic dienes.[6] An important unanswered question, however, is the relationship between the formation of the silylene and the formation of the rearranged conjugated silene.

b.   Silylene Addition to Allyl Ethers, Sulfides, and Halides

Although formally an insertion, the reaction of silylene with allyl ethers is almost certainly mechanistically an initial attack on an oxygen lone pair followed by direct rearrangement or by fragmentation to a radical pair followed by recombination.[1] Tortorelli and Jones have suggested the following reaction pathway for photogenic dimethylsilylene and alkyl allyl ethers.[74]

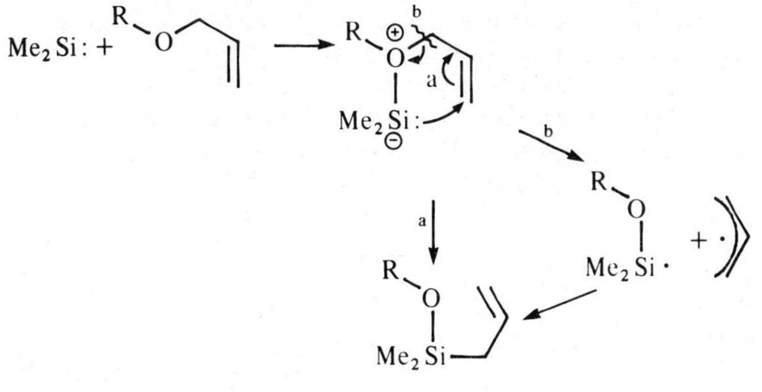

Tzeng and Weber have found that substituted allyl methyl ethers react regiospecifically, an observation that favors concerted rearrangement of the initial zwitterionic complex over a fragmentation-recombination radical mechanism.[75]

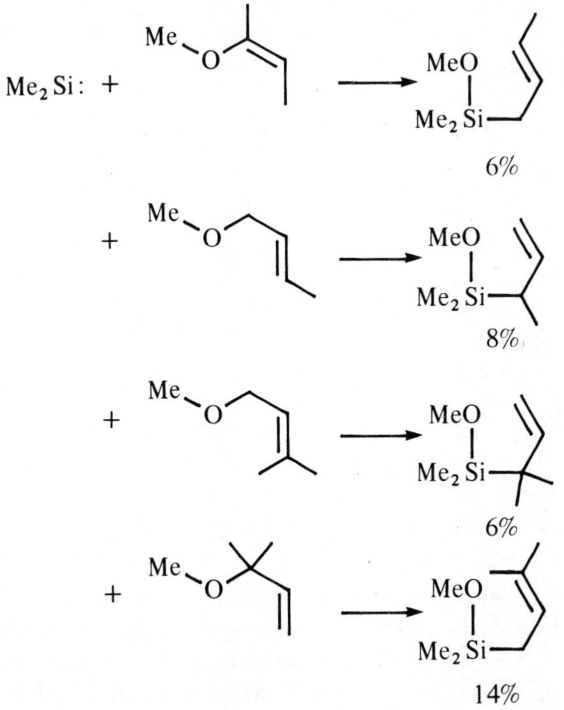

The same products were obtained in higher yields when dichlorodimethyl-silane and sodium or lithium metal dispersion were allowed to react with allyl methyl ether, α,α-dimethylallyl methyl ether, and α-methylallyl methyl ether. With *trans*-γ-methylallyl methyl ether, the product of formal insertion was obtained, however, and thus it seems likely that all these alkali-metal-induced reactions follow pathways initiated by reduction of the allyl ether followed by attack of the resulting radical anions on the chlorosilane.[6]

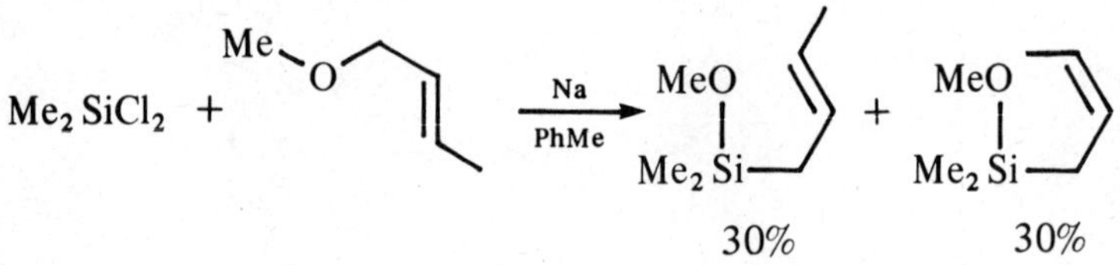

The reaction of *trans*-γ-methylallyl methyl sulfide with photogenic Me$_2$Si: apparently follows a similar mechanism.[76] The product of an Si–S bond insertion was also obtained, presumably by way of initial attack of the silylene on sulfur.

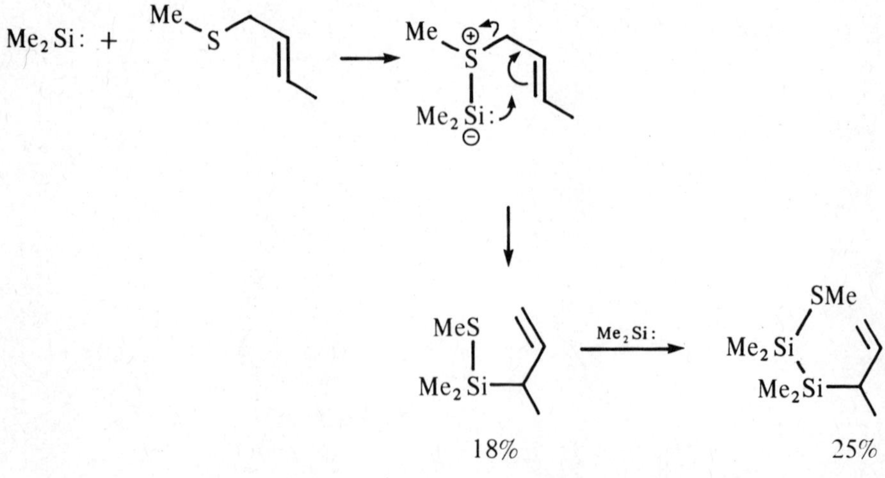

In contrast to all this evidence for attack at the heteroatom, Ishikawa, Naka-gawa, and Kumada have found that photochemically generated phenyltrimethyl-silylsilylene seems to attack the double bond of allyl ethyl ether to form a silacyclopropane that survives the reaction.[77] Addition of, first, dry hydrogen chloride, and then methanol yields the products appropriate for silirane ring opening followed by nucleophilic attack at the Si—Cl bond:

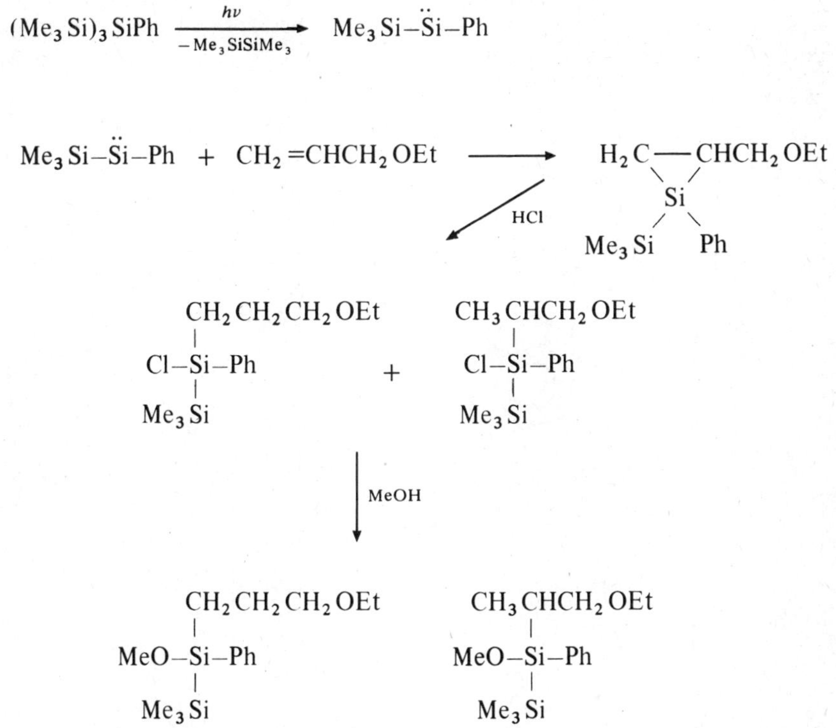

$$(Me_3Si)_3SiPh \xrightarrow[-Me_3SiSiMe_3]{h\nu} Me_3Si-\overset{\cdot\cdot}{Si}-Ph$$

$$Me_3Si-\overset{\cdot\cdot}{Si}-Ph \; + \; CH_2{=}CHCH_2OEt \longrightarrow$$

$$\begin{array}{c} CH_2CH_2CH_2OEt \\ | \\ Cl-Si-Ph \\ | \\ Me_3Si \end{array} \quad + \quad \begin{array}{c} CH_3CHCH_2OEt \\ | \\ Cl-Si-Ph \\ | \\ Me_3Si \end{array}$$

$$\Big\downarrow \text{MeOH}$$

$$\begin{array}{c} CH_2CH_2CH_2OEt \\ | \\ MeO-Si-Ph \\ | \\ Me_3Si \end{array} \qquad \begin{array}{c} CH_3CHCH_2OEt \\ | \\ MeO-Si-Ph \\ | \\ Me_3Si \end{array}$$

Direct methanolysis of the intermediate silirane gave a product of apparent concerted ring opening with elimination reminiscent of the reaction with ethyl vinyl ether (see below) together with the simple ring-opening product.

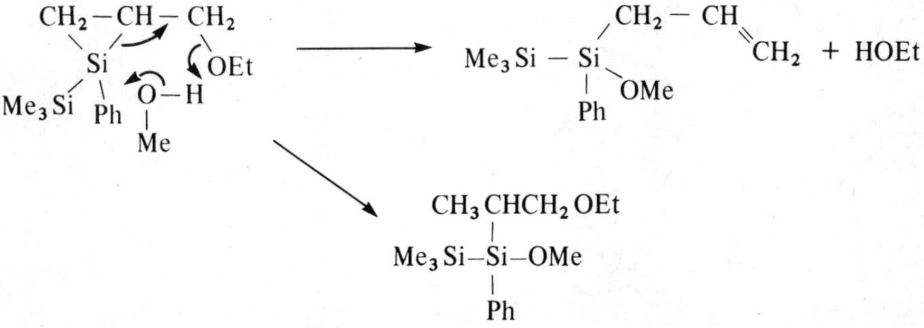

$$\begin{array}{c} CH_3CHCH_2OEt \\ | \\ Me_3Si-Si-OMe \\ | \\ Ph \end{array}$$

c.   Silylene Additions to Carbonyl Compounds

Ishikawa, Kumada, and co-workers obtained an adduct of dimethylsilylene and acetone in the course of investigating the photochemical generation of a silacyclopropene so stable that it tolerates the presence of oxygen and can be purified by recrystalization from ethanol![78]

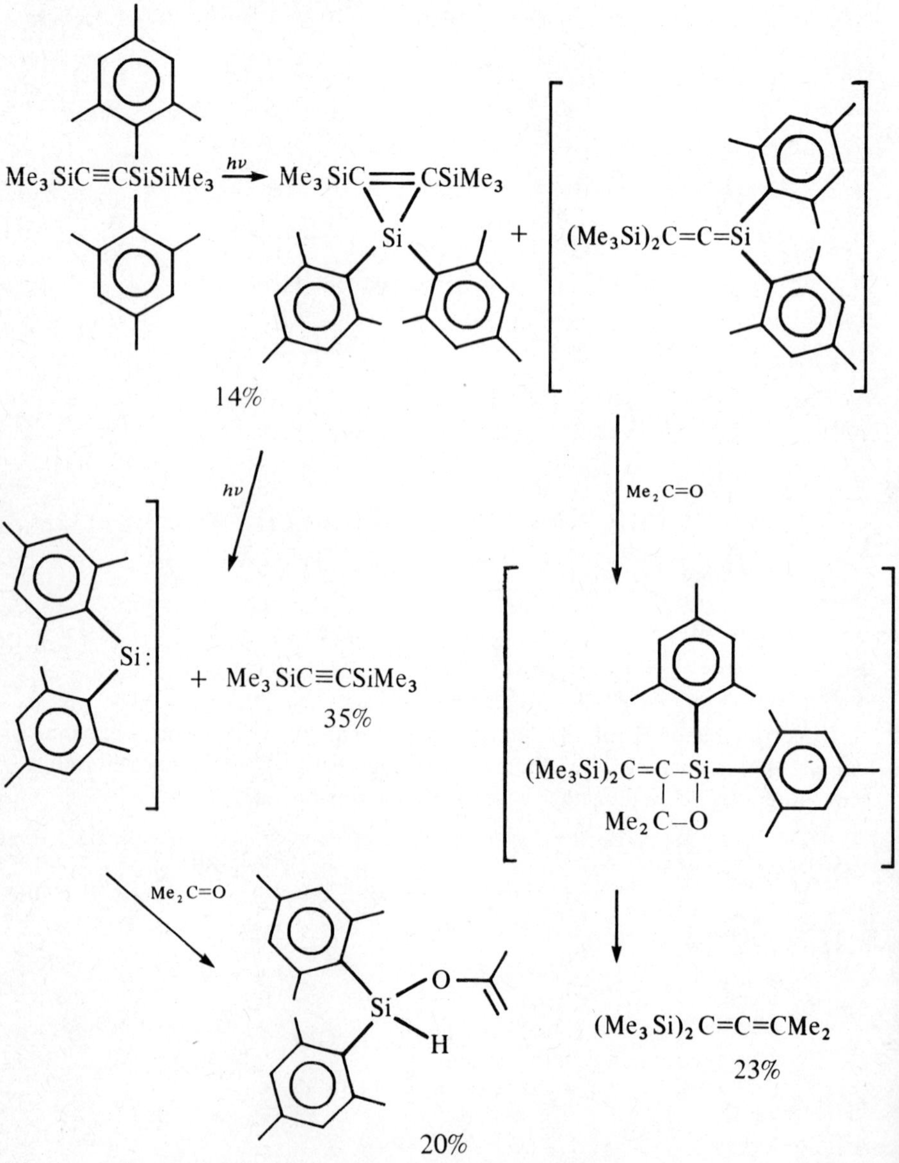

d. Silylene Addition to Acetylenes

The gas-phase addition to acetylenes of silylenes carrying labile substituents has been presented in Section II.E together with the rearrangement of the resulting silacyclopropenes to vinylsilylenes.

Halevi and West have discussed the addition of dimethylsilylene to acetylene in terms of the Orbital Correspondence Analysis in Maximum Symmetry formalism.[79] In this analysis, transition states of maximum symmetry are first considered. Molecular orbital correlation diagrams are constructed to reveal what reduction in symmetry is required to transform bonding-antibonding to bonding-bonding orbital correlations and thus convert "symmetry-forbidden" to "symmetry-allowed" reactions.

For the addition of dimethylsilylene to acetylene the approach of maximum symmetry is C_{2v} and is forbidden in the Woodward–Hoffmann sense. The OCAMS analysis points out an allowed pathway as one in which only the horizontal mirror plane of the C_{2v} approach is retained.

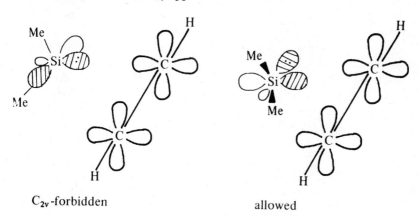

C_{2v}-forbidden   allowed

For this simple case, elementary frontier orbital theory leads to the same result with much less fuss, but this analysis is helpful for more elaborate examples like the dimerization of silacyclopropenes to disilacyclohexadienes. The sigma-dimerization route experimentally observed,[6] has gained instructive theoretical support.

e. Addition Reactions of Difluorosilylene

The finding that products containing more than one SiF_2 unit can be formed by secondary reactions of difluorosilylene adducts[1] has led literally to the rediscovery of the gas-phase chemistry of thermally generated SiF_2.[80]

Hwang and Liu have found that reaction of SiF_2 with butadiene gives a 30%

yield of a mixture in which the adduct containing a single SiF_2 predominated by more than 20–1.[80] Tang had obtained 1,1-difluoro-1-[31]Si-silacyclopent-3-ene as a product of nucleogenic difluorosilylene in 1972.[6]

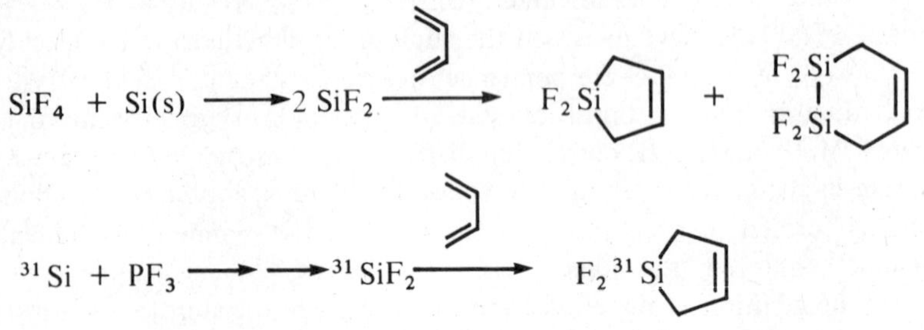

$$SiF_4 + Si(s) \longrightarrow 2\ SiF_2 \longrightarrow F_2Si \bigcirc\!\!| \quad + \quad \begin{matrix} F_2Si \\ | \\ F_2Si \end{matrix} \bigcirc\!\!|$$

$$^{31}Si + PF_3 \longrightarrow \longrightarrow ^{31}SiF_2 \longrightarrow F_2{}^{31}Si \bigcirc\!\!|$$

Tang and co-workers have recently demonstrated that recoiling silicon atoms acquire hydrogen and fluorine atoms in a stepwise manner.[17] When the $^{31}P(n, p)^{31}Si$ nuclear transformation was carried out in mixtures of PH_3, PF_3, and butadiene, the silacyclopentenes observed indicated that $^{31}SiH_2$, $^{31}SiHF$, and $^{31}SiF_2$ had all been formed. This implies that the hydrogen and fluorine atoms are obtained one at a time but does not distinguish between abstraction and insertion-dissociation, or a combination of both, as the operative mechanisms. The further implication that silynes are intermediates in the formation of silylenes in recoil reactions suggests that hot atom experiments will prove to be convenient and indeed essential in the exploration of the chemistry of these virtually unknown species.

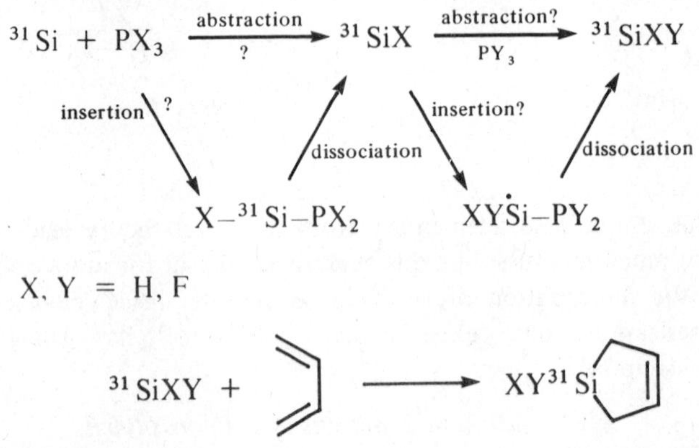

$$^{31}Si + PX_3 \xrightarrow[?]{\text{abstraction}} {}^{31}SiX \xrightarrow[PY_3]{\text{abstraction?}} {}^{31}SiXY$$

insertion? ?   dissociation   insertion?   dissociation

$$X{-}^{31}Si{-}PX_2 \qquad XY\dot{S}i{-}PY_2$$

$$X, Y = H, F$$

$$^{31}SiXY + \quad \diagup\!\!\diagdown \longrightarrow XY^{31}Si \bigcirc\!\!|$$

When nucleogenic difluorosilylene is formed in the presence of ethylene, the product is 1,1-difluoro-1-[31]Si-silacyclopentane, and a reasonable mechanism for its information would include a silirane intermediate.[81] There is precedent for reaction of a difluorosilirane with an olefin.[1]

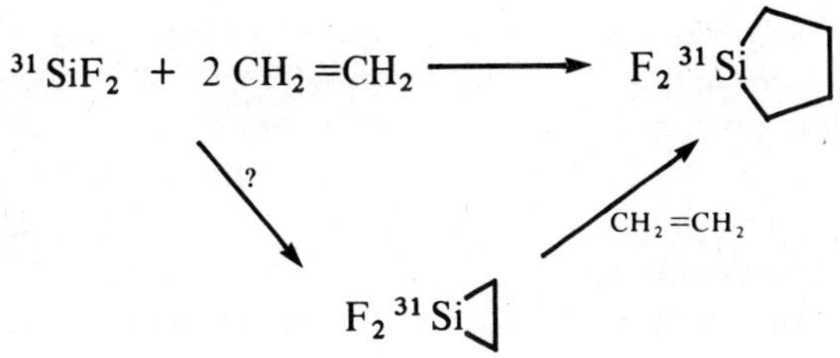

$$^{31}SiF_2 \ + \ 2\,CH_2{=}CH_2 \ \longrightarrow \ F_2{}^{31}Si\big\langle\bigcirc$$

(reaction scheme with $^{31}SiF_2$ path marked "?" leading to $F_2{}^{31}Si\triangleleft$, and $CH_2{=}CH_2$ addition)

Liu has reinvestigated the reactions of chemically generated SiF_2 and halo-ethylenes in light of the hypothesis that monomeric SiF_2 is the major attacking reagent.[82]

Reaction in the gas phase of SiF_2 with *cis*- or *trans*-1,2-difluoroethylene gave the same ratio of *cis*- and *trans*-2-fluorovinyltrifluorosilanes. In contrast with co-condensation experiments reported earlier, no products containing more than one silicon were observed.

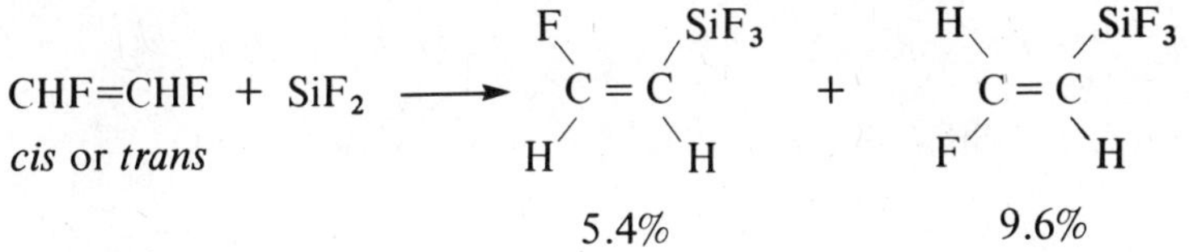

$$CHF{=}CHF \ + \ SiF_2 \ \longrightarrow \ \underset{\text{5.4\%}}{\overset{F}{\underset{H}{>}}C{=}C\overset{SiF_3}{\underset{H}{<}}} \ + \ \underset{\text{9.6\%}}{\overset{H}{\underset{F}{>}}C{=}C\overset{SiF_3}{\underset{H}{<}}}$$

cis or *trans*

Although these reactions are formally C–F insertions, the nonstereospecificity was explained by suggesting initial attack at the π-bond, followed by rearrangement of the intermediate silirane by way of a diradical with lifetime sufficient for rotation about the C–C bond.

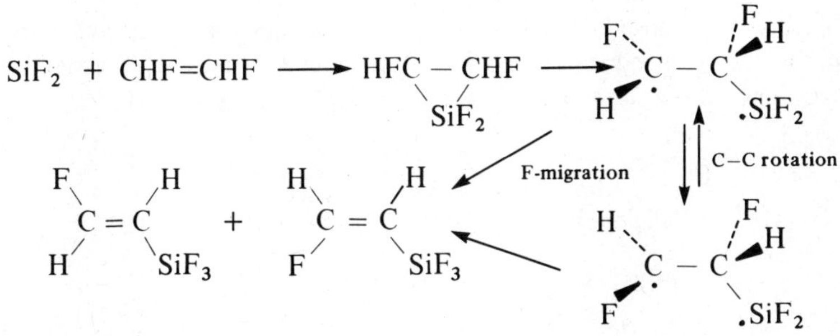

$$SiF_2 \ + \ CHF{=}CHF \ \longrightarrow \ HFC{-}CHF \atop \quad\ SiF_2$$

(reaction scheme showing F-migration, C–C rotation, and products)

$$\overset{F}{\underset{H}{>}}C{=}C\overset{H}{\underset{SiF_3}{<}} \ + \ \overset{H}{\underset{F}{>}}C{=}C\overset{H}{\underset{SiF_3}{<}}$$

Liu pointed out that silirane formation is sufficiently exoergic (about 91 kcal/mol) that rearrangement to the more stable vinylsilane is inevitable under the reaction conditions and that the vinylsilane is itself formed with so much excess energy (about 95 kcal/mol) that it too could undergo cis-trans isomerization. The product distribution may therefore reflect the thermodynamics rather than the kinetics of the reaction.

Reaction of SiF_2 with vinyl chloride in the gas phase also gives products of formal C–X insertion of monomeric SiF_2 rather than the mixture of cyclic and open-chain products formed upon cocondensation.[1] Some halogen exchange is apparently also occurring. The reaction with vinylfluoride is quite clean in the gas phase.

$$SiF_2 \ + \ CH_2{=}CHCl \ \longrightarrow \ CH_2{=}CHSiF_2Cl$$

$$2\%$$

$$+ \ CH_2{=}CHSiFCl_2 \ + \ CH_2{=}CHSiF_3$$

$$4\% \qquad\qquad 4\%$$

$$SiF_2 \ + \ CH_2{=}CHF \ \longrightarrow \ CH_2{=}CHSiF_3$$

$$\geqslant 30\%$$

From a repetition of the cocondensation experiments, including some in which alternate layers of $SiF_2 + SiF_4$ and reaction substrate were condensed, it was concluded that even in the cocondensation experiments some SiF_2 reacts in monomeric form yielding unstable siliranes that react further to give products containing more than one SiF_2 unit. The involvement of some SiF_2 oligomers in the cocondensation experiments was also inferred.

Liu ends his fine paper by asking whether triplet SiF_2 might be involved in these reactions. It should be pointed out that triplet carbenes react with 1,2-dihaloethylenes to produce diradical intermediates that undergo halogen migrations similar to those found by Liu, for example:[83]

$$Ph_2C{\uparrow}{\uparrow} \ + \ \underset{CH=CH}{\overset{Cl \qquad\ Cl}{\diagdown \qquad \diagup}} \ \longrightarrow \ Ph_2\dot{C}{-}CHCl{-}\dot{C}HCl \ \xrightarrow{\ \Omega\ }$$

$$Ph_2C{=}CH{-}CHCl_2$$

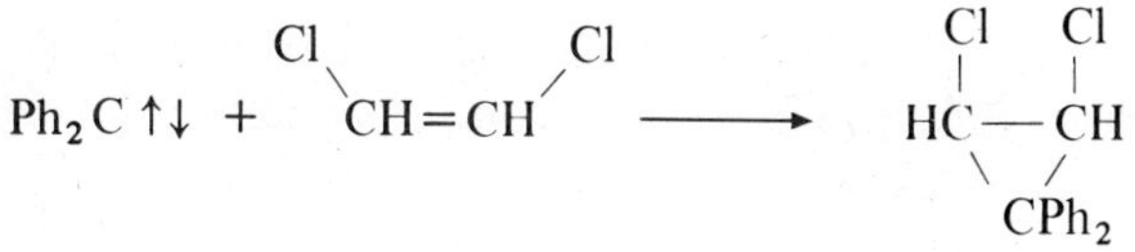

$$Ph_2C \uparrow\downarrow + \overset{Cl}{\underset{}{CH}}=\overset{Cl}{\underset{}{CH}} \longrightarrow \overset{Cl \quad Cl}{\underset{CPh_2}{HC-CH}}$$

Considering the high affinity of silicon for fluorine, perhaps the most puzzling aspect of these results is the apparent sluggishness of direct attack at F, which would be expected to give stereospecific insertion, probably by way of zwitterionic intermediate.

A mechanistic scheme similar to that of Liu has been suggested by Reynolds, Thompson, and Wright on the basis of cocondensation experiments in which SiF_2 was allowed to react with ethylene, propene, cis- and trans-2-butene, isobutylene, 2-methyl-2-butene, and tetramethylethylene.[84] No low molecular weight product containing only a single SiF_2 unit was found, but the polymer that comprised 60–90% by weight of the isolated products contained mainly of isolated SiF_2 units. The low molecular weight products consisted of 1,2-disilacyclobutanes and 1,2-disilacyclohexanes and open-chain isomers of the latter compounds. The disilacyclobutanes are formed stereospecifically and are therefore probably not formed by way of diradical pathways.

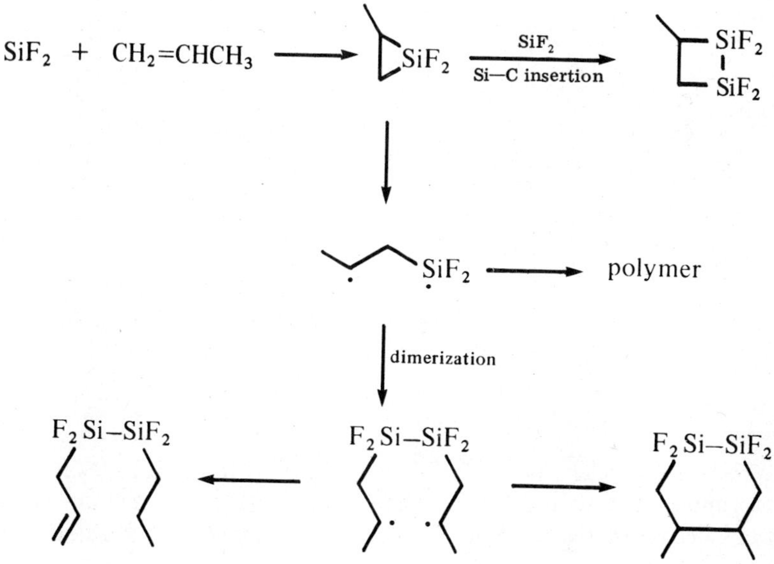

Liu and co-workers also have studied the reaction of SiF_2 and propene, both in the gas-phase and upon cocondensation.[85] The same volatile products were obtained in the cocondensation experiments as were found by Thompson.

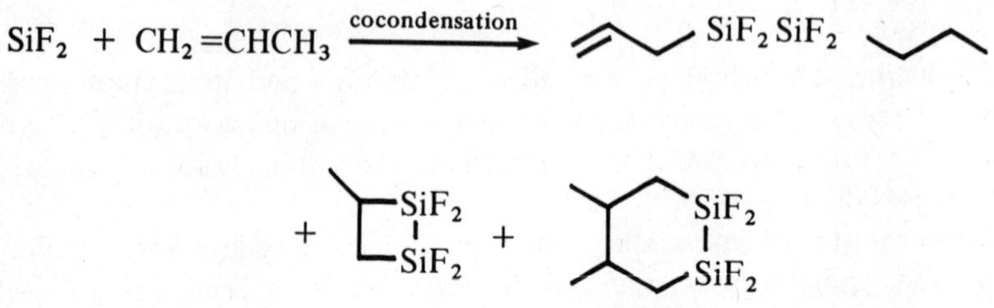

The reaction products produced in the gas phase contained only a single SiF_2 unit, but their formation clearly required intermolecular fluorine abstraction. The following mechanism was suggested:

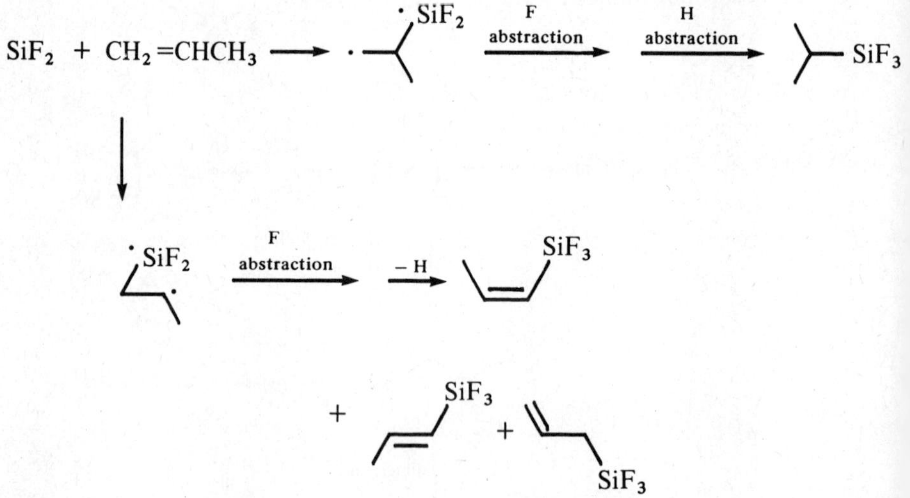

When the cocondensation of SiF_2 with cyclopentene and 1,3-cyclohexadiene was examined, the volatile products all contained more than one silylene unit, and the mechanism suggested by Liu and co-workers featured a diradical silylene dimer as the attacking reagent:[86]

387

Although several of the observed products could arise from siliranes that are SiF_2 adducts, only one product was attributed to monomeric SiF_2:

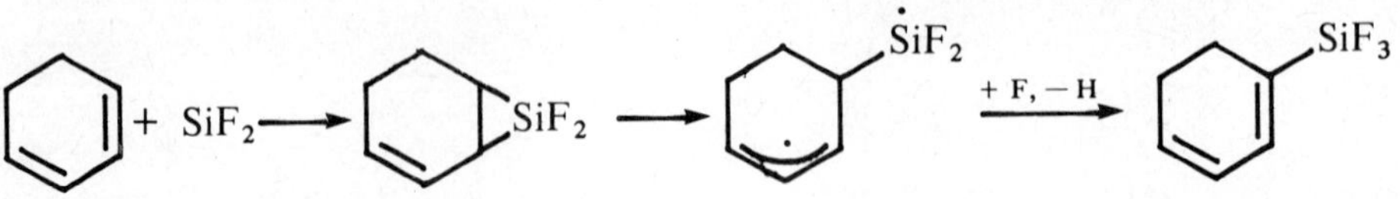

Cocondensation of difluorosilylene and *tert*-butylacetylene led to the formation of a disilacyclobutene.[87]

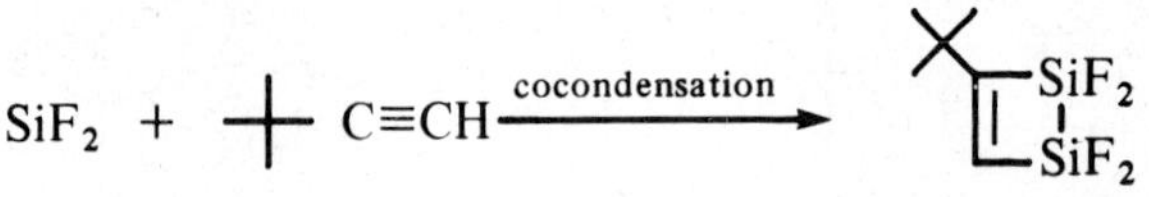

This product could clearly arise either from silirene formation followed by Si–C bond insertion, or by attack on the alkyne by a difluorosilylene dimer.

Although the interpretation of the chemistry of SiF_2 has veered sharply toward postulating primary reactions involving the monomeric silylene, none of the suggested primary products such as 1,1-difluoro-1-silacyclopropenes has as yet been isolated. Direct observation of these SiF_2 adducts is likely in the very near future, and they may turn out to be valuable synthetic intermediates.

C.   Abstraction by Silylenes

Alnaimi and Weber have found that sterically hindered dialkylsulfoxides undergo deoxygenation by photochemically generated dimethylsilylene more slowly than less encumbered sulfoxides.[88] This suggested that the mechanism is different from that for sulfoxide deoxygenation by dichlorocarbene. In the latter process acceleration by relief of steric strain is believed due to rate-determining decomposition of a carbene-sulfoxide adduct. Weber has considered two mechanisms: One is the same as that suggested for the carbene, with the first step slow, producing dimethylsilanone. In the second mechanism the first step is fast, but the resulting adduct undergoes reactions producing silanone oligomers without the intervention of free silanone.

Ando and co-workers have reported the deoxygenation of epoxides by photo-chemically generated dimesitylsilylene. The observed products were attributed to attack by an intermediate silanone on a second epoxide molecule to give **1**.[89]

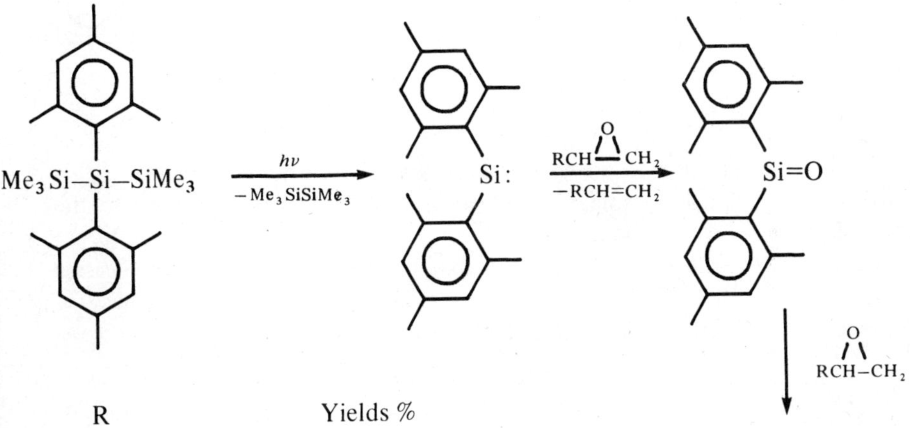

R	Yields %		
	Me₃SiSiMe₃	RCH=CH₂	**1**
H	100		24
Me	90	65	16
Et	92	70	9
Ph	82	39	11

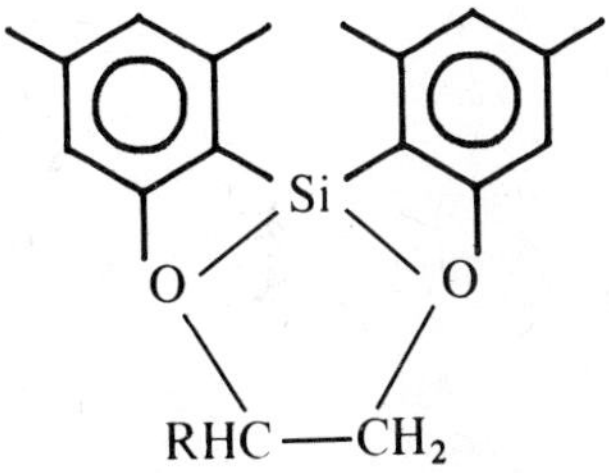

This evidence presented does not seem to demand the intermediacy of a silanone.

D.   Silylene Dimerization

The isolation and characterization by West, Fink, and Michl of a stable disi-lene formed by the dimerization of dimesitylsilylene was a major achievement.[7] The silylene was photochemically generated in hydrocarbon solution at room temperature.

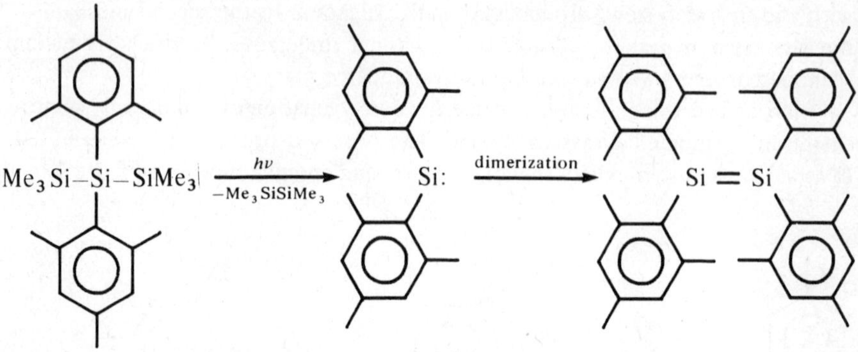

Although sensitive to oxygen, the bright orange-yellow disilene (λ_{max} = 420 nm) is thermally stable. This remarkable compound is a tribute to the synthetic utility of silylene reactions.

In our last review it was suggested that the formation of another pleasing molecule, octakis(trimethylsilyl)cyclotetrasilane, requires the intermediacy of the dimer of bis(trimethylsilyl)silylene.[1]

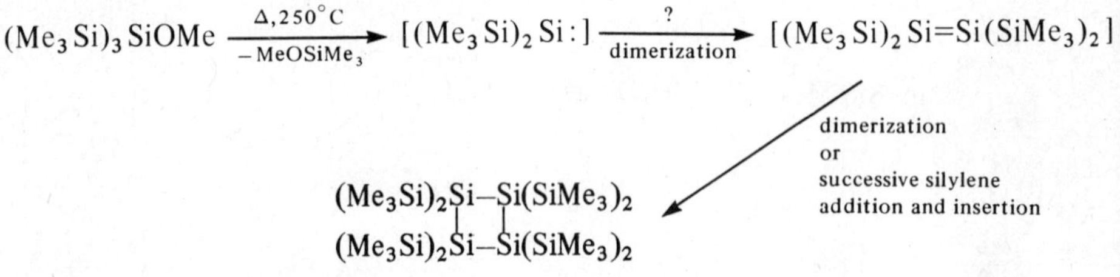

An X-ray crystal structure determination has revealed that this stable and rather inert cyclotetrasilane is planar,[90] in contrast to the highly folded structure found for 1,2,3,4-tetra-*tert*-butyl tetramethylcyclotetrasilane.[91]

Although the intermediacy of tetrakis(trimethylsilyl)disilene seems extremely likely, its source may be other than silylene dimerization.[90] When chlorotris(trimethylsilyl)silane or tris(trimethylsilyl)silane is employed as an alternate source of $(Me_3Si)_2Si$:, *no* cyclotetrasilane is obtained. Instead, products of silylene insertion into its precursor are found.

$$(Me_3Si)_3SiZ \xrightarrow[-Me_3SiZ]{\Delta,\,300^\circ C} (Me_3Si)_2Si: \xrightarrow{(Me_3Si)_3SiZ} (Me_3Si)_3SiSiZ(SiMe_3)_2$$

$Z = H, Cl$

$Z = H \quad 67\%$

$Z = Cl \quad 30\%$

Although this result may be due to the greater reactivity of Si–H and Si–Cl bonds toward silylene insertion, insertion may also occur with the methoxy compound (about 2% insertion product is found) followed by α- or β-elimination. The former seems more likely, and an example found by Sakurai of the suggested rearrangement is discussed in Section III.E.

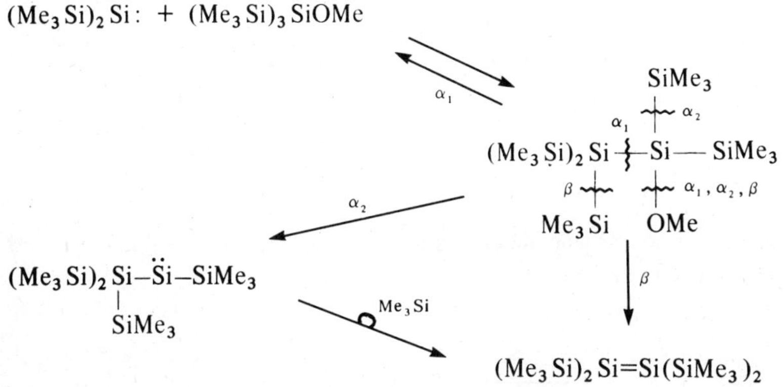

$$(Me_3Si)_2Si: \ + \ (Me_3Si)_3SiOMe$$

$$\alpha_1$$

$$SiMe_3$$

$$\alpha_1 \quad \alpha_2$$

$$(Me_3Si)_2Si \!-\! Si \!-\! SiMe_3$$

$$\beta \qquad \alpha_1, \alpha_2, \beta$$

$$Me_3Si \quad OMe$$

$$\alpha_2$$

$$(Me_3Si)_2Si\!-\!\overset{..}{Si}\!-\!SiMe_3$$
$$|$$
$$SiMe_3$$

$$Me_3Si$$

$$\beta$$

$$(Me_3Si)_2Si=Si(SiMe_3)_2$$

The moral of this yet incomplete story is that dimerization is not the only route from silylenes to disilenes.

E. Silylene Rearrangements and Other Reorganizations

There has been much work recently on the intramolecular reactions of silylenes, and many interesting results have been found, largely by the Barton group. Silylene-to-silylene rearrangements are grouped together with other reorganizations to include all the intramolecular processes not already discussed. Where applicable, comparison is made with the reactions of the analogous carbenes.

a. Ring Expansion by Carbon–Carbon Bond Shifts

In these reactions, regarded as intramolecular carbon–carbon bond insertions, relief of ring strain is made a driving force for thermally induced silylene-to-silene interconversion.

In the case of methylcyclopropylsilylene, which can be trapped prior to rearrangement, the 1-silacyclobut-1-ene intermediate can undergo ring opening by an "allowed" thermal electrocyclic reaction to a silabutadiene that undergoes self-trapping by dimerization:[92]

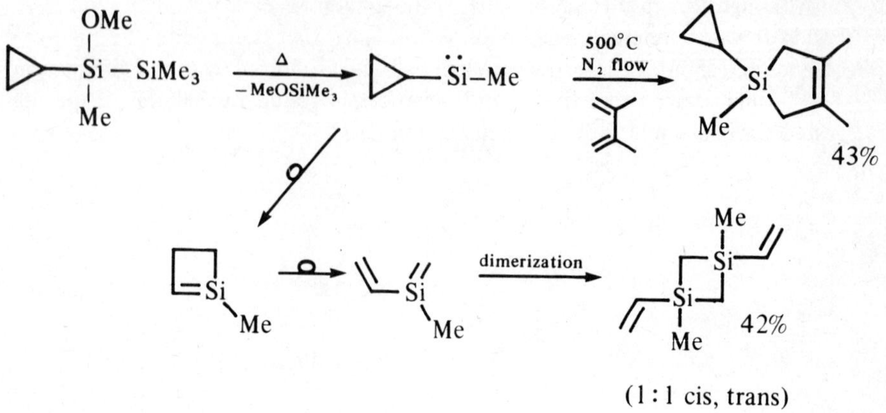

The analogous carbene undergoes the same reaction, stopping at the 1-methyl-cyclobut-1-ene in 92% yield.[93]

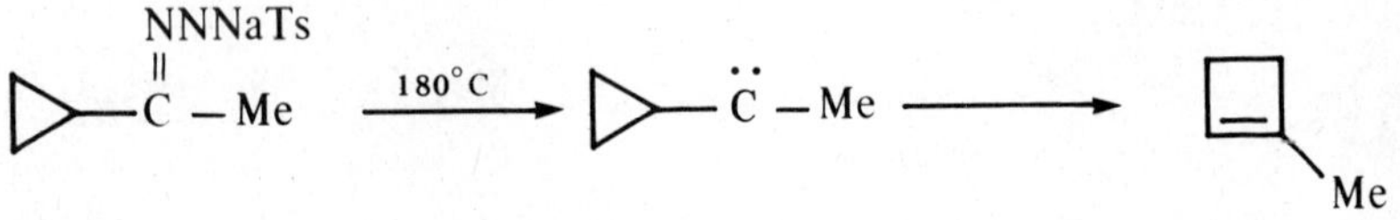

When the cyclopropylsilylene is generated with a reasonable migrating group on the silicon, the 1-silacyclobut-1-ene rearranges to a new cyclic silylene.[94]

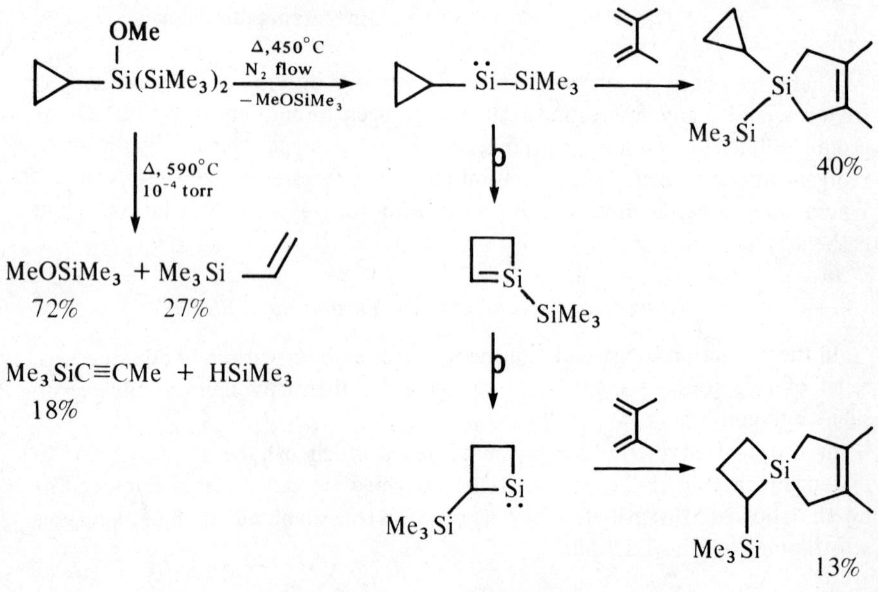

Although both silylenes can be trapped, there are two products formed, tri-methylvinylsilane and trimethyl(1-propynyl)silane, whose origins are obscure.

Barton suggested a somewhat far-fetched route to the vinyl silane by way of silylene ring contraction analogous to a well-known carbene reaction.[93, 95]

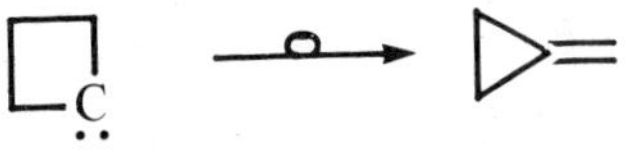

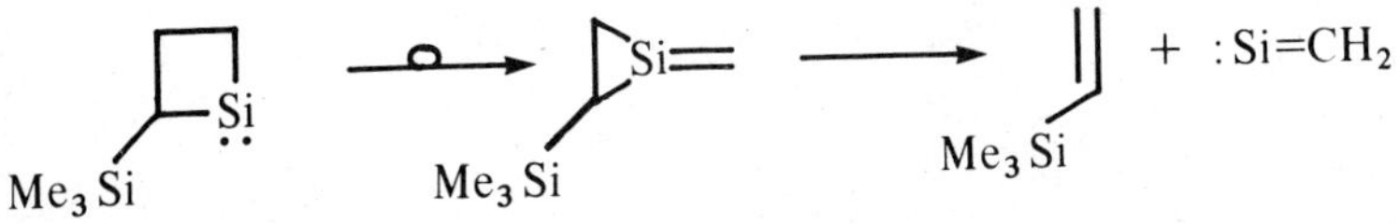

Silylidene $:Si=CH_2$ has been calculated by Gordon and Pople as being the most stable CH_2Si isomer, about 45 kcal/mol lower in energy than silyne $HC\equiv SiH$, with a less than 10 kcal/mol barrier separating the two from the higher energy side.[96]

Barton's mechanism would also have been expected to yield ethylene, not reported, and trimethylsilylsilylidene.

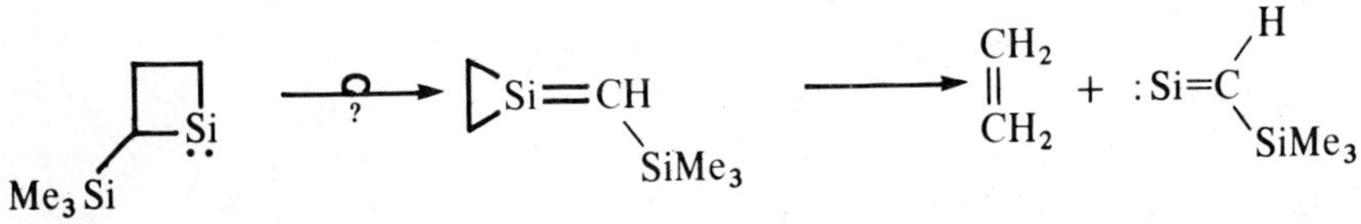

Any reader supplying a stamped, self-addressed envelope will receive an even more bizarre mechanism for formation of $Me_3SiC\equiv CMe$.

Ando and co-workers have trapped the 2-sila-1,3-butadiene formed from the rearrangement of cyclopropylphenylsilylene.[97]

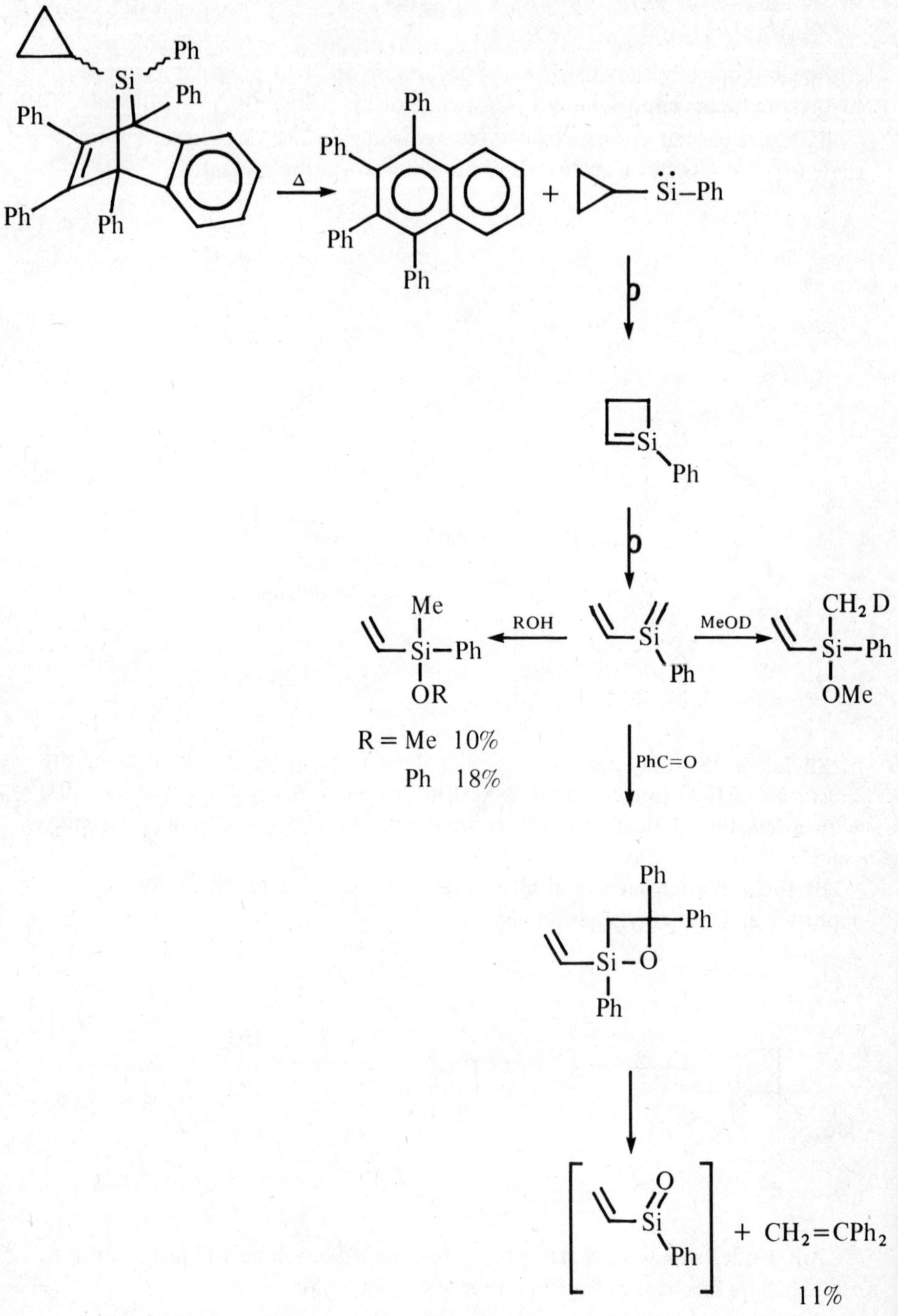
Ph
Si
Ph
Ph
Ph
Ph
Δ
Ph
Ph
Ph
Ph
+
Si–Ph
Si
Ph
Me
ROH
Si–Ph
OR
R = Me 10%
Ph 18%
Si
Ph
MeOD
CH2 D
Si–Ph
OMe
PhC=O
Ph
Ph
Si – O
Ph
O
Si
Ph
+ CH2=CPh2
11%

b. Intramolecular π-Addition

Methylvinylsilylene undergoes rearrangement to ethynylmethylsilane.[92] Since ring opening of 1-silacycloprop-2-enes has previously been suggested as a route to ethynylsilanes,[6] Barton has proposed an intramolecular π-addition leading to a 1-silacycloprop-1-ene followed by a hydrogen migration that is a "forbidden" 1,3-sigmatropic rearrangement. Other mechanisms involving intramolecular C–H insertions yielding the 1-silacycloprop-2-ene directly or the ethynylsilane by way of a 1-sila-1,2-propadiene seem less likely.

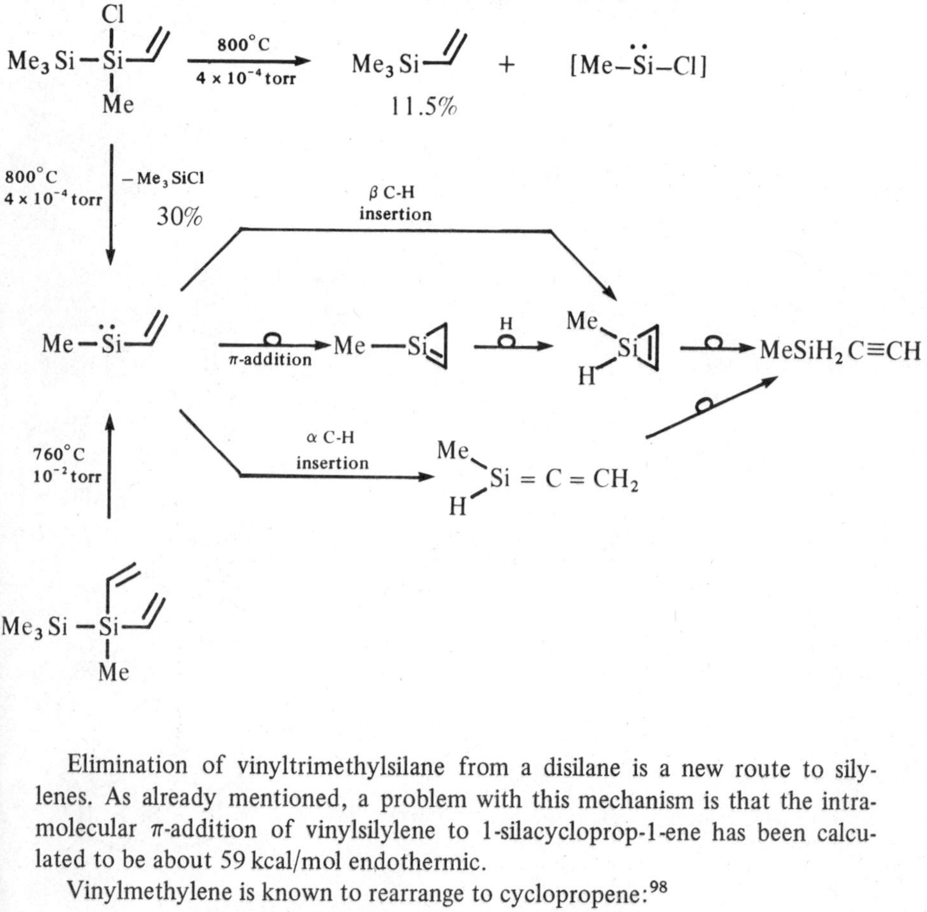

Elimination of vinyltrimethylsilane from a disilane is a new route to silylenes. As already mentioned, a problem with this mechanism is that the intramolecular π-addition of vinylsilylene to 1-silacycloprop-1-ene has been calculated to be about 59 kcal/mol endothermic.

Vinylmethylene is known to rearrange to cyclopropene:[98]

To test for the intermediacy of a 1-silacycloprop-1-ene, Barton and Burns equipped a vinylsilylene with a migrating group.[99] The trimethylsilylvinylsilylene could be trapped, as could the silylene resulting from intramolecular π-addition followed by trimethylsilyl migration. The latter silylene also seems able to stabilize itself by intramolecular C—H insertion followed by rearrangement of the disilahousane.

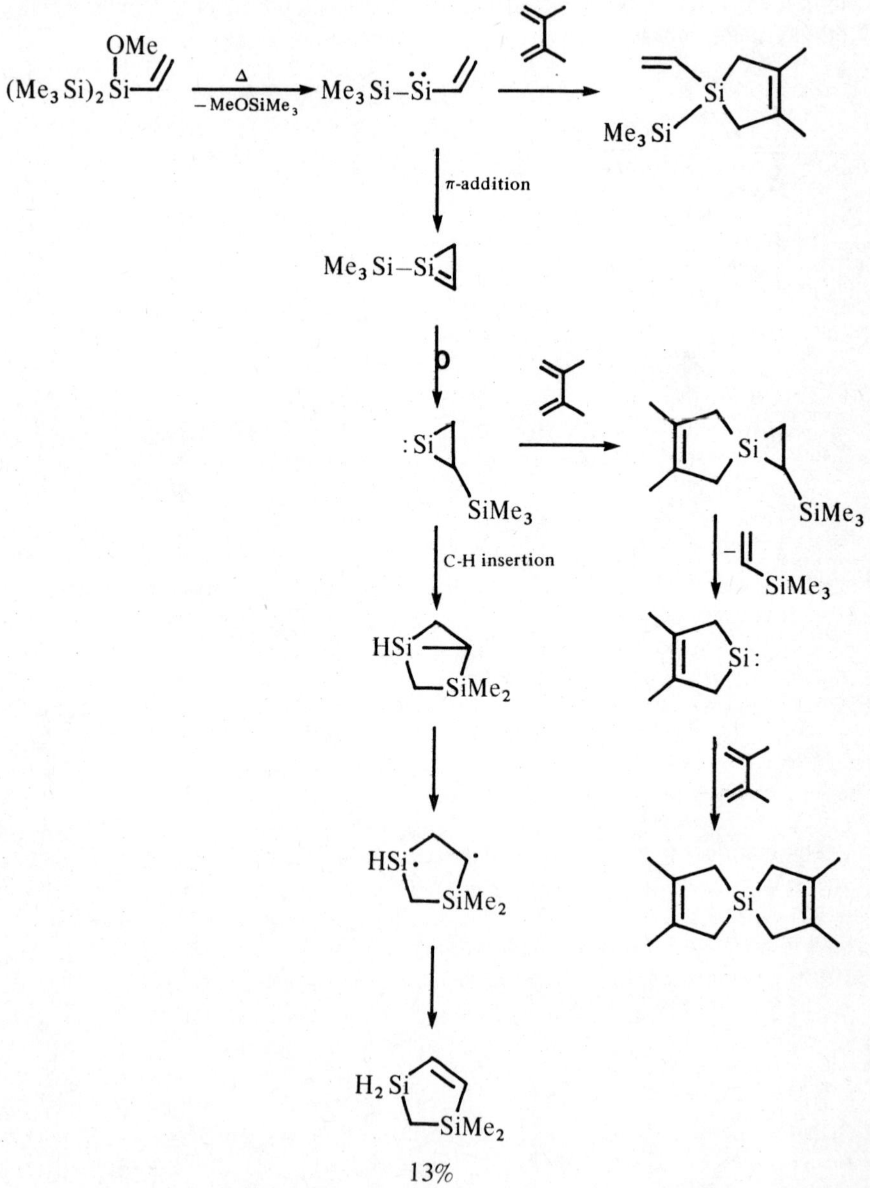

13%

The formation of trimethyl(vinyl)silane in the absence of dimethylbutadiene was attributed to silicon atom loss from the 1-silacyclopropyl-1-idene, but it is more likely, however, that this silylene inserts into an available Si–O bond prior to fragmentation of the silacyclopropane ring.

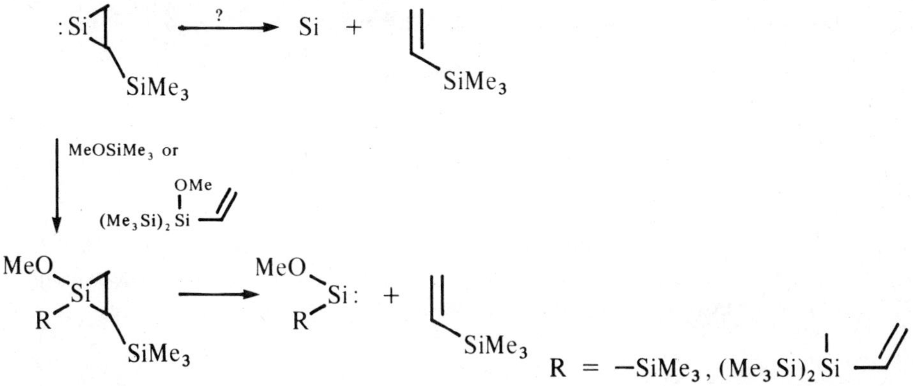

Methyl(2-propenyl)silylene can be trapped, but in the absence of trapping agent it yields a silacyclobutene whose formation was tentatively suggested to involve intramolecular π-addition, rearrangement to the more stable silacyclopropene, ring opening to a vinylsilylene (see Section II.E), and finally, intramolecular C–H insertion.[100]

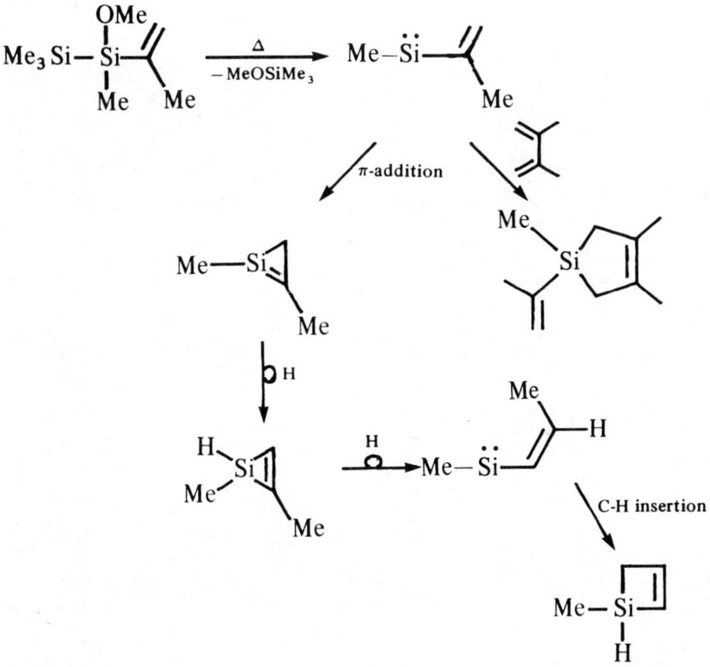

An alternative mechanism takes the silylene to an exomethylenesilacyclopropane by way of intramolecular allylic C—H insertion and several higher-energy intermediates, and is therefore somewhat less likely.

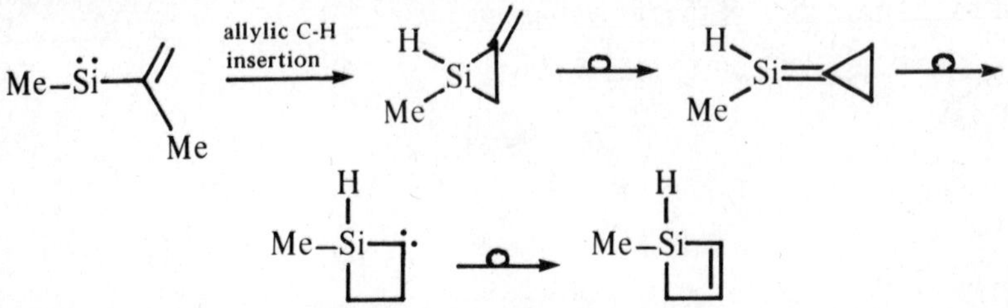

When an allylsilylene undergoes reorganization by way of intramolecular π-addition a silabicyclobutane is formed that can undergo further rearrangement to silylenes whose 2,3-dimethylbutadiene adducts were isolated:[94]

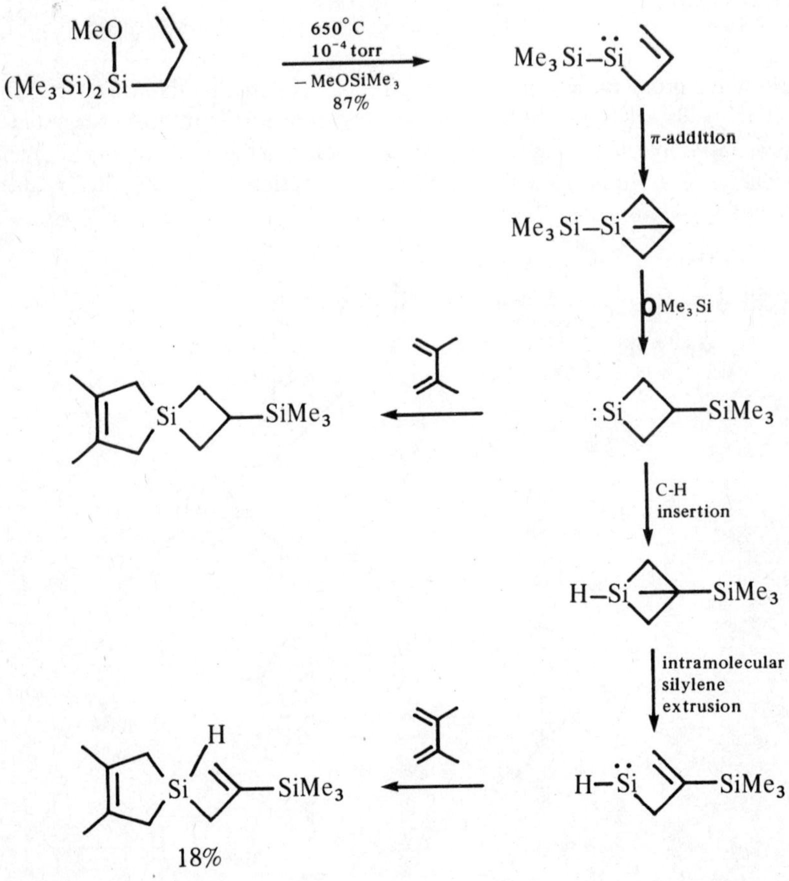

The formation of trimethylvinylsilane from this precursor was again rationalized by a silacyclobutanylidene ring contraction, and the unbiquitous product $Me_3SiC\equiv CCH_3$ (18% yield) was left as a puzzle for the reader (see offer on p. 393).

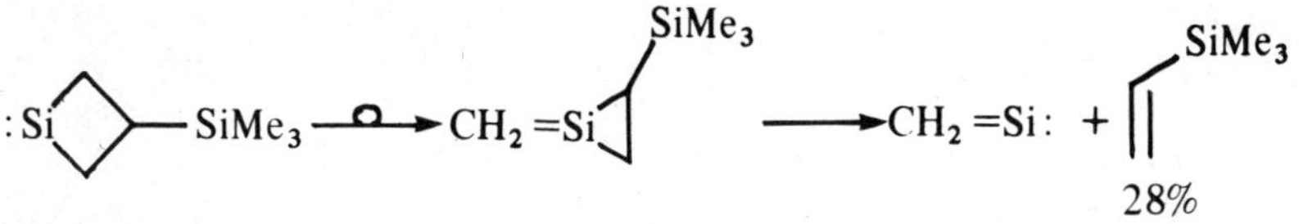

Curiously, generation of propenyltrimethylsilylene leads to the formation of the same two products, trimethylvinylsilane and trimethyl(1-propynyl)silane, as obtained from cyclopropyltrimethylsilylene and allytrimethylsilylene.[100]

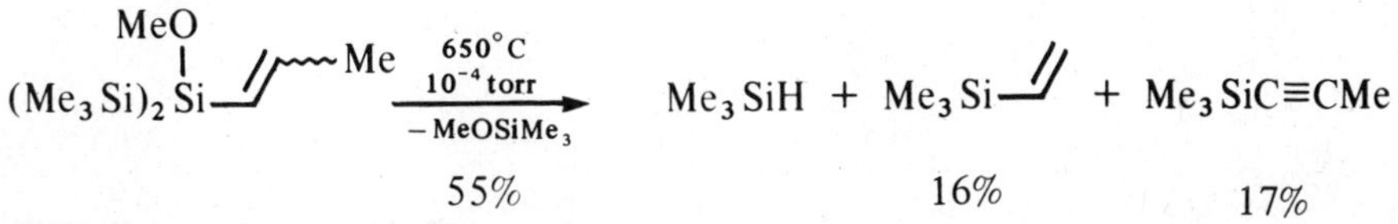

Although Barton proposes that this reaction is triggered by an intramolecular silylene C–H insertion, it is presented here because the suggested sequence also includes a silylene intramolecular π-addition and seems to lead to the same potential surface reached from the isomeric cyclopropyl- and allyltrimethyl-silylsilylenes.

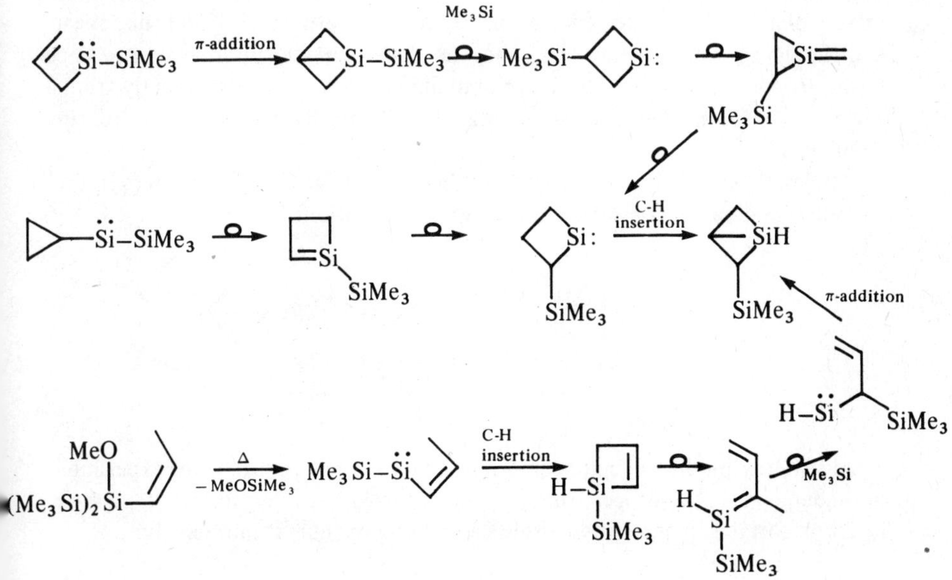

When allylmethylsilylene is generated under similar conditions, a 25% yield of 1-methyl-1-silacyclobut-2-ene is obtained.[100] Again intramolecular π-addition followed by rupture of the bridging bond is suggested.

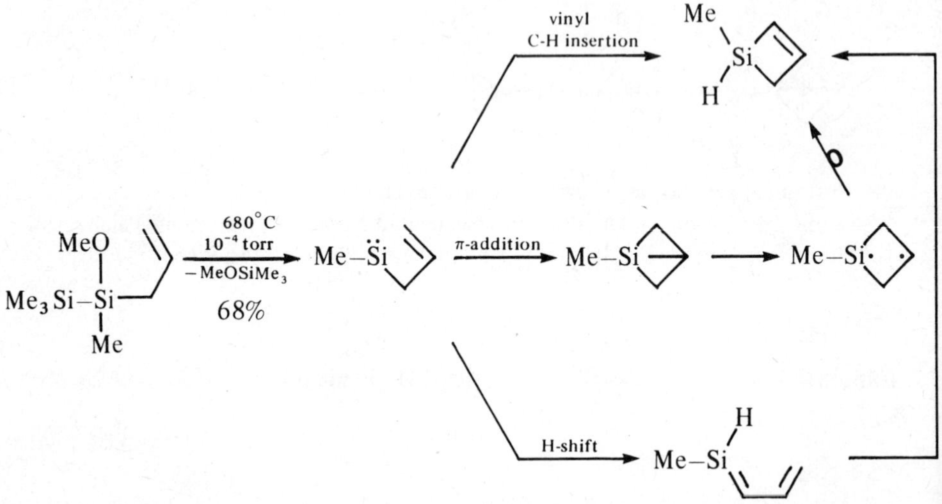

Alternative routes involving C—H insertions were considered less likely. It is curious that the methylsilacyclobutene survives this reaction and its formation from methyl(2-propenyl)silylene, but the trimethylsilylsilacyclobutene intermediate proposed to arise from propenyltrimethylsilylsilylene disappears completely by ring opening. This may be due to greater internal energy since in the trimethylsilyl case the silacyclobutene is believed to arise directly from a silylene C—H insertion and hence may be chemically activated toward ring opening.

Allylmethylene is known to undergo intramolecular C—H insertion (a hydrogen shift) in competition with intramolecular π-addition:[101]

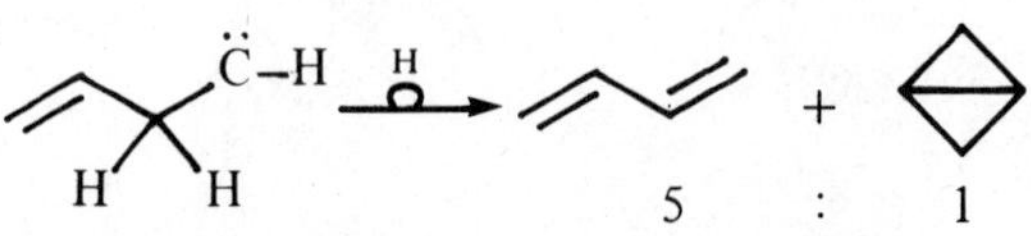

Allylmethylmethylene has been proposed on the basis of a labeling experiment to undergo a 1,4-hydrogen shift,[102] but, as suggested by Barton,[100] a C—H insertion forming a cyclobutene followed by ring opening is more likely:

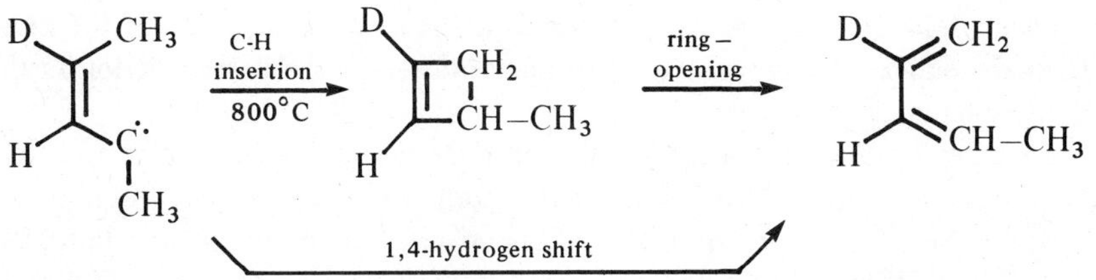

C-H insertion 800°C   ring-opening

1,4-hydrogen shift

In the case of allylsilylene with a choice between an intramolecular C–H insertion that produces a silicon–carbon double bond and an intramolecular π-addition yielding a strained bicyclic, the C–H insertion is disfavored both by the greater activation energy and the enthalpic advantage of forming two carbon–silicon σ-bonds over the formation of a carbon–silicon π-bond together with an Si–H bond.

The products obtained from the generation of methyl(2,4-pentadienyl)-silylene suggest that intramolecular π-addition to the terminal double bond occurs followed by competing rearrangements of the bicyclic intermediate:[103]

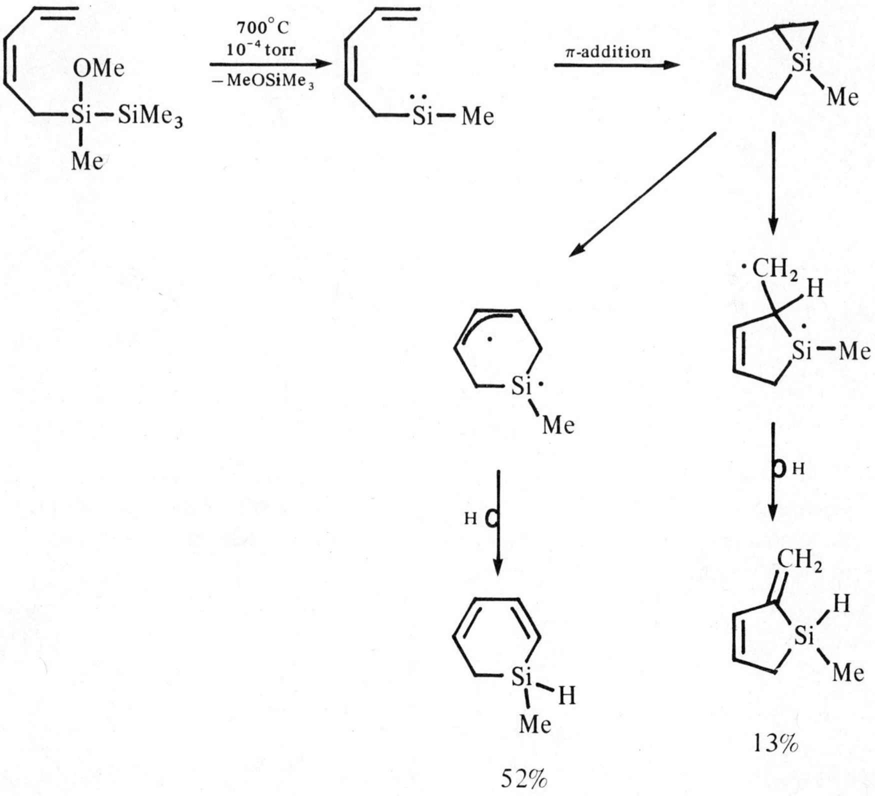

700°C
10^{-4} torr
– MeOSiMe₃

π-addition

52%

13%

Formation of the six-membered ring is favored by the breaking of the more strained bond in the bicyclic intermediate and the allylic delocalization in the resulting biradical.

Lengthening the pentadienyl chain by a methyl group seems not to affect the initial intramolecular silylene π-addition, but the bicyclic intermediate apparently undergoes preferential C–C bond cleavage of the silacyclopropane ring.[103] This process has precedent in the intramolecular addition of dimethylsilylene to 1,3-cyclopentadiene.[6]

c. Ring Contraction (Intramolecular C–C Insertion)

Two cases of rearrangement of trimethylsilyl-substituted 1-silacyclobutan-1-ylidenes to 1-methylene-1-silacyclopropanes have been suggested to account for the products from cyclopropyltrimethylsilylsilylene and allyltrimethylsilylsilylene in the immediate preceding sections:[94]

d.   Intramolecular Carbon–Hydrogen Insertion

Several examples of this reaction have already been presented (Section III.E.2)

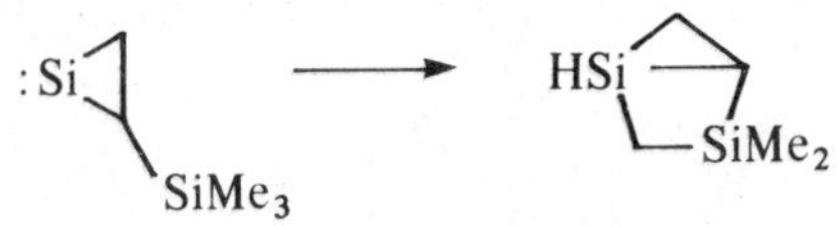

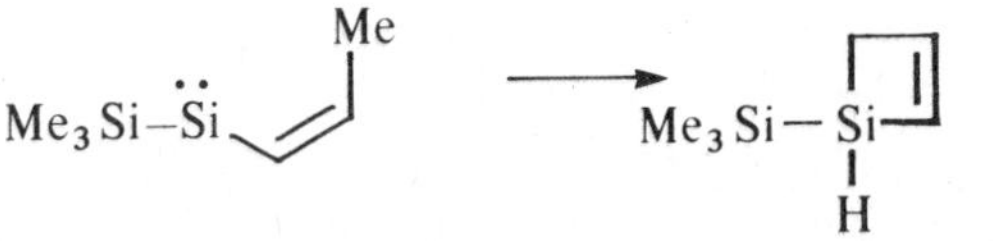

Intramolecular C–H insertions of ω-trimethylsilylalkyl(methyl)silylenes (p. 360),[54] and 1-silacyclopentan-1-ylidene (p. 353)[41] have already been presented, as has the controversial interconversion of alkylsilylenes and silenes by 1,2-hydrogen shift that can be regarded as an intramolecular C–H insertion (pp. 351–352).[38–40,43,44] The probable nonrearrangement of $H_2C=Si:$ to $HC\equiv SiH$ has also been touched upon (p. 393).[96]

A number of cases of intramolecular allylic C–H insertions by silylenes has recently been noted.

The high yield of 1-methyl-1-silacyclopent-3-ene obtained from 3-butenyl-(methyl)silylene was attributed to allylic C–H insertion followed by a vinyl-silacyclopropane-to-silacyclopentene rearrangement.[103] π-Addition followed by rearrangement and vinyl C–H insertion would be expected to produce the 2-ene isomer.

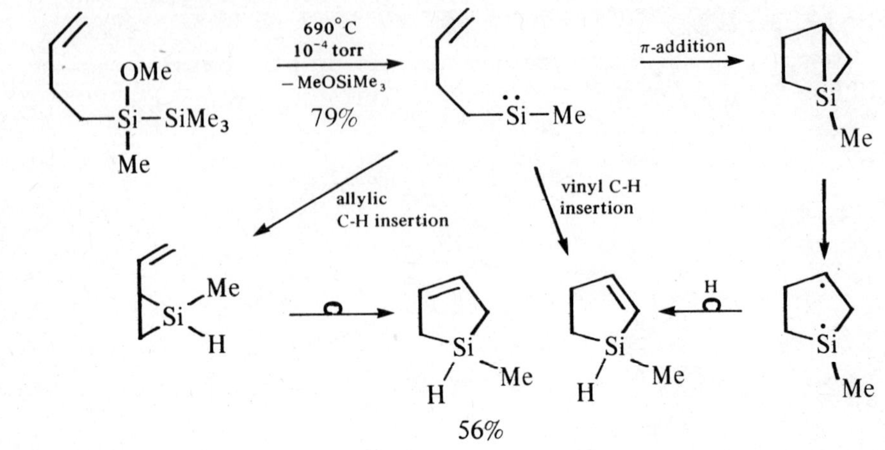

Silacyclopentenes can also arise by direct intramolecular allylic insertions:[103]

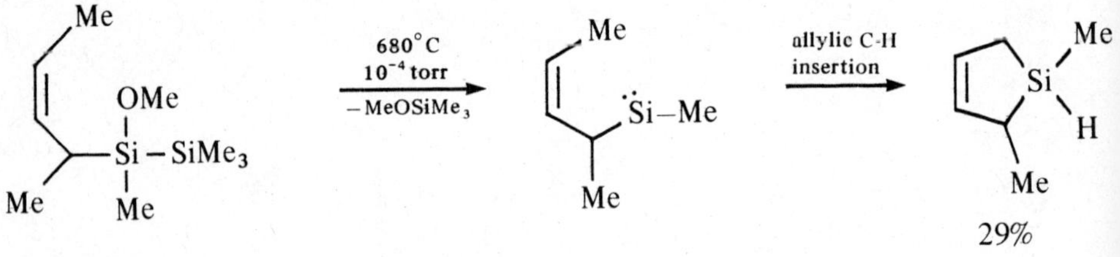

Another case of silacyclobutene formation by direct silylene allylic C–H insertion is[100]

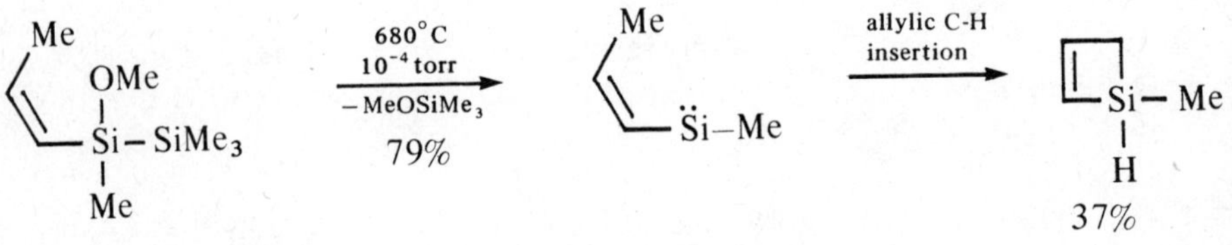

Given a choice between allylic C–H insertion to form a silacyclobutane and a π-addition to give a bicyclic, the following silylene seems to prefer the insertion.[100] Some of the silacyclobutene expected as a rearrangement product of the silabicyclobutane is also formed. However, since the exomethylenesilacyclo-butane undergoes efficient rearrangement to the silacyclobutene, this is believed by Barton to be its sole source.

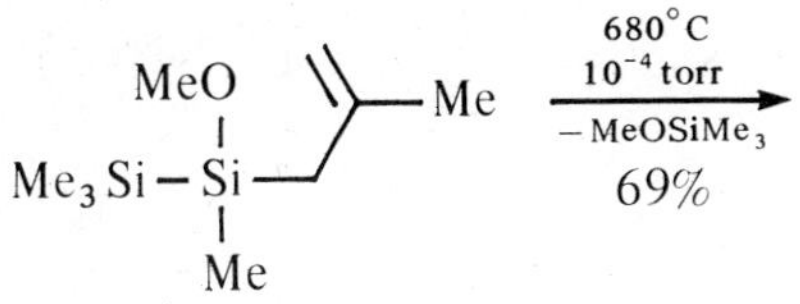

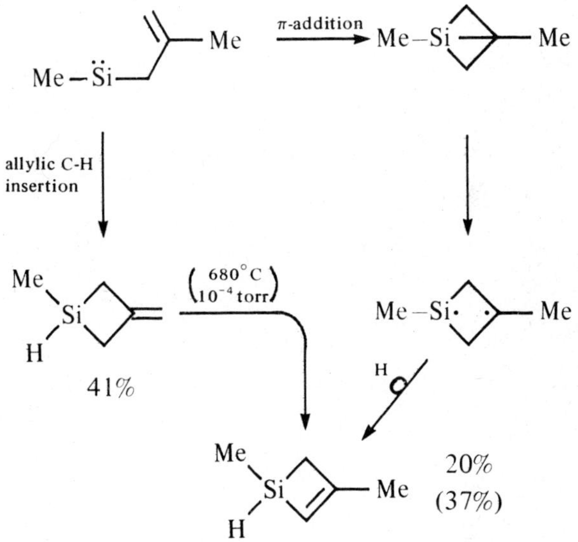

$$\text{Me}_3\text{Si}-\underset{\underset{\text{Me}}{|}}{\overset{\overset{\text{MeO}}{|}}{\text{Si}}}-\overset{}{\diagup}\!\!\diagdown\!\!-\text{Me} \xrightarrow[\substack{-\text{MeOSiMe}_3 \\ 69\%}]{\substack{680°\text{C} \\ 10^{-4}\,\text{torr}}}$$

$$\text{Me}-\ddot{\text{Si}}-\diagup\!\!\diagdown\!\!-\text{Me} \xrightarrow{\pi\text{-addition}} \text{Me}-\text{Si}\triangleleft\!\!\square\!\!-\text{Me}$$

allylic C-H insertion

$$\underset{\text{H}}{\overset{\text{Me}}{\diagdown}}\text{Si}\square\!\!=\;\; \xrightarrow{\left(\substack{680°\text{C} \\ 10^{-4}\,\text{torr}}\right)} \quad \text{Me}-\text{Si}\cdot\;\square\;\cdot-\text{Me}$$

41%

$$\underset{\text{H}}{\overset{\text{Me}}{\diagdown}}\text{Si}\square\!\!-\text{Me} \qquad \begin{matrix}20\% \\ (37\%)\end{matrix}$$

Intramolecular C–H insertion to generate a 1-silacyclopenta-1,3-diene by way of 1,2-hydrogen shift may be the process that converts 1-silacyclopent-3-en-1-ylidene to 5-silacyclopenta-1,3-diene. It was suggested in 1974 that the major product from the gas-phase addition of high-energy silicon atoms to 1,3-butadiene is unsubstituted silole.[104] Further evidence has been found in the reduction of the no-carrier-added compound to 1-silacyclopent-3-ene.[105]

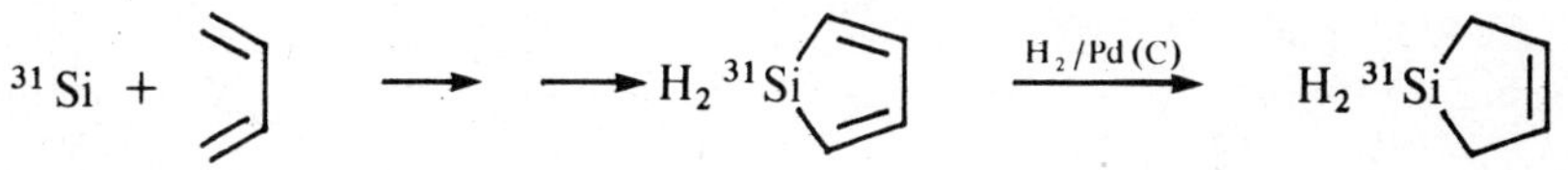

$$^{31}\text{Si} + \diagup\!\!\diagdown\!\!\diagup \longrightarrow \;\longrightarrow \text{H}_2\,^{31}\text{Si}\!\!\bigcirc\!\! \xrightarrow{\text{H}_2/\text{Pd(C)}} \text{H}_2\,^{31}\text{Si}\!\!\bigcirc$$

It has recently been found that generation of 1-silacyclopent-3-en-1-ylidene in the gas phase by a conventional chemical method[10] leads to a dimeric species whose spectroscopic properties suggest that it is the silole Diels—Alder self-dimer.[106] The following mechanism is suggested:

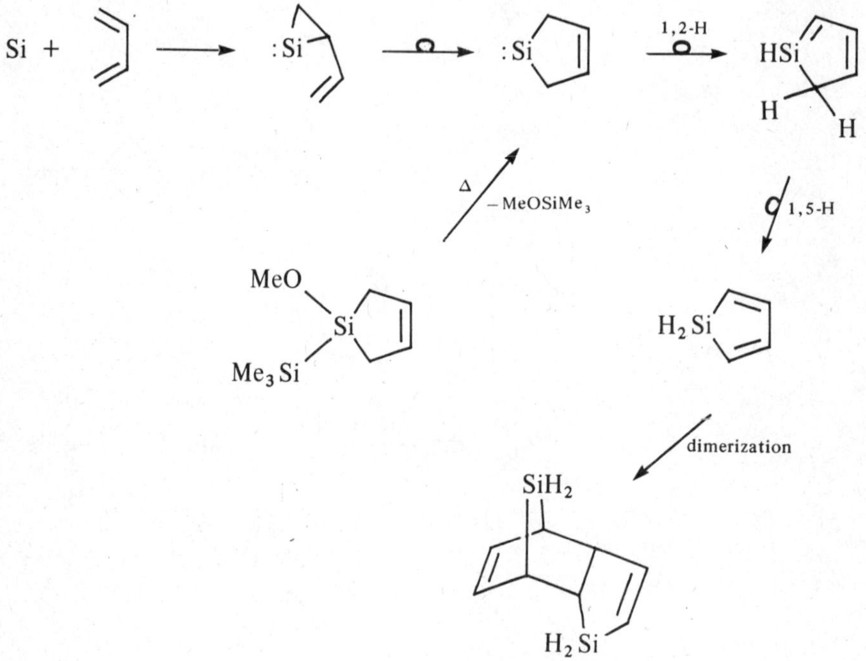

e.   Intramolecular Si—Si Insertion (Silyl Shift)

The possible rearrangement of trimethylsilyl[tris(trimethylsilyl)silyl]silylene to the tetrakis(trimethylsilyl)disilene has already been considered (Section III.D).[90]

Sakurai and co-workers have presented clear evidence for this process.[107] Extrusion of methyl(pentamethyldisilanyl)silylene from a bicyclic precursor in the presence of 2,3-dimethylbutadiene led to the trapping of both the silylene and its rearrangement product, permethylated trisilapropene. When generated in the presence of anthracene, the silylene undergoes rearrangement in competition with dimerization, with only a trace of the trapping product of the dimer being formed.

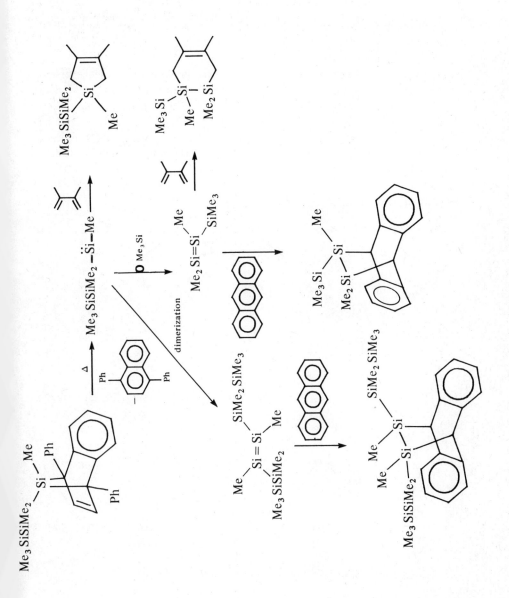

f.   *Intramolecular Si–C Insertion (Alkyl Shift)*

Although the rearrangement of tetramethyldisilene to methyl(trimethylsilyl)-silylene was reported by Barton and co-workers in 1978,[1] evidence for the reverse process has only recently been found.

When 1,1,1,3,3,3-hexamethyltrisilane is pyrolyzed in the presence of trimethylsilane in the gas phase, the compound formed in second-highest yield is the product of intermolecular Si–H insertion by a silylene that has undergone a transposition of α- and β-substituents.[47] The other products can be rationalized by the rearrangement proposed by Barton[1] in which α-silylsilylenes rearrange to β-silylsilylenes by intramolecular C–H insertion and α-elimination.

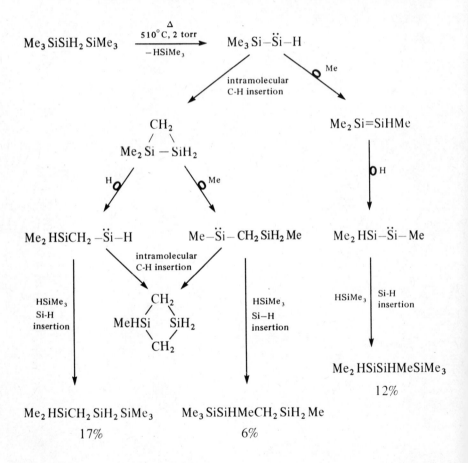

The intermediacy of trimethyldisilene was confirmed by trapping with butadiene.[47]

$$Me_2Si=SiHMe \; + \; \text{(diene)} \longrightarrow \begin{matrix} Me_2Si \\ | \\ MeHSi \end{matrix} \text{(ring)}$$

Since quantum mechanical calculations suggest that unsubstituted disilene and silylsilylene are nearly isoenergetic (see Section IV), rearrangement in either direction seems equally feasible. Which species is trapped will of course depend on the relative reactivity of the trapping agent toward silylsilylene and disilene and on the rate of their interconversion relative to the rates of trapping processes. It will be interesting to see what the predicted barrier is for this interconversion, considering the high barrier calculated for the $CH_2=SiH_2 \rightleftharpoons CH_3-\ddot{S}i-H$ process.

Having examined however briefly a vast array of silylene rearrangements and reorganizations, one may well ask whether any generalizations are as yet possible. Certainly many of the suggested mechanisms are highly tentative and cry out for further investigation. π-Addition and C–H insertion are the most common intramolecular rearrangement processes thus far encountered, with allylic C–H bonds particularly prone to attack. The formation of three-membered rings seems to be the favored process despite the instability of the resulting siliranes.

The complex matrix of silylene rearrangements speaks for the necessity of supporting our already vast body of empirical knowledge of silylene chemistry with basic mechanistic studies. Unless a greater effort is made to extract fundamental principles of silylene reactivity, our collection of new reactions, however interesting, will soon become unmanageable.

IV.   THEORETICAL CALCULATIONS

In considering the applications of theoretical calculations to silylene chemistry it is important to keep in mind Schaefer's warning that "overly simplistic theoretical models can yield very unrealistic theoretical predictions."[38] He gave the example of the singlet-triplet energy separation for the simplest silene $CH_2=SiH_2$. Minimal basis set self-consistent field calculations suggested that the triplet lies 1.4 kcal/mol *below* the singlet. Using two basis functions for each atomic orbital, a so-called double-zeta calculation, gave what Schaefer called "a more plausible value" of 13.7 kcal/mol, with the ordering reversed, singlet now lower in energy than triplet. Finally the inclusion of d orbitals and extensive configuration-interaction gave "the first reliable" energy separation of 38.5 kcal/mol.

Schaefer and co-workers have calculated the energies of methylsilylene CH_3-Si-H, silaethylene $CH_2=SiH_2$, and silylmethylene SiH_3-C-H at various levels of theory.[39] The following results have been obtained for the *singlet species* (all entries are in Kilocalories per mole):

	DZ SCF	DZ CI	DZ + d SCF	DZ + d CI
$E_{SiH_3-\ddot{C}-H} - E_{SiH_2=CH_2}$	63.3	76.6	54.7	69.0
Barrier $SiH_3-\ddot{C}-H \xrightarrow{\ \circ\ } SiH_2=CH_2$	13.1	3.2	8.4	1.9
Barrier $SiH_2=CH_2 \xrightarrow{\ \circ\ } SiH_3-\ddot{C}-H$	76.4	79.8	63.1	70.9
$E_{SiH_2=CH_2} - E_{CH_3-\ddot{Si}-H}$	11.6	2.3	4.9	0.4
Barrier $SiH_2=CH_2 \xrightarrow{\ \circ\ } CH_3-\ddot{Si}-H$	45.0	43.4	44.6	41.0
Barrier $CH_3-\ddot{Si}-H \xrightarrow{\ \circ\ } SiH_2=CH_2$	56.6	45.7	49.5	41.4

Both a near degeneracy and a large barrier for interconversion of silaethylene and methylsilylene are predicted, and although the energy difference shrinks with inclusion of polarization functions and extensive configuration interaction, the barrier for interconversion is not significantly altered. As might be expected, the transition states for the interconversion of silaethylene to methylsilylene and silylmethylene were both found to approximate the motion of a hydrogen from the MH_3 group toward the "empty" p orbital of the divalent atom. Yet further refinement of the calculation did not lower the barrier for the $SiH_2=CH_2 \xrightarrow{\ \circ\ } CH_3-\ddot{Si}-H$ interconversion beyond 40.6 kcal/mol.[40]

The suggestion[40] that such a calculated barrier throws doubt on the experimental evidence for silene-to-silylene conversions (see Section II.E) is somewhat shortsighted, however, since the thermally induced reactions have initial steps with much higher barriers.

Gordon and Koob have carried out self-consistent field calculations with large basis sets on 15 C_2SiH_4 isomers, including vinylsilylene, $CH_2=CH-\ddot{Si}-H$, 1-silacyclopropanylidene

and ethylidenesilylene $CH_3CH=Si:$.[46] Both vinylsilylene and silacyclopropanylidene were predicted to be slightly more stable than silacyclopropene

$$\overset{SiH_2}{\underset{\triangle}{}}$$

but all three of these species were found to lie some 13–25 kcal/mol above the most stable C_2SiH_4 isomer, silylacetylene $SiH_3C{\equiv}CH$. The species containing silicon—carbon triple bonds were found to be some 60 kcal/mol above silylacetylene, as was the 1-silacycloprop-1-ene, as already noted.

Lien and Hopkinson have also carried out SCF calculations at the double-zeta basis set level, on 10 C_2SiH_4 isomers.[108] When polarization functions were included for all atoms

$$\text{SiH}_2 \ (15.2) \quad CH_2 = CH - \ddot{S}iH \ (15.4) \quad \text{Si} \ (17.2)$$

and $CH_3CH=Si$: (18.0) were again found to be similar and remarkably near in relative energy to silylacetylene (0). Sila-allenes were much higher in energy, 30–50 kcal/mol above silylacetylene, and silicon triple-bonded species yet higher.

Köhler and Lischka have carried out SCF calculations with large basis sets and configuration interaction on the lowest singlet and triplet states of the $CSiH_4$, Si_2H_4, and C_2H_4 isomers.[109] In an earlier communication SCF results obtained with the use of an extended basis set were reported for SiH_2SiH_2 and SiH_3SiH:.[110] The singlet ethylene calculations could be compared with experiment, and good agreement was found, for example, $r_{C=C} = 1.336$ Å calculated, versus 1.339 Å experimental. Calculated force constants and vibrational frequencies agreed to within about 10% with experimental values.

The energy differences for singlets $E_{SiH_3\ddot{C}H} - E_{SiH_2=CH_2} = 67.0$ and $E_{SiH_2=CH_2} - E_{CH_3\ddot{S}iH} = -0.6$ kcal/mol were similar to those predicted by Schaefer,[39] but now silaethylene is slightly lower in energy than methylsilylene. The corresponding triplet state energy differences calculated by Köhler and Lischka were 10.2 and −11.1 kcal/mol respectively. Disilene is predicted to be more stable than silylsilylene by 0.4 kcal/mol in the singlet manifold and 4.8 kcal/mol in the triplet. No energy barriers were calculated.

Earlier, Poirer and Goddard had reported geometries and relative energies of the SiH_2SiH_2 and SiH_3SiH molecules based on their SCF CI calculations.[111] Results on SiH_2 were also included, with a predicted triplet-singlet separation of 10.6 kcal/mol, the smallest yet reported. Planar and trans-bent singlet disilene are predicted to be nearly equal in energy. Singlet silylsilylene was predicted to be from 0.1 to 12.4 kcal/mol more stable than disilene by SCF calculations using various basis sets, but the inclusion of correlation effects by configuration interaction reversed the prediction, making $SiH_2=SiH_2$ 10.1 kcal/mol more stable than SiH_3SiH.

Even more elaborate calculations by Krogh—Jesperson predict that trans-bent singlet disilene is more stable than its planar form by 0.9 kcal/mol, with silylsilylene 5 kcal/mol higher in energy.[112]

The calculations of Gordon and Pople on $:Si=CH_2$ and $HSi\equiv CH$ have already been mentioned (see p. 393).[96] An energy difference of about 45 kcal/mol favoring the silylene was predicted with a small energy barrier in the exoergic direction.

The singlet and triplet energy surfaces for $CSiH_2$ have also been examined by Hopkinson, Lien, and Csizmadia.[113,114] An energy difference of 205.4 kJ/mol (49.1 kcal/mol) between $HC\equiv SiH$ and $H_2C=Si$: was predicted for the singlet states and 130.1 kJ/mol (31.1 kcal/mol) in the triplet manifold. There is, however, a sizeable barrier (115.9 kJ/mol = 27.7 kcal/mol) predicted for the isomerization of *triplet* $HC\equiv SiH$ to $H_2C=Si$:.

Several silabutadienes and their silylene isomers have been treated by Trinquier and Malrieu at the double-zeta SCF level.[115] The following relative energies were obtained.

	Relative energies (kcal/mol)	
	SCF Calculation	After Correlation Correction
$CH_3-\ddot{S}i-CH=CH_2$	0	5
$CH_2=SiH-CH=CH_2$	2.18	0
$SiH_2=CH-CH=CH_2$	8.01	6
$H-\ddot{S}i-CH_2-CH=CH_2$	16.92	22

It is interesting that π-conjugation seems to stabilize a silene nearly as much as it does an olefin, with charge transfer in the 2-silabutadiene from the $C=Si$ to the $C=C$ π-bond.

Trinquier and Malrieu also report results on the silaethylene and methylsilylene singlet states, in reasonable agreement with the calculations of other workers reported before.

The structures of four isomeric forms of the potential silylenoid SiH_2LiF have been investigated by SCF molecular orbital calculations.[116] The most stable form at all levels of theory was the ion pair $SiH_2Li^+F^-$, with the $H_2Si\cdots FLi$ silylene complex only 1.3 kcal/mol higher in energy in the most extensive calculations. Another silylene–lithium fluoride complex $H_2Si\cdots LiF$ and the tetrahedral SiH_2LiF both are predicted to be considerably less stable.

In this paper,[116] as indeed in all the other *ab initio* studies reported in this section, detailed geometries and charge distributions, as well as absolute energies are reported. We have concentrated on relative energies because they are most directly pertinent to current experimental investigations of silylene reactions. The wealth of structural predictions contributed by theory will receive more attention when our meagre hoard of spectroscopic information about silylenes expands.

V.   PHYSICAL MEASUREMENTS

From the measurement of the rates of reaction in the gas phase as a function of temperature of iodine with a series of silanes, Walsh has been able to obtain an important set of Si–H and C–H bond dissociation energies.[3] These in turn allowed the derivation of the heats of formation of a number of reactive silicon species. Walsh gives a $\Delta H_f°$ of 26 kcal/mol for Me_2Si and 23 ± 5 kcal/mol for its isomer $MeHSi=CH_2$, thus indicating that these two molecules, whose inter-conversion has been so controversial, are nearly isoenergetic.

Walsh has noted that the *second* bond dissociation energy for SiH_4, $SiCl_4$, and SiF_4, that is, $D(SiZ_2-Z)$ is the smallest of all, and he associates this with the stabilization of a silylene by the s-character of the lone-pair orbital. He defines a lone-pair stabilization energy as the difference between the first and second dissociation energies, and these values (kcal/mol) are 26 for SiH_2, 45 for $SiCl_2$ and 37 for SiF_2, and a further value of 27 for $SiMe_2$ was derived and believed to be valid also for other nonpolar organosilylenes.

The lone-pair stabilization energy is equal to the exothermicity of the silyl-radical disproportionation to silylenes:

$$2 \cdot SiZ_3 \rightleftharpoons SiZ_4 + :SiZ_2$$

A general conclusion by Walsh pertinent to mechanistic discussions in organo-silicon chemistry is that: "In reactions where the choice of pathways seems to lie between those involving silylene intermediates and those involving bi-radicals, the former seem to offer a definite energetic advantage."

In surprising contrast alike with Walsh's kinetic-measurement-based estimates of the heats of formation of $Me_2Si:$ and $MeHSi=CH_2$, with the predictions of theoretical calculations, and with the experiments of the Conlin and West–Michl groups, Pau, Pietro, and Hehre have deduced heats of formation for $MeHSi=CH_2$ and Me_2Si of 18 and 46 kcal/mol, respectively.[117] This places the silene 28 kcal/mol *lower* in energy than the silylene and makes the silene-to-silylene conversion sufficiently endoergic that its rate at low temperature would be fairly slow. Reversible interconversion at high temperatures would of course favor the silene.

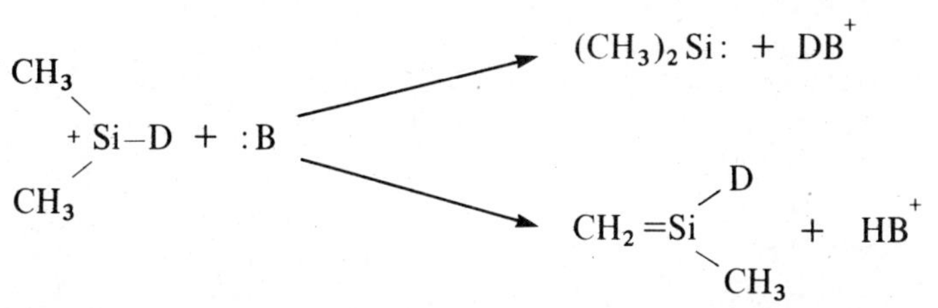

The technique used by Hehre and co-workers was to measure the onset of the two different deprotonation reactions producing silylene and silaolefin, using bases of known proton affinity. These gas-phase reactions were monitored by ion-cyclotron resonance spectroscopy.

The proton affinity data provided the enthalpy changes in the two processes, and these together with the heat of formation of the starting silylenium cation lead to the heats of formation of the two neutral products. The source of the discrepancy between these measurements and all other data on the relative and absolute stability of the silaolefin is unclear.

Electron-diffraction experiments have provided direct structural information on $SiCl_2$ and $SiBr_2$ generated in the following reactions:[118]

$$Si(solid) \; + \; Si_2Cl_6 \, (gas) \; \rightarrow \; SiCl_2 \, (gas)$$

$$Si(solid) \cdot + \; SiBr_4 \, (gas) \; \rightarrow \; SiBr_2 \, (gas)$$

The following parameters were deduced:

	$SiCl_2$	$SiBr_2$
$r(Si–X)$ Å	2.083 ± 0.004	2.243 ± 0.005
$\angle$ X–Si–X, deg	102.8 ± 0.6	102.7 ± 0.3

The spectrosopic detection of SiH_2 in the infrared multiphoton decomposition of SiH_4 (p. 339),[22] and matrix isolation spectroscopic studies of SiHMe, SiClMe, and $SiMe_2$ (p. 354)[44] as well as HSiOH and HSiF (p. 339)[19,20] have been mentioned earlier.

VI.   SILYLENE-TRANSITION METAL COMPLEXES

The potential importance of silylene-transition metal complexes as "silylenoid" synthetic reagents and as precursors for the photochemical generation of silylenes has yet to be realized. A few recent examples of the formation and reactions of these metal complexes are given here. A magisterial survey of the corresponding μ-carbene complexes has appeared in the past year.[119]

Stone and co-workers have obtained μ-$SiMe_2$ complexes from reactions of dimethylbis(phenylethynyl)silane with bis(ethylene)(tertiaryphosphine)-platinum complexes.[120]

$$Me_2Si(C{\equiv}CPh)_2 \ + \ (C_2H_4)_2Pt(PR_3)$$

$$PR_3 \ = \ P(cyclo\text{-}C_6H_{11})_3$$
$$PMe(t\text{-}Bu)_2$$
$$PPh(i\text{-}Pr)_2$$

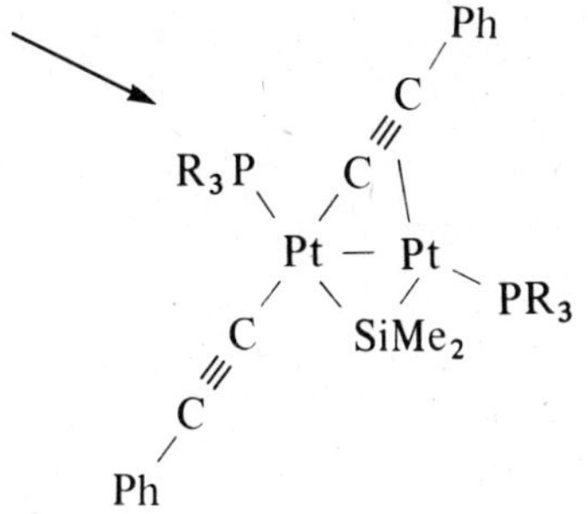

μ-Silylene complexes have also been obtained with a hydrogen atom bridging a silicon–platinum bond.[121] These are obtained both from decomposition of triorganosilyl-metal complexes and, in much higher yield, from reaction of dihydridodiorganosilanes with mononuclear platinum–ethylene complexes, the latter process believed to be a free radical reaction.

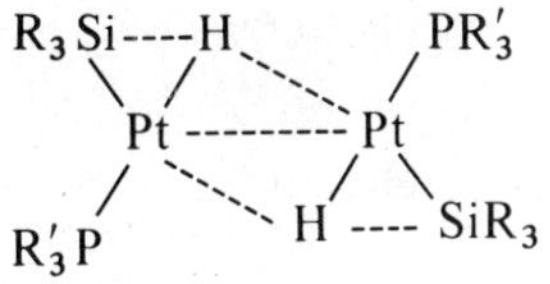

$$R_3Si \ = \ SiMe_2Ph$$
$$R'_3P \ = \ P(cyclo\text{-}C_6H_{11})_3$$

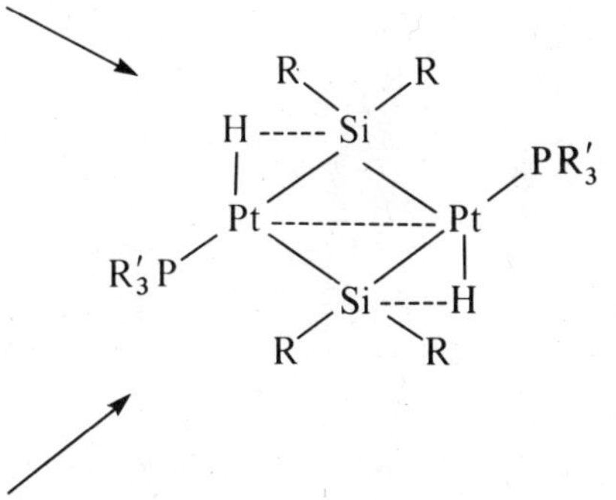

$$[Pt(C_2H_4)_2(PR'_3)] \ + \ H_2SiR_2$$

$$PR'_3 \ = \ P(cyclo\text{-}C_6H_{11})_3, \ PMe(t\text{-}Bu)_2, \ PPh(i\text{-}Pr)_2, \ PPh_3$$
$$H_2SiR_2 \ = \ H_2SiMe_2, \ H_2SiPh_2$$

Corriu and Moreau have obtained μ-silylene iron complexes in photochemical reactions.[122]

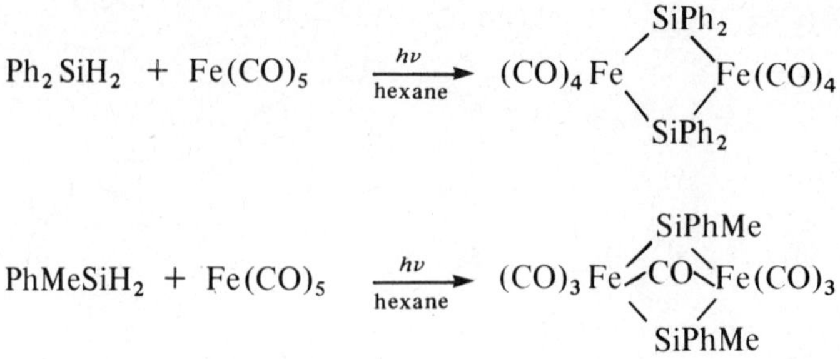

$$Ph_2SiH_2 + Fe(CO)_5 \xrightarrow[\text{hexane}]{h\nu} (CO)_4Fe \underset{SiPh_2}{\overset{SiPh_2}{\diagup\diagdown}} Fe(CO)_4$$

$$PhMeSiH_2 + Fe(CO)_5 \xrightarrow[\text{hexane}]{h\nu} (CO)_3Fe \underset{SiPhMe}{\overset{SiPhMe}{\diagup}} CO \underset{}{\diagdown} Fe(CO)_3$$

The diphenylsilylene complex has been employed as a synthetic reagent to form disilametallacyclopentenes that in turn show promising reactions like the conversion of nitriles to aldehydes.

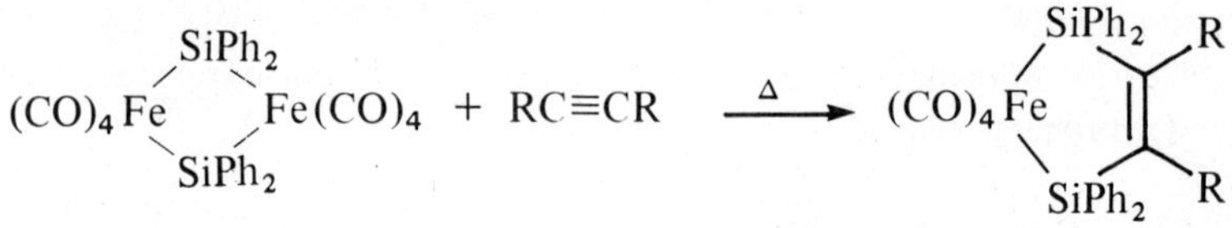

$$(CO)_4Fe \underset{SiPh_2}{\overset{SiPh_2}{\diagup\diagdown}} Fe(CO)_4 + RC\equiv CR \xrightarrow{\Delta} (CO)_4Fe \underset{SiPh_2}{\overset{SiPh_2}{\diagup}} \begin{matrix} R \\ \| \\ R \end{matrix}$$

$$\begin{array}{ll} R = Ph & 60\% \\ Et & 65\% \\ Me & 30\% \end{array}$$

$$(CO)_4Fe \underset{SiPh_2}{\overset{SiPh_2}{\diagup}} \begin{matrix} R \\ \| \\ R \end{matrix} + R'CH_2C\equiv N \longrightarrow R'C=CH-N \underset{SiPh_2}{\overset{SiPh_2}{\diagup}} \begin{matrix} R \\ \| \\ R \end{matrix}$$

$$\begin{array}{lc} R = Et, R' = H & 84\% \\ R = Et, R' = Ph & 75\% \\ & (cis{:}trans = 55{:}45) \\ R = R' = Me & 71\% \\ & (cis{:}trans = 58{:}42) \end{array}$$

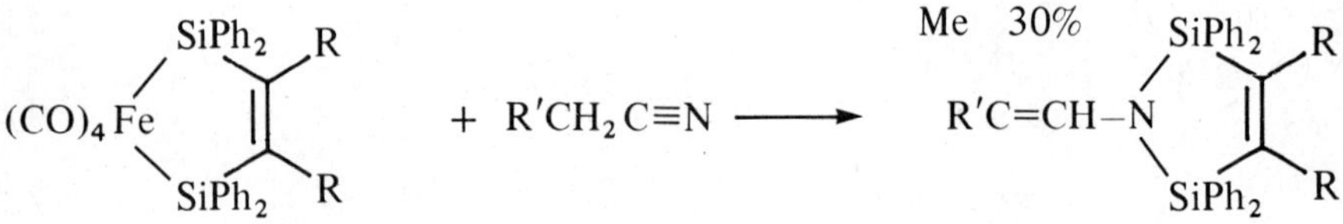

$$R'C=CH-N \underset{SiPh_2}{\overset{SiPh_2}{\diagup}} \begin{matrix} R \\ \| \\ R \end{matrix} \xrightarrow{\text{hydrolysis}} R'CH-C\underset{H}{\overset{O}{\diagup}}$$

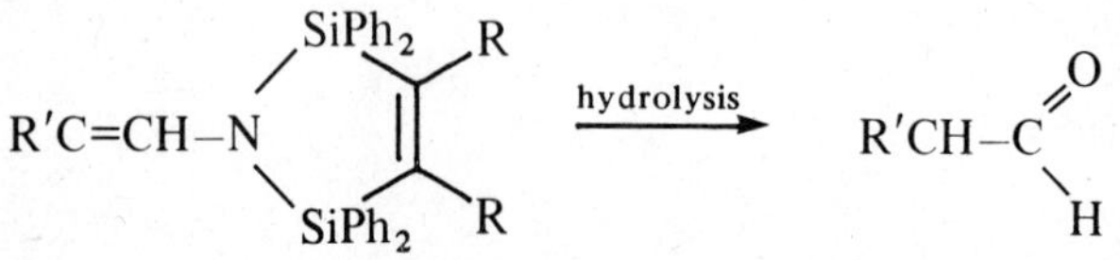

The following reaction schemes have been proposed, without benefit of evidence to account for the formation of bis(μ-silylene) iron and cobalt complexes by reactions of 1,1,2,2-tetramethyldisilane with nonacarbonyldiiron and octacarbonyldicobalt, respectively.[123]

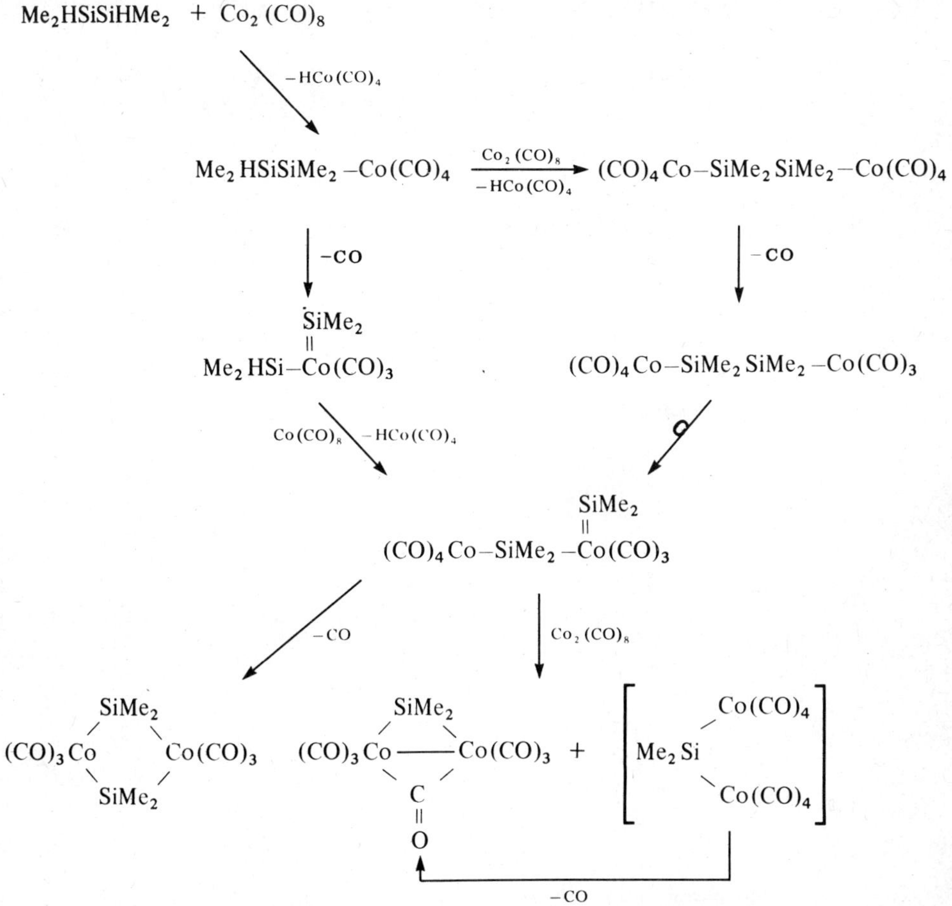

$Me_2HSiSiHMe_2$ + $Co_2(CO)_8$

$\xrightarrow{-HCo(CO)_4}$

$Me_2HSiSiMe_2-Co(CO)_4$ $\xrightarrow[-HCo(CO)_4]{Co_2(CO)_8}$ $(CO)_4Co-SiMe_2SiMe_2-Co(CO)_4$

$\downarrow -CO$ $\qquad\qquad\qquad\qquad\qquad\qquad\qquad\qquad\downarrow -CO$

$$\overset{\overset{\textstyle SiMe_2}{||}}{Me_2HSi}-Co(CO)_3$$

$(CO)_4Co-SiMe_2SiMe_2-Co(CO)_3$

$\xrightarrow[Co(CO)_8 \quad -HCo(CO)_4]{}$

$$\overset{\overset{\textstyle SiMe_2}{||}}{(CO)_4Co-SiMe_2-Co(CO)_3}$$

$\swarrow -CO \qquad\qquad \downarrow Co_2(CO)_8$

$$(CO)_3Co\overset{\diagup SiMe_2\diagdown}{\underset{\diagdown SiMe_2\diagup}{}}Co(CO)_3$$

$$(CO)_3Co-\overset{\diagup SiMe_2}{\underset{\underset{\overset{||}{O}}{C}}{}}Co(CO)_3 \quad + \quad \left[Me_2Si \overset{\diagup Co(CO)_4}{\diagdown Co(CO)_4} \right]$$

$\xrightarrow{-CO}$

$Me_2HSiSiHMe_2$ $\xrightarrow{-Fe(CO)_5}$ $Me_2HSiSiMe_2-FeH(CO)_4$
+ $Fe_2(CO)_9$

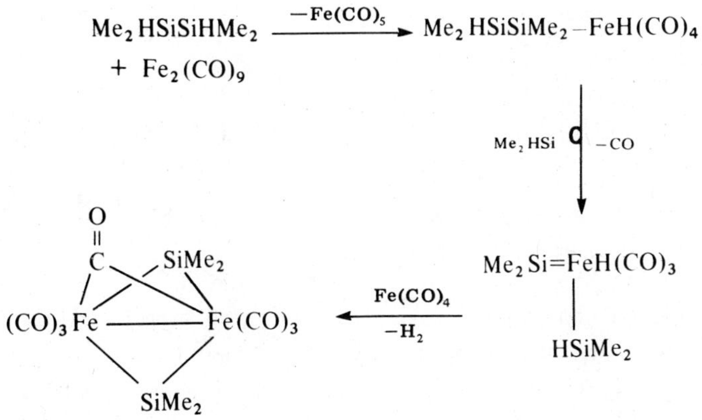

$\downarrow$ Me_2HSi $-CO$

$Me_2Si=FeH(CO)_3$

$$\overset{\overset{\textstyle O}{\overset{\textstyle ||}{\textstyle C}}}{(CO)_3Fe}-Fe(CO)_3 \quad\xleftarrow[-H_2]{Fe(CO)_4}\quad$$
with SiMe_2 bridges

$HSiMe_2$

417

$Me_2Si[Co(CO)_4]_2$ was also believed formed, but could not be isolated, from the reaction of an iron–silyl complex with octacarbonyldicobalt.[124]

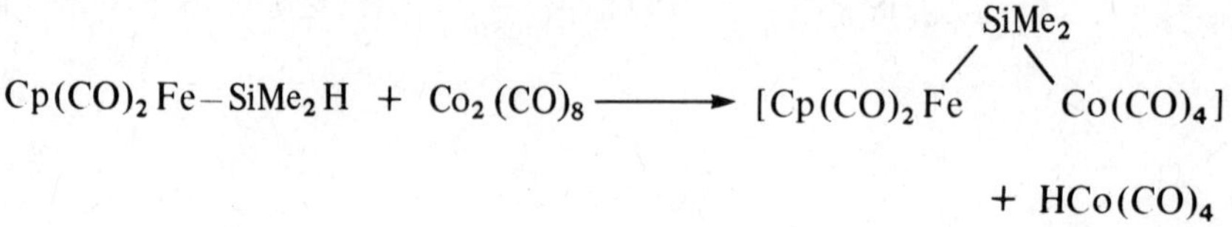

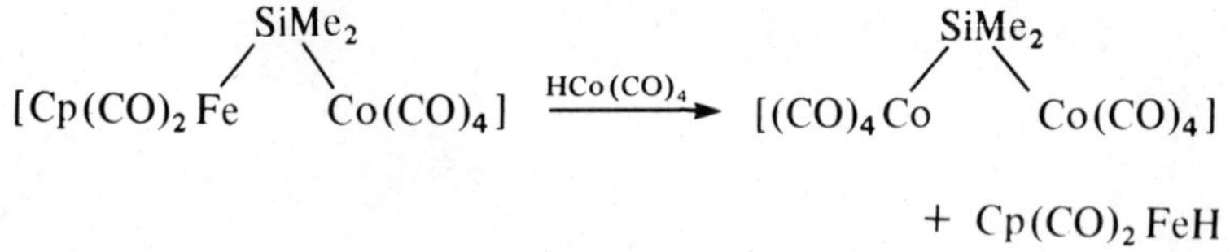

$$Cp(CO)_2\,Fe-SiMe_2\,H \ + \ Co_2\,(CO)_8 \longrightarrow [Cp(CO)_2\,Fe \overset{\displaystyle SiMe_2}{\diagup \diagdown} Co(CO)_4]$$

$$+ \ HCo(CO)_4$$

$$[Cp(CO)_2\,Fe \overset{\displaystyle SiMe_2}{\diagup \diagdown} Co(CO)_4] \xrightarrow{\ HCo(CO)_4\ } [(CO)_4\,Co \overset{\displaystyle SiMe_2}{\diagup \diagdown} Co(CO)_4]$$

$$+ \ Cp(CO)_2\,FeH$$

Malisch made the interesting suggestion that elimination of dimethylsilylene may have occured:[124]

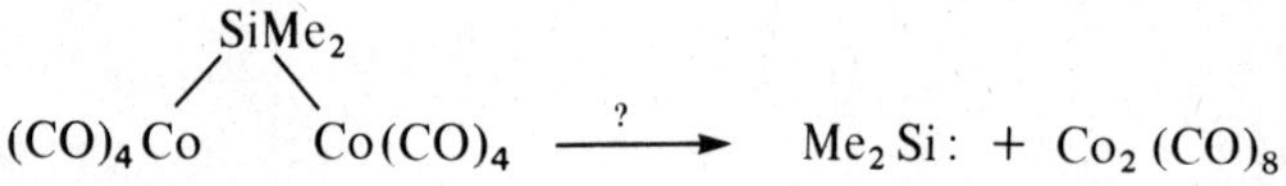

$$(CO)_4\,Co \overset{\displaystyle SiMe_2}{\diagup \diagdown} Co(CO)_4 \xrightarrow{\ ?\ } Me_2\,Si: \ + \ Co_2\,(CO)_8$$

A stronger iron–silicon bond was believed to permit the isolation of a heteronuclear complex in the following reaction:

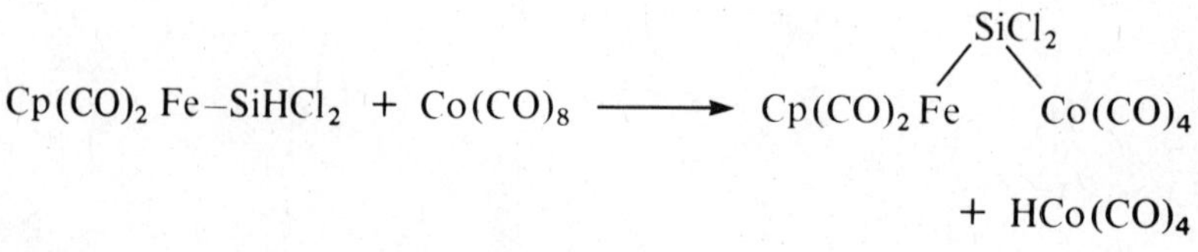

$$Cp(CO)_2\,Fe-SiHCl_2 \ + \ Co(CO)_8 \longrightarrow Cp(CO)_2\,Fe \overset{\displaystyle SiCl_2}{\diagup \diagdown} Co(CO)_4$$

$$+ \ HCo(CO)_4$$

This dichlorosilylene complex could be converted into the corresponding difluoro complex by treatment with $AgBF_4$. When two or three silicon–hydrogen bonds are present on the initial silyl–iron complex, silyne and silicon atom complexes are obtained:

$$L_n M - SiH_2 Me \; + \; 2\, Co_2\,(CO)_8 \quad \xrightarrow[-CO]{} \quad$$

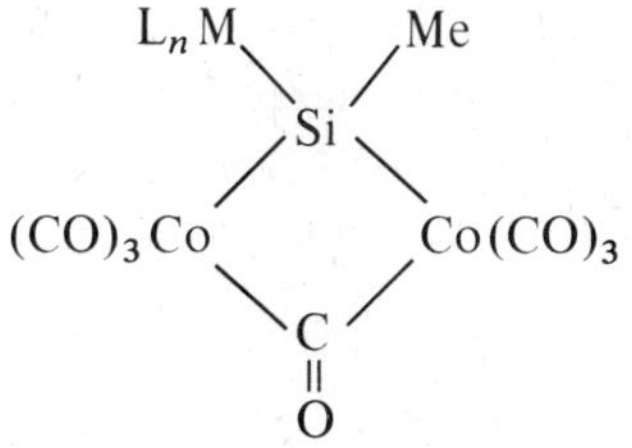

$$L_n M \; = \; Cp(CO)_2\, Fe$$

$$trans\text{-}Cp(CO)_2\,(PMe_3)W$$

$$Cp(CO)_2\, Fe - SiH_3 \; + \; \tfrac{3}{2}\, Co(CO)_8 \quad \xrightarrow[-3/2\ H_2]{} \quad$$

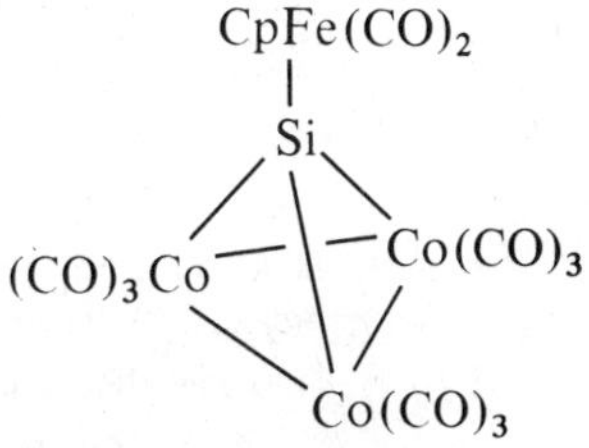

In the near future we can expect to see much more of the chemistry and spectroscopy of silylene–metal complexes, both for their own interest, and in relation to silylene chemistry.

VII.   FUTURE DEVELOPMENTS

Despite the vigorous activity evident in the last three years, our conceptual framework for the understanding of silylene chemistry is lagging behind the accretion of interesting and useful phenomena. It is therefore a hope, rather than a prediction, that mechanistic studies will begin catching up with the discovery of new reactions, or at least keep pace.

Time-resolved spectroscopic studies of the reactions of photochemically generated silylenes are likely to contribute to our understanding. The era of the laser has reached silylene chemistry, and we shall feel its impact soon.

With the wide application of tetravalent silicon reagents in organic synthesis, the distinctive uses of divalent silicon in the design of synthetic strategy are bound to emerge.

The richness of silylene rearrangements in producing molecules, reactions, and ideas will continue functioning as the cornucopia of silylene chemistry.

VIII.   ACKNOWLEDGMENTS

The fellowship of workers in the field should again be mentioned. For the sharing of facts and thoughts I am grateful to Tom Barton, Phil Boudjouk, John

Bookstaver, Rob Conlin, Iain Davidson, Dewey Holten, Mait Jones, Josef Michl, Peter Potzinger, Morey Ring, Robin Walsh, Bill Weber, and Bob West. My co-workers Bong-Hyun Boo, Sarangan Chari, Stephen Chiarello, Yue-Shen Chen, Ashim Ghosh, Stanislaw Konieczny, Eric Ma, Siu-Hong Mo, Marcy North, Eric Suchanek, and Dan Svoboda have contributed dedicated and thoughtful effort and good cheer to the pursuit and understanding of silylenes. This chapter was prepared with financial assistance from the United States Department of Energy. This is technical report COO-1713-116.

IX.   REFERENCES

1.   P. P. Gaspar, "Silylenes," in M. Jones, Jr. and R. A. Moss, Eds., *Reactive Intermediates,* Vol. 2, Wiley, New York, 1981, pp. 335–385.

2.   Y.-N. Tang, "Reactions of Silicon Atoms and Silylenes," R. A. Abramovitch, Ed., *Reactive Intermediates,* Vol. 2, Plenum, New York, 1982, Chapter 4, pp. 297–366.

3.*   R. Walsh, "Bond Dissociation Energy Values in Silicon-Containing Compounds and Some of Their Implications," *Acc. Chem. Res.,* **14,** 246 (1981).

4.   M. Ishikawa and M. Kumada, "Photochemistry of Organopolysilanes," *Adv. Organomet. Chem,* **19,** 51 (1981).

5.   M. Kumada, "Szerves Polisilánok Fotolizise Reaktiv Szilícium-Köztitemétek Keletkezése és Reakciói," *Kém. Közl.* **52,** 347 (1979).

6.   P. P. Gaspar, "Silylenes," in M. Jones, Jr., and R. A. Moss, Eds., *Reactive Intermediates,* Vol. 1, Wiley, New York, 1978, pp. 229–277.

7.*   R. West, M. J. Fink, and J. Michl, "Tetramesityldisilene, A Stable Compound Containing A Silicon-Silicon Double Bond," *Science,* **214,** 1343 (1981).

8.   J. Dzarnoski, S. F. Rickborn, H. E. O'Neal, and M. A. Ring, "Shock-Induced Kinetics of the Disilane Decomposition and Silylene Reactions with Trimethylsilane and Butadiene," *Organometallics,* **1,** 1217 (1982).

9.   T. J. Barton, S. A. Burns, P. P. Gaspar, and Y-S. Chen, "2,2-Dimethoxyhexamethyltrisilane. A 'One-Pot' Silicon Atom Synthon," *Synthesis and Reactivity in Inorganic and Metal-Organic Chemistry,* in press.

10.   P. P. Gaspar, Y.-S. Chen, A. P. Helfer, S. Konieczny, E. C.-L. Ma, and S.-H. Mo, "1-Silacyclopent-3-en-1-ylidene, A Cyclic Silylene From the Reactions of Silicon Atoms, and A Silicon Atom Synthon," *J. Am. Chem. Soc.,* **103,** 7344 (1981).

11.   T. J. Barton and S. A. Jacobi, "Radical Rearrangements in the Pyrolysis of Allyldisilanes," *J. Am. Chem. Soc.,* **102,** 7979 (1980).

12.* T. J. Barton, S. A. Burns, and G. T. Burns, "Thermochemistry of Di- and Trisilanyl Radicals, Observations and Comments on Silene-to-Silylene Rearrangements," *Organometallics,* **1,** 210 (1982).

13.   P. P. Gaspar, B. D. Pate, and W. C. Eckelman, "Reaction of Recoiling Silicon Atoms with Phosphine and Silane," *J. Am. Chem. Soc.,* **88,** 3878 (1966).

14.* P. P. Gaspar, S. Konieczny, and S.-H. Mo, "The Formation of Thermalized Singlet Silylene in the Reactions of Recoiling Silicon Atoms," *J. Am. Chem. Soc.,* submitted for publication.

15. P. P. Gaspar, "Recoil Chemistry and Mechanistic Studies with Polyvalent Atoms," in J. W. Root and K. A. Krohn, Eds., *Short-Lived Radionuclides in Chemistry and Biology*, Advances in Chemistry Series, No. 197, American Chemical Society, 1981, Chapter 1, pp. 3–31.

16. P. P. Gaspar and J. W. Root, "Chemical Kinetics and Hot Atom Chemistry – Present Status and Future Prospects," *Radiochim. Acta*, **28**, 191 (1981).

17.* E. E. Siefert, S. D. Witt, K.-L. Loh, and Y.-N. Tang, "On the Mechanisms of Silicon Atom Abstraction Reactions," *J. Organometal. Chem.*, submitted for publication.

18. S.-H. Mo, J. D. Holten III, S. Konieczny, E. C.-L. Ma, and P. P. Gaspar, "Silylene Rearrangements in the Reactions of Recoiling Silicon Atoms with Trimethylsilane," *J. Am. Chem. Soc.*, **104**, 1424 (1982).

19. Z. K. Ismail, R. H. Hauge, L. Fredin, J. W. Kauffman, and J. L. Margrave, "Matrix Isolation Studies of the Reactions of Silicon Atoms: I. Interaction with Water: The Infrared Spectrum of Hydroxysilylene HSiOH," *J. Chem. Phys.*, **77**, 1617 (1982).

20. Z. K. Ismail, L. Fredin, R. H. Hauge, and J. L. Margrave, "Matrix Isolation Studies of the Reactions of Silicon Atoms: II. Infrared Spectrum and Structure of Matrix-Isolated Fluorosilylene: HSiF," *J. Chem. Phys.*, **77**, 1626 (1982).

21. C. George and R. D. Koob, "Excited-State Chemistry of 1,1-Dimethylsilacyclobutane. The Role of Singlet and Triplet States at 254 nm," *Organometallics*, **2**, 39 (1983).

22.* J. F. O'Keefe and F. W. Lampe, "Spectroscopic Detection of Silylene in the Infrared Multiphoton Decomposition of Silane," *Appl. Phys. Lett.*, **42**, 217 (1983).

23. P. A. Longeway and F. W. Lampe, "Infrared Multiphoton Decomposition of Monosilane," *J. Am. Chem. Soc.*, **103**, 6813 (1981).

24. M. Ishikawa, H. Sugisawa, K. Yamamoto, and M. Kumada, "Photochemically Generated Silicon–Carbon Double-Bonded Intermediates. XII. Some Properties of Silacyclopropenes and Silapropadienes Produced by Photolysis of Ethynylpolysilanes," *J. Organometal. Chem.*, **179**, 377 (1979).

25. M. Ishikawa, D. Kovar, T. Fuchikami, K. Nishimura, M. Kumada, T. Higuchi, and S. Miyamoto, "Photochemically Generated Silicon–Carbon Double-Bonded Intermediates. 13. The Formation and Reactions of 1,2-Disilacyclobutanes," *J. Am. Chem. Soc.*, **103**, 2324 (1981).

26. M. Ishikawa, H. Sugisawa, T. Fuchikami, M. Kumada, T. Yamabe, H. Kawakami, K. Fukui, Y. Ueki, and H. Shizuka, "Photolysis of Organopolysilanes. Photochemical Behavior of Phenylethynyldisilanes," *J. Am. Chem. Soc.*, **104**, 2872 (1982).

27. P. Boudjouk, J. R. Roberts, C. M. Golino, and L. H. Sommer, "Photochemical Dehydrosilylation of Pentaphenylmethyldisilane. Generation and Trapping of an Unstable Intermediate Containing a Silicon–Carbon Double Bond or Its Equivalent," *J. Am. Chem. Soc.*, **94**, 7926 (1972).

28. H. Shizuka, Y. Sata, M. Ishikawa, and M. Kumada, "Evidence That the Fluorescence of Phenyldisilanes Involves the $^1(2p\pi, 3d\pi)$ Intramolecular Charge-Transfer State," *J. Chem. Soc., Chem. Commun.*, **1982**, 439.

29.* S. Masamune, Y. Hanzawa, S. Murakami, T. Bally, and J. F. Blount, "Cyclotrisilane $(R_2Si)_3$ and Disilene Systems: Synthesis and Characterization," *J. Am. Chem. Soc.*, **104**, 1150 (1982).

30.* R. T. Conlin and D. L. Wood, "Evidence for the Isomerization of 1-Methylsilene to Dimethylsilylene," *J. Am. Chem. Soc.*, **103**, 1843 (1981).

31.*   R. T. Conlin and R. S. Gill, "Thermal Fragmentation of Silacyclobutane. Formation of Silylene, Methylsilylene and Silene," *J. Am. Chem. Soc.,* **105,** 618 (1983).

32.*   L. E. Gusel'nikov, V. M. Sokolova, E. A. Volnina, Z. A. Kerzina, N. S. Nametkin, N. G. Komalenkova, S. A. Bashkirova, and E. A. Chernyshev, "Role of 1,1-Dichloro-1-silaethylene and Dichlorosilylene in the Mechanism of Thermal Decomposition of 1,1-Dichloro-1-silacyclobutane," *Dokl. Akad. Nauk. SSSR,* **260,** 348 (1981).

33.*   D. Seyferth, D. C. Annarelli, S. C. Vick, and D. P. Duncan, "Hexamethylsilirane I. Preparation, Characterization and Thermal Decomposition," *J. Organometal. Chem.,* **201,** 179 (1980).

34.*   D. Seyferth, D. C. Annarelli, and D. P. Duncan, "Hexamethylsilirane. 3. Dimethylsilylene-Transfer Chemistry," *Organometallics,* **1,** 1288 (1982).

35.   B. Mayer and W. P. Neumann, "7-Silanorbornadienes and Their Thermal Cycloeliminations," *Tetrahedron Lett.,* 4887 1980.

36.   T. J. Barton, W. F. Goure, J. L. Witiak, and W. D. Wulff, "Observations and Comments on the Thermal Behavior of 7-Silanorbornadienes," *J. Organometal. Chem.,* **225,** 87 (1982).

37.   Reference 31, footnote 15.

38.   H. F. Schaefer III, "The 1,2-Hydrogen Shift: A Common Vehicle for the Disappearance of Evanescent Molecular Species," *Acc. Chem. Res.,* **12,** 288 (1979).

39.   J. D. Goddard, Y. Yoshioka, and H. F. Schaefer III, "Methylsilylene, Silaethylene, and Silylmethylene: Energies, Structures and Unimolecular Reactivities," *J. Am. Chem. Soc.,* **102,** 7644 (1980).

40.   Y. Yoshioka and H. F. Schaefer III, "Theoretical Studies of the 1,2-Hydrogen Shift. 11. The Controversial Barrier Between Silaethylene and Methylsilylene," *J. Am. Chem. Soc.,* **103,** 7366 (1981).

41.   T. J. Barton, G. T. Burns, and D. Gschneider, "New Silene Rearrangements from a Study of the Mechanism of Silacyclopentene Formation From [3.3] Silaspirocycloheptane," *Organometallics,* **2,** 8 (1983).

42.   N. S. Nametkin, L. E. Gusel'nikov, V. Y. Orlov, R. L. Ushakova, O. V. Kuzin, and V. M. Vdovin, "Thermal Decomposition of 4-Silaspiro[3.3]heptane," *Dokl. Akad. Nauk. SSSR,* **211,** 106 (1973).

43.   T. J. Drahnak, J. Michl, and R. West, "Photoisomerization of Dimethylsilylene to 2-Silapropene and Thermal Reversion to Dimethylsilylene," *J. Am. Chem. Soc.,* **103,** 1845 (1981).

44.*   H. P. Reisenauer, G. Mihm, and G. Maier, "Reversible Photoisomerization between Silaethenes and Methylsilylenes," *Agnew. Chem. Int. Ed. Engl.,* **21,** 854 (1982).

45.*   T. J. Barton, S. A. Burns, and G. T. Burns, "Silacyclopropene to Vinylsilylene Isomerization in Reactions of Silyl- and Hydridosilylenes with Acetylenes," *Organometallics,* **2,** 199 (1983).

46.   M. S. Gordon and R. D. Koob, "Relative Stability of Multiple Bonds to Silicon: An *ab initio* Study of C_2SiH_4 Isomers," *J. Am. Chem. Soc.,* **103,** 2939 (1981).

47.*   B. H. Boo, P. P. Gaspar, A. K. Ghosh, J. D. Holten, C. R. Kirmaier, and S. Konieczny, "Transposition, A New Silylene Rearrangement and the Laser Photolysis of Polysilanes," paper delivered at the *17th Organosilicon Symposium,* Fargo, ND, June 4, 1983, to be published.

48.   V. N. Bochkarev, A. N. Polivanov, T. F. Slyusarenko, A. A. Bernadskii, N. N.

Silkina, and B. N. Klimentov, "Formation of Dichlorosilene Under Electron Impact," *Zh. Obshch. Khim.,* **51,** 824 (1981).

49.   B. Chapman, *Glow Discharge Processes,* Wiley, New York, 1980.

50.   M. J. Vasile and F. A. Stevie, "Reaction of Atomic Fluorine with Silicon: The Gas-Phase Products," *J. Appl. Phys.,* **53,** 3799 (1982).

51.   J. A. Mucha, D. L. Flamm, and V. M. Donnelly, "Chemiluminescent Reaction of SiF_2 with Fluorine and the Etching of Silicon by Atomic and Molecular Fluorine," *J. Appl. Phys.,* **53,** 4533 (1982).

52.   B. A. Scott, M. H. Brodsky, D. C. Green, P. B. Kirby, R. M. Plecenik, and E. E. Simonyi, "Glow Discharge Preparation of Amorphous Hydrogenated Silicon From Higher Silanes," *Appl. Phys. Lett.,* **37,** 725 (1980).

53.   O. M. Nefedov and P. S. Skell, "Formation and Reactions of Dimethylsilylene and Dimethylgermylene in the Gas Phase," *Dokl. Akad. Nauk. SSSR,* **259,** 377 (1981).

54.   L. E. Gusel'nikov, E. Lopatnikova, Yu. P. Polyakov, and N. S. Nametkin, "Generation and Reactions of Methyl(trimethylsilylalkyl)silylenes $Me_3Si(CH_2)_nSiMe$ in the Gas Phase," *Dokl. Akad. Nauk. SSSR,* **253** 1387 (1980).

55.   R. T. Conlin and L. L. Peterson, "Alkali Metal Promoted Transfer of Dimethylsilylene from Dodecamethylcyclohexasilane to Silanes and Dienes," *J. Organometal. Chem.,* **232,** C71 (1982).

56.   M. S. Gordon, "*Ab Initio* Study of the Insertions of CH_2 and SiH_2 into H_2," *J. Chem. Soc., Chem. Commun.,* 890 (1981).

57.   P. John and J. H. Purnell, "Arrhenius Parameters for Silene Insertion Reactions," *J. Chem. Soc., Faraday Trans. 1,* **69,** 1455 (1973).

58.   T. N. Bell, "The Role of the Triplet State in Insertion Reactions of Silylenes," *J. Chem. Soc., Chem. Commun.,* 1046 (1980).

59.   K. Raghavachari, J. Chandrasekhar, and M. J. Frisch, "*Ab Initio* Study of Silylene Intermediates," *J. Am. Chem. Soc.,* **104,** 3779 (1982).

60.   K. P. Steele and W. P. Weber, "Kinetic Isotope Effects for Silylene Insertions into Oxygen-Hydrogen and Silicon-Hydrogen Bonds," *Inorg. Chem.,* **20,** 1302 (1981).

61.*   K. P. Steele and W. P. Weber, "Solvent Modified Reactivity of Dimethylsilylene," *J. Am. Chem. Soc.,* **102,** 6095 (1980).

62.   K. P. Steele, D. Tzeng, and W. P. Weber, "Insertion Reactions of Dimethylsilylene: Relative Reactivity Towards Oxygen—Hydrogen, Silicon—Hydrogen and Silicon—Alkoxy Bonds," *J. Organometal. Chem.,* **231,** 291 (1982).

63.   T. Y. Gu and W. P. Weber, "Insertion Reactions of Dimethylsilylene into Si–H and Si–OR Bonds," *J. Organometal Chem.,* **195,** 29 (1980).

64.*   I. M. T. Davidson, F. T. Lawrence, and N. A. Ostah, "Kinetics of Insertion Reactions of Dimethylsilylene," *J. Chem. Soc., Chem. Commun.,* 859 (1980).

65.*   I. M. T. Davidson and N. A. Ostah, "Gas-Phase Photochemical Reactions of Dodecamethylcyclohexasilane with Silicon Compounds. Kinetics of Some Insertion Reactions of Dimethylsilylene," *J. Organometal. Chem.,* **206,** 149 (1981).

66.   J. Blazejowski and F. W. Lampe, "Monosilylphosphine Formation by Rapid Silylene Insertion in the IR Photochemistry of SiH_4–PH_3 Mixtures," *J. Photochem.,* **20,** 9 (1982).

67.   M. D. Sefcik and M. A. Ring, "Relative Insertion Rates of Silylene and Evidence for Silylsilylene Insertion into Silicon—Hydrogen and Silicon—Silicon Bonds," *J. Am. Chem. Soc.,* **95,** 5168 (1973).

68.   A. Chihi and W. P. Weber, "Insertion Reactions of Dimethylsilylene into Silicon–Sulfur and Sulfur–Sulfur Single Bonds", *J. Organometal. Chem.*, **210**, 163 (1981).

69.* M. Ishikawa, K.-I. Nakagawa, S. Katayama, and M. Kumada, "Photolysis of Organopolysilanes. The Reaction of Photochemically Generated Trimethylsilylphenylsilylene with Alkyl Chlorides," *J. Organometal Chem.*, **216**, C48 (1981).

70.   H. Schnöckel, "Matrixreaktionen von SiO mit F_2. IR–Spektroskopischer Nachweis von Molekularem $OSiF_2$," *J. Mol. Structure*, **65**, 115 (1980).

71.   A. V. Novoselova and Yu. V. Azhikina, "Equilibrium in a System Containing Beryllium Oxide, Silicon Dioxide, and Sodium Fluoroberyllate," *Izv. Akad. Nauk. SSSR, Neorganicheski Materialy*, **2**, 1064 (1966).

72.* V. J. Tortorelli, M. Jones, Jr., S.-H. Wu, and Z.-H. Li, "Stereospecific Additions of Dimethylsilylene and Diphenylsilylene to Olefins," *Organometallics*, **2**, 759 (1983).

73.   M. Ishikawa, K-I. Nakagawa, S. Katayama, and M. Kumada, "Photolysis of Organopolysilanes. Reactions of Trimethylsilylphenylsilylene with Functional-Group Substituted Olefins," *J. Am. Chem. Soc.*, **103**, 4170 (1981).

74.   V. J. Tortorelli and M. Jones, Jr., "Reaction of Dimethylsilylene with Allyl Ethers," *J. Chem. Soc., Chem. Commun.*, 785 (1980).

75.   D. Tzeng and W. P. Weber, "Regiospecific Synthesis of Allylic Dimethylmethoxysilanes," *J. Org. Chem.*, **46**, 693 (1981).

76.   A. Chihi and W. P. Weber, "Reactions of Dimethylsilylene with Allylic Methyl Sulfides," *Inorg. Chem.*, **20**, 2822 (1981).

77.   M. Ishikawa, K.-I. Nakagawa, and M. Kumada, "Photolysis of Organopolysilanes. Reactions of Trimethylsilylphenylsilylene with Allylic Halides and Allyl Ethyl Ether," *J. Organometal. Chem.*, **214**, 277 (1981).

78.   M. Ishikawa, K. Nishimura, H. Sugisawa, and M. Kumada, "Photolysis of Organopolysilanes. The Synthesis and Reactions of Stable Silacyclopropenes," *J. Organometal. Chem.*, **194**, 147 (1980).

79.   E. A. Halevi and R. West, "Orbital Symmetry Analysis of the Reactions of Silylenes with Acetylenes and the Dimerization of 1-Silacyclopropenes," *J. Organometal. Chem.*, **240**, 129 (1982).

80.* T.-I. Hwang and C.-S. Liu, "Rediscovery of the Gas-Phase Chemistry of Difluorosilylene Generated by Thermal Reduction," *J. Am. Chem. Soc.*, **102**, 385 (1980).

81.   E. E. Siefert, S. D. Witt, and Y.-N. Tang, "Reactions of Monomeric Difluorosilylene with Ethylene," *J. Chem. Soc., Chem. Commun.*, 217 (1981).

82.* T.-L. Hwang, Y.-M. Pai, and C.-S. Liu, "Reactions of Difluorosilylene with Halogen-Substituted Ethylenes. A Reinvestigation of the Reaction Mechanism," *J. Am. Chem. Soc.*, **102**, 7519 (1980).

83.   P. P. Gaspar, B. L. Whitsel, M. Jones, Jr., and J. B. Lambert, "Addition of Diphenylmethylene to 1,2-Dichloroethylenes. New Chemical Evidence for a Carbene Singlet-Triplet Equilibrium," *J. Am. Chem. Soc.*, **102**, 6108 (1980).

84.   W. F. Reynolds, J. C. Thompson, and A. P. G. Wright, "A Systematic Investigation of the Reaction of SiF_2 with Ethylene and Six Methyl-substituted Ethylenes. I. Product yields and Mechanistic Conclusions," *Can. J. Chem.*, **58**, 419 (1980).

85.   C.-C. Shiau, T.-I. Hwang, and C.-S. Liu, "The Reaction of Propene with Difluorosilylene," *J. Organometal. Chem.*, **214**, 31 (1981).

86.   Y.-M. Pai, C.-K. Chen, and C.-S. Liu, "Reactions of Difluorosilylene with Cyclopentene and Cyclohexa-1,3-diene," *J. Organometal. Chem.*, **226**, 21 (1982).

87. C.-P. Chin and C.-S. Liu, "Cycloaddition Reactions of 1,1,2,2-Tetrafluoro-1,2-Disilacyclobutene with Aldehydes and Ketones," *J. Organometal. Chem.*, **235**, 7 (1982).

88. I. S. Alnaimi and W. P. Weber, "Deoxygenation of Dialkyl Sulfoxides by Dimethylsilylene. Steric Requirements," *J. Organometal. Chem.*, **241**, 171 (1983).

89. W. Ando, M. Ikeno, and Y. Hamada, "Reactions of Dimesitylsilylene with Epoxides. Silanone-Epoxide Adducts," *J. Chem. Soc., Chem. Commun.*, **1981**, 621.

90.* Y.-S. Chen and P. P. Gaspar, "Octakis(trimethylsilyl)cyclotetrasilane. A Stable Cyclotetrasilane from a Silylene Precursor," *Organometallics*, **1**, 1410 (1982).

91. C. J. Hurt, T. C. Calabrese, and R. West, "Cyclic Polysilanes VIII. The Crystal Structure of 1,2,3,4-Tetra-*tert*-butyltetramethylcyclotetrasilane," *J. Organometal. Chem.*, **91**, 273 (1975).

92.* T. J. Barton, G. T. Burns, W. F. Goure, and W. D. Wulff, "Silylene to Silene Thermal Rearrangement. Generation and Rearrangement of Cyclopropylsilylene and Vinylsilylene," *J. Am. Chem. Soc.*, **104**, 1149 (1982).

93. L. Friedman and H. Shechter, "Rearrangement and Fragmentation Reactions in Carbenoid Decomposition of Diazo Hydrocarbons," *J. Am. Chem. Soc.*, **82**, 1002 (1960).

94.* S. A. Burns, G. T. Burns, and T. J. Barton, "Orchestrated Silylene to Silene to Silylene Rearrangement. The Unusual Behavior of C_3H_5 Silylsilylenes," *J. Am. Chem. Soc.*, **104**, 6140 (1982).

95. J. Meinwald, J. W. Wheeler, A. A. Nimetz, and J. S. Liu, "Synthesis of Some 1-Substituted 2,2-Dimethyl-3-isopropylidenecyclopropanes," *J. Org. Chem.*, **30**, 1038 (1965).

96. M. S. Gordon and J. A. Pople, "Structure and Stability of a Silicon–Carbon Triple Bond," *J. Am. Chem. Soc.*, **103**, 2945 (1981).

97. W. Ando, Y. Hamada, and A. Sekiguchi, "Silanediyl to Sila-alkene Conversions. Generation and Rearrangement of Cyclopropylsilanediyl," *J. Chem. Soc., Chem. Commun.*, 787 (1982).

98. W. Kirmse, *Carbene Chemistry*, 2nd ed., Academic, New York, 1971, pp. 328–332.

99.* T. J. Barton and G. T. Burns, "Vinylsilylene Rearrangement. A Possible Silacyclopropylidene Intermediate," submitted for publication.

100.* G. T. Burns and T. J. Barton, "Silacyclobutene Synthesis via Intramolecular Cyclization of Unsaturated Silylene," submitted for publication.

101. D. M. Lemal, F. Menger, and G. W. Clark, "Bicyclobutane," *J. Am. Chem. Soc.*, **85**, 2529 (1963).

102. J. D. Perez and G. I. Yranzo, "Evidence for a 1,4-hydrogen Shift in a Deuterium-Labeled Vinyl Carbene Intermediate in the Formation of 1,3-Pentadiene from 3,5-Dimethylpyrazole," *J. Org. Chem.*, **47**, 2221 (1982).

103.* T. J. Barton and G. T. Burns, "Intramolecular Reactions of Alkenylsilylenes," *Organometallics*, **2**, 1 (1983).

104. P. P. Gaspar, R.-J. Hwang, and W. C. Eckelman, "Reactions of Recoiling Silicon Atoms with Phosphine and Butadiene, and the Addition of Silylene to Butadiene," *J. Chem. Soc. Chem. Commun.*, 242 (1974); R.-J. Hwang and P. P. Gaspar, "Reactions of Recoiling Silicon Atoms with Phosphine-Diene Mixtures and the Question of Silylene Intermediates," *J. Am. Chem. Soc.*, **100**, 6626 (1978).

105.   E. E. Siefert, K.-L. Loh, R. A. Ferrieri, and Y.-N. Tang, "Formation of 1-Silacyclo-penta-2,4-diene Through Recoil Silicon Atom Reactions," *J. Am. Chem. Soc.,* **102,** 2285 (1980).

106.   Y. S. Chen and P. P. Gaspar, unpublished work (1981).

107.*   H. Sakurai, H. Sakaba, and Y. Nakadaira, "Facile Preparation of 2,3-Benzo-1,4-diphenyl-7-silanorbornadiene Derivatives and the First Clear Evidence of Silylene to Disilene Thermal Rearrangement," *J. Am. Chem. Soc.,* **104,** 6156 (1982).

108.   M. H. Lien and A. C. Hopkinson, "An *Ab Initio* Molecular Orbital Study of Ten Isomers on the C_2SiH_4 Energy Surface," *Chem. Phys. Lett.,* **80,** 114 (1981).

109.   H. -J. Köhler and H. Lischka, "A Systematic Investigation of the Structure and Stability of the Lowest Singlet and Triplet States of Si_2H_4 and SiH_3SiH and the Analogous Carbon Compounds SiH_2CH_2, SiH_3CH, CH_3SiH, C_2H_4, and CH_3CH", *J. Am. Chem. Soc.,* **104,** 5884 (1982).

110.   H. Lischka and H. -J. Köhler, "On the Structure and Stability of Singlet and Triplet Disilene and Silylsilylene," *Chem. Phys. Lett.,* **85,** 467 (1982).

111.   R. A. Poirier and J. D. Goddard, "The Isomers of Si_2H_4: Disilene and Silylsilylene," *Chem. Phys. Lett.,* **80,** 37 (1981).

112.   K. Krogh-Jespersen, "Geometries and Relative Energies of Singlet Silysilylene and Singlet Disilene," *J. Phys. Chem.,* 1492 (1982).

113.   A. C. Hopkinson and M. H. Lien, "*Ab initio* Calculations on Tautomers of $CSiH^-$, $CSiH_2$ and $CSiH_3^+$. Examples of the Tendency of Silicon to Avoid Formation of Multiple Bonds with Carbon," *J. Chem. Soc., Chem. Commun.,* 107 (1980).

114.   A. C. Hopkinson, M. H. Lien, and I. G. Csizmadia, "*Ab Initio* Calculations on the Singlet and Triplet Energy Surfaces for $CSiH_2$," *Chem. Phys. Lett.,* **95,** 232 (1983).

115.   G. Trinquier and J.-P. Malrieu, "Silabutadienes and Their Silylene Isomers. An *ab Initio* Study", *J. Am. Chem. Soc.,* **103,** 6313–6319 (1981).

116.   T. Clark and P. v. R. Schleyer, "The Isomeric Structures of SiH_2LiF," *J. Organometal. Chem.,* **191,** 347 (1980).

117.   C. F. Pau, W. J. Pietro, and W. J. Hehre, "Relative Thermochemical Stabilities of 1-Methylsilaethylene and Dimethylsilylene by Ion Cyclotron Double Resonance Spectroscopy," *J. Am. Chem. Soc.,* **105,** 16 (1983).

118.*   I. Hargittai, G. Schultz, J. Tremmel, N. D. Kagramanov, A. K. Maltsev, and O. M. Nefedov, "Molecular Structure of Silicon Dichloride and Silicon Dibromide From Electron Diffraction Combined with Mass Spectrometry," *J. Am. Chem. Soc.,* **105,** 2895 (1983).

119.   W. A. Herrmann, "The Methylene Bridge," *Adv. Organometal. Chem.,* **20,** 159 (1982).

120.   M. Ciriano, J. A. K. Howard, J. L. Spencer, F. G. A. Stone, and H. Wadepohl, "Reactions of Bis(ethylene)(tertiary phosphine)platinum Complexes with Phenylethynyl Derivatives of Titanium and Silicon; Crystal Structure of (μ-Dimethyl-silanediyl) (σ-phenylethynyl) [μ-(1-σ:1-2-η-phenylethynyl)] -bis-(tricyclohexylphosphine)diplatinum (Pt–Pt)," *J. Chem. Soc., Dalton Trans.,* 1749 (1979).

121.   M. Auburn, M. Ciriano, J. A. K. Howard, M. Murray, N. J. Pugh, J. L. Spence, F. G. A. Stone, and P. Woodward, "Synthesis of Bis-μ-diorganosilanediyl-af-dihydro-bis-(triorganophosphine)diplatinum Complexes, Crystal and Molecular Structure of $[\{PtH(\mu–SiMe_2)[P(C_6H_{11})_3]\}]$," *J. Chem. Soc., Dalton Trans.,* 659 (1980).

122.   R. J. P. Corriu and J. J. E. Moreau, "Organosilyl Iron Carbonyl Complexes: Synthesis and Reactivity Towards Alkynes and Nitriles," *J. Chem. Soc., Chem. Commun.,* 278 (1980).

123.   R. C. Kerber and T. Pakkanen, "Reactions of 1,1,2,2-Tetramethyldisilane with Group VIII Metal Carbonyls", *Inorg. Chim. Acta,* **37,** 61 (1979).

124.   W. Malisch, H.-U. Wekel, I. Grob, and F. H. Köhler, "Synthese and Reaktivität von Silicium-Ubergangsmetallkomplexen, XIV. Ümsetzung von Cp(CO)$_2$Fe-substituierton Silicium-Wasserstoff-Verbindungen mit Dicobaltoctacarbonyl: Synthese von Siliciumverbindungen mit zwei, drei and vier Übergansmetall-Liganden," *Z. Naturforsch.,* **376,** 601 (1982).

125.*   M. Ishikawa, K.-I. Nakagawa, R. Enokida, and M. Kumada, "Photolysis of Organopolysilanes. The Reaction of Photochemically Generated Mehtylphenylsilylene with Conjugated Dienes," *J. Organometal. Chem.,* **201,** 151 (1980).

INDEX